Human Factors and Ergonomics in Consumer Product Design

Uses and Applications

Ergonomics Design and Management: Theory and Applications

Series Editor
Waldemar Karwowski
Industrial Engineering and Management Systems
University of Central Florida (UCF) – Orlando, Florida

Published Titles

Aircraft Interior Comfort and Design
Peter Vink and Klaus Brauer

Ergonomics and Psychology: Developments in Theory and Practice
Olexiy Ya Chebykin, Gregory Z. Bedny, and Waldemar Karwowski

Ergonomics in Developing Regions: Needs and Applications
Patricia A. Scott

Handbook of Human Factors in Consumer Product Design, 2 vol. set
Waldemar Karwowski, Marcelo M. Soares, and Neville A. Stanton

Volume I: Methods and Techniques
Volume II: Uses and Applications

Human–Computer Interaction and Operators' Performance: Optimizing Work
Design with Activity Theory
Gregory Z. Bedny and Waldemar Karwowski

Trust Management in Virtual Organizations: A Human Factors Perspective
Wiesław M. Grudzewski, Irena K. Hejduk, Anna Sankowska, and Monika Wańtuchowicz

Forthcoming Titles

Ergonomics: Foundational Principles, Applications and Technologies
Pamela McCauley-Bush

Knowledge Service Engineering Handbook
Jussi Kantola and Waldemar Karwowski

Manual Lifting: A Guide to the Study of Simple and Complex Lifting Tasks
Daniela Colombiani, Enrico Ochipinti, Enrique Alvarez-Casado, and Thomas R. Waters

Neuroadaptive Systems: Theory and Applications
Magalena Fafrowicz, Tadeusz Marek, Waldemar Karwowski, and Dylan Schmorrow

Organizational Resource Management: Theories, Methodologies, and Applications
Jussi Kantola

Human Factors and Ergonomics in Consumer Product Design

Uses and Applications

Edited by
Waldemar Karwowski
Marcelo M. Soares
Neville A. Stanton

CRC Press
Taylor & Francis Group
Boca Raton   London   New York

CRC Press is an imprint of the
Taylor & Francis Group, an **informa** business

CRC Press
Taylor & Francis Group
6000 Broken Sound Parkway NW, Suite 300
Boca Raton, FL 33487-2742

Printed in the United States of America on acid-free paper
10 9 8 7 6 5 4 3 2 1

International Standard Book Number: 978-1-4200-4624-3 (Hardback)

Visit the Taylor & Francis Web site at
http://www.taylorandfrancis.com

and the CRC Press Web site at
http://www.crcpress.com

Contents

SECTION II   Design for Usability

SECTION III   Case Studies

Preface

Human factors and ergonomics (HF/E) is a unique and far-reaching discipline that focuses on the nature of human–artifact interactions, which are viewed from a unified perspective on science, engineering, design, technology, and management of human-compatibility systems (Karwowski 2005). The HF/E discipline promotes a holistic, human-centered approach that considers physical, cognitive, social, organizational, environmental, and other design-relevant factors. As such, HF/E aids designers by raising their awareness of the full scope of knowledge required when designing consumer products, and plays an important role in facilitating a better performance of consumer products in general. HF/E-based design of products encompasses a wide variety of consumer preferences, and accounts for differences in such preferences due to factors such as age, gender, or health issues.

Every day, we interact with thousands of consumer products. As users, we expect these products, no matter how simple or complex, to perform their expected functions in a safe, reliable, and efficient manner. Unfortunately, this is not always the case, as designing consumer products that satisfy human needs and expectations is not an easy task. The design process that involves the application of HF/E principles and knowledge strives to achieve the above goals and, at the same time, reduce the risk of product malfunction or failure, reduce the potential for accidents, and contribute to overall product acceptance and utility, all while reducing the total product life cycle cost.

The goal of the human-centered design paradigm as applied to consumer products is to improve levels of user satisfaction, efficiency of use, increase comfort, and assure safety under normal use as well as foreseeable misuse of the product. It is in this context that we are very pleased to present the second volume of the *Handbook of Human Factors and Ergonomics in Consumer Product Design*. The motivation to produce this *Handbook* was to facilitate wider acceptance of HF/E as an effective body of knowledge for improving quality of life and safety for millions of users of consumer products with a variety of needs and expectations. In this *Handbook*, consumer products are defined as those goods used by the general public without any special training, skills, or supervision. Consumers are individuals of any age, gender, or physical condition with varying educational, cultural, and economic backgrounds. The consumer products are usually used in or around the home, in a social setting, rather than in a workplace environment with commercial needs.

Currently, there is substantial and convincing evidence that the application of HF/E knowledge can improve critical product features. These features include: ease of use, learning, efficiency, comfort, safety, and adaptability, all of which meet the needs and contribute to consumer satisfaction. Therefore, this *Handbook* aims to offer a comprehensive review of the HF/E state of the art relevant to design, development, testing, evaluation, usability, and use of consumer products. The *Handbook* also aims to provide a comprehensive source of information regarding new methods, techniques, and software applications for consumer product design.

The second volume, *Human Factors and Ergonomics in Consumer Product Design: Uses and Applications*, contains 29 chapters divided into three sections. Section I contains seven chapters that discuss challenges and opportunities in the design for product safety. Among other topics, these chapters consider such issues as consumer risks and hazards and tools for their assessment, accident analysis, user welfare, design of consumer product warnings, design of technology for safe patient handling, and ergonomics of consumer packaging. Section II, which contains eight chapters, focuses on the critical aspects of human-centered design for usability. The chapters in this section discuss user perspective on design, user empathy, and other aspects of user research, as well as assessment and testing of the usability of various systems, consumer e-health, and the organizational context of design.

Section III contains 14 carefully selected case studies of human-centered design. These chapters involve the discussion of the analysis, design, and development of a media player, mobile phones, headgear, supermarket checkout, an interactive kiosk, self-assembling products, office chairs, school furniture, and personal hygiene products. The presented case studies include application of a variety of innovative approaches that incorporate HF/E principles, standards, and best practices of user-centered design, cognitive psychology, participatory macro-ergonomics, and mathematical modeling. These chapters also identify many unique aspects of new product development projects, which have adopted a user-centered design paradigm as a way to attend to user requirements.

We hope that this second volume will also be useful to a large number of professionals, students, and practitioners who strive to incorporate HF/E principles and knowledge in the design of consumer products in a variety of applications. As was the case with the first volume of the *Handbook*, we hope that the knowledge presented in this volume will ultimately lead to an increased appreciation of the benefits of the HF/E discipline by ordinary consumers of the myriad of products used every day, and increase the HF/E literacy (Karwowski 2007) of citizens around the world.

Waldemar Karwowski
Orlando, Florida, USA

Marcelo M. Soares
Recife, Brazil

Neville A. Stanton
Southampton, England

REFERENCES

Karwowski, W. 2005. Ergonomics and human factors: The paradigms for science, engineering, design, technology, and management of human-compatible systems. *Ergonomics* 48 (5): 436–63.
———. 2007. Toward an HF/E-literate society. *Bulletin of the Human Factors and Ergonomics Society* 50 (2): 1–2.

Acknowledgments

The editors would like to take this opportunity to express their sincere appreciation to Madelda Thompson for her invaluable help and assistance in editing the final draft of the *Handbook of Human Factors and Ergonomics in Consumer Product Design.* Marcelo Soares would also like to acknowledge the CNPq – Brazilian National Council of Research and Development, which sponsored his post doctorate research conducted at the Department of Industrial Engineering and Management Systems at the University of Florida, Orlando, USA, and Nalva and Gabriel Soares for their love and unfailing encouragement, support, and patience.

Editors

Waldemar Karwowski, PE, is currently professor and chairman of the Industrial Engineering and Management Systems Department at the University of Central Florida, Orlando. Florida. He holds an MS (1978) in production engineering and management from the Technical University of Wroclaw, Poland, and a PhD (1982) in industrial engineering from Texas Tech University, USA. He was also awarded the DSc (dr hab.) postgraduate degree in management science, by the Institute for Organization and Management in Industry (ORGMASZ), Warsaw, Poland (2004). He is a recipient of honorary doctorate degrees, including those from the South Ukrainian State University of Odessa, Ukraine (2004), the Technical University of Koscie, Slovakia (2006), and the MIRA Technical University of Moscow, Russia (2007). Dr. Karwowski is a board certified professional ergonomist (BCPE). His research, teaching, and consulting activities focus on human systems integration, work systems compatibility, human–computer interaction, prevention of work-related musculoskeletal disorders, manufacturing enterprises and management ergonomics, and theoretical aspects of ergonomics science. He is past president of the International Ergonomics Association (2000–2003), and of the Human Factors and Ergonomics Society, USA (2006–2007). Dr. Karwowski currently serves as editor of *Human Factors and Ergonomics in Manufacturing* (John Wiley), and the editor-in-chief of *Theoretical Issue in Ergonomics Science* (TIES) (Taylor & Francis Group, London).

Marcelo M. Soares, PhD, is currently a professor in the Department of Design and the Department of Industrial Engineering at the Federal University of Pernambuco, Brazil. He was an invited lecturer at the Technical University of Lisbon, Portugal, and the University of Guadalaraja, Mexico. He was also a visiting scholar and lecturer at the University of Central Florida, USA. He holds an MS in production engineering from the Federal University of Rio de Janeiro, Brazil. He was also awarded his PhD at the Loughborough University in England. Dr. Soares is a professional certified ergonomist from the Brazilian Ergonomics Association (ABERGO). He was president of this organization for seven years. He has also provided leadership in Human Factors and Ergonomics Latin America and internationally as a member of the executive committee of the International Ergonomics Association. He is currently the chairman of IEA 2012 (the Triennial Congresses of the International Ergonomics Association), which will be held in Brazil. His research, teaching, and consulting activities focus on manufacturing ergonomics, usability, product design, and information ergonomics. Dr. Soares currently serves on the editorial board of *Theoretical Issues in Ergonomics Science* (TIES), *Human Factors and Ergonomics in Manufacturing*, and several other publications in Brazil. He has also done significant research and consulting for several companies in Brazil.

Neville A. Stanton, PhD, was appointed chairman in human factors in the School of Civil Engineering and the Environment at the University of Southampton in February 2009. Prior to that, he held a chair in human factors at Brunel University (since September 1999). Previously, he held a lectureship and then readership in engineering psychology at the University of Southampton (since September 1993). Professor Stanton was also a visiting fellow at Cornell University during 1998. He has published over 140 peer-reviewed journal papers (including papers in *Nature* and *New Scientist*) and 18 books on human factors and ergonomics. In 1998, he was awarded the Institution of Electrical Engineers Divisional Premium Award for a co-authored paper on engineering psychology and system safety. The Ergonomics Society awarded him the Otto Edholm medal in 2001 and the President's Medal in 2008 for his contribution to basic and applied ergonomics research. In 2007, The Royal Aeronautical Society awarded him the Hodgson Medal and Bronze Award with colleagues for their work on flight-deck safety. He also acted as an expert witness for Network

Rail in the civil litigation following the Ladbroke Grove rail accident. He has undertaken research work for the Ministry of Defence, and received grant funding from ESRC, EPSRC, EU, DTI, Ford, Jaguar, and National Grid. Professor Stanton is an editor of the journal *Ergonomics* and is on the editorial board of *Theoretical Issues in Ergonomics Science*. Professor Stanton is a fellow and chartered occupational psychologist registered with The British Psychological Society, and a fellow of The Ergonomics Society. He has a BSc (Hons) in occupational psychology from the University of Hull and a PhD in human factors from Aston University in Birmingham.

Contributors

Erminia Attaianese
Laboratorio di Ergonomia Applicata e
    Sperimentale
Naples University
Naples, Italy

Evelyn Rodrigues Azevedo
The Brazilian Society of Information Design
Curitiba, Brazil

Roger Ball
School of Design
The Hong Kong Polytechnic University
Kowloon, Hong Kong, People's Republic of
    China

Marina de Lima N. Barros
Department of Design
Federal University of Pernambuco
Recife, Brazil

Mariana P. Bezerra
Department of Design
Federal University of Pernambuco
Recife, Brazil

Róber Dias Botelho
School of Design
State University of Minas Gerais
Belo Horizonte, Brazil

Ralph Bruder
Institute of Ergonomics
Darmstadt University of Technology
Darmstadt, Germany

Giuseppe Di Bucchianico
IDEA Department
University of Chieti-Pescara
Pescara, Italy

Jairo José Drummond Câmara
School of Design
State University of Minas Gerais
Belo Horizonte, Brazil

Richard I. Cook
Department of Anesthesia and Critical Care
University of Chicago
Chicago, Illinois

Walter Franklin M. Correia
Department of Design
Federal University of Pernambuco
Recife, Brazil

Sarah Davies
Department of Industrial Design, Coventry
    School of Art and Design
Coventry University
Coventry, United Kingdom

Muriel Didier
Institute of Ergonomics
Darmstadt University of Technology
Darmstadt, Germany

Ken Eason
Department of Ergonomics (Human
    Sciences)
Loughborough University
Loughborough, United Kingdom

Matthias Göbel
Department of Human Kinetics and
    Ergonomics
Rhodes University
Grahamstown, South Africa

Lia Buarque de Macedo Guimarães
Industrial Engineering Graduate Program
Federal University of Rio Grande do Sul
Porto Alegre, Brazil

David Hitchcock
David Hitchcock Limited
Hugglescote, United Kingdom

John Jansen
Philips Design
Eindhoven, the Netherlands

Daniel P. Jenkins
Sociotechnic Solutions
St. Albans, United Kingdom

Kenneth R. Laughery
Department of Psychology
Rice University
Houston, Texas

Michael G. Lenné
Monash University Accident Research Centre
Monash University
Victoria, Australia

Steve Love
Human-Centred Design Institute
Brunel University
Uxbridge, United Kingdom

Matthew Lyons
Boots
Nottingham, United Kingdom

Elaine Mackie
Department of Industrial Design, Coventry
    School of Art and Design
Coventry University
Coventry, United Kingdom

Tiago Costa Maia
Tuiuti University of Paraná
Curitiba, Brazil

Christopher B. Mayhorn
Department of Psychology
North Carolina State University
Raleigh, North Carolina

Eve Mitsopoulos-Rubens
Monash University Accident Research Centre
Monash University
Victoria, Australia

Louise Moody
Department of Industrial Design, Coventry
    School of Art and Design
Coventry University
Coventry, United Kingdom

Christopher P. Nemeth
Klein Associates Division of Applied Research
    Associates
Evanston, Illinois

Jan M. Noyes
Department of Experimental Psychology
University of Bristol
Bristol, United Kingdom

James O'Malley
Department for Work and Pensions
Sheffield, United Kingdom

Luis Carlos Paschoarelli
Department of Industrial Design, Faculty of
    Architecture, Arts and Communication
Universidade Estadual Paulista
Bauru, Brazil

José Carlos Plácido da Silva
Department of Industrial Design, Faculty of
    Architecture, Arts and Communication
Universidade Estadual Paulista
Bauru, Brazil

Lilia R. Prado-León
Ergonomic Research Center – Art,
    Architecture and Design Center
University of Guadalajara
Guadalajara, Mexico

David Ravnik
Rehabilitation Center
Kranj, Republic of Slovenia
University Clinic of Respiratory and Allergic
    Diseases
Golnik, Republic of Slovenia

Sinja Röbig
Institute of Ergonomics
Darmstadt University of Technology
Darmstadt, Germany

Peter Rogerson
Jobcentre Plus
Sheffield, United Kingdom

Paul M. Salmon
Monash University Accident Research Centre
Monash University
Victoria, Australia

Marcelo M. Soares
Department of Design
Federal University of Pernambuco
Recife, Brazil

Carla Galvao Spinillo
Department of Design
Federal University of Parana
Curitiba, Brazil

Neville A. Stanton
School of Civil Engineering and the
    Environment
University of Southampton
Highfield, United Kingdom

Bruce Thomas
Philips Design
Eindhoven, the Netherlands

Margaret J. Trotter
Monash University Accident Research Centre
Monash University
Victoria, Australia

Andrea Vallicelli
IDEA Department
University of Chieti-Pescara
Pescara, Italy

Ramon van de Ven
Philips Design
Eindhoven, the Netherlands

Guy H. Walker
School of the Built Environment
Heriot Watt University
Edinburgh, United Kingdom

Wen-Chia Wang
Human-Centred Design Institute
Brunel University
Uxbridge, United Kingdom

Stephen J. Ward
Industrial Design Program, Faculty of the Built
    Environment
University of New South Wales
Sydney, Australia

Thomas R. Waters
National Institute for Occupational Safety and
    Health
Cincinnati, Ohio

Damien J. Williams
School of Medicine
University of St Andrews
St Andrews, United Kingdom

Michael S. Wogalter
Department of Psychology
North Carolina State University
Raleigh, North Carolina

Mark S. Young
Human-Centred Design Institute
Brunel University
Uxbridge, United Kingdom

Aleksandar Zunjic
Faculty of Mechanical Engineering
University of Belgrade
Belgrade, Serbia

Section I

Design for Product Safety

1 Reducing the Risk to Consumers: Implications for Designing Safe Consumer Products

Damien J. Williams and Jan M. Noyes

CONTENTS

1.1 INTRODUCTION

The term "consumer product" refers to a wide variety of products (Hedge 2001), ranging from relatively simple products (e.g., cigarette lighters, washing-up liquid, and cosmetics) through to "white goods" (e.g., toasters, refrigerators, and washing machines) and "brown goods" (e.g., televisions, DVD players, and camcorders), to more complex technologies (e.g., aircraft and cars). In general, these products serve to protect, support, and/or replace particular activities, or extend consumer capabilities (Kanis 1998) and ultimately improve quality of life. While consumer products usually have very specific purposes or functions (Bonner 2001), they may also be used in unintended ways (i.e., the intended use of washing-up liquid is to wash dishes, but it is often used as a general cleaner; van Veen, van Engelen, and van Raaij 2001), which can result in an increased risk of injury or death. Indeed, each year, millions of injuries and thousands of fatalities can be attributed to consumer products (Rider et al. 2000; Gagg 2005). For instance, Baber and Mirza (1997) identified the small but significant risk that white goods pose to users. Consequently, while consumer product designers must strive for usability, given the importance of product design in consumer safety (Gagg 2005) and the expectation among consumers with regard to the safety of the diverse range of products they encounter daily (van Duijne, Kanis, and Green 2002), there is also a need to minimize as far as possible any hazards or risks (Benedyk and Minister 1997) and to produce safe products.

An accident resulting from an inadequately designed consumer product could have far-reaching implications. First, and most importantly, it can have an impact on the consumer directly (physical, psychological, financial harm). Second, it can have an indirect impact on the manufacturer. For instance, the recall of a product that is labeled "unsafe" may damage the manufacturer's reputation. As an example, on June 22, 2010, the Trading Standards Institute announced the recall of the 8-Sheet Diamond Cut Shredder purchased from Asda as a result of a potential safety issue identified in testing, which showed "there may be a fault in the plug that could cause an electric shock" (see http://www.tradingstandards.gov.uk/advice/advice-recall-list.cfm). Further, on June 24, 2010, following 47 reports of incidents/injuries (including a child who was found unconscious, eight reports of scratches and bruises, and one child who sustained a broken collarbone), the Consumer Product Safety Commission in cooperation with Jardine Enterprise Ltd. announced the recall of various models of drop-side cribs due to entrapment, suffocation, and fall hazards (see http://www.cpsc.gov/cpscpub/prerel/prhtml10/10275.html). These events can also have financial implications through lost income, criminal prosecution, or a (successful) product liability claim (Page 1997). Therefore, not only is there a moral requirement to assess the safety and safe usage of consumer products, but it is also in the best interests of the designers and manufacturers to undertake a process that reduces the occurrence of accidents involving products.

The aim of the current chapter is to identify how a consideration of the risk posed to consumers through consumer product use can be achieved and subsequently applied to the design of consumer products. To begin, the importance of risk in human behavior is identified; however, the way in which risk is conceptualized in the design process can influence how risk is approached and, ultimately, the designers understanding of the risk(s) faced by the consumer. Consequently, the issue of risk conceptualization (objective vs. subjective) will be addressed, followed by ways in which these different views of risk can be investigated in order to inform consumer product design (objective: accident statistics and exposure estimates vs. subjective: psychometric approach and usage-centered approach). These methodologies will then be considered within the typical approach to safety design (a hazard control hierarchy) and finally some possible barriers (risk transference and risk compensation) to the implementation of safety measures in consumer product design will be discussed.

1.2   RISK AND BEHAVIOR

Accidents involving consumer products are often attributed directly to consumer's risky behaviour (Ryan 1987). The perception of risk plays an important role in decision making and subsequent behavior (Williams 2007a): those who make erroneous assessments of risk are also likely to make sub-optimal decisions, which could lead to unsafe behavior and human error (Rundmo 2001). Such behavior may be initiated through inaccurate assessments of risk based on ineffective product design (i.e., physical characteristics and accompanying information and warnings); however, it is argued that when undertaking routine tasks (e.g., the use of consumer products), people may not worry about risks, nor even think about them (Wagenaar 1992). Consequently, under these circumstances, accidents may not be attributable to the (mis)perception of risk (Wagenaar 1992). Even from this perspective, given the number of accidents associated with consumer product use, it becomes evident that consumer product design plays an important role (Gagg 2005). Moreover, an understanding of risk perception is considered relevant to safety design as it *may* affect behavior, which can influence the likelihood of an accident (Rundmo 1996).

While an understanding of the risks faced by consumers (be they real or perceived or indeed, if they are raised to a level of conscious awareness at all) would facilitate the implementation of safety measures and mitigate the risk to which the consumer was exposed, the way in which risk is conceptualized during design could influence the understanding of the potential risk faced by the consumer. The following section will briefly discuss the issues surrounding conceptualization of risk and identify the implications for consumer product design.

1.2.1   CONCEPTUALIZATION OF RISK

A sharp distinction has been made between experts (i.e., risk assessors) and non-experts in terms of what constitutes a risk (Williams 2006). The traditional approach to the conceptualization of risk, particularly in scientific risk analysis, is to define risk as quantifiable and objective (Naoe 2008)—a mathematical value (Barki, Rivard, and Talbot 1993; Hoegberg 1998)—known as the probabilistic approach. With reference to consumer products, risk (of injury) is typically represented as a ratio between an accident or injury rate and a measure of exposure (Weegels and Kanis 2000). It is believed that these probabilistic risk assessments are based on "real risks" characterized as objective, analytic and rational, whereas anyone who does not subscribe to this approach is said to have an erroneous "risk perception" that is, it is subjective, hypothetical, emotional and irrational assessments of risk by Slovic (2007).

In questioning the suitability of the concept of real risk, Hansson (2005) identified seven pertinent myths:

1. Risk must have a single, well-defined meaning.
2. The severity of risks should be judged according to the probability-weighted averages of the severity of their outcomes.
3. Decisions on risk should be made by weighing total risks against total benefits.
4. Decisions on risk should be taken by experts rather than laymen (*sic*).
5. Risk-reducing measures in all sectors of society should be decided according to the same standards.
6. Risk assessments should be based only on well-established scientific facts.
7. If there is a serious risk, then scientists will find it if they look for it.

Indeed, a number of limitations have been identified with the probabilistic concept of real risk. First, the value derived by an expert merely considers past events (Rehmann-Sutter 1998); thus, a risk with a small probability gives no assurance that a statistically low event will not happen in the near future. With regard to accidents involving consumer products, Hayward (1996) indicated that there are not enough fatalities, for example, to deliver a reliable measure of risk. Secondly, not only is there a large degree of variation between experts' judgments of the same risk (Brehmer 1994), probabilistic risk assessments often differ from actual frequencies by at least two orders of magnitude (Shrader-Frechette 1991). Thirdly, the probabilistic view typically defines the physical impact, which neglects other important consequences (i.e., financial, psychological, and social) and issues that are not easily quantifiable (e.g., human element in exposure rate and human error). Finally, real risk neglects the large number of factors that influence those individuals facing the risks (Rehmann-Sutter 1998; Bohnenblust and Slovic 1998). Consequently, there would appear to be two limitations associated with a reliance on objective risk: (1) the degree of uncertainty with regard to the reliability of its use; and (2) the apparent neglect of a number of factors that are important in enabling a complete understanding of risk.

A large body of research from the psychometric paradigm, largely initiated by the seminal paper by Fischhoff et al. (1978), has demonstrated the complex, value-laden nature (Cross 1992) of non-experts' view of risk (see also Morgan et al. 1985; Kraus and Slovic 1988; Mullett et al. 1993; Sjöberg and Torell 1993). Through the use of factor analytic techniques, a number of underlying factors of risk have been identified (e.g., voluntariness, timelessness, knowledge about the risk, controllability, newness, catastrophic consequence, and dread). These factors have been grouped into four dimensions:

1. "Dread risk," which includes "perceived lack of control, dread, catastrophic potential, fatal consequence, and the inequitable distribution of risks and benefits" (Slovic 1987, 283).
2. "Unknown risks," which are "unobservable, unknown, new, and delayed in their manifestation of harm" (Slovic 1987, 283).

3. The number of people exposed to a given risk.
4. "Unnatural and immoral risk," which includes notions of tampering with nature and moral issues (Sjöberg 2000).

Despite the apparent limitations of the psychometric paradigm (see Millstein and Halpern-Felsher 2002 for a discussion) such findings provide the beginnings of a psychological classification system for risks (Slovic, Fischhoff, and Lichtenstein 1984) that may help improve our understanding of what constitutes a risk (Williams 2006) from the perspective of the consumer.

Hence, in order to develop safe consumer products, it is essential that designers and manufacturers recognize the complexity of the consumer, because even if they could calculate accurately the (objective) risk involved, it is their subjective perceptions that motivates behavior (Mitchell 1999). This is not meant to imply that probabilities are not important, but rather to indicate that numerous other subjective factors, beyond the objective scientific measures associated with risk of injury, are fundamentally important to the consumer's conceptualization of risk and subsequent behavior. Bernstein (1998) noted that risk-averse choices are based on a consideration of consequences without regard to the probability, while "foolhardy" choices are based solely on probability without regard to the consequences. Hence, any consideration of risk in the design process should necessarily incorporate the two elements of probability and consequence (Bernstein 1998) that correspond to the different ways in which risk is conceptualized; as a synonym for danger or threat (subjective risk) or a statistical value (objective risk) (Oppe 1988).

1.2.2   RISK IN CONSUMER PRODUCTS

1.2.2.1   Objective Risk

There are two primary ways in which the risk to consumers can be identified through objective methods—analysis of accident statistics and estimates of exposure. Each will be dealt with in turn.

1.2.2.1.1   Accident Statistics

One approach to identify the risk associated with consumer products is through the review of accident statistics. Such data are often held to be a key element in risk assessments (Rider et al. 2000). Three recent papers will be considered that report analyses of accident statistics and the implications for the design of safe consumer products.

1. Eye injuries—A broad range of consumer products are a significant cause of eye injuries. Recently, McGwin and colleagues utilized data from the Consumer Product Safety Commission's (CPSC) National Electronic Injury Surveillance System to identify all consumer product-related eye injuries in children (aged under 12 years) between 1997 and 2006 (Moren Cross et al. 2008) and adults between 1998 and 2002 (McGwin et al. 2006). There were apparent differences between the sexes and across the age spectrum in both studies. Nonetheless, the top 10 products among children were: sports equipment, household cleaning chemicals, toys, furniture, desk supplies, "other" chemicals and compounds, swimming pools and related equipment, toy guns, tools/hardware, and lawn and garden equipment. In adults, the top 10 products were: welding equipment, household cleaners, basketball (activity, clothing, or equipment), workshop/grinders/buffers/polishers, saws, adhesives, lawn mowers, bleaches, hair brushes/combs/curlers, and (ironically) eye protection devices. Moren Cross et al. note that the findings provide guidance concerning which categories of consumer product should be the target of preventative measures; however, McGwin et al. noted that many factors play a role in the occurrence of eye injuries, including the product's inherent risk to cause injury, behaviors to minimize the risks, and the frequency with which a product is used, which are not available when considering accident statistics. Indeed, while analysis of the accident statistics identifies the clear need for

manufacturers to add or strengthen safety measures (i.e., warnings, safety guards, etc.) in order to minimize the potential for eye injury, it does not provide any specific information to indicate what measure(s) should be used or how to implement them.

2. Scooter injuries—In recent times there has been an increase in injuries related to the use of unpowered scooters in the United States. Parker, O'Shea, and Simon (2004) reviewed 469 unpowered scooter-related injury reports compiled by the CPSC between January 1995 and June 2001. Most of the injuries were found to occur in older children and young adolescents; however, there were instances of injury in young children and adults. A broad spectrum of injuries was reported with the most frequent injuries including lacerations (26%), fractures (22%), and contusions (16%). Of these injuries, 24 (5.1%) required hospitalization and 15 were fatal. These findings led the authors to conclude that there was a need for improved safety awareness, with a particular emphasis on the use of protective equipment. Once again, however, no specific recommendations (i.e., content of safety awareness campaigns, or the type of protective equipment to be used) could be made for reducing the risk of injury posed by the use of unpowered scooters based solely on a consideration of the accident statistics.

3. Baby walker injuries—In response to the high number of baby walker injuries treated in U.S. hospital emergency departments associated with falls down stairs, the CPSC worked with industry to develop a voluntary standard to address the stair-fall risk. In order to evaluate the effectiveness of this standard, Rodgers and Leland (2008) conducted a thorough analysis of annual baby walker emergency department injury rates between 1981 and 2005. It was found that the requirements in the standard introduced in 1997 reduced the baby walker injury rate by approximately 60%, and the stair-fall injury rate by as much as 75%–80%. Moreover, a cost-benefit analysis indicated that the standard resulted in an overall benefit of $173 per walker, as a result of the reductions in the costs of injuries. While the study clearly demonstrates the success of the standard, which resulted in a relatively simple, effective, and inexpensive product modification, in terms of substantially reducing the risk of stair-fall injuries (Rodgers and Leland 2008), there were still a large number of injuries attributable to baby walkers and specifically stair falls for which the standard did not address. Moreover, there was no means of identifying the cause of the remaining accidents, and hence, the steps required to further reduce this risk.

The three examples reported here demonstrate the utility of accident statistics in evaluating the effectiveness of the real-world implementation of safety measures and the identification of risks in product use (van Duijne, Kanis, and Green 2002) albeit after accidents/incidents have occurred. Thus, the data can provide warnings about the presence of previously unidentified risks (van der Sman, Rider, and Chen 2007) and the need for intervention in order to improve the safety of products; however, there are a number of limitations in terms of possible safety recommendations based on accident statistics. First, the studies only report injuries treated in hospitals, which neglects those injuries treated in other settings, thereby potentially underestimating the risk associated with some products (Moren Cross et al. 2008). Secondly, it is not always possible to specify the cause of the accident or to locate the precise details about the accident (see van Duijne, Kanis, and Green 2002), which means that recommendations for interventions are very broad (i.e., safety awareness). For instance, one of the safety recommendations made by McGwin et al. (2006) was the use of eye protection while using certain consumer products; however, they found that eye protection devices were a significant cause of eye injuries. Consequently, a more detailed follow-up analysis is required following the identification of risk to consumers through accident statistics. Finally, the very nature of accident statistics means that accidents/incidents have already occurred, thereby requiring a reactive approach to "fix" the problem. What is required is a preventative approach that would reduce the risk of an event occurring in the first place.

1.2.2.1.2   Exposure Estimates

Exposure to the risk posed by consumer products is an issue that consumers can readily control; they can choose whether to allow the product into their home, how/where it is stored, and how/where/when it is used (Girman, Hogdson, and Wind 1987). Nonetheless, as outlined earlier, an important step in any (objective) risk analysis is the quantification of exposure to a given product (van Leeuwen and Hermans 1995; Hayward 1996). Hayward illustrated this through an investigation of 76 consumer products (kitchen, do-it-yourself, and household appliances) that were known from accident data to be involved in a high number of accidents. The overall measure of risk adopted for this study was the number of hospital-treated injuries per hour that the product was in use (number-of-injuries-per-hour-of-use). A number of limitations were identified with this approach. First, it is difficult to generate a parameter that accounts for injury severity (i.e., fatal or varying degrees of non-fatal injuries). Secondly, in this instance, the measure of exposure was calculated by multiplying the number of occasions per year on which the product was used and how long the product was used on the most recent occasion. Consequently, this is reliant on the most recent use of the product being representative of the general use of the product. Further, it may not be possible to pinpoint exactly the duration of use. For instance, the task of vacuuming may be broken down and spread across an entire day. Finally, given the importance of the (intended and unintended) actual use of consumer products in determining risk of injury, such a simple model does not account for additional factors that influence the risk associated with a product.

A large variety of everyday consumer products expose the consumer to various chemicals (e.g., textiles, televisions, computer equipment, and cosmetics) (Peters 2003). Sanderson et al. (2006) identify three routes by which contact with chemicals can occur as a result of intended and unintended usage, including: inhalation following emission into the air, dermal contact through spillage, and oral contact as a result of mouthing and subsequent ingestion (see also Steenbekkers 2001). Indeed, information about chemical constituents and exposure data play an important role in risk reduction (Steinemann 2009). Consequently, in order to ensure the safety of these products, risk assessments need to consider both the intrinsic hazard of constituents (Loretz et al. 2005) as well as exposure levels (i.e., concentrations) and the dose (i.e., milligram per day or milligram per kilogram per day) of any chemical that could potentially reach humans (van Leeuwen and Hermans 1995); however, Steinemann (2009) found that there is relatively little information available on the chemical constituents of many products, particularly fragranced consumer products (e.g., air fresheners, laundry supplies, personal care products, and cleaners). In addition to a consideration of the type of product and the emission and distribution of the chemicals, an exposure assessment should also take into account the influence of human behavior on contact with the chemicals (Lioy 1990; van Veen 1996). Indeed, it has been acknowledged that the consideration of consumer behavior is as important in exposure assessments as measures of exposure concentrations and the type of product (Weegels and van Veen 2001; Steenbekkers 2001).

One group of consumer products that rely on reliable exposure information in risk assessment includes cosmetic and personal care products (see Loretz et al. 2005, 2006, 2008). Indeed, millions of consumers use cosmetic and personal care products on a daily basis and human systemic exposure to their ingredients can rarely be prevented (Nohynek et al. 2010). A number of techniques have been employed to generate product exposure data (most usefully, the amount of product applied, and the frequency of use) to assess consumer use practices that would be useful in safety assessment. One method is the completion of detailed diaries of usage. For instance, Loretz and colleagues required participants to keep diaries recording detailed daily usage information over a period of 2 weeks for widely used cosmetic products (facial cleanser, hair conditioner, and eye shadow; see Loretz et al. 2008) and personal care products (spray perfume, hairspray, liquid foundation, shampoo, body wash, solid antiperspirant; see Loretz et al. 2006). In addition, the products were weighed at the start and on completion of the study in order to determine the total amount of product used. Nevertheless, this approach is somewhat limited in terms of generating a valid view of behavior as people do not always accurately or honestly report their behavior. By contrast, Weegels and van Veen (2001) utilized diaries, in-home observations, and direct measurements to study exposure to chemicals in a number of consumer products (dishwashing detergents, cleaning products, and hairstyling products). The mixed methods approach enabled an investigation of the type of actual contact, duration, frequency, and amount of use, and provided a clearer understanding of product usage.

In general, these and other similar studies (e.g., Weegels 1997; Riley et al. 2001; Loretz et al. 2005) have identified a large intra- and inter-individual variation in the use of the products; however, there was also evidence that people had developed a particular routine in the use of some products. Nonetheless, Weegels and van Veen (2001) concluded that by utilizing a number of different methods of data collection, they were able to get a more valid insight into product usage and therefore consumer exposure (see also Steenbekkers 2001). For instance, they noted that diary accounts were useful in collecting coarser information (i.e., frequency, duration, purpose, and amount of products used), while observations and home visits enable the collection of finer information (i.e., specific information regarding product use, and the specific measurements regarding amount used and duration). While the use of observational data enables a more valid account of product usage than the diary method, the behavior displayed may still differ from "real-world" behavior (see Garland and Noyes [2006] for a critique of the use of observation as a research methodology). Consequently, no matter what method is utilized, there may be a degree of inescapable uncertainty regarding the validity of the resulting data.

Due to the paucity of direct exposure data, such information is often estimated from mathematical models that describe the emission and distribution process and exposure factors (e.g., product use), which can provide a basis for conducting risk assessments (McNamara et al. 2007; van Veen, van Engelen, and van Raaij 2001; van Veen 1996; Riley et al. 2001; Matoba and van Veen 2005; van Veen et al. 1999). For instance, McNamara et al. developed a predictive model of European exposure to seven cosmetic products (body lotion, shampoo, deodorant spray, deodorant non-spray, facial moisturizer, lipstick, and toothpaste) by analyzing the usage profile (generated from market information databases and a controlled product use study) of frequency of use and amount of product used for a large European sample (44,100 households and 18,057 habitual users, comprising males and females). While such models are more complex than the one used by Hayward, in terms of accounting for user behavior and other compensatory actions, there are still issues with the calculation of the necessary parameters. For instance, while the emission and distribution process can be constant for a given product, exposure factors can be highly variable. In addition to the intra- and inter-individual variability in product use, other factors, including compensatory actions such as room ventilation, cannot be considered constant (either in terms of duration of use, positioning, or the power of the equipment). With regard to the intra- and inter-individual variation in product usage, Weegels and van Veen (2001) suggested that the quantification of this variability could be used to define "worst-case" or more extreme uses of a product for use in exposure assessments. Nonetheless, the lack of consideration of the variability in behavior represents a simplified view of exposure. Thus, risk assessment is still in search of methods, data, and models that can ultimately account for the large variability in human behavior (van Veen, van Engelen, and van Raaij 2001).

Deriving precise measures of exposure is a difficult task because the process is different for the various types of consumer products and because of the number of factors that influence exposure. In particular, the general lack of information with regard to actual product use severely hampers exposure estimation for consumer products (Weegels and van Veen 2001). Moreover, there appears to be a general lack of consideration for the variability in the parameters used to calculate exposure (i.e., duration of use, nature of use, injury severity, compensatory actions [ventilation], etc.). This creates a simplified view of the risk associated with a product, often at the level of worst-case scenario or the more extreme end of the usage spectrum.

1.2.2.2 Subjective Risk

Consumer behavior with a given product is determined largely by their perceptions of the risks associated with that product (Young, Brelsford, and Wogalter 1990; Wogalter, Desaulniers, and Brelsford 1986, 1987). Thus, safety solutions may fail if they do not take into account the perceptions and attitudes of those affected (Williamson et al. 1997). Furthermore, the possibility that consumers may incorrectly perceive the level of risk associated with a given product highlights the importance of understanding the perceived riskiness of the product (see Slovic, Fischhoff, and Lichtenstein 1979; Brems 1986). For instance, if a consumer perceives a lower degree of risk than is actually present in a given product, this could lead them unwittingly to act in an inappropriate/unsafe manner. Consequently, steps need to be taken during product design in order to overcome such misperceptions. Two general approaches exist to the study of risk perception associated with consumer products—the psychometric approach and the usage-centered approach. These will be dealt with in turn.

1.2.2.2.1 Psychometric Approach

Empirical research into risk perception of consumer products has largely been undertaken through a psychometric approach, which attempts to uncover the underlying components that influence product risk/hazard perceptions through the analysis of the outcomes (often made on a Likert scale) of a variety of questions using a number of statistical techniques (e.g., regression analysis, factor analysis, principal component analysis, analysis of variance, and correlation analysis). For instance, early work by Slovic, Fischhoff, and Lichtenstein (1979) demonstrated that while a small but measurable subjective component of risk influenced perceptions of risk, the objective component ultimately determined people's perceptions of risk associated with various products and activities. By contrast, Wogalter, Desaulniers, and Brelsford (1986) revealed that the subjective component (in particular severity of injury) played the primary role in shaping people's perceptions of risk. Similarly, Young and Wogalter (1998) showed that injury severity information accounted for the greatest proportion (78%) of the variance in perceived hazard.

Despite the apparent contradictions in the research findings reported in the literature, a number of studies have suggested that both objective and subjective risk components may influence perceptions of risk. For instance, Vaubel and Young (1992) examined the underlying dimensions associated with perceived risk for 40 consumer products (e.g., hair dryers, lighter fluid, whirlpool/hot tubs, and bicycles). Analysis of participant responses to 17 questions for each product, using principal component analysis, revealed that both subjective (i.e., degree to which potential risks were known and the immediacy of their onset) and objective (i.e., magnitude of the potential harm) aspects of risk play a role in risk perception, along with the person's familiarity with the product. Indeed, the relationship between familiarity and perceived hazardousness has been found previously (Godfrey et al. 1983; Godfrey and Laughery 1984; Wogalter, Desaulniers, and Brelsford 1986); however, it is highlighted that the relationship is more complex than intuition might suggest. Furthermore, Vaubel and Young found that each component was significantly related to participants' ratings of intention to act cautiously with a product.

The extent to which the outcomes from the psychometric approach can inform consumer product design has been questioned. For instance, it is argued that the psychometric approach is only suitable for uncovering opinions on risk as the analysis of questionnaire responses provides a very restricted interpretation of people's risk perception (van Duijne, Green, and Kanis 2001; Weegels and Kanis 2000). Hayward (1996) noted that perceptions of danger are an unreliable guide to the actual risk of injury associated with a product as it is driven by an over-reliance on characteristics of the products (i.e., the sharpness or power of the products) rather than on an assessment of the type of situations that can arise during product use, and the likelihood of occurrence. Consequently, it could be argued that the outcomes of the psychometric approach offer little in the way of usable information to support the design of safe consumer products as it does not enable any insight into risk perception associated with actual product usage.

Although the limitations identified with the psychometric approach are valid, it must be noted that none of the researchers have suggested that the outcomes of the psychometric approach could be informative for actual product design. Rather, the implications of the findings have often been directed at enabling an understanding of consumer perception of product risks in order to provide insight into the type of information that might be useful in the design and provision of warnings. Indeed, warnings that communicate appropriate and inappropriate product usage through labels and instruction manuals are a common approach to communicating risk information (Lesch 2005, 2006; Hancock, Fisk, and Rogers 2005) and improving safety (Hedge 2001). There is, however, evidence that warnings (and other product information) are not often noticed (see Trommelen 1997; Trommelen and Akerboom 1994) or understood by consumers (Hancock, Fisk, and Rogers 2005) or are ineffective (lack of appropriate information; Presgrave et al. 2008). Wogalter (2006a, 4) highlights the fundamental role of adequately designed warnings in the design process, stating that: "if a warning is absent or defective, then the product may be considered defective." Consequently, in instances where warnings are deemed necessary, it is not sufficient simply to provide a warning; it must be meaningful and communicate the correct information (Williams 2007b). Indeed, there is substantial human factors literature on the issues surrounding the provision of salient, well-designed warnings that attract attention and convey the appropriate risk information (Williams and Noyes 2007; Wogalter 2006b; Rogers, Lamson, and Rousseau 2000; Edworthy and Adams 1996; Wogalter, Dejoy, and Laughery 1999; Laughery, Wogalter, and Young 1994; Wogalter, Young, and Laughery 2001), which has a role in reducing the risk to users of consumer products.

1.2.2.2.2 Usage-Centered Approach

In consumer product use, numerous factors influence the consumer's perception of risk. In order to gain insight into the process of risk perception associated with consumer products, and provide designers with information that will assist in the development of safe products, it is necessary to consider the interaction of the consumer and the product. Kanis (1998) referred to this as usage-centered design.

One area of particular interest in product safety is the influence of featural and functional characteristics of a product on the way it is used (Weegels and Kanis 2000; van Duijne, Green, and Kanis 2001; Norman 1988; Laughery 1993). This relates to the concept of an open and obvious risk in which the design or function of a product communicates the risk (Laughery 1993). For instance, the teeth of a hand saw conveys the risk of injury through being cut, and the red glow emitted by a cooker hob when in use conveys the risk of being burnt; however, Laughery notes that numerous products do not convey such information and the risks faced by consumers are hidden (whether it is in the actual design or in the ineffective warning labels or instruction manuals). In such cases, consumers may not take proper precautions when using the product, which will increase the risk of injury (Williams, Kalsher, and Laughery 2006). For instance, Weegels and Kanis (2000) cite an example of an accident involving a metallic wastepaper basket. It was reported that the rim of the basket did not feel sharp, but when a user moved their hand along the rim while emptying the basket, they sustained a deep cut. It was concluded that the user might have emptied the basket in another way, had the risk posed by the rim not been hidden. Consequently, as usability problems can evoke unsafe situations (van Duijne et al. 2007), the investigation of the relationship between user activities and featural and functional product characteristics can reveal the role of risk perception in consumer product use. This will subsequently enable the identification of focal points for consideration in the design of safe consumer products (Kanis 2002).

A study by Weegels and Kanis (2000) utilized in-depth interviews with victims of consumer product accidents (involving kitchen utensils, do-it-yourself products and personal care items) and accident reconstructions in order to provide information about the risks perceived by users during product use. A number of interesting issues were identified. First, the majority of participants were simply unaware of the risk that had contributed to the accident. Secondly, product features did not adequately reflect the product's actual condition, which led participants to believe that there was no risk involved in the use of the product. Weegels and Kanis noted that these findings identify the influential role that featural and functional characteristics can have on consumer product use and consumer safety; however, they note that only those people involved in accidents with consumer products participated in the study, which limits the understanding of how the product is used by different people, such as those not involved in accidents. Further, the findings were based on interviews and accident reconstructions carried out a number of days after the actual accident. Consequently, participants could have simply forgotten some important details about the accident or they may have been selective in what they told the interviewer or fabricated information (i.e., false memories, Loftus and Pickrell [1995]; see Mazzoni and Vannucci [2007] for a review of the factors that can influence the recollection of autobiographical memories) so as not to appear to have caused the accident through their own negligence or stupidity. Thus, while the methods employed could provide insight into the role of risk in product use and the causes of accidents, there is a strong possibility that some relevant material was missed.

An alternative approach employs a common tool in consumer product evaluation—user trialing (Baber and Mirza 1997). This enables the identification of the variation in the actual usage of a given product by potential consumers (Kanis and Vermeeren 1996). For instance, in a study by van Duijne, Kanis, and Green (2002), observations of the use of a chip pan and a food blender were compared to the accident statistics for each of the products. In general, there was a degree of similarity between the two sources of information; however, it was noted that as a result of the necessary interpretation of the circumstances of the accidents, the descriptions from observed usage appeared to be more consistent with potential consumer actions. Furthermore, user trialing was able to show how risk perceptions were related to product characteristics. For instance, it was reported that when using the food blender, participants were aware of the risk of hot food being ejected at speed, but overlooked the more serious risk associated with putting their hand into the blender to remove food from the blades.

In an earlier study by van Duijne, Green, and Kanis (2001), participants were observed (video-recorded) and subsequently interviewed about the use of a number of common do-it-yourself products (hand saw, a chisel and hammer, an electric drill, and a screwdriver) while making a halving joint. This combination of observed product use and participants' previous experience provided interesting insights into how participants' perceived and evaluated risk when using the tools. First, it was noted that risk perception was difficult to observe in tool usage; it was suggested that this may be because the risk did not reach conscious awareness. Secondly, it was inferred through participants' personal experience that there was acceptance of a degree of risk and the occurrence of minor injuries.

The studies reported here (van Duijne, Green, and Kanis 2001; van Duijne, Kanis, and Green 2002; see also van Duijne et al. 2008) identified the same two issues: first, most participants were unaware of the risks posed by the products; secondly, safety was not necessarily the primary motivator in the use of the products, rather that most user actions were guided by the products afforded "ease of use" and the goals of the task. Consequently, in order to reduce the occurrence of even minor injuries, it is necessary to make the risks associated with a given product open and obvious, through either the provision of appropriate warnings or appropriate use cues (Kanis, Rooden, and Green 2000). For instance, van Duijne et al. (2008, 118) conclude that "by emphasising the characteristics of a product that trigger a perception of risk, users may become more alert to risk in usage."

In conclusion, knowledge about actual consumer activities is a useful aid in the (re)design of safe products and the prevention of accidents (see van Duijne et al. 2007). Indeed, a usage-centered approach enables a consideration of the effect of (potential) consumers' perceptions and evaluations of risk on product use (van Duijne et al. 2008). Further, such an approach can be undertaken during the design process (see McRoberts 2005) using related products, models, or prototypes (van Duijne, Kanis, and Green 2002). The knowledge gained can reveal subtle ways in which featural and functional product characteristics influence consumers, which would provide designers and manufacturers with insights into the requirements necessary to generate design solutions that assimilate the variation in user activities in order to prevent risks to consumer safety.

One limitation with the body of work associated with the usage-centered approach is that it involved mainly Dutch participants. Therefore, it is possible that the outcomes were specific to that subset of consumers. Consequently, given the global market of many consumer products, in order to gain real insight into potential use and misuse of products, it would be necessary to consider different types of consumers who might come in contact with the product (i.e., experienced and inexperienced, the elderly, the young, different cultures, physically and mentally impaired, etc.). Nonetheless, this work highlights the utility of the usage-centered design approach in reducing risk to consumers.

1.3   APPROACH TO SAFETY DESIGN

Increased attention has been given to the design of artifacts (products, equipment, environments, and services) that individuals encounter (Laughery and Hammond 1999) as a result of the moral concern for their welfare and because of the increase in litigation and personal injury claims in the developed world. A standard approach to dealing with risks and enhancing safety is through the process prescribed by a "hazard control hierarchy" (Sanders and McCormick 1993). Many such schemes exist (e.g., Laughery and Hammond 1999; Lehto and Clark 1990), which share the same fundamental approach to creating a product/environment that is inherently safe through the removal of hazards or their reduction to a level that is insufficient to cause harm (Clark and Lehto 2006). Dejoy, Cameron, and Della (2006) indicated that each of the steps in a hazard control hierarchy lie on a passive–active continuum. Passive measures provide automatic protection for the user through the removal, control, or alteration of the hazard through design/re-design or other steps that prevent exposure, such as the implementation of safety guards (Leonard 1999; Rogers, Lamson, and Rousseau 2000). By contrast, active measures require the user to engage or refrain from certain behavior(s) in order to avoid possible harm, which are implemented through training and warning people. It is not always possible or even viable to design-out risks as this can reduce the utility of the product/environment (Clark and Lehto 2006); accordingly, passive measures are becoming more commonplace and, in particular, warnings are frequently used to inform users of potential risks (Lesch 2005). Consequently, there is a particular need for both approaches to subjective risk, as outlined here: the outcomes of the usage-centered approach would identify the target functional and featural characteristics of a product that would need passive measures, while the psychometric approach could facilitate the design of warnings and enable the provision of active measures.

Despite the importance of warnings and the ease with which they can be provided (Dejoy, Cameron, and Della 2006), they are first and foremost injury control interventions (Laughery 2006) and not a remedy for poor design (Wogalter 2006a). Consequently, it is essential that passive measures are carefully implemented where possible in order to eliminate or control hazards, and that active measures, in the form of warnings, should be used as a supplement rather than a substitute (Edworthy, Stanton, and Hellier 1995; Rogers, Lamson, and Rousseau 2000; Trommelen 1997; Frantz, Miller, and Lehto 1991; Lehto and Salvendy 1995; Lehto 1992).

1.4   BARRIERS TO DESIGNING FOR PRODUCT SAFETY

A successful consumer product will facilitate a balance between exposing the consumer to an acceptable level of risk and depriving the consumer of useful services and features (Hecht 2003). Within risk management, the idea of acceptable risk as the basis for design (Vrijling, van Hengel, and Houben 1998; Naoe 2008) is based on the *de minimis* principle in which risks are reduced to a level that is insufficient to cause harm (Clark and Lehto 2006), meaning there is (effectively) no risk at all (Sandin 2005). Comar (1979) stated that the *de minimis* approach should promote understanding about how to deal with risk in the real world; encourage the provision of risk estimates; focus attention on actions that can effectively improve health and welfare; and avoid attempts to reduce small risks while neglecting larger ones. In this sense, the *de minimis* approach is primarily concerned with the probability of an outcome: a *de minimis* risk has a sufficiently small probability to make it negligible (Peterson 2002). There are a number of ways through which the *de minimis* approach can be implemented.

1. The *specific number view* in which a risk is considered *de minimus* if its probability falls below a certain, arbitrary number.
2. The *non-detectability view* through which a risk is *de minimus* if there is no scientific confirmation of its manifestation.
3. The *natural-occurrence view* considers a risk *de minimis* if its probability does not exceed the "natural" background levels of the type of risk.

Although, the *de minimis* principle has largely focused on objective risk levels consistent with the expert, probabilistic approach, Peterson (2002) acknowledged that there is nothing to exclude negligible risks being discussed in subjective terms. The usage-centered approach could provide a way in which to identify negligible risk, as was the case in the study by van Duijne, Kanis, and Green (2002), who identified an acceptance of the occurrence of minor injuries as a necessary part of the task. Further, Davey and Dalgetty (2007) noted that some product characteristics are an inherent part of design due to their function or purpose, and the risks associated with the characteristics are well known and accepted (e.g., hot water from the hot tap, hot steam emitted from a boiling kettle, and a sharp blade of a kitchen knife). Consequently, the *de minimis* principle could theoretically be applied to all consumer products in order to understand and manage both objective and subjective risks; however, it is possible that through the re-distribution of risk (Keeney 1995) any safety measure would merely transfer the risk to another source or action. In particular, measures to manage objective risk might be affected by the process of risk transference, while measures to manage subjective risk might be affected by the process of risk compensation. These two issues are briefly dealt with in the following sections.

1.4.1 RISK TRANSFERENCE

Within the field of risk regulation, it has been reported that attempts to control one risk unavoidably increases another risk (Cross 1998). While the concept of risk transference does not relate specifically to consumer products, it does identify the possibility that attempts to manage mainly objective risks could simply re-distribute the risk. Cross identifies three routes through which risk transference can occur. These are

1. *Substituting or using alternative substances or activities.* For instance, Cross (1996) noted how the change from gas fueled cars to electric cars was seen as a way of tackling the growing risk of pollution; however, this action simply transferred the source of pollution from the car to the electric generating plant, which was also suggested to increase the risk of lead exposure.
2. *Foregone benefits.* For instance, Cross (1996) noted that risk to health associated with pesticide residue could cause consumers to avoid fresh produce, which would be far more risky to their health.
3. *Remediation efforts.* For instance, Graham and Wiener (1995) noted that as a result of government requirements for minimum mileage requirements, manufacturers were required to produce smaller cars, which were associated with an increase in road deaths.

While risk transference is associated with risk regulation, which is often based on objective risk estimates, it is also evident that the re-distribution of risk associated with risk transference is applicable to consumer products. Considering the possibility that consumers deliberately put their hand in a food blender to remove food stuck around the blade (see van Duijne, Kanis, and Green [2002] for observational data that support the occurrence of this action), Baber and Mirza (1997) identified two potential measures to prevent this action. The first involved fitting interlocks on a cover over the blender aperture in order to prevent consumers putting their hand into the aperture. The second involved the standardization of the depth and size of the blender aperture to prevent consumers from reaching down to the blade. Baber and Mirza noted that while the first solution would be more expensive to implement, the second might lead consumers to transfer the risk to another behavior through the use of other objects such as a wooden spoon, thereby risking injury from flying splinters if the wooden spoon came in contact with the rotating blade. Consequently, while concerted efforts may be made to reduce the risk to consumers, it is possible that the safety measures may merely act to re-distribute the risk (Keeney 1995). Thus, it is necessary that the designer(s) and manufacturer(s) consider the consequences of implementing safety measures, and design with these in mind.

1.4.2   RISK COMPENSATION

The implicit belief in risk management is that risk can be reduced through the use of appropriate safety measures (Stanton and Glendon 1996). Despite this, empirical evidence from the driver-safety literature suggests that individuals have an innate, predetermined level of acceptable risk, and when confronted with a risk that is lower or higher, they will act in manner to maintain their desired level. One way in which this will occur, which is consistent with the idea of risk transference, is for the risk to be transferred from "regulated" to "unregulated" behavior (Jackson and Blackman 1994). A potential consequence of this re-distribution of risk is that the effect of any safety measures may be negated (Weegels and Kanis 2000).

Specific explanations for the consistent finding that drivers respond to the addition of safety measures (e.g., anti-lock brake system, seat belts, and air bags) or a change in legislation (i.e., mandatory seat belt use or helmet use on motorcycles/bicycles) with risky behavior (see Jackson and Blackman 1994; Horswill and Coster 2002; Stanton and Pinto 2000; Hoyes et al. 1996; cf. Lewis-Evans and Charlton 2006; Scott et al. [2007] who failed to find compensatory behavior in other risky domains) have been attempted via two models. First, Pelzman (1975) proposed the risk compensation theory (RCT), which simply suggested that when the risks associated with a certain behavior decrease, individuals would compensate by taking greater risks through another behavior. For instance, when individuals are provided with a protective device (e.g., bicycle helmet or seat belt), they will act in a riskier manner because of the sense of increased protection, thereby nullifying the protection afforded by the device (Thompson, Thompson, and Rivara 2002). However, the theory does not indicate the degree of compensation. By contrast, risk homeostasis theory (RHT; Wilde 1982, 1988, 1989) explicitly predicts that any reduction in the level of risk (real or perceived) following the implementation of safety measures, would cause the individual to re-distribute the risk around other behaviors (Stewart 2004), thereby returning to the original combination of risk and reward (Hoyes 1994; Dulisse 1997). The consequences of this are that any attempt to manage risk would be continually thwarted (Stewart 2004).

While the main body of work on RHT and RCT has been largely associated with driving behavior, Noyes (2001) notes that the possibility of some compensatory mechanism presents disturbing theoretical implications if people are found to operate in a similar way in other aspects of their life. There is some evidence that people do behave in a manner consistent with RHT within other domains (see Hoyes 1994; Desmond and Hoyes 1996; Hersch and Viscusi 1990). Viscusi and Cavallo (1994) reported a field study investigating the influence of a lighter with child-resistant features on risk beliefs and precautions. It was found that the inclusion of child-resistant features reduced perceptions of risk and there was some degree of compensatory behavior; however, the child-resistant features were found to reduce fire-related injuries by much more than any diminished precaution taking (i.e., allowing children to play more freely with the lighter than before the safety features were added). This suggests that some degree of compensatory behavior *may* occur when safety measures are added to consumer products (see also Viscusi 1984; cf. Walton 1982). Indeed,

Wilde (1998, 91) states that "the mechanisms involved in risk homeostasis are probably universal" and would therefore apply across domains.

Compensatory behaviors could be considered a natural reaction through which people attempt to regain control of their environment (Stanton and Glendon 1996); however, Hedlund (2000, 87) stated that "I believe the evidence is overwhelming that every safety law or regulation is not counterbalanced by compensating behavior." Nonetheless, the paucity of research concerning the effect of reducing risk associated with consumer products through safety measures limits the extent to which reliable conclusions can be formed. Moreover, despite evidence supporting the occurrence of a mild degree of compensatory behavior (see Stetzer and Hofmann 1996; Viscusi and Cavallo 1994) contrary to Wilde's (1998) assertion of the generality of the homeostatic mechanisms, recent evidence shows that no such compensation occurs in a number of domains (see Lewis-Evans and Charlton 2006; Scott et al. 2007), suggesting that it may actually be domain specific. Nonetheless, until the theory is robustly tested in the domain of consumer product design, there is a theoretical possibility that consumers *might* react to safety measures in a compensatory fashion such that alternative risky behaviors result from perceptions that a product has become safer. In this case, it would be necessary to consider these barriers to safety design to ensure that risk is dealt with in an appropriate manner to safeguard the consumer and those potentially affected by consumer product use.

1.5   DISCUSSION

The most effective consumer products enhance usability, in the sense that they are easy to use as they assimilate consumer behavior and, at the same time, prevent injuries and accidents (see Stanton 1997). Considering the risk to consumers is a particularly important step in designing safe products. To do this, it is first essential to consider the conceptualization of risk (objective vs. subjective) employed during design as neglecting the role of either type would necessarily mean that important safety issues will be overlooked. For instance, the consideration of objective risk is necessary in identifying products that pose a risk to consumers (accident statistics) and the extent of the risk (exposure data). Given the large number of chemicals that can be emitted by consumer products, exposure data are a crucial element in any risk assessment. Nonetheless, there is increasing acknowledgment of the importance of the human element (compensatory actions and product use) in estimates of exposure. Consequently, consumers' perceptions and evaluations of risk and their resulting behavior are an important element in the consideration of both subjective and objective risk. Hence, the different approaches to risk provide different kinds of information applicable to the improvement of consumer product safety.

While there is a substantial body of literature relating to issues relevant to the risk (and safety) debate, these are distributed across a vast number of domains. For instance, while risk transference and risk compensation are theoretically important in consumer product use and design, most of the work is concentrated in other domains. Consequently, given the role that risk plays in the use of consumer products, it is necessary that research addresses the relevant risk issues that may influence the design of consumer products.

In conclusion, the most effective approach to designing usable consumer products is to implement human factors principles as early as possible, rather than as reactive measures to usability problems. This should also be the approach in designing for safety, wherein a preventative approach is adopted such that risk and safety issues are given appropriate consideration, prior to product release. Given the difficulties experienced by the human factors discipline in terms of designers and manufacturers implementing appropriate usability considerations throughout the design life cycle, it is likely that the preventative approach to safety design will face the same lack of acceptance; however, given the potential financial risks associated with the release of an unsafe product, designers and manufacturers might be encouraged to adopt the principle of preventative safety design with a little more haste.

REFERENCES

Baber, C., and Mirza, M.G. 1997. Ergonomics and the evaluation of consumer products: Surveys of evaluation practices. In *Human Factors in Consumer Products*, ed. N. Stanton, 91–103. London: Taylor & Francis.

Barki, H., Rivard, S., and Talbot, J. 1993. Toward an assessment of software development risk. *Journal of Management Information Systems* 10 (2): 203–25.

Benedyk, R., and Minister, S. 1997. Evaluation of product safety using the BeSafe method. In *Human Factors in Consumer Products*, ed. N. Stanton, 55–74. London: Taylor & Francis.

Bernstein, P.L. 1998. *Against the Gods: The Remarkable Story of Risk*. New York: John Wiley.

Bohnenblust, H., and Slovic, P. 1998. Integrating technical analysis and public values in risk-based decision making. *Reliability Engineering and System Safety* 59:151–59.

Bonner, J.V.H. 2001. Human factors design tools for consumer-product interfaces. In *International Encyclopedia of Ergonomics and Human Factors*, ed. W. Karwowski, 1839–42. Florence, KY: Taylor & Francis.

Brehmer, B. 1994. Some note on psychological research related to risk. In *Future Risks and Risk Management*, eds. B. Brehmer and N.E. Sahlin, 79–91. Dordrecht: Kluwer Academic.

Brems, D.J. 1986. Risk estimation for common consumer products. In *Proceedings of the Human Factors Society 30th Annual Meeting*, 556–60. Santa Monica, CA: Human Factors Society.

Clark, D.R., and Lehto, M.R. 2006. Information design: Warning signs and labels. In *International Encyclopedia of Ergonomics and Human Factors*, ed. W. Karwowski, 1152–55. London: CRC Press.

Comar, C.L. 1979. Risk: A pragmatic *de minimis* approach. *Science* 203:319.

Commission, U.S. Consumer Product Safety. 2007. *News for CPSC. Release #07-147* 2007, http://www.cpsc.gov/cpscpub/prerel/prhtml07/07147.html (accessed April 26, 2007).

Cross, F.B. 1992. The risk of reliance on perceived risk. Review of reviewed item. *Risk: Health, Safety & Environment*. Available from http://www.piercelaw.edu/risk/vol3/winter/cross.htm

———. 1996. Paradoxical perils of the precautionary principle. *Washington and Lee Law Review* 53:851–925.

———. 1998. Facts and values in risk assessment. *Reliability Engineering & System Safety* 59 (1): 27–40.

Davey, C., and Dalgetty, I. 2007. Questions about product hazards and product characteristics. Paper presented at the 3rd Meeting of the EuroSafe Working Group on Risk Assessment. Vienna.

Dejoy, D.M., Cameron, K.A., and Della, L.J. 2006. Postexposure evaluation of warning effectiveness: A review of field studies and population-based research. In *Handbook of Warnings*, ed. M.S. Wogalter, 35–48. Mahwah, NJ: Lawrence Erlbaum Associates.

Desmond, P.A., and Hoyes, T.W. 1996. Workload variation, intrinsic risk and utility in a simulated air traffic control task: Evidence for compensatory effects. *Safety Science* 22 (1–3): 87–101.

Dulisse, B. 1997. Methodological issues in testing the hypothesis of risk compensation. *Accident Analysis and Prevention* 29 (3): 285–92.

Edworthy, J., and Adams, A. 1996. *Warning Design: An Integrative Approach to Warnings Research*. London: Taylor & Francis.

Edworthy, J., Stanton, N., and Hellier, E. 1995. Warnings in research and practice. *Ergonomics* 38 (11): 2145–54.

Fischhoff, B., Slovic, P., Lichtenstein, S., Read, S., and Coombs, B. 1978. How safe is safe enough? A psychometric study of attitudes toward technological risks and benefits. *Policy Sciences* 9:127–52.

Frantz, J.P., Miller, J.M., and Lehto, M.R. 1991. Must the context be considered when applying generic safety symbols: A case-study in flammable contact adhesives. *Journal of Safety Research* 22 (3): 147–61.

Gagg, C. 2005. Domestic product failures – Case studies. *Engineering Failure Analysis* 12 (5): 784–807.

Garland, K.J., and Noyes, J.M. 2006. Observation. In *International Encyclopedia of Ergonomics and Human Factors*, ed. W. Karwowski, 3285–88. London: Taylor & Francis.

Girman, J.R., Hogdson, A.T., and Wind, M.L. 1987. Considerations in evaluating emissions from consumer products. *Atmospheric Environment* 21:315–20.

Godfrey, S.S., Allender, L., Laughery, K.R., and Smith, V.L. 1983. Warning messages: Will the consumer bother to look? In *Proceedings of the Human Factors Society 27th Annual Meeting*, 950–54. Santa Monica, CA: Human Factors Society.

Godfrey, S.S., and Laughery, K.R. 1984. The biasing effect of product familiarity on consumers' awareness of hazard. In *Proceedings of the Human Factors Society 28th Annual Meeting*, 483–86. Santa Monica, CA: Human Factors Society.

Graham, J.D., and Wiener, J.B. 1995. *Risk vs. Risk: Tradeoffs in Protecting Health and the Environment*. Cambridge, MA: Harvard University Press.

Hancock, H.E., Fisk, A.D., and Rogers, W.A. 2005. Comprehending product warning information: Age-related effects and the roles of memory, inferencing, and knowledge. *Human Factors* 47 (2): 219–34.

Hansson, S.O. 2005. Seven myths of risk. *Risk Management: An International Journal* 7 (2): 7–17.

Hayward, G. 1996. Risk of injury per hour of exposure to consumer products. *Accident Analysis and Prevention* 28 (1): 115–21.

Hecht, M. 2003. The role of safety analyses in reducing products liability exposure in "smart" consumer products containing software and firmware. In *Proceedings of the Annual Reliability and Maintainability Symposium*, 153–58. http://ieeexplore.ieee.org/xpls/abs_all.jsp?arnumber=1181918

Hedge, A. 2001. Consumer product design. In *International Encyclopedia of Ergonomics and Human Factors*, ed. W. Karwowski, 888–91. Florence, KY: Taylor & Francis.

Hedlund, J. 2000. Risky business: Safety regulations, risk compensation, and individual behavior. *Injury Prevention* 6:82–89.

Hersch, J., and Viscusi, W.K. 1990. Cigarette smoking, seatbelt use, and differences in wage-risk tradeoffs. *The Journal of Human Resources* 25 (2): 202–27.

Hoegberg, L. 1998. Risk perception, safety goals and regulatory decision-making. *Reliability Engineering and System Safety* 59:135–39.

Horswill, M.S., and Coster, M.E. 2002. The effect of vehicle characteristics on drivers' risk-taking behaviour. *Ergonomics* 45 (2): 85–104.

Hoyes, T.W. 1994. Risk homeostasis theory: Beyond transportational research. *Safety Science* 17 (2): 77–89.

Hoyes, T.W., Dorn, L., Desmond, P.A., and Taylor, R. 1996. Risk homeostasis theory, utility and accident loss in a simulated driving task. *Safety Science* 22 (1–3): 49–62.

Jackson, J.S.H., and Blackman, R. 1994. A driving-simulator test of Wilde's risk homeostasis theory. *Journal of Applied Psychology* 79 (6): 950–58.

Kanis, H. 1998. Usage centred research for everyday product design. *Applied Ergonomics* 29 (1): 75–82.

———. 2002. Can design-supportive research be scientific? *Ergonomics* 45 (14): 1037–41.

Kanis, H., Rooden, M.J., and Green, W.S. 2000. Use cues in the Delft design course. In *Contemporary Ergonomics 2000*, ed. M.A. Hanson, 365–69. London: Taylor & Francis.

Kanis, H., and Vermeeren, A.P.O.S. 1996. Teaching user involved design in the Delft curriculum. In *Contemporary Ergonomics 1996*, ed. S.A. Robertson, 98–103. London: Taylor & Francis.

Keeney, R.L. 1995. Understanding life-threatening risks. *Risk Analysis* 15:627–38.

Kraus, P.P., and Slovic, P. 1988. Taxonomic analysis of perceived risk: Modeling individual and group perceptions within homogeneous hazard domains. *Risk Analysis* 8:435–55.

Laughery, K.R. 1993. Everybody knows: Or do they? *Ergonomics in Design* July: 8–13.

———. 2006. Safety communications: Warnings. *Applied Ergonomics* 37 (4): 467–78.

Laughery, K.R., and Hammond, A. 1999. Overview. In *Warnings and Risk Communication*, eds. M.S. Wogalter, D.M. Dejoy, and K.R. Laughery, 3–13. London: Taylor & Francis.

Laughery, K.R., Wogalter, M.S., and Young, S.L. (eds). 1994. *Human Factors Perspectives on Warnings*. Vol. 1. Santa Monica, CA: Human Factors and Ergonomics Society.

Lehto, M.R. 1992. Designing warning signs and warning labels: Part I – Guidelines for the practitioner. *International Journal of Industrial Ergonomics* 10 (1–2): 105–13.

Lehto, M.R., and Clark, D.R. 1990. Warning signs and labels in the workplace. In *Workspace, Equipment and Tool Design*, eds. W. Karwowski and A. Mital, 303–44. Amsterdam: Elsevier.

Lehto, M.R., and Salvendy, G. 1995. Warnings – A supplement not a substitute for other approaches to safety. *Ergonomics* 38 (11): 2155–63.

Leonard, S.D. 1999. Does color of warnings affect risk perception? *International Journal of Industrial Ergonomics* 23 (5–6): 499–504.

Lesch, M.F. 2005. Remembering to be afraid: Applications of theories of memory to the science of safety communication. *Theoretical Issues in Ergonomics Science* 6 (2): 173–91.

———. 2006. Consumer product warnings: Research and recommendations. In *Handbook of Warnings*, ed. M.S. Wogalter, 137–45. Mahwah, NJ: Lawrence Erlbaum Associates.

Lewis-Evans, B., and Charlton, S.G. 2006. Explicit and implicit processes in behavioural adaptation to road width. *Accident Analysis & Prevention* 38 (3): 610–17.

Lioy, P.J. 1990. Assessing total human exposure to contaminants: A multidisciplinary approach. *Environmental Science & Technology* 24 (7): 938–45.

Loftus, E.F., and Pickrell, J.E. 1995. The formation of false memories. *Psychiatric Annals* 25:720–25.

Loretz, L.J., Api, A.M., Babcock, L., Barraj, L.M., Burdick, J., Cater, K.C., Jarrett, G., et al. 2008. Exposure data for cosmetic products: Facial cleanser, hair conditioner, and eye shadow. *Food and Chemical Toxicology* 46 (5): 1516–24.

Loretz, L., Api, A.M., Barraj, L., Burdick, J., Davis, D.A., Dressler, W., Gilberti, E., et al. 2006. Exposure data for personal care products: Hairspray, spray perfume, liquid foundation, shampoo, body wash, and solid antiperspirant. *Food and Chemical Toxicology* 44 (12): 2008–18.

Loretz, L.J., Api, A.M., Barraj, L.M., Burdick, J., Dressler, W.E., Gettings, S.D., Hsu, H.H., et al. 2005. Exposure data for cosmetic products: Lipstick, body lotion, and face cream. *Food and Chemical Toxicology* 43 (2): 279–91.

Matoba, Y., and van Veen, M.P. 2005. Predictive residential models. In *Occupational and Residential Exposure Assessment*, eds. E. Franklin and J. Worgan, 209–42. Chichester: John Wiley.

Mazzoni, G., and Vannucci, M. 2007. Hindsight bias, the misinformation effect, and false autobiographical memories. *Social Cognition* 25 (1): 203–20.

McGwin, G., Jr., Hall, T.A., Seale, J., Xie, A., and Owsley, C. 2006. Consumer product-related eye injury in the United States, 1998–2002. *Journal of Safety Research* 37:501–6.

McNamara, C., Rohan, D., Golden, D., Gibney, M., Hall, B., Tozer, S., Safford, B., Coroama, M., Leneveu-Duchemin, M.C., and Steiling, W. 2007. Probabilistic modelling of European consumer exposure to cosmetic products. *Food and Chemical Toxicology* 45 (11): 2086–96.

McRoberts, S. 2005. Risk management of product safety. In *Proceedings of the 2005 IEEE Symposium on Product Safety Engineering*. http://ieeexplore.ieee.org/search/srchabstract.jsp?tp=&arnumber=1529524&queryText%3DRisk+management+of+product+safety%26openedRefinements%3D*%26searchField%3DSearch+All

Millstein, S.G., and Halpern-Felsher, B.L. 2002. Perceptions of risk and vulnerability. *Journal of Adolescent Health* 31 (1): 10–27.

Mitchell, V-W. 1999. Consumer perceived risk: Conceptualisations and models. *European Journal of Marketing* 33 (1/2): 163–95.

Moren Cross, J., Griffin, R., Owsley, C., and McGwin G., Jr. 2008. Pediatric eye injuries related to consumer products in the United States, 1997–2006. *Journal of American Association for Pediatric Ophthalmology and Strabismus* 12 (6): 626–28.

Morgan, M.G., Slovic, P., Nair, I., Geisler, D., MacGregor, D., Fischhoff, B., Lincoln, D., and Florig, K. 1985. Powerline frequency electric and magnetic fields: A pilot study of risk perception. *Risk Analysis* 5:139–49.

Mullett, E., Duquesnoy, C., Raiff, P., Fahrasmane, R., and Namur, E. 1993. The evaluative factor of risk perception. *Journal of Applied Social Psychology* 23 (19): 1594–605.

Naoe, T. 2008. Design culture and acceptable risk. In *Philosophy and Design: From Engineering to Architecture*, eds. P. Kroes, P.E. Vermaas, A. Light, and S.A. Moore, 119–30. Netherlands: Springer.

Nohynek, G.J., Antignac, E., Re, T., and Toutain, H. 2010. Safety assessment of personal care products/cosmetics and their ingredients. *Toxicology and Applied Pharmacology* 243:239–59.

Norman, D.A. 1988. *The Psychology of Everyday Things*. Cambridge, MA: MIT Press.

Noyes, J.M. 2001. *Designing for Humans*. Hove: Psychology Press.

Oppe, S. 1988. The concept of risk – a decision theoretic approach. *Ergonomics* 31 (4): 435–40.

Page, M. 1997. Consumer products: More by accident than design? In *Human Factors in Consumer Products*, ed. N. Stanton, 127–46. London: Taylor & Francis.

Parker, J.F., O'Shea, J.S., and Simon, H.K. 2004. Unpowered scooter injuries reported to the Consumer Product Safety Commission: 1995–2001. *American Journal of Emergency Medicine* 22 (4): 273–75.

Pelzman, S. 1975. The effects of automobile safety regulation. *Journal of Political Economics* 83:677–725.

Peters, R.J.B. 2003. *Hazardous Chemicals in Consumer Products* (Report no. 34629). TNO Nederlands Organisation for Applied Scientific Research.

Peterson, M. 2002. What is a *de minimis* risk? *Risk Management: An International Journal* 4 (2): 47–55.

Presgrave, R.D., Alves, E.N., Camacho, L.A.B., and Boas, M.H.S.V. 2008. Labelling of household products and prevention of unintentional poisoning. *Ciencia & Saude Coletiva* 13:683–88.

Rehmann-Sutter, C. 1998. Involving others: Towards an ethical concept of risk. Review of reviewed item. *Risk: Health, Safety & Environment*. Available from http://www.piercelaw.edu/risk/vol9/spring/Rehman.pdf

Rider, G., Milkovich, S., Stool, D., Wiseman, T., Doran, C., and Chen, X. 2000. Quantitative risk analysis. *Journal for Injury Control and Safety Prevention* 7 (2): 115–33.

Riley, D.M., Fischhoff, B., Small, M.J., and Fischbeck, P. 2001. Evaluating the effectiveness of risk-reduction strategies for consumer chemical products. *Risk Analysis* 21 (2): 357–69.

Rodgers, G.B., and Leland, E.W. 2008. A retrospective benefit-cost analysis of the 1997 stair-fall requirements for baby walkers. *Accident Analysis & Prevention* 40 (1): 61–68.

Rogers, W.A., Lamson, N., and Rousseau, G.K. 2000. Warning research: An integrative perspective. *Human Factors* 42 (1): 102–39.

Rundmo, T. 1996. Associations between risk perception and safety. *Safety Science* 24 (3): 197–209.

———. 2001. Employees images of risk. *Journal of Risk Research* 4 (4): 393–404.

Ryan, R.P. 1987. Consumer behaviour considerations in product design. In *Proceedings of the Human Factors Society 31st Annual Meeting*, 1236–40. Santa Monica: Human Factors Society.

Sanders, M., and McCormick, E. 1993. *Human Factors in Engineering and Design*, 7th ed. New York: McGraw-Hill.

Sanderson, H., Counts, J.L., Stanton, K.L., and Sedlak, R.I. 2006. Exposure and prioritization – Human screening data and methods for high production volume chemicals in consumer products: Amine oxides a case study. *Risk Analysis* 26 (6): 1637–57.

Sandin, P. 2005. Naturalness and de minimis risk. *Environmental Ethics* 27 (2): 191–200.

Scott, M.D., Buller, D.B., Andersen, P.A., Walkosz, B.J., Voeks, J.H., Dignan, M.B., and Cutter, G.R. 2007. Testing the risk compensation hypothesis for safety helmets in alpine skiing and snowboarding. *Injury Prevention* 3:173–77.

Shrader-Frechette, K.S. 1991. *Risk and Rationality*. Berkeley, CA: University of California.

Sjöberg, L. 2000. Factors in risk perception. *Risk Analysis* 20 (1): 1–11.

Sjöberg, L., and Torell, G. 1993. The development of risk acceptance and moral valuation. *Scandinavian Journal of Psychology* 34:223–36.

Slovic, P. 1987. Perception of risk. *Science* 236:280–85.

———. 1997. Trust, emotion, sex, politics and science. In *Environment, Ethics and Behaviour*, eds. M.H. Bazerman, D.M. Messick, A.E. Tenbrunsel, and K.A. Wade-Benzoni, 277–313. San Francisco: Lexington Press.

Slovic, P., Fischhoff, B., and Lichtenstein, S. 1979. Rating the risks. *Environment* 21 (3): 14–20.

———. 1984. Behavioural decision theory perspectives on risk and safety. *Acta Psychologia* 56:183–203.

Stanton, N., and Glendon, I. 1996. Risk homeostasis and risk assessment. *Safety Science* 22 (1–3): 1–13.

Stanton, N.A., and Pinto, M. 2000. Behavioural compensation by drivers of a simulator when using a vision enhancement system. *Ergonomics* 43 (9): 1359–70.

Steenbekkers, L.P.A. 2001. Methods to study everyday use of products in households: The Wageningen mouthing study as an example. *Annals of Occupational Hygiene* 45 (1001): s125–29.

Steinemann, A.C. 2009. Fragranced consumer products and undisclosed ingredients. *Environmental Impact Assessment Review* 29 (1): 32–38.

Stetzer, A., and Hofmann, D.A. 1996. Risk compensation: Implications for safety interventions. *Organizational Behavior and Human Decision Processes* 66 (1): 73–88.

Stewart, A. 2004. On risk: Perception and direction. *Computers & Security* 23 (5): 362–70.

Thompson, D.C., Thompson, R.S., and Rivara, F.P. 2002. Risk compensation theory should be subject to systematic reviews of the scientific evidence. *Injury Prevention* 8:e1.

Trommelen, M. 1997. Effectiveness of explicit warnings. *Safety Science* 25 (1–3): 79–88.

Trommelen, M., and Akerboom, S.P. 1994. Explicitness in warnings provided with child-care products. Paper read at Proceedings of the International Symposium of Public Graphics. Lunteren.

van der Sman, C., Rider, G., and Chen, X. 2007. Questions about (mechanism of) reported injuries and complaints. Paper presented at the 3rd Meeting of the EuroSafe Working Group on Risk Assessment. Vienna.

van Duijne, F.H., Green, W.S., and Kanis, H. 2001. Risk perception: Let the user speak. In *Contemporary Ergonomics 2001*, ed. M.A. Hanson, 291–96. London: Taylor & Francis.

van Duijne, F.H., Hale, A., Kanis, H., and Green, B. 2007. Design for safety: Involving users' perspectives. Redesign proposals for gas lamps using a pierceable cartridge. *Safety Science* 45:253–81.

van Duijne, F.H., Kanis, H., and Green, B. 2002. Risks in product use: Observations compared to accident statistics. *Injury Control and Safety Promotion* 9 (3): 185–91.

van Duijne, F.H., Kanis, H., Hale, A.R., and Green, B.S. 2008. Risk perception in the usage of electrically powered gardening tools. *Safety Science* 46:104–18.

van Leeuwen, C.J., and Hermans, J.L.M. 1995. *Risk Assessment of Chemicals: An Introduction*. Dordrecht: Kluwer.

van Veen, M.P. 1996. A general model for exposure and uptake from consumer products. *Risk Analysis* 16:331–38.

van Veen, M.P., Fortezza, F., Bloemen, H.J.T.H., and Kliest, J.J. 1999. Indoor air exposure to volatile compounds emitted by paints: Model and experiment. *Journal of Exposure Analysis and Environmental Epidemiology* 9 (6): 569–74.

van Veen, M.P., van Engelen, J.G.M., and van Raaij, M.T.M. 2001. Crossing the river stone by stone: Approaches for residential risk assessment for consumers. *Annals of Occupational Hygiene* 45:S107–18.

Vaubel, K.P., and Young, S.L. 1992. Components of perceived risk for consumer products. In *Proceedings of the Human Factors 36th Annual Meeting*, 494–98. Santa Monica, CA: Human Factors Society.

Viscusi, W.K. 1984. The lulling effect: The impact of child-resistant packaging on aspirin and analgesic ingestions. *American Economic Review* 74 (2): 323–27.

Viscusi, W.K., and Cavallo, G.O. 1994. The effect of product safety regulation on safety precautions. *Risk Analysis* 14 (6): 917–30.

Vrijling, J.K., van Hengel, W., and Houben, R.J. 1998. Acceptable risk as a basis for design. *Reliability Engineering and System Safety* 59:141–50.

Wagenaar, W.A. 1992. Risk taking and accident causation. In *Risk-Taking Behavior*, ed. J.F. Yates, 257–81. Chichester: John Wiley.

Walton, W.W. 1982. An evaluation of the poison prevention packaging act. *Pediatrics* 69 (3): 363–70.

Weegels, M.F. 1997. Exposure to chemicals in consumer product use. Technical Report. Delft University of Technology, Faculty of Industrial Design Engineering, The Netherlands.

Weegels, M.F., and Kanis, H. 2000. Risk perception in consumer product use. *Accident Analysis and Prevention* 32:365–70.

Weegels, M.F., and van Veen, M.P. 2001. Variation of consumer contact with household products: A preliminary investigation. *Risk Analysis* 21 (3): 499–511.

Wilde, G.J.S. 1982. The theory of risk homeostasis: Implications for safety and health. *Risk Analysis* 2:209–25.

———. 1988. Risk homeostasis theory and traffic accidents: Proposition, deductions and discussion of dissension in recent reactions. *Ergonomics* 31 (4): 441–68.

———. 1989. Accident countermeasures and behavioral compensation: The position of risk homeostasis theory. *Journal of Occupational Accidents* 10 (4): 267–92.

———. 1998. Risk homeostasis theory: An overview. *Injury Prevention* 4:89–91.

Williams, D.J. 2006. Conceptualization of risk. In *International Encyclopedia of Ergonomics and Human Factors*, ed. W. Karwowski, 301–3. London: CRC Press.

———. 2007a. Risk and decision making. In *Decision Making in Complex Systems*, eds. M.J. Cook, J.M. Noyes, and Y. Masakowski, 43–53. Aldershot, UK: Ashgate.

———. 2007b. An investigation of risk perception and decision making, Unpublished PhD thesis, University of Bristol.

Williams, K.J., Kalsher, M.J., and Laughery, K.R. 2006. Allocation of responsibility for injuries. In *Handbook of Warnings*, ed. M.S. Wogalter. Mahwah, NJ: Lawrence Erlbaum Associates.

Williams, D.J., and Noyes, J.M. 2007. Effect of risk perception on decision-making: Implications for the provision of risk information. *Theoretical Issues in Ergonomics Science* 8 (1): 1–35.

Williamson, A.N., Feyer, A., Cairns, D., and Biancotti, D. 1997. The development of safety climate: The role of safety perceptions and attitudes. *Safety Science* 25 (1–3): 15–27.

Wogalter, M.S. 2006a. Purposes and scope of warnings. In *Handbook of Warnings*, ed. M.S. Wogalter, 3–9. Mahwah, NJ: Lawrence Erlbaum Associates.

Wogalter, M.S. (ed.) 2006b. *Handbook of Warnings*. Mahwah, NJ: Lawrence Erlbaum Associates.

Wogalter, M.S., Dejoy, D.M., and Laughery, K.R. (eds.) 1999. *Warnings and Risk Communication*. London: Taylor & Francis.

Wogalter, M.S., Desaulniers, D.R., and Brelsford, J.W. 1986. Perceptions of consumer products: Hazardousness and warning expectations. In *Proceedings of the Human Factors Society 30th Annual Meeting*, 1997–2201. Santa Monica, CA: Human Factors Society.

———. 1987. Consumer products: How are the hazards perceived. In *Proceedings of the Human Factors Society 31st Annual Meeting*, 615–19. Santa Monica, CA: Human Factors Society.

Wogalter, M.S., Young, S.L., and Laughery, K.R. (eds.) 2001. *Human Factors Perspectives on Warnings*. Vol. 2. Santa Monica, CA: Human Factors and Ergonomics Society.

Young, S.L., Brelsford, J.W., and Wogalter, M.S. 1990. Judgments of hazard, risk and danger: Do they differ? In *Proceedings of the 34th Annual Meeting of the Human Factors Society*, 503–7. Santa Monica, CA: Human Factors Society.

Young, S.L., and Wogalter, M.S. 1998. Relative importance of different verbal components in conveying hazard-level information in warnings. In *Proceedings of the Human Factors and Ergonomics Society 42nd Annual Meeting*, 1063–67. Santa Monica, CA: Human Factors and Ergonomics Society.

2 Consumer Product Risk Assessment

Aleksandar Zunjic

CONTENTS

2.1 INTRODUCTION

Product risk assessment is performed in order to assess the safety of products in relation to the consumer. According to the Technical University of Liberec, risk is a probability of accident occurrence and the possible consequences that might occur under certain circumstances or during a certain period (Duffey and Saull 2008). Unlike risk assessment of technical systems, assessment of risk in the system man - product is based primarily on the prediction of the effects of hazards in relation to humans. In this regard, according to the UK Department of Health, risk is the probability, high or low, that somebody (or something) will be harmed by a hazard, multiplied by the severity of the potential harm (Duffey and Saull 2008).

Data indicating the number of accidents while using different products are alarming. For example, the U.S. Consumer Product Safety Commission (CPSC) points to the fact that in 1999, 205,850 playground equipment-related injuries were treated in U.S. hospital emergency rooms (CPSC 2001). In the United States, approximately 8700 injuries from home exercise equipment occur annually in children younger than 5 years (Abbas, Bamberger, and Gebhart 2004). The top ten consumer products involved in domestic work accidents in the European Union (EU) (for the period 2002–2004) are: kitchen knife, household ladder, drinking glass, lawn mower, stick, wheeled shopping bag, chair, slicing machine, fats and oils, and tins (Zimmermann and Bauer 2006). In addition, the top ten consumer products involved in play and leisure activity (for the period 2003–2005) are: bicycle and accessories, equipment in playground, door, rolling sports equipment, chair (bench), toys, table, bed, ball, and gymnastic and body-building equipment (Angermann et al. 2007). Injuries that occur during the use of consumer products are in the range from minor cuts to fatal outcomes.

In the United States, concerns about hazardous consumer products have led to the passage of the new Consumer Product Safety Improvement Act of 2008 (Rider et al. 2009). Similarly, in order to build a system for controlling product safety, the EU has adopted Directive 2001/95/EC on general product safety (GPSD), which establishes a community rapid information system (RAPEX) for the fast exchange of information between the member states and the commission on measures and actions concerning consumer products that pose a serious risk for the health and safety of consumers (EC 2004).

2.2   CONSUMER PRODUCT RISK ASSESSMENT METHODS

The selection of an appropriate procedure that can be applied for the assessment of risks is vital for consumer product risk assessment. There are dozens of methods that can be applied for the assessment of risk in technical systems. However, these methods have very limited application in the system man - product because they do not include humans. Such methods are concentrated on the detection of possibilities for the occurrence of failure, as well as the risk assessment in relation to technical systems.

A relatively small number of methods are applicable to the system man - product. In the EU, the official method for consumer product risk assessment is RAPEX. This method is described in detail in DTI (2007), as well as in the growing number of EU Commission documents. Practical application of this method is presented in Floyd et al. (2006), where some weaknesses of this method are also observed.

In order to overcome the shortcomings of the existing methods for risk assessment, the ergonomic product risk assessment (EPRAM) method has been developed. This new method will be presented here for the first time.

2.2.1   ERGONOMIC PRODUCT RISK ASSESSMENT METHOD

The first step in the implementation of the EPRAM method for product risk assessment consists of gathering the necessary information that will be used as a basis for the risk assessment. Information that should be collected primarily relates to

- Analytical and/or statistical data related to the defective parts of products or some of their functions
- Data on the number of accidents that occurred as a result of defective products
- Information regarding the circumstances under which the accident occurred
- Other information that may be of importance for the risk assessment

The analytical data relating to the malfunction of some parts of the product can be obtained by applying some of the many methods for the risk assessment of technical systems, such as HAZOP, FMEA, FTA, and others. In fact, data collected by these methods can be used as a basis for identifying hazards in the system man - product. It should be noted that none of this methods in practice are used for risk assessment in relation to humans.

Application of methods for the risk assessment of technical systems is not necessary for the implementation of the EPRAM method if there are statistics related to the malfunction of certain products. For example, it is known from practice and the available statistical data that the airbags on a particular type of vehicle do not deploy because of a technical failure. In this case, the collection of statistical data related to this phenomenon is sufficient to identify risks in the system man - product. Moreover, the collected statistical data about the malfunctioning of products reveal defects in the products that methods for the risk assessment of technical systems previously failed to detect (e.g., in the product design phase).

It is desirable to collect data concerning the number of accidents that occur as a result of defective products. In fact, these data reveal the dimension of the problem. However, such data are often difficult to find because, in most cases, there is no developed system for the collection of such data of general social significance. Still, there are a small number of organizations dealing with the collection of such information in a systematic manner.

Collection of data regarding the circumstances under which the accident occurred may be important for two main reasons. First of all, the collection of these data is of interest to identify hazards. The product can function correctly under certain conditions. However, under certain circumstances that have not been taken into account in the design phase, the product can become defective and endanger consumers. In addition, the collection of this data is relevant for determining the probability of occurrence of risk events. Other circumstances and environmental conditions that have had an impact on the occurrence of an accident also belong here. These data can also be unavailable.

The second step in the implementation of the EPRAM method consists of the formal identification of all the risks associated with the use of the products and descriptions of the scenarios that lead to the possibilities for the occurrence of accidents. One product may be defective for several reasons. For example, a car may have a faulty airbag, but also the brakes can block under certain circumstances. Therefore, in this case there are two identified risks. For each of the identified risks, a scenario should be written that relates to the occurrence of risky events. For example, a side airbag is not activated in a case of strikes to the car from the opposite side. In addition, the brake remains blocked if a person acts on the pedal with great force. Therefore, for each identified risk it is necessary to create a scenario that describes the possibility of accident appearance and injuring of users. This scenario should also include a description, i.e., the kind of injuries that a consumer sustains when using the product. A description of the injury may be based on existing statistical data related to the use of the product, or it may be based on the evaluation by assessors of the type of injury that may occur as a result of the use of the product in a way that is described in the scenario.

In order to identify hazards, in addition to the implementation of analytical methods for the risk assessment and collection of statistical data, expert assessment can be used. In practice, risk assessors often perform a risk assessment based on the expert identification of risks. In this case, in order to identify hazards, it is necessary to consider in particular the following general sources of risks associated with the use of products:

- Risk of fire
- Risk of explosion
- Risk of falling due to slippery and uneven surfaces
- Risk of electric shock
- Risk of mechanical injury
- Risk of acceleration effects
- Risk of effects of noise
- Risk of effects of high or low temperature
- Risk of chemical hazards and toxicity
- Risk of evaporation
- Other risks associated with use of the products in an inadequate way

It is essential that for each potential risk, the evaluator identifies the hazards that can lead to the occurrence of an accident. For this purpose, it is necessary that the evaluator considers all possible conditions under which the product can be used. Therefore, it is essential that the evaluator identifies the hazards and describes a scenario that could lead to the occurrence of an accident and injury.

The third step in the implementation of the EPRAM method determines the probability of occurrence of injury during the predicted life of a product. There are several factors that influence the probability of injury. These are

P_1: The probability that the product is defective and dangerous, or will become defective and dangerous during use, or it can be used in such a way that the risk can lead to user injury.

P_2: The probability that the product does not have adequate safeguards, which may prevent the occurrence of injuries.

P_3: The probability that a hazard is not obvious while using the product.

P_4: The probability that the product does not have adequate warnings to alert the user to danger.

P_5: The probability that vulnerable persons (children, elderly, handicapped, disabled) can use the product.

P_6: The probability of product usage (exposure to the risk).

P_7: The probability that there are other environmental factors that can affect the prevention of injury (e.g., persons who can help in preventing the injury from happening, the existence of other circumstances that may prevent or reduce an injury).

The total probability P for the occurrence of injury is equal to the product of probabilities that contribute to the occurrence of injury, i.e.,

$$P = P_1 \times P_2 \times P_3 \times P_4 \times P_5 \times P_6 \times P_7. \tag{2.1}$$

According to this formula, the probability of occurrence of an injury is equal to 1 (100%), if

- The assessor estimates that the product is defective or it will become defective during its working life
- The product does not have adequate safeguards that may prevent the occurrence of injury
- The hazard is not apparent to the user
- The product does not have adequate warnings to alert the user to danger
- It is certain that vulnerable persons will use the product
- The user is exposed to the hazard every day (on a continual basis)
- There are no other circumstances that may contribute to the prevention of injuries

The fourth step in the implementation of the EPRAM method is determining the severity of possible injuries. It is necessary to classify the kind of injury described in the scenario into a particular group of injuries, depending on the severity of injury. For the purpose of the EPRAM method, injuries are classified into three groups.

The first group consists of *slight* injuries. This group comprises injuries that are characterized with <2% incapacity. These injuries are usually reversible and do not require hospital treatment. The slight group of injuries consists of

- Minor cuts
- Minor fractures

The second group consists of *serious* injuries. This group of injuries comprises injuries that are characterized with 2%–15% incapacity. These injuries are usually irreversible and require hospital treatment. The group of serious injuries includes:

- Serious cuts
- Loss of finger or toe
- Damage to sight
- Damage to hearing

The third group consists of *very serious* injuries. This group includes injuries that are characterized with >15% incapacity. These injuries are usually irreversible. The group of very serious injuries includes:

- Serious injury to internal organs
- Loss of limbs
- Loss of sight
- Loss of hearing

This classification of injuries is adopted to ensure consistency with the classification of injuries adopted by the EU within the RAPEX method for product risk assessment. Such a classification also facilitates a comparison of results of product risk assessment obtained by using the EPRAM and RAPEX methods.

The fifth step in the implementation of the EPRAM method is the final assessment of the risks of using the product. This assessment is performed on the basis of the calculated probability for the occurrence of injury P and performed classification of injuries according to the severity. In order to assess the risk, the diagram shown in Figure 2.1 is used.

Figure 2.1 shows that there are nine areas (indicated by numbers from 1 to 9) for the purpose of the assessment of risk. These areas are

1. Unacceptable risk
2. Very high risk
3. High risk
4. Increased risk
5. Moderate risk
6. Low risk
7. Very low risk
8. Extremely low risk
9. Risk is negligible

The vertical axis of the diagram indicates the probability of injuries, which can range from 0 to 1. On the horizontal axis, there are three groups (slight, serious, and very serious) in relation to which an injury is classified. Risk assessment is performed by pulling the vertical line depending on the severity of injury to the area that corresponds to the probability of occurrence of injury.

Depending on the assessed risk, it is necessary to undertake appropriate measures for eliminating or reducing the risk. The risk that corresponds to area 1 in Figure 2.1 is of the type that requires immediate withdrawal of products from the market and suspension of sales. Risks in areas 2 and 3 indicate the necessity of taking urgent measures to reduce the risk, within a certain short period of time, proportional to the risk. The risk reduction in area 4 is less urgent, but it is necessary. The risk in area 5 on the diagram does not require urgent resolving. However, measures to reduce risks in this area are recommended in a certain medium period of time. Risks that belong to areas 6 and 7 in Figure 2.1 are at a very low level. Reducing the risks in these areas is recommended in a long period of time. The community or its competent authority makes the decision about the necessity of risk reduction in this area. Risks in areas 8 and 9 do not require special treatment. Their elimination is optional and has a voluntary character.

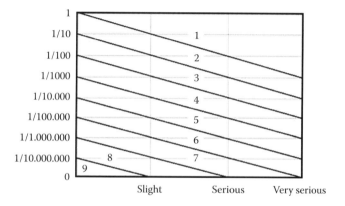

FIGURE 2.1 Diagram of the consumer product risk assessment.

2.3   EXAMPLE OF CONSUMER PRODUCT RISK ASSESSMENT

The application of the EPRAM method will be presented with an example. In order to compare the results of risk assessment, the RAPEX method will also be used.

2.3.1   EXAMPLE OF PRODUCT RISK ASSESSMENT BY THE ERGONOMIC PRODUCT RISK ASSESSMENT METHOD

2.3.1.1   Step 1

According to the National Electronic Injury Surveillance System (NEISS), over two-thirds (67%) of all injuries involving home equipment in the United States occurred on swings (CPSC 2001). This finding indicates that the swing is a potentially risky product. Therefore, it is necessary to assess the risk that comes from using swings.

2.3.1.2   Step 2

Two main hazards with using swings have been identified. The data indicate that a number of accidents in the United States emerged as a result of the chain breaking on swings during use. The resulting fall of the children from a swing caused injuries related to serious cuts and fractures. Additionally, use of a heavy seat may cause injury when a child is hit by a moving swing. In this case, bruising is the type of injury that is a result of this danger.

2.3.1.3   Step 3

Foremost, it is necessary to first determine the probability of occurrence of an injury as a result of the chain breaking. For this purpose, it is necessary to determine component probabilities that affect the overall probability of occurrence of injury. In this case we have

- The probability of the chain breaking is not great. Therefore, the probability that the product is defective and dangerous, or will become defective and dangerous during use is $P_1 = 0.05$.
- Given that the swings do not have special protection preventing the chain breaking, the probability that the product does not have adequate safeguards that can prevent injury is then $P_2 = 1$.
- Given that the chain breaking is not a predictable phenomenon for the user, the probability that the hazard is not obvious while using the product is $P_3 = 1$.
- Given that swings do not have a warning on the occurrence of chain breaking, the probability that the product does not have an adequate warning so that the user is aware of the danger is $P_4 = 1$.
- Bearing in mind that swings are primarily intended for children, the probability that the product can be used by vulnerable persons is $P_5 = 1$.
- The estimated probability of exposure to this kind of danger for the average consumer is $P_6 = 0.55$.
- There are some mitigating circumstances that may affect the reduction of injury. For example, the swing is positioned on grass instead of on concrete ground. Taking this into account, the probability that there are no other environmental factors that can affect the prevention of injuries is $P_7 = 0.8$.

Probability of occurrence of injuries during the predicted life of the product is

$$P = 0.05 \times 1 \times 1 \times 1 \times 1 \times 1 \times 0.55 \times 0.8 = 0.022.$$

Now it is necessary to determine the probability of occurrence of an injury as a result of the child being hit by a heavy seat. In this case, the component probabilities that affect the overall probability of injury occurrence are

- The probability that the swing can be used in a risky manner (so that the seat of the swing can hit the child) is $P_1 = 1$.
- The probability that the swing does not have adequate safeguards, which can prevent the hazard of hitting is $P_2 = 1$.
- Children are, to some extent, intuitively aware of the danger of a moving swing seat. Therefore, the estimated probability that the danger while using the product is not obvious is $P_3 = 0.3$.
- Given that swings do not have a warning in relation to this type of hazard, the probability that the product does not have an adequate warning that can warn the user to the danger is $P_4 = 1$.
- The probability that the product can be used by vulnerable persons is $P_5 = 1$.
- The estimated probability of exposure to this kind of hazard for the average consumer (a child in this case) is $P_6 = 0.55$.
- There are certain environmental factors that can affect the prevention of injuries. For example, an adult may be placed in the vicinity and monitor use of the swing. In this sense, the adult can prevent the child from entering into the path of movement of the seat (dangerous zone), thereby preventing injuries. Taking this into account, the probability that there are no other environmental factors affecting the prevention of injuries is $P_7 = 0.5$.

Probability of occurrence of injuries during the predicted lifetime of the product in this case is

$$P' = 1 \times 1 \times 0.3 \times 1 \times 1 \times 0.55 \times 0.5 = 0.0825.$$

2.3.1.4 Step 4

In relation to the type of injuries that are described in the scenario, their classification will be conducted. Since, in the first case (breaking of the chain) the child may suffer serious cuts and fractures, these injuries are classified as serious. In the second case (stroke of the seat), bruises are injuries that are classified in the group of slight injuries.

2.3.1.5 Step 5

The final risk assessment for both cases will be performed by using the calculated values of P and P' as well as the diagram shown in Figure 2.1. By drawing a vertical line from the serious injury location up to the value of $P = 0.022$, we get a point located in zone 1 of the diagram. Therefore, the risk that originates from the chain breaking is assessed as an unacceptable risk. In the other case, by pulling the vertical line from the location of slight injury up to the value $P' = 0.0825$, we get a point located in zone 2 of the diagram. Therefore, the risk that originates from the heavy seat is assessed as a very high risk.

Given that the risk from the chain breaking is estimated as unacceptable, it is necessary to promptly withdraw the product from the market and to suspend sales. This conclusion is consistent with the procedure of the U.S. CPSC. In 2000, the CPSC recalled 7000 play sets due to chain swings breaking during use.

In the second case, the risk of hitting that originates from the swing chair is assessed as very high risk. Therefore, it is necessary to take urgent measures to reduce the risk. Let us now consider the extent to which some of the activities for risk reduction could affect the risk assessment in this case. One of the possible measures that can be taken is to set up warning signs. By adequate marking of a risk zone in which a second child should not be in during swing usage, and placing appropriate labels or warning signs, the probability, P_4, would amount to 0.01 instead of 1. In this case, as a result of effective warning, the probability of occurrence of injury would be equal to $P_1' = 0.000825$. Based on the implementation of this measure, the risk would be significantly reduced. In that case, the risk would match zone 3 of the diagram (high risk).

In the next iteration, it is necessary to consider taking additional measures to reduce the risk. One of the possible measures consists in educating children about the dangers related to the use of swings. By way of effective implementation of this measure, the probability that there are no other environmental factors that can affect the prevention of injury would amount to 0.05 instead of 0.5. In this case, as a result of environmental factor changes, the probability of occurrence of injury would be reduced to $P_2 = 0.0000825$. Accordingly, the risk would be further reduced. In that case, it would match zone 4 of the diagram (increased risk).

In the next iteration, the risk could be reduced by using some design changes. In this sense, the use of lighter materials for the production of seats and their coating with soft material (e.g. rubber) would affect the reduction of injury severity. Two possible cases will be considered, bearing in mind that not all serious injuries are of equal weight and all slight injuries are not equally serious.

In the first case, by using the lighter material, weight of injuries is to some extent reduced, but injuries can still be classified as serious. Now suppose that the severity of injuries is located in the middle of the interval between the vertical lines, with signs slight and serious in Figure 2.1 (which essentially correspond to the limits of the intervals). By drawing the vertical line from the point on the horizontal axis of the diagram, which is located in the middle of the mentioned interval, up to the area that corresponds to the probability of injury P_2', we get a point on the chart that is located in area 5 (moderate risk). This kind of risk does not require urgent resolving.

In the second case, we can suppose that the effective design solution significantly reduced the seriousness of injuries (e.g., by coating the seat with rubber). Now suppose that the injuries are at the level that corresponds to the middle of the interval, which is in the range from the vertical axis at the beginning of the coordinate system, up to the vertical line that corresponds to the border of slight injuries. By drawing the vertical line from the point that is located at the mentioned location up to the probability P_2', we get a point on the chart that is located in area 6 (low risk). Deciding on the need for further risk reduction is the responsibility of the community or its competent authority. This level of risk is practically tolerated in most cases.

2.3.2 Example of Product Risk Assessment by the Community Rapid Information System Method

For the same example, the risk assessment will be carried out by the RAPEX method. First, the risk assessment for the case where the chain may break will be carried out.

As a result of a child falling due to the chain breaking, injuries occur (serious cuts and fractures) that, based on the table for assessing the severity of injury, correspond to the group of serious injuries.

We can estimate that the criterion "probability of hazardous product" has a value <1%. We estimate that the criterion "probability of health/safety damage from regular exposure to hazardous product" has the value "hazard may occur under one improbable or two possible conditions." Based on these two values, from the table for assessing the gravity of the outcome, we read that the probability of harm has a "low" value.

Based on the first table for risk assessment, the value "overall gravity of outcome" also has a "low" value. We adopt that children from 3 to 11 years are among the vulnerable group of users. Based on the table for the final risk assessment, taking into account the values for the overall gravity of outcome and the existence of a vulnerable group of users, we get the value "moderate risk" as the final risk assessment.

If the competent institution made a decision about taking action in this case on the basis of the application of the RAPEX method for risk assessment, then they would come to the conclusion that it is not necessary to take rapid action. Such a decision would be contrary to the CPSC decision on the recall of 7000 swings.

Now, using the RAPEX method, we can evaluate the risk in the case of hitting a child with the seat. Injuries that occur as a result of hitting a child with the seat can be characterized as a bruising. Based on the table for assessing the severity of injury, this kind of injury belongs to the group of slight injuries.

The value for the criterion "probability of hazardous product" in this case is 100%. We estimate that the criterion "probability of health/safety damage from regular exposure to hazardous product" has the value "hazard may occur under one improbable or two possible conditions." Based on these two values, from the table for assessing the gravity of outcome, we can read that the probability of harm has a "high" value.

Based on the first table for the risk assessment, the value "overall gravity of outcome" has a "moderate" value. We adopt that children from 3 to 11 years are among the vulnerable group of users. Based on the table for the final risk assessment, taking into account the values for the overall gravity of outcome and the existence of the vulnerable group of users, we get the value "serious risk" as the final risk assessment.

Although we cannot consider that, using the RAPEX method, the final risk assessment of hitting a child with the seat is generally overestimated, this assessment seems overrated in relation to the assessment "moderate risk," regarding the case of the chain breaking. Reversed values of risk assessment results when using the RAPEX method in these two cases would, to some extent, more realistically fit.

2.4 CONCLUSION

Consumers are often not aware of the fact that most products that they use daily are potentially hazardous. A product that worked correctly can become defective, thereby endangering the consumer. In addition, a product can be used in a risky manner. Designers should bear this in mind when designing the product.

In order to assess the risk that originates from a product, methods for risk assessment are commonly used. For every product, it is necessary to make a risk assessment in order to determine the level of vulnerability of consumers. Providing such information is of particular importance not only for consumers, but also for manufacturers and sellers. If the conclusion of a risk assessment is that the product generates unacceptable risk, it is necessary to initiate the procedure for its withdrawal from the market.

There are several methods that can be applied to consumer product risk assessment. The method that has very broad application is the RAPEX method. This method is applied in the member countries of the EU. In some ways, this method includes the requirements prescribed by the European Commission regarding risk assessment. However, methodologically, these requirements (criteria) are not adequately covered in all cases. One of the main disadvantages of this method is that during the risk assessment, it is not possible to include at the same time the risks that come from using the products by the vulnerable persons, and the risks that come from the lack of adequate warning or protection (and the risks that occur when the danger is not obvious). In other words, according to this method, when vulnerable people use the product, the risk does not depend on the fact that the product has appropriate protection and warning signs, or that the hazard may be obvious (e.g., open flame). If the protection is adequate and works in 100% of cases, then the risk is zero. However, this fact could not be taken into account if the risk assessment for vulnerable persons was performed by the RAPEX method.

In this chapter, the new EPRAM method for consumer product risk assessment is described. This method includes all the requirements prescribed by the European Commission regarding consumer product risk assessment. In addition, by using the criteria of the environment (surrounding), this method also includes other possible factors that have a potential impact on the risk assessment. It does not have the shortcomings that the RAPEX method possesses. Also, the application of this method enables quantifying of each criterion that can affect the risk assessment. If all numerical data that affect the risk assessment are exact and known, its accuracy is theoretically absolute. If all the data needed for the risk assessment are not known, this method allows performing an expert evaluation of these values (as with other methods of risk assessment). However, it is more sensitive than the RAPEX method, since, instead of several levels that are available to the qualitative assessment of some criteria that affect the risk, the EPRAM method allows a virtually infinite number of levels for the evaluation of some criteria (according to the range of P values). In addition, its use is relatively simple.

REFERENCES

Abbas, M.I., Bamberger, B.H., and Gebhart, R.W. 2004. Home treadmill injuries in infants and children aged to 5 years: A review of consumer product safety commission data and an illustrative report of case. *JAOA* 104 (9): 372–376.

Angermann, A., Bauer, R., Nossek, G., and Zimmermann, N. 2007. *Injuries in the European Union: Statistics Summary 2003–2005*. Vienna: Austrian Road Safety Board.

CPSC. 2001. Special study: Injuries and deaths associated with children's playground equipment. Consumer Product Safety Commission, Washington.

DTI. 2007. The general product safety regulations 2005: Notification guidance for local authorities. Department of Trade and Industry, London.

Duffey, R.B., and Saull, J.W. 2008. *Managing Risk: The Human Element*. Chichester: John Wiley.

EC. 2004. Corrigendum to Commission Decision 2004/418/EC of 29 April 2004 laying down guidelines for the management of the Community Rapid Information System (RAPEX) and for notifications presented in accordance with Article 11 of Directive 2001/95/EC. *Official Journal of the European Union* L 208/73.

Floyd, P., Nwaogu, A.T., Salado, R., and George, C. 2006. Establishing a comparative inventory of approaches and methods used by enforcement authorities for the assessment of the safety of consumer products covered by Directive 2001/95/EC on general product safety and identification of best practices. Final Report prepared for DG/SANCO, European Commission by Risk & Policy Analysts Ltd., UK, London.

Rider, G., van Aken, D., van de Sman, C., Mason, J., and Chen, X. 2009. Framework model of product risk assessment. *International Journal of Injury Control and Safety Promotion* 16 (2): 73–80.

Zimmermann, N., and Bauer, R. 2006. Injuries in the European Union: Statistics summary 2002–2004. Austrian Road Safety Board, Vienna.

3 Hazard Control Hierarchy and Its Utility in Safety Decisions about Consumer Products

Kenneth R. Laughery and Michael S. Wogalter

CONTENTS

3.1 INTRODUCTION

In general, the public at large expects the consumer products that they purchase to be relatively safe. In order to meet this expectation and to avoid injuries and product damage, manufacturers need to take steps in bringing products to the marketplace to ensure that the products meet people's beliefs about safety.

There is a concept in safety, as well as in human factors engineering and other disciplines, known as the hazard control hierarchy, or alternatively as simply the safety hierarchy (National Safety Council 1989; Sanders and McCormick 1993). This concept is a prioritization scheme for dealing with hazards. The basic sequence of priorities in the hierarchy consists of three approaches: first is to design the hazard out; the second is to guard against the hazard; and the third is to warn.

If a hazard exists with a product, the first step is to try to eliminate or reduce it through an alternative design. If a non-flammable propellant in a can of hairspray can be substituted for a flammable carrier and still adequately serve its function, then this alternative design would be preferred. Eliminating sharp edges on product parts or pinch points on industrial equipment are additional examples of eliminating hazards. However, safe alternative designs are not always available.

The second approach to dealing with product hazards is guarding. The purpose of guarding is to prevent contact between people and the hazard. Guarding procedures can be divided into two categories: physical guards and procedural guards. Personal protective equipment such as rubber gloves and goggles, barricades on the highway, and bed rails on the side of an infant's crib are examples of physical guards. Designing a task so as to prevent people from coming into contact with a hazard is a procedural guard. An example would be the controls on a punch press that require the operator to simultaneously press two switches, one with each hand, a sequence of activities that ensures fingers will not be under the piston when it strokes. Another example is a physician's prescription for a medication. Without it, the medication cannot be obtained.

However, guarding, like alternative designs, is not always a feasible solution for dealing with hazards. One cannot design out all the hazards of a lawnmower even though the shell or cover of the mower physically guards against certain kinds of contact with the blade, and a so-called dead-man's switch at the handle provides a procedural guard that stops the engine when the handle is released from a grip. The protection that alternative designs and guarding can provide can be incomplete and serve only to reduce the hazard, not completely eliminate it or serve as a complete barrier to hazards, e.g., there may be some residual hazards given the design alternatives and guarding employed.

In cases where there are still hazards associated with the product after design and guarding have been implemented, warnings may be used as a third line of defense. Warnings can be thought of as safety communications. One of the purposes of a warning is to provide people with the information needed to make informed decisions about how to use a product safely, including the choice of whether to use it at all. Warnings are third in the priority sequence because they are generally less reliable than design or guarding solutions. Even the best warnings are not likely to be 100% effective. People at risk may not see or hear a warning, or they may not understand it. Further, even warnings that are understood may not be successful in motivating compliance because the message does not fit well with people's beliefs and attitudes. It is these and other reasons and difficulties that place warnings as the third strategy in hazard control, behind design and guarding.

There are other approaches to dealing with product hazards, such as training (influencing how the product is used), personnel selection (influencing who uses it), and administrative controls (employer/supervisor sets and enforces rules). In the context of dealing with product hazards, these approaches are viewed as similar to warnings in that they mostly involve efforts intended to inform and influence behavior.

3.2 ISSUES ASSOCIATED WITH THE HIERARCHY

Numerous questions or issues may arise when applying the safety hierarchy. A starting point, of course, is to have a good understanding of the product hazards. While it is not within the scope of this chapter to discuss the goals and methods of hazard analysis, there are two noteworthy points worth mentioning. The first point is that there are formal analytic procedures and/or tools for carrying out a product hazard analysis (Frantz, Rhoades, and Lehto 1999). Examples of such procedures are fault-tree analysis and failure modes and effects analysis. Such procedures are widely recognized and practiced. A second point to note is that hazard analysis is, or should be, viewed as part of the design stage of product development. Hazard analysis of the product ought to be carried out before it is made available to consumers. A product hazard that is not recognized until the product has been in the marketplace can be costly both financially and with regard to safety outcomes. Recalls and retrofits are not a good substitute for timely and competent hazard analyses. After the product is in the marketplace and being used by consumers, it is also necessary to conduct ongoing analysis of consumer injury data from sources such as government agencies and customer service departments. If data suggest a problem with the product, post-sale warnings and recalls can be used for hazard control. Also, those data can serve as input into future designs.

Whether from hazard analysis during product development or through feedback after the product has been marketed, the hazard control hierarchy comes into play. The hierarchy's role is to aid in decision making about how to address the hazards. Some of the issues involved in such decisions are discussed in the following sections.

3.3 ALTERNATIVE DESIGNS

A general rule of thumb for when to implement an alternative design is when it is technologically and economically feasible. However, the decision process is more complex than that. Clearly, alternatives must be technically possible, such as whether non-flammable carriers in hairsprays can be produced or whether there is a way to reduce automotive tire deterioration due to aging processes. But decisions about alternative designs must include consideration of other aspects such as reliability and adequate function. If the alternative detracts from the effectiveness of the hairspray or causes the tire tread to wear faster, the alternative may not be an acceptable option, even though it addresses the hazard that led to its consideration.

It is also necessary to take into account economic feasibility in considering alternative designs. If the cost of eliminating a hazard with an alternative design is prohibitively expensive, it may not be an acceptable fix. Here again, however, the economically feasible decision may be considerably more complex than meets the eye. It might create another hazard elsewhere. Thus, a complex evaluation is needed, not just at the product level but also in a more global scope, as a part of a system of interacting components. Such considerations are not within the scope of this chapter, but one factor that is sometimes suggested or considered, rightly or wrongly, is the potential cost of defending lawsuits based on safety issues associated with the product.

When hazard elimination is feasible on both technical and economic dimensions through some alternative design, it should be examined with respect to the possibility of creating a new and worse hazard. An example would be a non-flammable carrier for hairspray that is extremely toxic if it gets into the eyes. Likewise, the harm could be to the environment, which could indirectly cause adverse health effects on users and others. The carrier in hairsprays used to be chlorofluorocarbons (CFCs), but its use was found to negatively affect the ozone layer and greenhouse gases, and they were banned from use in the United States and some other countries. Clearly, one should avoid using an alternative design that creates a worse hazard. Any new hazard that is created to eliminate another requires deliberate consideration about tradeoff acceptability. Thus, alternative designs that create as many or more hazards as they solve is not the intent of the safety hierarchy. The decision to ban CFCs was made to reduce a societal, environmental hazard, but it resulted in an increased personal-use hazard.

3.4 FACTORS THAT INFLUENCE SAFETY DECISIONS

In the previous section on alternative designs, a few factors were described that influence decisions on how to address product hazards. Technological and economic feasibility and the potential creation of other hazards were noted. There are other factors that can play a role in deciding how to address hazards. One factor is what the consumer wants or will accept; or, alternatively, what the manufacturer believes the consumer wants or will accept. An example of this issue in the context of a consumer product will help make the point. Most vehicles marketed in the United States have front seats that can be reclined to a nearly horizontal position. (Pickup trucks with bench seats are an exception.) It is generally agreed that it is hazardous for a passenger to have the seat significantly reclined to where the shoulder belt is not in contact with the torso while the vehicle is moving. The problem is that when the occupant is in the reclined position, the restraint system loses its effectiveness. Vehicle manufacturers do not even test restraint effectiveness with dummies in a reclined seat. There have been people in accidents who were reclined in passenger seats who were ejected or partially ejected and are now dead or with high level spinal fractures resulting in quadriplegia. Virtually all manufacturers now warn in the vehicle owner's manual not to recline the seat while the vehicle is in motion. While the quality of such warnings varies, the warning approach has been chosen to address the hazard—the third line of defense in the safety hierarchy. Studies show that most people are unaware of this hazard, although when called to their attention, people understand it (Leonard 2006; Leonard and Karnes 1998; Paige and Laughery 2003; Rhoades and Wisniewski 2004). Laughery and Wogalter (2008) have explored the use of warnings to address this hazard.

An alternative approach exists for addressing the seat recline hazard. It is technically and economically feasible to design the seat so that it cannot recline to an unsafe angle. According to the safety hierarchy, this would be a preferred solution compared to a warning approach. Part of the reason is that people do not read, and do not have the opportunity to read, a vehicle owner's manual before using it, as in the case of rentals.

Vehicle manufacturers have taken into account at least two factors in deciding to address the seat recline hazard with warnings. First, they considered a marketing factor based on the belief that customers want the seat recline feature. A second cited factor is that in circumstances where the driver is experiencing fatigue, it will be possible to rest by stopping and reclining the seat, a safety consideration.

A guarding approach has also been proposed for addressing the seat recline hazard. Here, the vehicle cannot be driven from a stopped condition if the seat is reclined beyond some safe angle, and if the engine is running, the seat will not recline. Note that this guarding solution permits the fatigued driver to stop the vehicle, recline the seat, and rest. They can still get the benefit of being able to recline the seat. Like the above design alternative, it is likely to be more successful than warnings in dealing with the seat recline hazard. Note that there may be other design solutions, such as designing the restraint system so it works while in a reclined position.

3.5 WARNING VERSUS ALTERNATIVE DESIGN VERSUS GUARDING

The above seat recline example illustrates a product where the hazard is understood and there are options to deal with it. More specifically, there is a choice between a technologically and economically feasible alternative design or guarding or warnings. Note that to be successful the design and guarding options need to be fail-safe, unless of course there is some kind of structural failure or successful effort to override the kill switch. The effectiveness of a warning option depends on the communications successfully informing and motivating the occupant not to recline the seat in the moving vehicle. The differences in effectiveness, of course, illustrate the underlying value or purpose of the safety hierarchy.

Another example of a consumer product where the safety hierarchy could or should come into play is a turkey fryer. The base or stand for such a fryer, or cooker, is shown in Figure 3.1a. A large aluminum pot sits on top of the propane-fueled base shown in the figure. A typical application or use of the product would be to put cooking oil, such as peanut oil, in the pot and cook turkey parts or other meat.

A considerable hazard associated with this product is that it is unstable and can tip over if intentionally or unintentionally bumped or moved. The resulting hot oil spill can result in severe or catastrophic burns. Such incidents have occurred in situations such as outdoor picnics or similar events where children or animals may be active in the vicinity of the cooker.

The cooker comes with an owner's manual. The manual contains a warning that includes a statement that the hot oil can cause severe burns and advising to keep children and pets away. Note that the instruction to keep children and pets away is an example of a warning recommending a guarding solution. While the adequacy or inadequacy of the warning could be a concern, the manufacturer of the product should explore how to deal with the tip over hazard from the perspective of the hazard control hierarchy. As stated earlier, design alternatives are preferred over guarding or warning.

There are several design aspects of the turkey fryer that contribute to its instability. Included among these characteristics are: the width of its base, the height of its center of gravity, and the fact that it has only three legs. In terms of alternatives, these are design features that can be improved in ways that result in a significant increase in stability. For example, adding a fourth leg, lowering the center of gravity by shortening the legs, or adding a ring at the base of the legs, as shown in Figure 3.1b, are examples of design alternatives that are readily achievable.

There are numerous examples of the different ways that the hazard control hierarchy is used for any given product, person, and context of use. Take the example that Karnes, Lenorovitz, and Leonard (2010) discuss with respect to personal water craft (PWC). There is a hazard of orifices injuries caused by water jets used to propel PWC. For many years, manufacturers used warnings

(a)

(b)

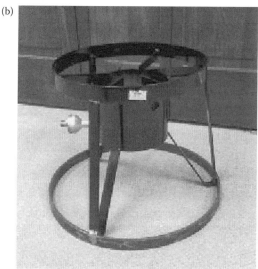

FIGURE 3.1   (a) Poultry roaster, (b) modified poultry roaster.

as the means of hazard control. Research indicates that people did not see the warning, and even if they had, they would not be able to carry out the warning due to the high cost of compliance of having to wear a wetsuit if riding as a passenger. A warning like this is not going to protect people if wetsuits are not readily available. A better solution is to design the PWC so that when a person falls off the unit, they do not end up in the path of the jets; perhaps by covering the top of the jet nozzles or shaping the back of the PWC so that people fall in directions away from the jets.

In the example above, the solution of redesigning the back of the PWC came after the warning method had been considered and used. The warning was not working. Also note that there is another related situation where a manufacturer decides, for whatever reason, not to warn. Both instances call for a recursive step, a return to consideration of design alternatives, perhaps some of which were not considered in the first round of hazard control analysis. Thus, these would be examples of cases in which the design–guard–warn hierarchy was considered but none of the methods looked promising for the various reasons already discussed in this chapter, followed by a step to relook at design alternatives and guarding methods to see if they can be accomplished, perhaps differently and in a different light given the preceding analyses.

3.6   FINAL COMMENTS

The examples of the vehicle seat recline hazard and the turkey fryer tip over hazard were presented as a context for exploring some of the issues encountered in deciding how to address product hazards. The hazard control hierarchy provides some principles and/or guidelines based on what is likely to be most effective; that is, the design, guard, and warn priority scheme. But, as indicated with the seat recline example, decisions about whether to seek solutions based on alternative design, guarding, or warning may be complex. In addition to technological and economic feasibility, there are other factors that can come into play, such as secondary safety effects and customer preferences.

Sometimes, the decision-making process may be relatively straightforward, as in the case with the turkey fryer. Clearly, it does not require a revision of Newton's laws of physics to come up with a more stable cooker by what would appear to be some simple design changes that would likely amount to only a modest increase in cost to produce the product. Certainly, in comparison to a warning that recommends a guarding solution (keep children and pets away), the design alternative that increases stability would appear to be more effective. The point, however, is not to suggest that children and pets need not be monitored around the fryer or that a warning spelling out the potential severe burn consequences of a tip over is not appropriate. These aspects are important and should be included. Rather, the point is that guarding and warnings should be viewed as a complement to better, safer design, not as a substitute for it.

A few additional comments are worthwhile at this point. Influencing human behavior is often difficult and seldom foolproof. Concerns about the reliability of warnings should not be regarded as a basis for not warning when it is appropriate to do so. Warnings are one of several tools available to product manufacturers and designers to facilitate product safety, and they have an appropriate role in the safety hierarchy.

A final comment on the complimentary aspects of the design, guard, and warn safety hierarchy is worth mentioning. The hierarchy should not be viewed as a prioritization scheme consisting of three options from which a selection can/must be made. Rather, it defines a preference scheme based on what is likely to be most effective from a safety perspective. It is not meant to imply some sort of exclusion principle; for example, if you guard (such as putting up a fence around a power station), that there is no need to warn (hang a warning sign on the fence that emphasizes danger and not to enter). Instead, the matter may be better thought of as: even with a better design, it may still be appropriate and necessary to guard or warn, or both.

Future warnings may do a better job in fulfilling their role to protect against hazards as technology allows warnings that are triggered by sensors and that display tailored warning messages. Nevertheless, alternative designs and guarding will likely remain the main means to keep hazards away from people and property.

REFERENCES

Frantz, J.P., Rhoades, T.P., and Lehto, M.R. 1999. Practical considerations regarding the design and evaluation of product warnings. In *Warnings and Risk Communication*, eds. M.S. Wogalter, D.M. DeJoy, and K.R. Laughery, 291–311. Philadelphia, PA: Taylor & Francis.

Karnes, E.W., Lenorovitz, D.R., and Leonard, S.D. 2010. Reliance on warnings as the sole remedy for certain hazards: Some circumstances where that just doesn't work. In *Proceedings of the Applied Human Factors and Ergonomics Conference*, 1017–27. Miami, FL. Boca Raton: Taylor & Francis/CRC.

Laughery, K.R., and Wogalter, M.S. 2008. On the symbiotic relationship between warnings research and forensics. *Human Factors* 50 (3): 329–33.

Leonard, S.D. 2006. Who really knows about reclining the passenger seat? In *Proceedings of the Human Factors and Ergonomics Society 50th Annual Meeting*, 855–59. Santa Monica, CA: Human Factors and Ergonomics Society.

Leonard, S.D., and Karnes, E.W. 1998. Perception of risk in automobiles: Is it accurate? In *Proceedings of the Human Factors and Ergonomics Society 42nd Annual Meeting*, 1083–87. Santa Monica, CA: Human Factors and Ergonomics Society.

National Safety Council. 1989. *Accident Prevention Manual for Industrial Operation,* 5th ed. Chicago, IL: National Safety Council.

Paige, D.L., and Laughery, K.R. 2003. Risk perception: The effects of technical knowledge – or lack of it. In *Proceedings of the XVth Triennial Congress of the International Ergonomics Association,* 953–56. Seoul, Korea: International Ergonomics Association.

Rhoades, T.P., and Wisniewski, E.C. 2004. Judgments of risk associated with riding with a reclined seat in an automobile. In *Proceedings of the Human Factors and Ergonomics Society 48th Annual Meeting,* 1136–39. Santa Monica, CA: Human Factors and Ergonomics Society.

Sanders, M.S., and McCormick, E.J. 1993. *Human Factors in Engineering and Design,* 7th ed. New York: McGraw-Hill.

4 Communication-Human Information Processing Stages in Consumer Product Warnings

Michael S. Wogalter, Kenneth R. Laughery,
and Christopher B. Mayhorn

CONTENTS

4.1 INTRODUCTION

Research on warnings has grown considerably over the last three decades (e.g., see Laughery and Wogalter 2006; Miller and Lehto 2001; Wogalter and Laughery 2005). During this time period, researchers have investigated a wide variety of variables. The communication-human information processing (C-HIP) model provides a framework to organize and structure the seemingly disparate research literature by bringing coherence to the field. It also reveals needed research to fill gaps in knowledge (Wogalter, DeJoy, and Laughery 1999a). Most previous descriptions of C-HIP have focused on its broad generality. Some descriptions of the model demonstrate particular applicability to other more specific situations such as warning signs in the workplace (Conzola and Wogalter 2001) or for one specific category of consumer products such as pharmaceuticals or beverage alcohol (Wogalter and Sojourner 1999; Wogalter and Young 1998). No previous review of C-HIP has specifically focused on consumer product warnings. C-HIP has applicability to a wide assortment of consumer products.

In describing C-HIP and its component stages, this chapter reviews research of some of the influential factors found at each stage. After going through the stages of the model, another benefit of the C-HIP model is described, namely, it can serve as an investigative tool for helping determine why a warning failed to be effective.

The C-HIP model has two major sections, each with several component stages. A representation of the model can be seen in Figure 4.1. The first section of the framework employs the basic stages of a simple communication model. McGuire (1980) provides a detailed description of communication theory with respect to warnings. Here, the model focuses on a warning message being sent from one entity to another, i.e., sent by a source (sender) through some channel(s) to a receiver.

The second major section of the model focuses on the receiver and how people internally process information. This section interfaces with the first through effective delivery of the warning to individuals who are part of the target audience. When warning information is delivered to the receiver, processing may be initiated and, if not blocked in some way, will continue through several stages: from attention switch, attention maintenance, comprehension and memory, beliefs and

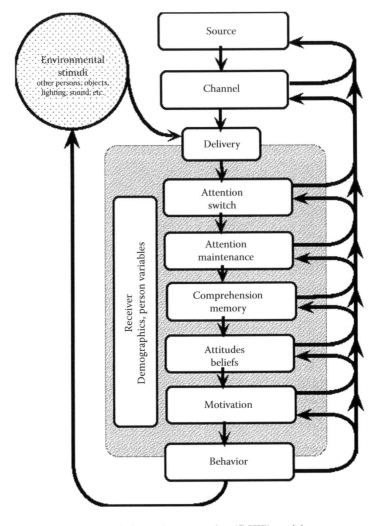

FIGURE 4.1   Communication-human information processing (C-HIP) model.

attitudes, motivation, and possibly ending in behavior. Similar information processing models have been discussed by others (Lehto and Miller 1986; Rogers, Lamson, and Rousseau 2000). Cameron and DeJoy (2006) and Lehto (2006) have reviewed other process models with respect to warnings.

4.1.1 How the Communication-Human Information Processing Model Has Evolved

One of the main benefits of the C-HIP model is that it serves as a guiding framework for organizing diverse findings in the warning research literature. Over the years, the body of research has grown to the extent that it now requires fairly substantial books to describe and summarize the literature (e.g., Wogalter, DeJoy, and Laughery 1999b; Wogalter 2006a). This chapter gives an overview of research findings relevant to each stage of C-HIP with specific focus on consumer products and their associated warnings. The purpose of the present chapter is to demonstrate that the C-HIP model is a useful conceptualization about warning processing across a wide area of consumer products. Both Wogalter et al. (1999) and Wogalter (2006a) have individual detailed chapters on most of the model's stages. The model has evolved over time. The model that pre-dated the C-HIP (Wogalter and Laughery 1996) simply presented some of the main human information processing stages (i.e., in the receiver section); in other words, only the second section of the stages of the eventual C-HIP model. The Wogalter et al. (1999) version of C-HIP added the first section from communication theory (source and channel). The most recent model from Wogalter (2006b) (i.e., Figure 4.1) is different in four ways from Wogalter et al.'s (1999) C-HIP model. First, in the current model the attention stage is split into two separate stages, attention switch and attention maintenance. The reason for the split is that these two stages are different (and often confused), and they are affected by different variables. The second major difference in the models is that there is now the stage of delivery (Williamson 2006). Delivery refers to the point of warning reception where information is provided to the receiver via one or more channels. The third change in the current model is an explicit reference to the influence of other environmental stimuli. Environmental influences are aspects other than the product warning itself that could affect how the warning is processed. They are extrinsic to the warning. Environmental influences can include other information on the product label, the product itself, other people's involvement, other warnings, and other aspects in the environment including illumination and background noise (Vredenburgh and Helmick-Rich 2006). The fourth major change from the Wogalter et al. (1999) C-HIP model to the current model is greater emphasis on the receiver's personal characteristics (e.g., demographics) and task involvment (Smith-Jackson 2006, 2007; Wogalter and Usher 1999). Both the third and the fourth changes serve to emphasize how context (outside the person and warning, and internal aspects of the target person) can influence the processsing of warning content.

4.2 HOW THE COMMUNICATION-HUMAN INFORMATION PROCESSING MODEL WORKS

The C-HIP model is both a stage model and a process model. The model is useful in describing a general sequencing of stages and the effects warning information might have as it is processed. If information is successfully processed at a given stage, the information "flows through" to the next stage. If processing at a stage is unsuccessful, it can produce a bottleneck, blocking the flow of information from getting to the next stage. If a person does not initially notice or attend to a warning, then processing of the warning goes no further. However, even if a warning is noticed and attended to, the individual may not understand it, and as a consequence, no additonal processing occurs beyond that point. Even if the message is understood, it still might not be believed, thereby causing a blockage to occur at this point. If the person believes the message, then low motivation (to carry out the warning's instructed behavior) could cause a blockage. If all of the stages are successful, the warning process ends in safety behavior (compliance) attributable to the warning information. While the processing of the warning may not make it all the way to the behavioral compliance stage, it can still be effective at earlier stages. For example, a warning might enhance understanding and beliefs but not change behavior.

Although the model tends to emphasize a linear sequence from source to behavior, there are feedback loops from later stages in the process that can impact earlier stages of processing, as illustrated on the right side of Figure 4.1. For example, when a warning stimulus becomes habituated from repeated exposures over time, less attention is given to it on subsequent occasions. A more specific example could be given in terms of over-the-counter (OTC) pharmaceuticals (Cheatham and Wogalter 2002, 2003). If a new hazard is added to a warning, people may not notice it if they have read the previous warning version and used the drug many times in the past. Here, memory affects an earlier stage, attention. A second example of feedback effects concerns the influence of beliefs on attention. Some individuals may not believe that a given product is hazardous, and as a result not think about looking for a warning. Thus, if people believe that a common and familiar analgesic can cause no harm, they will be less likely to read a warning that accompanies the drug. Thus, a later stage, beliefs and attitudes, affects an earlier stage of attention.

In the following sections, factors affecting each stage of the C-HIP model are described. The first three sections concern the communication features of C-HIP from the source via some channel(s) to the receiver. Later sections concern analysis of information processing factors that are internal to the receiver.

4.2.1 Source

The source is the initial transmitter of the warning information. The source can be a person or an organization (e.g., company, government). With respect to consumer products, the source is usually the manufacturer (although in cases of imported products, the importer/distributor in the United States may be responsible). One critical role that the source assumes is to determine if there are hazards present that necessitate a warning. Such a determination requires some form of hazard analysis (Frantz, Rhoades, and Lehto 1999; Young, Frantz, and Rhoades 2006). If a hazard is identified, the source must first determine if there are better methods of controlling it than the use of warnings, such as eliminating or designing out the hazard or guarding against it by using design and engineering procedures (see Laughery and Wogalter 2006). There are several general principles to guide when to employ a warning:

1. There is a hazard that cannot be designed out or guarded
2. The hazard, consequences, and appropriate safe modes of behavior are not known to persons at risk
3. The hazards are not open and obvious; that is, the appearance of the product or environment does not clearly expose the hazards
4. A reminder is needed to promote awareness of the hazard at the proper time

There are other considerations, such as the specific characteristics of the consumer product involved. Some products are inherently more dangerous than others. For instance, a manufacturer of drain cleaner will have a different role to play than a manufacturer of orange juice. Even relatively safe products such as orange juice can have hazards. It is the responsibility of the manufacturer to mitigate potential consumer risks, which might include the use of warnings.

If the need for a warning exists, then the source (generally the manufacturer) needs to determine how consumers should be warned, e.g., what channel(s) to use (see section below) and the warning's intrinsics characteristics. In addition, the perceived characteristics of the source can influence people's beliefs, credibility, and relevance (Cox 1999; Wogalter, Kalsher, and Rashid 1999). Information from a reliable, expert source is usually given greater credibility. It is generally assumed that the manufacturer is expert with regard to the product they produce. It is expected that they know or seek to learn about hazards and keep them at bay. That is their role. If the source does not carry out its role satisfactorily, persons can be injured, and in some cases, depending on the country and legal jursidiction, the manufacturer can be sued, fined, and the product recalled. Additional information on the source stage is given in Cox and Wogalter (2006). Note that some research concerning the source is properly classified as beliefs and attitudes and will be discussed further in that section of the C-HIP model.

4.2.2   CHANNEL

The channel is the medium and modality in which information is transmitted from the source to one or more receivers. Consumer product warnings can be transmitted in many ways. Warnings can be presented in labels directly on the product, on product containers, in product manuals, in package inserts, on posters/placards, in brochures, and as part of audio-video presentations in various media such as the internet. Most commonly, warnings use the visual (text and symbols) and auditory (alarms and voice) modalities as opposed to the other senses. There are exceptions, e.g., an odor added to petroleum-based gases to enable leak detection by the olfactory sense, and the rough vibration of a product that is not mechanically functioning well, which would give tactual, kinesthetic, and haptic sensation (Mazis and Morris 1999; Cohen et al. 2006).

4.2.2.1   Media and Modality

There are two dimensions of the channel. The first concerns the media in which the information is embedded (e.g., label, video). The second dimension is the sensory modality of the receiver (visual, auditory). Some media involve one modality (e.g., product manual involves the visual sense) and others involve two modalities (e.g., videos often have both visual and auditory). Visual presentation can be composed of both or either text and symbols. Auditory presentation can be non-verbal (noise, beeps, buzzers) and verbal (voice/speech) sounds. For example, traditional smoke alarms produce non-verbal signals whereas "talking" smoke alarms, such as those depicted in Figure 4.2, produce speech warnings.

   Research comparing the effectiveness of language-based warnings presented visually (text) versus auditorily (speech) is conflicting (Cohen et al. 2006). One can be better in certain circumstances with the reverse being true in other circumstances (e.g., video presentation of visual print is better than speech in terms of comprehension and memory, while audio presentation of voice is better

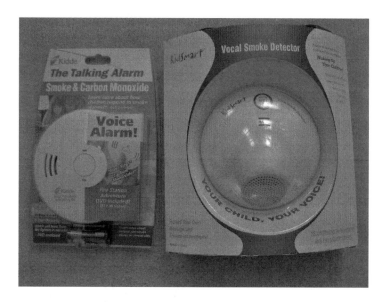

FIGURE 4.2   "Talking" smoke alarms that deliver verbal warnings.

than some signage in open environments to capture attention). However, most published research results are fairly consistent in showing that presentation in either modality is better than no warning presentation whatsoever (Edworthy and Hellier 2006). Also, warnings presented in two or more modalities are generally more effective than those presented in a single modality. This finding is applicable for the design of warnings associated with television and internet advertising as is done with prescription pharmaceuticals in the United States (Wogalter et al. 2002). Thus, a video-based warning is better if the words are shown on a screen compared to giving the same information orally or just visually (Barlow and Wogalter 1993; Kalsher and Wogalter 2007).

Multi-modal warnings provide redundancy. If an individual is not watching a visual display, he/she can still hear it (Barlow and Wogalter 1993; Wogalter and Young 1991). If the individual is blind or deaf, the information is available in the other modality. In addition, if an individual sees and hears warning information, there is a greater likelihood that the message will be delivered to otherwise vulnerable receivers (e.g., both deaf and blind persons will be satisfied and persons over-loaded in one modality could receive it in another modality). Also, there is a well-supported theory in cognitive psychology and education that multi-modal presentation enhances learning because the information is richer and may link to greater or better internal representational nodes (Paivio 1971).

Longer, more complex messages may be better presented visually because reading language is generally faster and allows easier review and re-review of the material. However, shorter, less complex messages presented auditorily can be more effective than the same messages presented visually. Also, the presentation of an auditory signal is generally better for switching attention (a stage described below). An implication from this analysis is that a short auditory warning, pointing to more detailed information accessible elsewhere would be beneficial for capturing attention as well as enabling the processing of longer and more complex information. An example demonstration of this is the "talking box" used in Conzola and Wogalter (1999).

4.2.2.2   Warning System

As the above discussion suggests, the conceptualization of warnings solely as static labels is too narrow a view of how warning information may be transmitted for a consumer product (Laughery and Wogalter 1997; Wogalter and Mayhorn 2005). For many consumer products, warnings may be transmitted by manufacturers via many media and be received at different times. Warning systems for a particular product may consist of a number of components. For example, a warning system for a prescription acne medication, such as Accutane, may consist of several components: a printed statement on the outside packaging or box, on a bottle or blister pack, and a sheet insert (Mayhorn and Goldsworthy 2007, 2009). Television advertisements for prescription drugs in the United States also may contain warnings (Vigilante, Wogalter, and Mayhorn 2007). The manufacturer's web site and other web sites may have warnings (Hicks, Wogalter, and Vigilante 2005; Vigilante and Wogalter 2005) or replacement product manuals that are available for consumers. An example web page with downloadable manuals is shown in Figure 4.3. The physician who prescribed it and the pharmacist who fills the prescription are other potential sources of warnings. Organizations including government agencies such as the U.S. Food and Drug Administration and the U.S. Consumer Product Safety Commission and consumer and trade groups such as Consumers Union and Underwriters Laboratories could provide additional materials.

The purpose and content of the components of a warning system are not necessarily identical. For example, some components may be designed for the purpose of capturing attention and direct the person to another component containing more information for comprehension or to affect beliefs and attitudes, or may be intended for different target audiences. The multiple components of the warning system can provide the advantages (e.g., redundancy) of multiple media and modalities described above.

4.2.2.2.1   Direct and Indirect Communications

The distinction between direct and indirect effects of warnings concerns the routes by which information gets to the target person (Wogalter and Feng in press). A direct effect occurs as a result of the

NATIONAL MOWER COMPANY

SINCE 1919

Maintenance Manuals	Acrobat Reader PDF file
68 Triplex-CE owners manual	Complete file (1.47MB)
8400 Owners manual	Complete file (2.17MB)
8400 Owners manual (Pages 01–10)	Pages 01–10 (459K)
8400 Owners manual (Pages 11–20)	Pages 11–20 (679K)
8400 Owners manual (Pages 21–30)	Pages 21–30 (426K)
8400 Owners manual (Pages 31–40)	Pages 31–40 (399K)
8400 Owners manual (Pages 41–52)	Pages 41–52 (430K)
84 Triplex owners manual	Complete file (1.65MB)
84 Triplex owners manual (Pages 01–11)	Pages 01–11 (489K)
84 Triplex owners manual (Pages 12–23)	Pages 12–23 (390K)
84 Triplex owners manual (Pages 24–35)	Pages 24–35 (490K)
84 Triplex owners manual (Pages 36–45)	Pages 36–45 (424K)
Hydro 70 owners manual	Complete file (1,029K)
Hydro 70 owners manual (Pages 01–12)	Pages 01–12 (329K)

FIGURE 4.3   Dissemination of replacement owner's manuals via the internet.

person being directly exposed to the warning. Warnings can also be delivered indirectly. One example is learning about a hazard in a conversation with a family member or friend. To illustrate this point, Tam and Greenfield (2010) provided evidence that exposure to alcohol warnings may be instrumental in preventing incidences of drunk driving by others. Likewise, the employer or physician who reads warnings and then verbally communicates the information to employees or patients are also examples. Adults who have responsibility for the safety of children are another important category. Figure 4.4 is an illustration of a warning for infant caregivers concerning fall hazards associated with inappropriate use of a child seat. (Unfortunately it is not very salient so many caregivers might not notice it.)

FIGURE 4.4   Child seat with warning located on the rear.

Potentially, a warning put out by a manufacturer could be useful even if an individual does not see the warning if it is communicated via another person who viewed it. With respect to C-HIP, the material sent from the source (usually the manufacturer) to the receiver through some channnels provides the direct communication of warnings to the receiver. Indirect effects involve the delivery (discussed below) of that warning information by others, which according to the current C-HIP model is part of from the environment component shown in Figure 4.1.

4.2.3   Delivery

While the source may try to disseminate warnings in one or more channels, the warnings might not reach some of the targets at risk (Williamson 2006). Delivery refers to the point of reception where a warning arrives at the receiver. To emphasize its importance, it is shown as a separate stage in the current C-HIP model shown in Figure 4.1. A warning that a person sees or hears is a warning that has been delivered. A safety video that is produced but never reaches the individual would be delivery failure. The reason for the failure to deliver the warning to targeted individuals can be multifold. The video may be sitting in bulk boxes in a warehouse and not have been distributed. Or the distribution could be haphazard, reaching some intended persons and not others. But even if individuals receive the video, they may not receive the needed information. They may not have the time or playback equipment to see it. Of course, even if the person does see the video, it may not include the necessary warning. Thus, it may be necessary to distribute warning information in multiple ways to reach receivers at risk. As stated above, warnings disseminated by the souce can have indirect effects, e.g., the warning information from a disseminated safety video may be conveyed by someone who viewed it. The point is that if warnings given by a source do not reach the targets at risk either directly or indirectly, then the warning will have no or limited effects on the receiver.

4.2.4   Environmental Stimuli

Besides the subject warning, other stimuli are almost always simultaneously present. They may be other warnings or a wide assortment of non-warning stimuli. These stimuli compete with the warning for the person's attention (described further below). With respect to a given warning, these other stimuli may be described as "noise" that could potentially interfere with warning processing. Several examples can illustrate. A cellular phone ringing just when an individual begins to examine a warning may cause distraction and lead to the warning not being fully read. Likewise, a crying infant during mealtime may prevent a parent from comprehending the almost illegible warning information on the child seat illustrated in Figure 4.4. The environment can have other effects. The illumination can be too dim to read the warning. In such cases of distraction or legibility, another warning of greater salience could have the capability to attract and hold a person's focus instead.

Environmental influences can include other people. Awareness about what other persons are doing in the local environment and elsewhere can affect warning compliance positively or negatively. Seeing other people wearing safety helmets on bicycles and motorcycles suggests it is proper behavior to wear them. But seeing adverisements with persons not wearing goggles, gloves, or other needed protective equipment while apparently using a hazardous product can suggest that such protection is not needed, even though the product's warning requires its use. Such a disconnect between warning materials and advertisements located on packaging materials is apparent on the box pictured in Figure 4.5 of the aforementioned child seat. While the warning text located on the product states "never use on a raised surface," the packaging materials portray pictures of children at a birthday party, positioned on a table (raised surface) while sitting in the child seat. Arguably, this apparent inconsistency in safety information might be confusing to parents and may lead to an infant being injured. Clearly then, the environment can have effects on warning processing.

4.2.5 RECEIVER

The receiver is the person(s) or target audience to whom the warning is directed. For a warning to effectively communicate information and influence behavior, the warning must first be delivered. Then, attention must be switched to it and maintained long enough for the receiver to extract the necessary information. Next, the warning must be understood and must concur with the receiver's existing beliefs and attitudes. Finally, the warning must motivate the receiver to perform the directed behavior. The following sections are organized around these stages of information processing.

4.2.5.1 Attention Switch

An effective warning must initially attract attention. To do so, it needs to be sufficiently salient (conspicuous or prominent). Warnings typically have to compete with other stimuli in the environment for attention. Several design factors influence how well warnings may compete for attention (see Wogalter and Leonard 1999; Wogalter and Vigilante 2006).

Larger is generally better. Increasing the overall size of the warning, its print size and contrast generally facilitates warning conspicuousness. Context also plays an important role. It is not just the absolute size of the warning, but also its size relative to other displayed information matters.

FIGURE 4.5   Advertising photograph located on the packaging of a child seat.

Consider the can of hairspray depicted in Figure 4.6. Here, the warning text regarding the flammability hazard is considerably smaller than the advertising pronouncement that the buyer gets "33.5% more free" when this product is purchased.

For some products, the available surface area is limited, e.g., small product containers such as pharmaceuticals. Putting all of the hazards on the primary on-product (container) label could reduce the salience of the most critical information (e.g., by decreasing print size). Solutions include expanding the surface area, including the addition of tags, peel-off labels (Barlow and Wogalter 1991; Wogalter, DeJoy, and Laughery 1999b; Wogalter and Young 1994), or ancillary sheets.

Color is an important attribute that can facilitate attracting attention (Bzostek and Wogalter 1999; Laughery, Young et al. 1993). While there are potential problems with using color as the only method of conspicuity, such as color blindness in some individuals, color is a frequently used design component to attract attention. The ANSI Z535 (2002) warning standard uses color as one of several components of the signal word panel to attract attention. Other design components in the ANSI Z535 signal word panel include an alert symbol, the triangle/exclamation point, and one of three hazard connoting signal words (DANGER, WARNING, and CAUTION). Context again can play a role with respect to color as a salience feature. An orange warning on a product label located on an orange product will have less salience than the same warning conveyed using a different color. The color should be distinctive in the environment in which it is placed.

Symbols can also be useful for capturing attention. One example already mentioned is the alert symbol (triangle enclosing an exclamation point) used in the signal word panel (Bzostek and Wogalter 1999; Laughery et al. 1993). This symbol serves as a general alert. Bzostek and Wogalter (1999) found results showing people were faster in locating a warning when it was accompied by an icon. Other kinds of symbols may be used to convey more specific information. This latter purpose is discussed in the later comprehension section, but the point here is that a graphic configuration can also benefit the attention switch stage.

Warnings located proximal to the hazard, both temporally and physically, generally increase the likelihood of attention switch (Frantz and Rhoades 1993; Wogalter, Barlow, and Murphy

FIGURE 4.6   Warning on a can of hairspray.

1995). Warnings should be located to maximize the chance that they will be encountered. This aids in delivery. For instance, a parent interacting with a child who is sitting in the child seat depicted in Figure 4.4 is unlikely to encounter the warning located on the rear of the product. To further illustrate this point, a warning about carbon monoxide (CO) hazards on a gas-powered electrical generator is more likely to be effective than one located in a separate, sometimes displaced (e.g., in a file or possibly lost or never received) product manual (Mehlenbacher, Wogalter, and Laughery 2002). Generally, placement directly on the product or its primary container is preferred, particularly if the product is potentially highly dangerous (Wogalter et al. 1991; Wogalter, Barlow, and Murphy 1995). There may be exceptions to the proximity rule, such as where the warning is presented too close in location and/or time to the hazard, and the individual sees or hears it too late to avoid the hazard.

Repeated, long-term exposure to a warning may result in a loss of its ability to evoke an attention switch at later times (Thorley, Hellier, and Edworthy 2001). This process or state of habituation can eventually occur even with well-designed warnings; however, better designed warnings with salient features can slow the habituation process. Where feasible, changing the warning's appearance may be useful in reinvigorating attention switch previously lost due to habituation.

Tasks that the individual may be performing and other stimuli in the environment may absorb attention and may compete with the warning for attention capture (Wogalter and Usher 1999). Thus, the warning should have characteristics to make it highly salient in context.

4.2.5.2   Attention Maintenance

Individuals may notice the presence of a warning but not stop to examine it. A warning that is noticed but fails to maintain attention long enough for its content to be encoded may be of very little direct value. Attention must be maintained on the message for some length of time to extract meaning from the material. During this process, the information is encoded or assimilated with existing knowledge in memory.

With brief text or symbols, the warning message may be grasped very quickly, sometimes as fast as a glance. For longer, more complex warnings, attention must be held for a longer duration to acquire the information. To maintain attention in these cases, the warning needs to have qualities that generate interest, so that the person is willing to maintain attention to the material. The effort necessary to acquire the information should be as little as possible. Thus, a goal is to enable the information to be grasped as easily as possible. Some of the same design features that facilitate the switch of attention also help to maintain attention. For example, large print not only attracts attention, it also tends to increase legibility, which makes the print easier to read.

It is not difficult to find products with print on labels that is too small for older adults with age-related vision problems to read without a magnifying glass (Wogalter, DeJoy, and Laughery 1999b; Wogalter and Vigilante 2003). Not only might people not read a warning due to the effort involved, they may also believe that the material is relatively unimportant, otherwise the print would be larger.

Print legibility can be affected by numerous factors including choice of font, stroke width, letter compression and distance between them, resolution, and justification (see Frascara 2006). Although there is not much research to support an unequivocal preference for particular fonts, the general recommendation is to use relatively plain, familiar alphanumerics. It is sometimes suggested that sans serif font like Helvetica, Futura, and Univers for large text sizes and a serif font like Times, Times Roman, and New Century Schoolbook be used for smaller-sized text. A chart with print sizes for expected reading distances in good and degraded conditions can be found in the ANSI (2002) Z535.4 product warning standard.

Legibility is also benefitted by high contrast between objects, such as text lettering, relative to their background. Consider the poor contrast between the warning text on the vaporizer illustrated in the gray scale photo in Figure 4.7. Both the text and the background are in the same color, blue. In this instance, it is unlikely that consumers will notice let alone maintain their attention with this particular warning. Black on white or the reverse has the highest contrast, but legibility can be

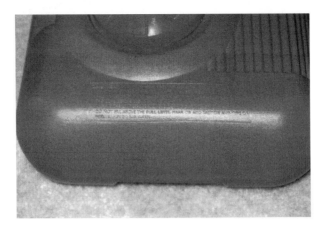

FIGURE 4.7   Warning on a vaporizer.

adequate with other combinations, such as black print on yellow (as in the ANSI Z535.4 "CAUTION" signal word panel) and white print on red (as in the ANSI Z535.4 "DANGER" signal word panel).

People will more likely maintain attention if a warning is well designed (i.e., aesthetic) with respect to formatting and layout. Research suggests that people prefer warnings that are in a list outline format as opposed to continous prose text (Desaulniers 1987). Also, text messages presented in all caps are worse than mixed-case text in glance legibility studies (Poulton 1967), and centered formatting is worse than left justified text (Hooper and Hannafin 1986). In terms of formatting, the warning text of the child seat illustrated in Figure 4.8 is poor with respect to several of these characteristics, and it is unlikely to maintain the attention of a parent using the product with his or her child. Moreover, visual warnings formatted with plenty of white space and containing organized information groupings are more likely to hold attention than a single chunk of dense text (Wogalter and Vigilante 2003, 2006). Research also shows that people like the added formatting, but a more important need for older adults was having adequate print size on labels so that they could read it (even if it loses the chunked structure provided because of the removal of white space). Younger readers do not have trouble reading smaller sizes, so formatting through white spacing is a useful add-on for this age group.

Because individuals may decide it is too much effort to read large amounts of text, structured formatting could be beneficial in lessening the mental load and perception of difficulty. Formatting can make the visual display aesthetically pleasing to help hold people's attention on the material.

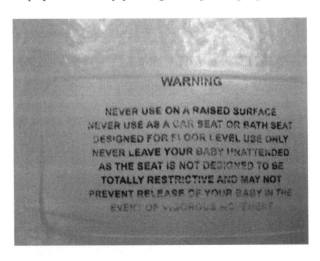

FIGURE 4.8   Warning text located on a child seat.

Formatting can help process the information by "chunking" it into smaller units. Formatting can also show the structure or organization of the material, thus making it easier to search for and assmililate the information into existing knowledge and memory (Hartley 1994; Shaver and Wogalter 2003). Figure 4.9 illustrates an example of the "Drug Facts" format used to communicate safety information on OTC drugs that is currently mandated by the U.S. Food and Drug Administration (U.S. FDA 2001).

4.2.5.3 Comprehension and Memory

Comprehension concerns understanding the meaning of something, in this case, the intended message of the warning. Comprehension may derive from several components: subjective understanding such as its hazard connotation, understanding of language and symbols, and an interplay with the individual's background knowledge. Background knowledge refers to relatively permanent long-term memory structure. The following sections contain short reviews of some major conceptual research areas with respect to warnings and the comprehension stage.

4.2.5.3.1 Signal Words

Aspects of a warning can convey a level of subjective hazard to the recipient. The ANSI (2002) Z535 standard recommends three signal words to denote different levels of hazard: DANGER, WARNING, or CAUTION (see also FMC Corporation 1985; Peckham 2006; Westinghouse Electric Corporation 1981). According to ANSI Z535, the DANGER panel should be used when serious injury or death *will* occur if the directive is not followed. A WARNING panel is used when serious injury or death *may* occur if the directive is not followed. The CAUTION panel is used when less severe personal injuries or property damage may occur if the directive is not followed. While the standard describes CAUTION and WARNING with different definitions, numerous empirical research studies indicate that people do not readily distinguish between the two. Although the term DEADLY has been shown in several research studies to connote significantly higher hazard than the standard's highest level DANGER, the use of DEADLY is not part of ANSI Z535 (e.g., see Hellier and Edworthy 2006; Wogalter, Kalsher et al. 1998; Wogalter and Silver 1990, 1995).

According to ANSI Z535, the signal word panels for DANGER, WARNING, and CAUTION are assigned specific colors: red, orange, and yellow, respectively. This assignment provides a form of redundancy due to the presence of more than one cue. However, most people do not reliably distinguish different levels of hazard associated with the colors orange and yellow (Chapanis 1994; Mayhorn, Wogalter, and Shaver 2004; Wogalter et al. 1998). The signal word panels also contain the alert symbol (triangle/exclamation point), which indicates it is a warning (Wogalter et al. 1998;

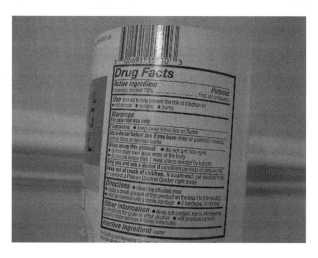

FIGURE 4.9   "Drug facts" formatting.

Wogalter, Jarrard, and Simpson 1994). Instead of the alert symbol, the older version of the ANSI Z535.2 (1991) standard had a different kind of shape cue co-occuring within the signal word panel (DANGER had a red, white, and black oval, and WARNING was surrounded by a hexagonal border).

4.2.5.3.2 Message Content

The content of the warning message should include information about the hazard, instructions on how to avoid the hazard, and the potential consequences if the hazard is not avoided (Wogalter et al. 1987). There are exceptions when the hazard is: (a) general knowledge; (b) known from previous experience; or (c) "open and obvious," i.e., apparent to everyone (except very young children).

a. *Hazard information.* At a minimum, the warning should identify the safety problem. Oftentimes, however, warnings might require more information regarding the nature of the hazard and the mechanisms that produce it.
b. *Instructions.* Warnings should instruct people about what to do or not do. The instructions should be specific inasmuch as reasonable to tell what exactly should be done or avoided. A classic non-explicit warning statement is "Use with adequate ventilation." Two others are "may be hazardous to health" or "maintain your tire pressure." By themselves these statements are inadequate to apprise people of what they should or should not do. In the case of "inadequate ventilation," does it mean to open a window, two windows, use a fan, or something more technical in terms of volume of air flow per unit time? The statement "may be hazardous to health" does not tell the mechanism by which injury may occur and the severity of the injury nor its probability. The statement "maintain your tire pressure" does not tell that there is an injury potential (as opposed to tread wear). In each case, without more information, users are left making inferences that may be partly or wholly incorrect (Laughery and Paige-Smith 2006; Laughery, Vaubel et al. 1993).
c. *Consequences.* Consequences information concerns what could result. It is not always necessary to state the consequences. However, one should be cautious in omitting it, because people may make the wrong inference.

A common shortcoming of warnings is that the consequences information is not explicit, i.e., lacking important specific details (Laughery and Paige-Smith 2006; Laughery et al. 1993). The statement "may be hazardous to your health" in the context of a toxic vapor hazard is insufficient by itself as it does not tell what kind of health problem could occur. The reader might believe it could lead to minor throat irritation not thinking that it could be something more severe, like permanent lung damage and perhaps death. To illustrate a poor example of consequence information communication via a warning, consider the depilatory product warning depicted in Figure 4.10. Here, the only consequence information regarding potential eye injuries states that "if irritation occurs" following eye contact, consumers should seek medical attention. From this, people might not readily infer that there is real potential for serious eye injury, possibly permanent blindness, resulting from this product. In a later section of this chapter, the specification of severe consequences is discussed as a factor in motivating compliance behavior.

4.2.5.3.3 Symbols

Safety symbols may also be used to communicate the above-mentioned information in lieu of or in conjunction with text statements (e.g., Dewar 1999; Mayhorn and Goldsworthy 2007, 2009; Mayhorn, Wogalter, and Bell 2004; Wolff and Wogalter 1998; Young and Wogalter 1990; Zwaga and Easterby 1984). Such symbols can contribute to understanding when illiterates or non-readers of the primary language are part of the target audience.

Comprehension is important for effective safety symbols (Dewar 1999). Symbols that directly represent concepts are preferred because they are usually better comprehended than more abstract symbols (Magurno et al. 1994; Wogalter et al. 2006; Wolff and Wogalter 1993). With abstract and

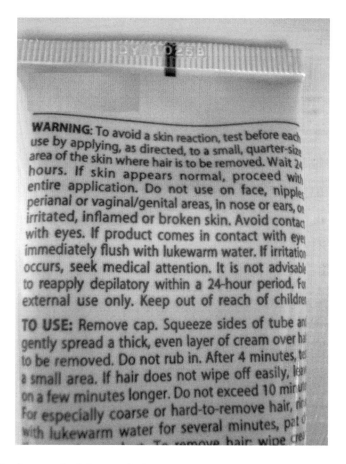

FIGURE 4.10 Warning on a depilatory product.

arbitrary symbols (Lesch 2003; Sojourner and Wogalter 1997, 1998; Wogalter, Sojourner, and Brelsford 1997), the meaning typically has to be learned via training.

What is an acceptable level of comprehension for safety symbols? In general, symbols should be designed to have the highest level of comprehension attainable. The ANSI (2002) Z535.3 standard suggests a goal of at least 85% comprehension using a sample of 50 individuals representative from the target audience for a symbol to be used without accompanying text. If 85% cannot be achieved, the symbol may still have utility (e.g., for attention capture) as long as it is not likely to be misinterpreted. According to the ANSI (2002) Z535.3 standard, an acceptable symbol must produce less than 5% critical confusions (opposite meaning or a meaning that would produce unsafe behavior). For example, the pharmaceutical warning shown in Figure 4.11 (circle/slash image superimposed over a pregnant female body) currently in use on the drug Accutane regarding the potential for birth

Do not

Get pregnant

FIGURE 4.11 Pharmaceutical warning on Accutane.

defects if the substance is taken during pregnancy might be wrongly interpreted as meaning that the drug is for birth control (Mayhorn and Goldsworthy 2007, 2009). ISO (2001) has similar comprehension criteria (see Deppa 2006; Peckham 2006).

Repeated exposure to an unchanged warning over time will not only result in it being less effective in switching attention, but also less effective for maintaining attention. As mentioned earlier, even a well-designed warning will eventually become habituated if repeatedly encountered. Fortunately, habituation as a memory concept implies that the person has learned some amount of information from the warning to "know" to ignore it. Unfortunately, only part of the warning may actually be known. Some techniques for reducing habituation include: (a) using salient features, and (b) periodically varying the warning's appearance (and content, if feasible and appropriate).

Although individuals may have knowledge about a hazard, they may not be aware of it at the time they are at risk. People have vast stores of knowledge in long-term memory based on an accumulation of experience in their lives. Despite this amazing memory storage space, at any given time only a small portion of it is consciously available. As people are doing their tasks in daily life and at work, their minds are not always actively accessing risk information. Thus, while a person may have some or an extensive store of risk knowledge, this information and related knowledge may not be activated unless there is an external cue to activate it. Consider the electrical hazard tag on hair dryers. Because of its presence, people are more likely to be reminded to keep away from water than if the tag were not secured to the electrical cord. Of course, seeing this tag every day results in habituation where it is infrequently noticed. But its presence is better than its absence, as for example it may serve as a reminder to some persons. So, despite habituation, the presence of a warning may serve to cue relevant hazard information. Some cues can activate large amounts of knowledge, so a single word or symbol may evoke much more than its literal interpretation. Without a reminder, known risk knowledge is less likely to be brought to mind.

In summary, information in long-term memory can be cued by the presence of a warning and bring forth related, previously dormant knowledge into conscious awareness. Reminders may be appropriate in situations: (a) where the hazard is infrequently encountered in which forgetting may be an issue, and (b) when there are foreseeable distractions or high task-load involvement that could pull attention away from normative hazard considerations.

4.2.5.3.4 Level of Knowledge

The levels of knowledge and understanding of the warning recipients should be taken into consideration. Three cognitive characteristics of receivers are important: language skill, reading ability, and technical knowledge.

With regard to pharmaceuticals, it is not unusual for consumers to be given textual warnings beyond their reading skill. In general, reading levels should be as low as feasible. For the general population, the reading level probably should be approximately the skill level of grades 4–6 (expected ability of 10 to 12-year-old readers). There are large numbers of functionally illiterate persons, even in some of the most technologically advanced countries. For example, in the United States there are estimates of over 16 million functionally illiterate adults. Thus, successful warning communication may require more than simply keeping reading levels to a minimum. The use of symbols, speech warnings, and special training programs may be beneficial adjuncts. Also, a related consideration is that different subgroups within a population may speak and read different languages. Because of increasing international trade and travel and the need to cross language barriers, this problem might require the use of multiple languages, graphics, and transmission through multiple methods (Lim and Wogalter 2003). An example is illustrated in Figure 4.12, which depicts the warning on a heat gun used to remove wall paper and paint. It shows a pictorial of a fire and text in both English and French, and further on the right slide is Spanish.

Despite considerations at the minimal end, reading levels should be consistent with the reading abilities of the receivers. A warning to trained health care professionals should use standard verbiage expected by that population. These technical experts have a more complete understanding

Photo file no. 98-4460244   Date taken: 12/5/00
Description: Heat gun
Copyright 2000 Richard M. Hansen & Associates, Inc.

FIGURE 4.12   Warning on a heat gun.

of domain-specific hazards and can perform their jobs better with area-appropriate technical data. A warning to the general consuming public does not have the luxury of knowing that the receivers have an extensive background. The short standardized text in the U.S. FDA's Drug Facts labels (see Figure 4.9) on OTC (or non-prescription) drugs is a simplified, less extensive description of the drug than a physician or other health professional may receive. Training depends on the type of occupation. A tire salesperson or tire buster (professional installer of tires) cannot be expected to have extensive training on the hazards and warnings associated with tire choice and installation. Any schooling or training on the topic is likely to be no more than a short course or two, and probably less, such as on-the-job training. Here, the warnings might not be much more different in level of difficulty than those transmitted to the public.

It is not usually necessary to give highly technical warning information to a general target audience of end users. Indeed, it can sometimes be counterproductive in the sense that encountering such information may result in little or no attention being given to the material. Reasons have already been discussed in the section on attention maintenance stage. Instead, pharmaceutical information directed to general consumers needs to give its indications for use, contraindications, side effects, and how to use it safely (i.e., hazard, consequences, and instructions as described above). When there are multiple groups of people with different characteristics, different parts of the warning system can be used to communicate to different groups.

4.2.5.4   Beliefs and Attitudes

Beliefs and attitudes is the next major stage of the C-HIP model. Beliefs refer to an individual's knowledge that is accepted as true (although some of it may not actually be true). It is related to the previous stage in that beliefs are formed from memory structure. In some respects, beliefs tend to be more global and overarching compared to specific memories. An attitude is similar to a belief except it includes more affect or emotional involvement.

People's benign experiences with a potentially hazardous product can produce beliefs that a product is safer than it is. This quickly changes after being involved in some way with (or seeing) a serious injury event. According to the C-HIP model, a warning will be successfully processed at the beliefs and attitudes stage if the message concurs (or at least is not discrepant) with the receiver's current beliefs and attitudes. However, if the warning information does not concur, then beliefs and attitudes may need to be altered before a person will be motivated to carry out the warning's directed behavior. The message and/or other information needs to be persuasive to override existing incorrect beliefs and attitudes. Methods of persuasion are commonly used in advertising and have been empirically explored in the social and cognitive psychology literatures. Sometimes, unequivocal and explicit statements can be used to persuade, but also the features of the warning may convey a higher level of importance. Such persuasion is important when a product is more hazardous than people believe. While changing people's beliefs may present some challenges, the task is even more difficult when other communications (e.g., through marketing and advertising, or simply poor news reporting) lead people to believe that the product is more safe than it is. For example, Figure 4.13 illustrates how advertising materials located on the packing materials of a child seat might invoke beliefs about product safety when it includes assertions about awards conferred by various organizations and language stating that physicians recommend it to improve the well being of a child. In the following paragraphs, several relevant and interrelated factors associated with the beliefs and attitudes stage: hazard perception, familiarity, prior experience, and relevance, are discussed (see DeJoy 1999; Riley 2006; Vredenburgh and Zackowitz 2006).

Hazard perceptions influence processing at the beliefs and attitudes stage. The greater the perceived hazard, the more responsive people will be to warnings, as in looking for, reading, and complying with them. The converse is also true. People are less likely to look for, read, or comply with a warning for products that they do not believe are hazardous. Perceived hazard is closely tied to beliefs about injury severity. People perceive a product is more hazardous and act more cautiously when injuries could be severe (Wogalter, Young et al. 1999). Interestingly, however, injury likelihood is a much less important factor in perceptions of risk or hazard for consumer products (Wogalter et al. 1991; Wogalter, Brems, and Martin 1993).

Familiarity beliefs are formed from past similar experiences stored in memory. It is the belief that almost everything that needs to be known about a product or situation is already known. A person believing that they are adequately familar with a product might assume that a different, but similar, product operates in the same way and has the same hazards (which may not be true), reducing the likelihood that he or she will look for or read a warning (Godfrey and Laughery 1984; Goldhaber and deTurck 1988; Wogalter et al. 1991). For example, women with prior tampon usage reported a reduced likelihood of reading a warning on more absorbant (and more hazardous) tampons (Godfrey and Laughery 1984).

Research indicates that hazard perception is more important than familiarity with respect to warnings (Wogalter et al. 1991). This is probably due to two factors. First, people more familiar with a situation or product may have more knowledge about the hazards and how to avoid them. Second, greater use also tends to increase exposure to warnings, which increases the opportunity to be influenced by them.

FIGURE 4.13   Advertising materials depicted on the packaging of a child seat.

Related to familiarity is prior experience. The concepts are somewhat different in that familiarity is a belief (that may or may not be true), and prior experience is an objective quantity that could potentially be measured. Prior experience can be influential in hazard perceptions. Having experienced some form of injury or having personal knowledge of someone else being injured enhances hazard perceptions (Wogalter, Brems, and Martin 1993). For instance, older adults who were personally familiar with the hazards associated with household products, such as cleaning solutions and small appliances, or who were aware of injuries to someone else were able to produce more effective hazard avoidance strategies (Mayhorn et al. 2004). Similarly, the lack of such experiences may lead to underestimating dangers, or not thinking about them at all. Warnings that give vivid explicit consequences may convince people to change beliefs when they have inappropriate low levels of perceived hazard. For instance, the Canadian cigarette warning illustrated in Figure 4.14 contains much more explicit information regarding the likelihood of nicotine addiction than is currently in cigarette warnings in the United States.

Perceived relevance is the belief that something is applicable to the person. If the individual does not believe that the warning is relevant to them, then the warning may fail to fulfill its intended purpose. The individual may instead attribute the warning as being directed to others and not to himself or herself. For example, men may utilize pharmaceutical substances such as Propecia (for male pattern baldness) that might cause birth defects if pregnant female family members come in contact with it. While men may be made aware of this property, they obviously will not believe that pregnancy warnings apply to them (Mayhorn and Goldsworthy 2007, 2009). One way to counter this is to personalize the warning so that it gets directed to relevant users and conveys facts that indicate that it is relevant (Wogalter et al. 1994).

A point related to beliefs and attitudes and more specifically, familiarity, concerns the problem of experts overestimating what lay persons know, which in turn may affect what kinds of warnings are produced (Laughery 1993). Experts in a domain can be so facile with their knowledge about a topic that they fail to realize that non-experts do not have similar knowledge. What is "obvious" to them may not be as obvious to end users. Without consumer input into the design of warnings, there may be a tendency to produce warnings that fail to meet the needs of end users.

4.2.5.5 Motivation

Motivation energizes the individual to carry out an activity. Some of the main factors that can influence the motivation stage of the C-HIP model are cost of compliance, severity of injury, social influence, and stress. These topics are discussed below.

Compliance generally requires that people take some action, and usually there are costs associated with doing so. The costs of complying may include time and effort to carry out the behavior (Wogalter et al. 1987; Wogalter, Allison, and McKenna 1989). When people perceive the costs of compliance to be too high, they are less likely to perform the safety behavior. This problem is

FIGURE 4.14   Canadian cigarette warning.

commonly encountered in warnings with instructions directing behaviors that are inconvenient, difficult, or occasionally impossible to carry out. One way to reduce cost is to make the directed behavior easier to perform. For example, if hand protection is required when using a product, the presence of gloves should be as simple, easy, and convenient as possible (Dingus, Hathaway, and Hunn 1991).

The costs of non-compliance can also exert a powerful influence on compliance motivation. With respect to warnings, a main cost for non-compliance is severe injury consequences. Previous research suggests that people report higher willingness to comply with warnings when they believe there is a high probability of incurring a severe injury (e.g., Wogalter et al. 1991, 1999; Wogalter, Brems, and Martin 1993).

Another motivator is social influence (Wogalter, Allison, and McKenna 1989; Edworthy and Dale 2000). When people see others comply with a warning, they are more likely to comply themselves. Likewise, seeing others not comply, lessens the likelihood of compliance. Other factors affecting motivation are time stress (Wogalter, Magurno et al. 1998) and mental workload (Wogalter and Usher 1999). Under high stress and workload, competing activities take resources away from processing warning information.

4.2.6 BEHAVIOR

The last stage of the sequential process is for individuals to carry out the warning-directed safe behavior. Behavior is one of the most important measures of warning effectiveness (Kalsher and Williams 2006; Silver and Braun 1999). Warnings do not always change behavior because of processing failures at earlier stages. Most research in this area focuses on the factors that affect compliance likelihood, including those that enhance safety behavior and those that do not.

Some researchers have used "intentions to comply" as the method of measurement because it is usually quite difficult to conduct behavioral tests. The difficulties include the following: (a) researchers cannot expose participants to real risks because of ethical and safety concerns; (b) events that could lead to injury are relatively rare; (c) the scenario must appear to have a believable risk, yet at the same time must be safe; and (d) running such research is costly in terms of time and effort. Nevertheless, compliance is an important criterion for determining which factors work better than others to boost warning effectiveness and, consequently, safe behavior. Additionally, many products are used inside homes where access to determine how a product is used and whether a warning was complied with is difficult. Virtual reality may play a role in allowing research to be conducted in simulated conditions that avoid some of the above problems (Duarte, Rebello, and Wogalter 2009). Also, compliance can be measured indirectly. For example determining whether protective gloves have been worn can be gleaned from whether they appear to be used or stretched in appearance (Wogalter and Dingus 1999; Kalsher and Williams 2006). Likewise, medication adherence to prescription pharmaceuticals can be assessed by using a hidden electronic chip in the cap that records each opening of the container lid (Park et al. 1992).

4.2.6.1 Receiver Variables

The receiver's characteristics and task workload can affect warning effectiveness (Young et al. 1999). Indeed, evidence supporting this has already been discussed. Person variables (Rogers, Lamson, and Rousseau 2000) such as the individuals' existing knowledge, beliefs, and language skill were noted in earlier sections as affecting whether and how a warning is processed. Mayhorn and Podany (2006) describe research findings showing age-related declines in sensory and cognitive processing that affect warning processing, particularly in attention switch and memory/comprehension stages. Although not much systematic warning research has been conducted with respect to children, Kalsher and Wogalter (2007) provide an overview of the existing research. In some studies, gender differences have been noted (e.g., see Laughery and Brelsford 1991; Smith-Jackson 2006) with women being somewhat more likely to look for and read warnings (e.g., Godfrey et al. 1983; LaRue and Cohen 1987; Young, Martin, and Wogalter 1989). Other research indicates that risk perception varies by ethnicity such that Latino farm workers reported higher risk perception associated with the use of pesticides than Americans of European descent (Smith-Jackson, Wogalter, and Quintela in press). Two other individual differences variables have been noted in the literature: self-efficacy (Lust, Celuch, and Showers 1993) and locus of control (Donner 1991; Laux and Brelsford 1989). It is not completely clear whether the relative paucity of research on personality variables and warning-related measures is due to the correlations being relatively small or that they simply have not attracted researchers as a topic of study (see also Lesch 2006).

Lastly, warning processing occurs in the context of other potential processing given other stimuli in the environment and the individual's ongoing and ever-changing task behavior. Whether and how a warning is processed can depend on mental workload (Wogalter and Usher 1999), time stress (Wogalter et al. 1998), and processing strategy (deTurk and Goldhaber 1988). An individual thinking about other information, under time pressure, and who is not in an information-seeking mode is less likely to fully process a warning compared to situations when not under those restraints. When such task loading can be anticipated (e.g., in emergency situations), the warning system may have to be highly salient to attract attention. For instance, people faced with televised warnings about impending natural hazards such as hurricanes or floods may be less likely to extract all pertinent protective action information when updates are transmitted (Mayhorn, Yim, and Orrock 2006). Because news tickers at the bottom of a screen may not be salient, attention must be directed to those updates, perhaps via the announcer occupying the fuller screen component.

4.3   SUMMARY AND UTILITY OF THE COMMUNICATION-HUMAN INFORMATION PROCESSING MODEL

The above review of the warning literature as applied to consumer products was organized around the C-HIP model. This model divides the processing of warning information into separate stages that must be successfully completed for compliance behavior to occur. A bottleneck at any given stage can hinder processing at subsequent stages. Feedback from later stages can affect processing at earlier stages. The model is valuable in describing some of the processes and organizing a large amount of research.

The C-HIP model can also be a valuable tool in systematizing the assessment process to help determine why a warning is not effective. It can aid in pinpointing where the bottlenecks in processing may be occurring and suggest solutions to allow processing to continue to subsequent stages. Warning effectiveness testing can be performed using methods similar to those used in research. Evaluations of the processing can be directed to any of the stages described in the C-HIP model: source, channel, environment, delivery, attention, comprehension, attitudes and beliefs, motivation, behavior, and receiver variables. Some of the methods for doing this evaluation are briefly described below.

Evaluating the source necessitates an attempt to determine whether the manufacturer has documented the potential hazards and has issued warnings. It is fundamental that manufacturers should analyze their product to determine whether there are foreseeable potential hazards associated with its use and misuse. When hazards are discovered, manufacturers have an obligation to employ methods to try to control the hazards to reduce personal injury and property damage. If a manufacturer is going to sell a product in which the hazard has not been eliminated through design or physical guarding, then it should provide effective warning(s) to consumers and users. One important question to address here is whether there is anything missing from the current warning that should be there? Hazard analysis is needed to answer this question (Young, Frantz, and Rhoades 2006).

Evaluating the channel mainly addresses questions relating to how warnings are sent to end users. One question to ask is what media and modalities are being used and are they adequate. Similarly, assessment regarding delivery asks whether end users receive the warnings. If not, other channels of distribution of warning materials may need to be considered.

To assess attention switch, the main question is whether end users see or hear the warnings. The answer could involve placing a warning on a product and having people carry out a relevant task and asking them later whether they saw it. Eye movement and response time paradigms can be used to measure what people tend to look at and how quickly.

To assess comprehension, there are several well-established methodologies involving memory tests, open-ended response tests, structured interviews, etc. These assessments can be valuable for determining what information was or was not understood and for suggesting revisions to warning text or symbols. To assess beliefs and attitudes, a questionnaire could be used to determine people's pre-existing beliefs on the topics of perceived hazard and familiarity with the product, task, or environment. For example, if people's perceived hazard is too low, greater persuasiveness may be needed.

To assess motivation, measures of behavioral intentions can be used. Low intentions to comply may indicate that consequence information should be enhanced (e.g., by being more explicit) or that cost of compliance should be reduced. To assess behavioral compliance, systematic observation can be used in both laboratory and field settings. As mentioned earlier, measurement of behavioral compliance is generally more difficult than any of the other methods; it may involve ethical issues such as participants' exposure to risk. However, in situations where the negative consequences are substantial, the effort and resources may be warranted. Sometimes behavioral intentions are measured as a proxy for overt behavioral compliance—however, some caution should be exercised, as noted earlier.

By using the above investigative methods (and others) in a systematic manner, the specific causes of a warning's failure may be determined. Resources would then be better directed at fixing the aspects that are limiting the warning's effectiveness.

In summary, the C-HIP model describes the processing of warnings in a series of stages that could block the processing of warnings. Although it has linear components from source to compliance behavior, there are feedback loops that account for later processing stages affecting earlier stages. The C-HIP model also serves as a useful framework in organizing the growing body of research in the area. Lastly, the model can be used as an investigative tool to determine why a warning is inadequately carrying out its intended purpose.

REFERENCES

ANSI. 1991. *Accredited Standards Committee on Safety Signs and Colors. Z535.1-5.* Arlington, VA: National Electrical Manufacturers Association.

———. 2002. *Accredited Standards Committee on Safety Signs and Colors. Z535.1-5.* Arlington, VA: National Electrical Manufacturers Association.

Barlow, T., and Wogalter, M.S. 1991. Increasing the surface area on small product containers to facilitate communication of label information and warnings. In *Proceedings of Interface 91*, 88–93. Santa Monica, CA: Human Factors Society.

———. 1993. Alcoholic beverage warnings in magazine and television advertisements. *Journal of Consumer Research* 20:147–55.

Bzostek, J.A., and Wogalter, M.S. 1999. Measuring visual search time for a product warning label as a function of icon, color, column, and vertical placement. *Proceedings of the Human Factors and Ergonomics Society* 43:888–92.

Cameron, K.A., and DeJoy, D.M. 2006. The persuasive functions of warnings: Theory and models. In *Handbook of Warnings,* ed. M.S. Wogalter, 301–12. Mahwah, NJ: Lawrence Erlbaum Associates.

Chapanis, A. 1994. Hazards associated with three signal words and four colours on warning signs. *Ergonomics* 37:265–75.

Cheatham, D.B., and Wogalter, M.S. 2002. Reported likelihood of reading over-the-counter (OTC) medication labeling and contacting a physician. *Proceedings of the Human Factors and Ergonomics Society* 46:1452–14.

———. 2003. Comprehension of over-the-counter drug label warnings regarding consumption of acetaminophen and alcohol. *Proceedings of the Human Factors and Ergonomics Society* 47:1540–44.

Cohen, H.H., Cohen, J., Mendat, C.C., and Wogalter, M.S. 2006. Warning channel: Modality and media. In *Handbook of Warnings,* ed. M.S. Wogalter, 123–34. Mahwah, NJ: Lawrence Erlbaum Associates.

Conzola, C.V., and Wogalter, M.S. 1999. Using voice and print directives and warnings to supplement product manual instructions. *International Journal of Industrial Ergonomics* 23:549–56.

————. 2001. A communication–human information processing (C-HIP) approach to warning effectiveness in the workplace. *Journal of Risk Research* 4:309–22.

Cox, E.P. III. 1999. Source. In *Warnings and Risk Communication*, eds. M.S. Wogalter, D.M. DeJoy, and K.R. Laughery, 85–97. London: Taylor & Francis.

Cox, E.P. III, and Wogalter, M.S. 2006. Warning source. In *Handbook of Warnings*, ed. M.S. Wogalter, 111–22. Mahwah, NJ: Lawrence Erlbaum Associates.

DeJoy, D.M. 1999. Beliefs and attitudes. In *Warnings and Risk Communication*, eds. M.S. Wogalter, D.M. DeJoy, and K.R. Laughery, 183–219. London: Taylor & Francis.

Deppa, S.W. 2006. U.S. and international standards for safety symbols. In *Handbook of Warnings*, ed. M.S. Wogalter, 477–86. Mahwah, NJ: Lawrence Erlbaum Associates.

Desaulniers, D.R. 1987. Layout, organization, and the effectiveness of consumer product warnings. *Proceedings of the Human Factors Society* 31:56–60.

deTurk, M.A., and Goldhaber, G.M. 1988. Consumers' information processing objects and effects of product warning. *Proceedings of the Human Factors Society* 32:445–49.

Dewar, R. 1999. Design and evaluation of graphic symbols. In *Visual Information for Everyday Use: Design and Research Perspectives*, eds. H.J.G. Zwaga, T. Boersema, and H.C.M. Hoonhout, 285–303. London: Taylor & Francis.

Dingus, T.A., Hathaway, J.A., and Hunn, B.P. 1991. A most critical warning variable: Two demonstrations of the powerful effects of cost on warning compliance. *Proceedings of the Human Factors Society* 35:1034–38.

Donner. 1991. Prediction of safety behaviors from Locus of Control statements. In *Proceedings of Interface '91*:94–98. Santa Monica, CA: Human Factors Society.

Duarte, M.E.C., Rebelo, F., and Wogalter, M.S. 2010. The potential of virtual reality (VR) for evaluating warning compliance. *Human Factors and Ergonomics in Manufacturing and Service Industries*, 20:526–37.

Edworthy, J., and Dale, S. 2000. Extending knowledge of the effects of social influence in warning compliance. In *Proceedings of the XIVth Triennial Congress of the International Ergonomics Association and 44th Annual Meeting of the Human Factors and Ergonomics Society*, vol. 4, 107–110. Santa Monica, CA: Human Factors and Ergonomics Society.

Edworthy, J., and Hellier, E. 2006. Complex nonverbal auditory signals and speech warnings. In *Handbook of Warnings*, ed. M.S. Wogalter, 199–220. Mahwah, NJ: Lawrence Erlbaum Associates.

FMC Corporation. 1985. *Product Safety Sign and Label System*. Santa Clara, CA: FMC Corporation.

Frantz, J.P., and Rhoades, T.P. 1993. A task analytic approach to the temporal placement of product warnings. *Human Factors* 35:719–30.

Frantz, J.P., Rhoades, T.P., and Lehto, M.R. 1999. Practical considerations regarding the design and evaluation of product warnings. In *Warnings and Risk Communication*, eds. M.S. Wogalter, D.M. DeJoy, and K.R. Laughery, 291–311. London: Taylor & Francis.

Frascara, J. 2006. Typography and the visual design of warnings. In *Handbook of Warnings*, ed. M.S. Wogalter, 385–406. Mahwah, NJ: Lawrence Erlbaum Associates.

Godfrey, S.S., Allender, L., Laughery, K.R., and Smith, V.L. 1983. Warning messages: Will the consumer bother to look? *Proceedings of the Human Factors Society* 27:950–54.

Godfrey, S.S., and Laughery, K.R. 1984. The biasing effect of familiarity on consumer's awareness of hazard. *Proceedings of the Human Factors Society* 28:483–86.

Goldhaber, G.M., and deTurck, M.A. 1988. Effects of consumer's familiarity with a product on attention and compliance with warnings. *Journal of Products Liability* 11:29–37.

Hartley, J. 1994. *Designing Instructional Text* (3rd ed.). London: Kogan Page/East Brunswick, NJ: Nichols.

Hellier, E., and Edworthy, J. 2006. Signal words. In *Handbook of Warnings*, ed. M.S. Wogalter, 407–17. Mahwah, NJ: Lawrence Erlbaum Associates.

Hicks, K.E., Wogalter, M.S., and Vigilante, W.J., Jr. 2005. Placement of benefits and risks in prescription drug manufacturers' web sites and information source expectations. *Drug Information Journal* 39:267–78.

Hooper, S., and Hannafin, M.J. 1986. Variables affecting the legibility of computer generated text. *Journal of Instructional Development* 9:22–28.

ISO. 2001. Graphical symbols–Test methods for judged comprehensibility and for comprehension. ISO 9186, International Organization for Standards.

Kalsher, M.J., and Williams, K.J. 2006. Behavioral compliance: Theory, methodology, and results. In *Handbook of Warnings*, ed. M.S. Wogalter, 289–300. Mahwah, NJ: Lawrence Erlbaum Associates.

Kalsher, M.J., and Wogalter, M.S. 2007. Hazard control methods and warnings for caregivers and children. In *Ergonomics for Children,* eds. R. Leuder and V. Rice, Ch. 14, 509–42. Boca Raton, FL: CRC Press.

LaRue, C., and Cohen, H. 1987. Factors influencing consumer's perceptions of warning: An examination of the differences between male and female consumers. *Proceedings of the Human Factors Society* 31:610–14.

Laughery, K.R. 1993. Everybody knows: Or do they? *Ergonomics in Design* July, 8–13.

Laughery, K.R., and Brelsford, J.W. 1991. Receiver characteristics in safety communications. *Proceedings of the Human Factors Society* 35:1068–72.

Laughery, K.R., and Paige-Smith, D. 2006. Explicit information in warnings. In *Handbook of Warnings*, ed. M.S. Wogalter, 419–28. Mahwah, NJ: Lawrence Erlbaum Associates.

Laughery, K.R., Vaubel, K.P., Young, S.L., Brelsford, J.W., and Rowe, A.L. 1993. Explicitness of consequence information in warning. *Safety Science* 16:597–613.

Laughery, K.R., and Wogalter, M.S. 1997. Risk perception and warnings. In *Handbook of Human Factors and Ergonomics* (2nd ed.) ed. G. Salvendy. New York: Wiley-Interscience.

———. 2006. Designing effective warnings. In *Reviews of Human Factors and Ergonomics*, Vol. 2, ed. R. Williges, 241–71. Santa Monica, CA: Human Factors and Ergonomics Society.

Laughery, K.R., Young, S.L., Vaubel, K.P., and Brelsford, J.W. 1993. The noticeability of warnings on alcoholic beverage containers. *Journal of Public Policy and Marketing* 12:38–56.

Laux, L.F., and Brelsford, J.W. 1989. Locus of control, risk perception, and precautionary behavior. In *Proceedings of Interface 89*, 121–24. Santa Monica, CA: Human Factors Society.

Lehto, M.R. 2006. Human factors models. In *Handbook of Warnings*, ed. M.S. Wogalter, 83–87. Mahwah, NJ: Lawrence Erlbaum Associates.

Lehto, M.R., and Miller, J.M. 1986. *Warnings: Volume 1. Fundamentals, Design and Evaluation Methodologies.* Ann Arbor, MI: Fuller Technical Publications.

Lesch, M.F. 2003. Comprehension and memory for warning symbols: Age-related differences and impact of training. *Journal of Safety Research* 34:495–505.

———. 2006. Consumer product warnings: Research and recommendations. In *Handbook of Warnings*, ed. M.S. Wogalter, 137–46. Mahwah, NJ: Lawrence Erlbaum Associates.

Lim, R.W., and Wogalter, M.S. 2003. Beliefs about bilingual labels on consumer products. *Proceedings of the Human Factors and Ergonomics Society* 47:839–43.

Lust, J.A., Celuch, K.G., and Showers, L.S. 1993. A note on issues concerning the measurement of self-efficacy. *Journal of Applied Social Psychology* 23:1426–34.

Magurno, A., Wogalter, M.S., Kohake, J., and Wolff, J.S. 1994. Iterative test and development of pharmaceutical pictorials. In *Proceedings of the 12th Triennial Congress of the International Ergonomics Association*, vol. 4:360–62.

Mayhorn, C.B., and Goldsworthy, R.C. 2007. Refining teratogen warning symbols for diverse populations. *Birth Defects Research Part A: Clinical and Molecular Teratology* 79 (6): 494–506.

———. 2009. "New and improved": The role text augmentation and the application of responses interpretation standards (coding schemes) in a final iteration of birth defects warnings development. *Birth Defects Research Part A: Clinical and Molecular Teratology* 85 (10): 864–71.

Mayhorn, C.B., Nichols, T.A., Rogers, W.A., and Fisk, A.D. 2004. Hazards in the home: Using older adults' perceptions to inform warning design. *Journal of Injury Control and Safety Promotion* 11 (4): 211–18.

Mayhorn, C.B., and Podany, K.I. 2006. Warnings and aging: Describing the receiver characteristics of older adults. In *Handbook of Warnings*, ed. M.S. Wogalter, 355–62. Mahwah, NJ: Lawrence Erlbaum Associates.

Mayhorn, C.B., Wogalter, M.S., and Bell, J.L. 2004. Are we ready? Misunderstanding homeland security safety symbols. *Ergonomics in Design* 12 (4): 6–14.

Mayhorn, C.B., Wogalter, M.S., and Shaver, E.F. 2004. What does Code Red mean? *Ergonomics in Design* 2 (4): 12.

Mayhorn, C.B., Yim, M.S., and Orrock, J.A. 2006. Warnings about potential natural and technological disasters and risks. In *Handbook of Warnings*, ed. M.S. Wogalter, 763–69. Mahwah, NJ: Lawrence Erlbaum Associates.

Mazis, M.B., and Morris, L.A. 1999. Channel. In *Warnings and Risk Communication*, eds. M.S. Wogalter, D.M. DeJoy, and K.R. Laughery, 99–121. London: Taylor & Francis.

McGuire, W.J. 1980. The communication-persuasion model and health-risk labeling. In *Banbury Report 6: Product Labeling and Health Risks*, eds. L.A. Morris, M.B. Mazis, and I. Barofsky, 99–122. Cold Spring Harbor, NY: Cold Spring Harbor Laboratory.

Mehlenbacher, B., Wogalter, M.S., and Laughery, K.R. 2002. On the reading of product owner's manuals: Perceptions and product complexity. *Proceedings of the Human Factors and Ergonomics Society* 46: 730–37.

Miller, J.M., and Lehto, M.R. 2001. *Warnings and Safety Instructions: Annotated and Indexed,* 4th ed. Ann Arbor, MI: Fuller Technical.

Park, D.C., Morrell, R.W., Frieske, D., and Kincaird, D. 1992. Medication adherence behaviors in older adults—Effects of external cognitive supports. *Psychology and Aging* 7 (2): 252–56.

Paivio, A. 1971. *Imagery and Verbal Processes.* New York: Holt, Rinehart and Winston.

Peckham, G.M. 2006. ISO design standards for safety signs and labels. In *Handbook of Warnings*, ed. M.S. Wogalter, 455–62. Mahwah, NJ: Lawrence Erlbaum Associates.

Poulton, E. 1967. Searching for newspaper headlines printed in capitals or lower-case letters. *Journal of Applied Psychology* 51:417–25.

Riley, D.M. 2006. Beliefs, attitudes, and motivation. In *Handbook of Warnings*, ed. M.S. Wogalter, 289–300. Mahwah, NJ: Lawrence Erlbaum Associates.

Rogers, W.A., Lamson, N., and Rousseau, G.K. 2000. Warning research: An integrative perspective. *Human Factors* 42:102–39.

Shaver, E.F., and Wogalter, M.S. 2003. A comparison of older v. newer over-the-counter (OTC) nonprescription drug labels on search time accuracy. *Proceedings of the Human Factors and Ergonomics Society* 47:826–30.

Silver, N.C., and Braun, C.C. 1999. Behavior. In *Warnings and Risk Communication*, eds. M.S. Wogalter, D.M. DeJoy, and K.R. Laughery, 245–62. London: Taylor & Francis.

Smith-Jackson, T. 2006. Culture and warnings. In *Handbook of Warnings*, ed. M.S. Wogalter, 363–72. Mahwah, NJ: Lawrence Erlbaum Associates.

Smith-Jackson, T.L. 2006. Receiver characteristics. In *Handbook of Warnings*, ed. M.S. Wogalter, 335–44. Mahwah, NJ: Lawrence Erlbaum Associates.

Smith-Jackson, T.L., and Wogalter, M.S. 2007. Application of mental models approach to MSDS design, *Theoretical Issues in Ergonomics Science* 8 (4): 303–19.

Smith-Jackson, T., Wogalter, M.S., and Quintela, Y. 2010. Safety climate and risk communication disparities for pesticide safety in crop production by ethnic group. *Human Factors and Ergonomics in Manufacturing and Service Industries* 20:481–83.

Sojourner, R.J., and Wogalter, M.S. 1997. The influence of pictorials on evaluations of prescription medication instructions. *Drug Information Journal* 31:963–72.

———. 1998. The influence of pictorials on the comprehension and recall of pharmaceutical safety and warning information. *International Journal of Cognitive Ergonomics* 2:93–106.

Tam, T., and Greenfield, T. 2010. Do alcohol warning labels influence men's and women's attempts to deter others from driving while intoxicated? *Human Factors and Ergonomics in Manufacturing and Service Industries* 20:538–46.

Thorley, P., Hellier, E., and Edworthy, J. 2001. Habituation effects in visual warnings. In *Contemporary Ergonomics 2001*, ed. M.A. Hanson, 223–28. London: Taylor & Francis.

U.S. FDA. 2001. *Format and Content Requirements for Over-the-Counter (OTC) Drug Product Labeling.* 21 C.F.R. § 201.66. Washington, DC: U.S. FDA.

Vigilante, W.J., and Wogalter, M.S. 2005. Assessing risk and benefit communication in direct-to-consumer medication web site advertising. *Drug Information Journal* 39 (1): 3–12.

Vigilante, W.J., Jr., Wogalter, M.S., and Mayhorn, C.B. 2007. Direct-to-consumer (DTC) prescription drug advertising on television and online purchases of medications. In *Proceedings of the Human Factors and Ergonomics Society 51st Annual Meeting*, 1272–76. Santa Monica, CA: Human Factors and Ergonomics Society.

Vredenburgh, A.G., and Helmick-Rich, J. 2006. Extrinsic nonwarning factors. In *Handbook of Warnings*, ed. M.S. Wogalter, 373–82. Mahwah, NJ: Lawrence Erlbaum Associates.

Vredenburgh, A.G., and Zackowitz, I.B. 2006. Expectations. In *Handbook of Warnings*, ed. M.S. Wogalter, 345–54. Mahwah, NJ: Lawrence Erlbaum Associates.

Westinghouse Electric Corporation. 1981. *Product Safety Label Handbook.* Trafford, PA: Westinghouse Printing Division.

Williamson, R.B. 2006. Fire warnings. In *Handbook of Warnings*, ed. M.S. Wogalter, 701–10. Mahwah, NJ: Lawrence Erlbaum Associates.

Wogalter, M.S. (ed.). 2006a. *Handbook of Warnings.* Mahwah, NJ: Lawrence Erlbaum Associates.

Wogalter, M.S. 2006b. Communication-human information processing (C-HIP) model. In *Handbook of Warnings*, ed. M.S. Wogalter, 51–61. Mahwah, NJ: Lawrence Erlbaum Associates.

Wogalter, M.S., Allison, S.T., and McKenna, N. 1989. Effects of cost and social influence on warning compliance. *Human Factors* 31:133–40.

Wogalter, M.S., Barlow, T., and Murphy, S. 1995. Compliance to owner's manual warnings: Influence of familiarity and the task-relevant placement of a supplemental directive. *Ergonomics* 38:1081–91.

Wogalter, M.S., Brelsford, J.W., Desaulniers, D.R., and Laughery, K.R. 1991. Consumer product warnings: The role of hazard perception. *Journal of Safety Research* 22:71–82.

Wogalter, M.S., Brems, D.J., and Martin, E.G. 1993. Risk perception of common consumer products: Judgments of accident frequency and precautionary intent. *Journal of Safety Research* 24:97–106.

Wogalter, M.S., DeJoy, D.M., and Laughery, K.R. 1999a. Organizing framework: A consolidated communication-human information processing (C-HIP) model. In *Warnings and Risk Communication*, eds. M.S. Wogalter, D.M. DeJoy, and K.R. Laughery, 15–24. London: Taylor & Francis.

Wogalter, M.S., DeJoy, D.M., and Laughery, K.R. (eds.) 1999b. *Warnings and Risk Communication*. London: Taylor & Francis.

Wogalter, M.S., and Dingus, T.A. 1999. Methodological techniques for evaluating behavioral intentions and compliance. In *Warnings and Risk Communication*, 53–82. London: Taylor & Francis.

Wogalter, M.S., and Feng, E. 2010. Indirect warnings/instructions produce behavioral compliance. *Human Factors and Ergonomics in Manufacturing and Service Industries* 20:500–10.

Wogalter, M.S., Godfrey, S.S., Fontenelle, G.A., Desaulniers, D.R., Rothstein, P.R., and Laughery, K.R. 1987. Effectiveness of warnings. *Human Factors* 29:599–612.

Wogalter, M.S., Jarrard, S.W., and Simpson, S.W. 1994. Influence of signal words on perceived level of product hazard. *Human Factors* 36:547–56.

Wogalter, M.S., Kalsher, M.J., Frederick, L.J., Magurno, A.B., and Brewster, B.M. 1998. Hazard level perceptions of warning components and configurations. *International Journal of Cognitive Ergonomics* 2:123–43.

Wogalter, M.S., Kalsher, M.J., and Rashid. R. 1999. Effect of signal word and source attribution on judgments of warning credibility and compliance likelihood. *International Journal of Industrial Ergonomics* 24:185–92.

Wogalter, M.S., and Laughery, K.R. 1996. WARNING: Sign and label effectiveness. *Current Directions in Psychology* 5:33–37.

———. 2005. Warnings. In *Handbook of Human Factors/Ergonomics* (3rd ed.). ed. G. Salvendy. New York: John Wiley.

Wogalter, M.S., and Leonard, S.D. 1999. Attention capture and maintenance. In *Warnings and Risk Communication*, eds. M.S. Wogalter, D.M. DeJoy, and K. R. Laughery, 123–48. London: Taylor & Francis.

Wogalter, M.S., Magurno, A.B., Dietrich, D., and Scott, K. 1999. Enhancing information acquisition for over-the-counter medications by making better use of container surface space. *Experimental Aging Research* 25:27–48.

Wogalter, M.S., Magurno, A.B., Rashid, R., and Klein, K.W. 1998. The influence of time stress and location on behavioral compliance. *Safety Science* 29:143–58.

Wogalter, M.S., and Mayhorn, C.B. 2005. Providing cognitive support with technology-based warning systems. *Ergonomics* 48:522–33.

Wogalter, M.S., Racicot, B.M., Kalsher, M.J., and Simpson, S.N. 1994. The role of perceived relevance in behavioral compliance in personalized warning signs. *International Journal of Industrial Ergonomics* 14:233–42.

Wogalter, M.S., and Silver, N.C. 1990. Arousal strength of signal words. *Forensic Reports* 3:407–20.

———. 1995. Warning signal words: Connoted strength and understandability by children, elders, and non-native English speakers. *Ergonomics* 38:2188–2206.

Wogalter, M.S., Silver, N.C., Leonard, S.D., and Zaikina, H. 2006. Warning symbols. In *Handbook of Warnings*, ed. M.S. Wogalter, 159–76. Mahwah, NJ: Lawrence Erlbaum Associates.

Wogalter, M.S., Smith-Jackson, T.L., Mills, B., and Paine, C. 2002. Effects of print format in direct-to-consumer prescription drug advertisements on risk knowledge and preference. *Drug Information Journal* 36:693–705.

Wogalter, M.S., Sojourner, R.J., and Brelsford, J.W. 1997. Comprehension and retention of safety pictorials. *Ergonomics* 40:531–42.

Wogalter, M.S., and Sojourner, R.J. 1999. Research on pharmaceutical labeling: An information processing approach. In *Processing of Medical Information in Aging Patients: Cognitive and Human Factors Perspectives*, eds. D.C. Park, R.C. Morrell, and K. Shifren, 291–310. Mahwah, NJ: Lawrence Erlbaum Associates.

Wogalter, M.S., and Usher, M. 1999. Effects of concurrent cognitive task loading on warning compliance behavior. *Proceedings of the Human Factors and Ergonomics Society* 43:106–10.

Wogalter, M.S., and Vigilante, W.J., Jr. 2003. Effects of label format on knowledge acquisition and perceived readability by younger and older adults. *Ergonomics* 46:327–44.

———. 2006. Attention switch and maintenance. In *Handbook of Warnings*, ed. M.S. Wogalter, 245–66. Mahwah, NJ: Lawrence Erlbaum Associates.

Wogalter, M.S., and Young, S.L. 1991. Behavioural compliance to voice and print warnings. *Ergonomics* 34:79–89.

———. 1994. Enhancing warning compliance through alternative product label designs. *Applied Ergonomics* 25:53–57.

———. 1998. Using a hybrid communication/human information-processing model to evaluate beverage alcohol warning effectiveness. *Applied Behavioral Sciences Review* 6:17–37.

Wogalter, M.S., Young, S.L., Brelsford, J.W., and Barlow, T. 1999c. The relative contribution of injury severity and likelihood information on hazard-risk judgments and warning compliance. *Journal of Safety Research* 30:151–62.

Wolff, J.S., and Wogalter, M.S. 1993. Test and development of pharmaceutical pictorials. In *Proceedings of Interface 93* (8): 187–92.

———. 1998. Comprehension of pictorial symbols: Effects of context and test method. *Human Factors* 40:173–86.

Young, S.L., Frantz, J.P., and Rhoades, T.P. 2006. Revisions of labeling for personal watercraft: Label development and evaluation. In *Handbook of Warnings*, ed. M.S. Wogalter, 723–38. Mahwah, NJ: Lawrence Erlbaum Associates.

Young, S.L., Laughery, K.R., Wogalter, M.S., and Lovvoll, D. 1999. Receiver characteristics in safety communications. In *The Occupational Ergonomics Handbook*, eds. W. Karwowski and W.S. Marras, 693–706. Boca Raton, FL: CRC Press.

Young, S.L., Martin, E.G., and Wogalter, M.S. 1989. Gender differences in consumer product hazard perceptions. In *Proceedings of Interface 89*, 73–78. Santa Monica, CA: Human Factors and Ergonomics Society.

Young, S.L., and Wogalter, M.S. 1990. Comprehension and memory of instruction manual warnings: Conspicuous print and pictorial icons. *Human Factors* 32:637–49.

Zwaga, H.J.G., and Easterby, R.S. 1984. Developing effective symbols or public information. In *Information Design: The Design and Evaluation of Signs and Printed Material,* eds. R.S. Easterby, and H.J.G. Zwaga, 277–97. New York: John Wiley.

5 Safety Design under the Perspective of Ergonomics in Consumer Products: Accident Studies in Users Welfare Promotion

Walter Franklin M. Correia, Marcelo M. Soares,
Marina de Lima N. Barros, and Mariana P. Bezerra

CONTENTS

5.1   INTRODUCTION

Nowadays, a large number of consumer products have reached a level of complexity and difficulty that is not easily accepted by consumers. Although the degree of sophistication and technology has provided a strong pull from the standpoint of marketing strategy, it can produce serious frustration for some users. It is possible that these products lack maturity and functional content that users really need. Consumers require products that provide safety, efficiency, comfort, and pleasure.

Products should be developed and designed to user needs. A product is supposed to make a user's life easier, assisting in carrying out actions that could not be performed without the help of a mechanical device or material. In theory, this presents itself as a major reason for consumer products to exist (Soares 1999). Consumer products may cause some kind of physical damage to the user as a result of lack of usability. The reality is that very little has been done in Brazil to change this situation.

It is extremely difficult to change this situation because the legal requirements to impose user standards have not been established for consumer products in Brazil. Today, there is more concern about accidents that occur with traffic and with manufacturing and construction industries.

In Brazil, there are no reliable statistics on the number of accidents involving consumer products. However, it is estimated that the percentage is as high or higher than those in European countries and the United States due to lack of governmental regulations and the general perception by the population regarding safety.

For a better understanding of the study, usability is defined by the International Standards Organizations (ISO) DIS 9241–11 as "the effectiveness, efficiency and satisfaction with which, specific users achieve specific goals in particular environments."

There is a difference between consumer products (or consumer goods) and capital (capital goods) (Iida 1990). The latter is usually machinery and equipment intended for industry in general and more specifically in production. Consumer products are those used in the home, i.e., appliances, furniture, electronics, etc. According to Cushman and Rosenberg (1991): "A consumer product may be understood as all tangible personal property, whether new or used, whose purposes are intended for use, usually personal, family or household." Table 5.1 shows the differences between consumer goods and capital goods (adapted from Iida 1990).

As one can observe from Table 5.2, the vast technological advances both to facilitate human life and to enhance certain capabilities have been dramatic in the last century. These advances obviously bring challenges for consumers. Salles and Wolff (1968), present a frame of dates with the evolution and the "jumps" of technology during certain periods is presented in Table 5.2.

The progress reflected in the products, particularly in products regarded as consumables, which are the main objects of the current study, results in a series of benefits and/or disadvantages. According to Jordan (1998), consumer products for home use or work are becoming increasingly complex in terms of features and functionality.

According to Jordan (1998, 34), "it is necessary and very important that designers, engineers and ergonomists responsible for product development take into account the requirements and limitations of users who will use the products. One truth is that this demand is already starting from users. There is more attention to what they are actually using and buying, making more companies do business and products more 'friendly' and 'ergonomically designed.'"

In addition, Soares (1999) emphasized the relationship of designers and ergonomists with the following statement:

TABLE 5.1
Basic Differences between Capital Goods and Consumer Goods

	Capital Goods	Consumer Goods
Goals	Defined by the manufacturer and company	Selected by the user and may vary
Operators	Enabled persons with special training	Big variety, without instructions or training
Failure costs	High, may causing a catastrophe	Dispersed, difficult to quantify
Use monitoring	Realized by skilled persons	Do not exist, specifically

TABLE 5.2
Advancing of Technology in Small Steps

Product	Time Interval
Photography	112 years (1727–1839)
Telephone	56 years (1820–1876)
Radio	35 years (1867–1902)
Radar	15 years (1925–1940)
Television	12 years (1922–1934)
Atomic bomb	6 years (1939–1945)
Transistor	5 years (1948–1953)
Integrated circuit	3 years (1958–1953)
Laser	2 years

The difficult relationship between designers and ergonomists has sometimes been mentioned by several authors. According to Meyer (1989) and Ward (1990), one of the main differences between designers and ergonomists, arises from the emphasis that each group puts on the methodology used to achieve their development goals. It is hoped that designers are always innovative and creative, always seek a different solution to a problem, considering the unique and intuitive way they work, trying a number of solutions and evaluating them later. Ergonomists, in turn, although sometimes use creative techniques, tend to analyze the problem and develop formulas or experiments that will lead to what they consider as the best solution or answer to this or that problem.

When Soares (1999, 28) states that consumer products that do not attain safety requirements may cause injury or death to users and be excluded from the market by preventive or repressive legislations, this is to highlight something that should be known by designers and planners. Professionals most frequently say that accidents occur from the misuse of the product by the user rather than from the designing and planning. However, manufacturing defects, poor design, and incompatibility with its function are also causes of accidents. When consumer products do not meet safety requirements, they can cause injury and even death.

This study attempted to demonstrate the relationship between accidents caused by consumer products and the analysis of the different aspects of product usage. In this chapter, consumer products are those used in or around the user's home. The study was carried out through questionnaires and direct surveys with users interacting with the products in their own homes. Tools for ergonomic analysis of accident situations were used, e.g., hazard assessment; checklists or cause taxonomy, questionnaires, scenario analysis, observations, analysis on-site, security analysis, anthropometric and biomechanics analysis, task analysis for error identification (TAFEI), and others.

5.2 FIELD STUDY WITH A SAMPLE OF USERS

A field study was conducted with a sample of users of consumer products in the city of Recife, Brazil. The main objectives of this study were to: (a) analyze the behavior of users with consumer products that they often buy and use; (b) check the incidence of accidents and how these users tend to act before certain difficulties in using such products, (c) check which products are more likely to cause accidents and which present the greatest difficulties during use and maintenance; and (d) identify possible difficulties in using the user manual.

5.2.1 STRATEGY OF THE FIELD STUDY

This field study required that a sample of users respond to a questionnaire. The sample was composed of 2100 questionnaires distributed via the internet of which 83 were completed and returned.

A team of three interviewers randomly picked 283 people at malls, bus stops, and on the street to complete the questionnaire, with a total of 320 completed and valid questionnaires for the study.

The Questionnaire for Usability Evaluation of Products of Domestic Use has been divided into three parts with a total of 16 questions, distributed as follows:

- Questions about the "characteristics and product usage," which referred to aspects such as brand, functionality, ease of cleaning, and maintenance.
- Questions about the "use of the user manual," which sought to identify dysfunctions between product use and the instruction manual.
- Questions about "product safety," which were related to: (a) possible accidents with consumer products and (b) issues of usability.

The sample of users who responded to these questionnaires was not intended to be representative of the entire population of users of consumer products in the city of Recife. Thus, this research was essentially exploratory. Its findings can be interpreted as a reflection of part of the population that may or may not represent the entire universe of users of consumer product. However, it should be emphasized that evidence from the research presented here is very significant.

The model of Spiegel (1985), separating the age groups in five-year increments, was adopted for purposes of statistical analysis of the participants' ages in the research sample. The Statistical Package for Social Sciences (SPSS), version 10.0 was used for the analysis and tabulation of the questionnaires. The sample was composed of an almost equal number of females (52%, $n = 165$) and males (48%, $n = 155$).

5.2.2   Analysis of Results of Field Study

After obtaining the respondents' opinion on the characteristics they considered most important in a consumer product, they were then asked to rank hierarchically the three products they considered to be most important of the 14 previously presented to them (Figure 5.1).

From the analysis of the questionnaires, the traits considered most important to users were quality, functionality and price. The item "security" was not considered as important, or as extreme importance to users. Norris and Wilson (1999) argue that a substantial number of products are developed by designers with no experience or by small groups of companies with no ergonomic support. This is a serious failure, considering that a product that does not take into account ergonomic principles concerning safety can be a potentially dangerous weapon.

According to the research, products that do not satisfy the user, in general, fall short especially in the previously mentioned requirements. Some example of products and their respective problems are listed below

- DVD player—cleaning and understanding of remote control (functions)
- Blender—cleaning and assembly
- Clothes iron—maintenance and durability
- Electric shower—different types of problems
- Pressure cooker—it cannot seal easily
- Others

Also according to those interviewed, products that appeared most frequently in complaints from users were

- Sound (micro system, radio, CD/tape player, etc.): problem of understanding certain functions.
- Computer chair: problem of discomfort for users often caused by a lack of proper adjustment.
- Electric coffee maker: problem of proper handling caused by inappropriate design for the activity.

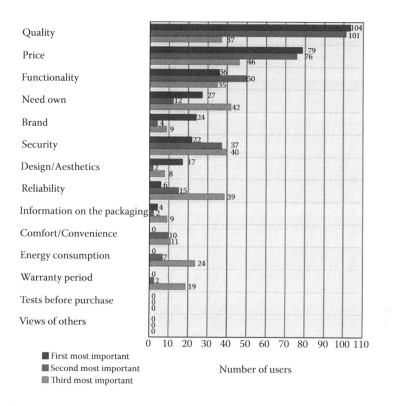

Quality
Price
Functionality
Need own
Brand
Security
Design/Aesthetics
Reliability
Information on the packaging
Comfort/Convenience
Energy consumption
Warranty period
Tests before purchase
Views of others

■ First most important
■ Second most important
▨ Third most important

Number of users

FIGURE 5.1 Order of importance for some factors by user's opinion.

- Iron: problems with fragile materials and bad components that result in low durability are major causes (according to users) of discomfort and accidents. The bad design of the handle and excessive weight are also presented as a cause of discomfort.
- Blender: problem with maintenance.
- Microwave: problem with the poor to very poor interface between the product and the user. The manuals were also cited as having poor informational quality.
- Cell phones, remote controls, and DVDs: problems caused by poor interface between user and product.
- Electric shower: problem with electric shock. Even though the product requires a high level of security, the users reported suffering an electric shock.

Almost half of the respondents (47%, $n = 149$) stated that they suffered accidents when using a consumer product. They were asked which product caused the accident and to give a brief description of the accident. The results are shown in Table 5.3.

The respondents were asked if they felt that the product or they themselves were the cause of the accident. The answers were almost evenly divided: 42% ($n = 62$) answered that it was the product's fault, 41% ($n = 61$) answered that they were the cause of the accident, and 17% ($n = 25$) said that they did not know.

When the respondents were asked whether they had read the manual or not, 65% of respondents stated that they had not read the instruction manual. The instruction manual may be considered as a key point to help users in obtaining the product functions and product usability. However, in the opinion of the users who responded to the survey, usually the manuals were not "user friendly." Many readers who did not read the manual cited that they did not understand the content, drawings, or organization of the manual in order to perform the tasks.

In general, it was noticed that the instruction manuals are not very precise with the context of information. According to the research, it is recommended to review the concepts used in the

TABLE 5.3
Products That Caused Accidents According to the Respondents with Cause and Frequency

Product (Frequency)	Accident	Product (Frequency)	Accident
Iron (16)	Electric shock and burn	Bed (3)	Contusion
Electric shower (13)	Electric shock	Computer (3)	Electric shock
Knife (12)	Cut	Beverage bottle (3)	Cut
Stove (9)	Cut and contusion	Computer monitor (3)	Electric shock
Blender (8)	Electric shock	Radio (3)	Electric shock
Refrigerator (5)	Electric shock	Television (3)	Electric shock
Food can (5)	Cut	Vacuum cleaner (3)	Electric shock
Fan (5)	Cut and contusion	Shave (2)	Cut
Opening can (4)	Cut	Coffee maker (2)	Burning
Screwdriver (4)	Cut	Glass (1)	Cut
Washing machine (4)	Contusion and cut	Air filter (1)	Electric shock and contusion
Microwave oven (4)	Contusion and cut	Electric drill (1)	Cut and contusion
Pressure cook (4)	Explosion	Videocassete recorder (1)	Electric shock
Plastic chair (3)	Contusion (fell down)		

current instruction manuals, using a language more user "friendly" without technical terms that have little interest to the consumer. Often, the product requires a lengthy and extensive manual, and it was found that in such cases the user does not read it. It is recommended that a more user "friendly" instruction manual be written to encourage consumers to read it because of the necessary information about the use, care, and safety issues regarding the product.

5.3   CASE STUDY: SEARCHING ACCIDENTS

The case study method was chosen because it provided the opportunity for an in-depth investigation of the problem in order to analyze in a systematic way the causes and consequence in a real-life context. A case study was conducted of eight accidents extracted from the previously applied questionnaires but only two will be detailed in this chapter.

5.3.1   STRATEGIES AND DESIGN OF THE CASE STUDY OF ACCIDENTS

Eight questionnaire respondents who had suffered an accident with a consumer product agreed to take part in the simulations. Each also agreed for the researcher to come to their residence to ask further questions and video record a simulation of the accident. Table 5.4 shows the products that were part of the study and the type of accidents. This chapter will focus only on the studies dealing with a fan and a pressure cooker.

The objectives of this case study were to: (a) analyze each accident based on a video-recorded simulation; (b) analyze the activity through a flowchart of each task, documenting comments from the simulations and the records; (c) perform a thorough analysis, seeking to find possible causes of the accidents; and (d) verify the level of usability of products according to user feedback.

The case study involved accident victims who were asked to

a. Respond to pre-research interviews regarding their report of the accident and how it happened
b. Perform a simulation of the accident, which was video recorded
c. Respond to a questionnaire, called the Usability Scale System (SUS), for each accident

The latter aimed at quantitatively evaluating the degree of usability of each of the products mentioned (Stanton and Young 1999). The collected data were analyzed and represented in a fault tree

TABLE 5.4
Products and Related Accidents That Were Part of the Study

Product	Type of Accident
1. Iron	Electric shock and burn
2. Stove	Cut
3. Drill	Cut and bruise
4. Electric shower	Electric shock
5. Screwdriver	Cut
6. Washing machine	Bruise
7. Pressure cooker	Explosion
8. Fan	Cut and bruise

for each type of accident. The conclusions from the case study were generated from the analysis using information from the fault tree.

5.3.2 PROCEDURES OF THE CASE STUDY

As previously stated, the case study was conducted in three stages: interviews, simulations, and analysis of usability. Initially, all the procedures of the case study were explained to each of the participants. The details of those steps are as follows.

5.3.2.1 Interviews with Users

An interview with each of the eight victims of some kind of accident was conducted following the model presented by Weegles (1996). Thus, it was possible to take into account several aspects of the accident. For example: (a) How did the accident occur? (b) What is the user opinion about the accident? (c) Which is the most relevant aspect related to the product and the place where the accident occurred? (d) What were the environmental conditions at the time of the event? and (e) What was the psychophysiological condition of the user at the time the accident occurred? All interviews were conducted where the accidents occurred.

5.3.2.2 Simulation of Accidents

Since the goal was to reconstruct the situation for future analysis, simulations of the accidents were made from the reports sent in by users. At the end of the simulation, the user was asked to describe the accident using as many details as possible in order to achieve as much accuracy as possible. Each video-recorded simulation included steps preceding the occurrence of each of the accidents. Images of these various stages are presented in the figures on the succeeding pages. Thus, it is possible not only to observe how the accident occurred, but also examine the steps leading to the accident using a fault tree to make an analysis of each of the simulations. Note that the considerations made were based on (i) the "voice of the user" in the statements regarding the descriptions and opinions about the accident, and (ii) the author's own insight to a reconstitution of the accident.

5.3.2.3 Usability Analysis

After the simulation of the accident, the study participants answered a questionnaire of usability (SUS) in order to get their opinion about the level of usability of the product that was involved in the accident.

5.3.3 ACCIDENT SIMULATION WITH PRODUCTS

The simulations and an analysis of the two products—pressure cooker and fan—are the focus of the next sections.

It may be noted that many times the users were reluctant to do the video-recording simulation because they were unable to remember the exact details of the accident. However, since the recordings were done in the residence of the user, recall became easier and did not hinder the integrity of the research.

5.3.3.1 Accident Simulation with a Pressure Cooker

For this simulation, from the eight volunteers, the person who had suffered an accident with the pressure cooker was selected. In this situation, the researcher requested that the artifact to be used in the simulation be similar to the product involved in the accident. Figures 5.2a through 5.3c show a simulation of

FIGURE 5.2   (a–c) Images of the pressure cooker closed, open, and in use by user.

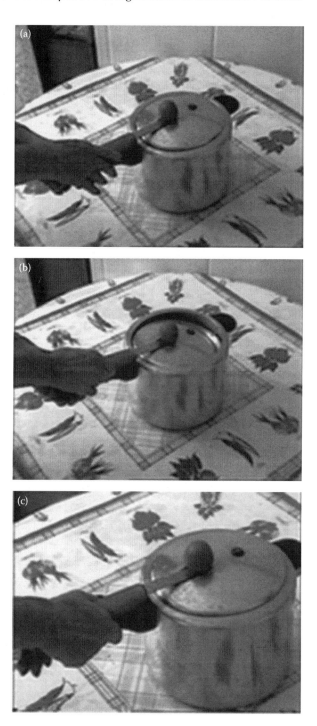

FIGURE 5.3   (a–c) Simulation of the accident with the pressure cooker.

the accident with the pressure cooker. It should be pointed out that the actual product involved in the accident was destroyed and, therefore, another product of the same brand and model was used.

The user reported that the pressure cooker exploded shortly after being placed on the stove to cook beans. She said she did not hear the noise of the exhaust valve. There was no physical damage, because the user was in another room. However, there was damage to the stove, ceiling, and the pot.

It should be mentioned that during the simulation, the user took some time (about 30 seconds) trying to put the cover of the pan in place. The simulation was held without any food inside the pot.

The flowchart of the task was prepared according to the explanation given by the user about her procedures for using a pressure cooker (Figure 5.4). The sequence of activities described by the user are those that are usually made when using an artifact like the one presented during the simulation.

During the face-to-face interview, it was necessary for the user to demonstrate the following steps in the use of the pressure cooker:

- Place the beans in the pan
- Secure the top
- Check to see if the top is secured correctly
- Take the pan to the stove
- Turn on the stove

The user reported that after a few minutes she heard an explosion and the cover of the pot had hit the ceiling.

The user also noted that she knew of other cases in which pressure cookers had exploded while in use. In this study, four users stated that they had suffered the same type of accidents when using pressure cookers. Two reported no physical damage to themselves, and the other two did not respond to the question.

It was found that the pressure cooker used in the simulation was difficult to seal completely because the cover of the pot had a side opening between it and the pot. Several attempts at closing the pot were necessary to have a secure seal. Figure 5.5a and b show this defect in the product design.

To reach conclusions about the real cause of the accident, a technical report would be required. However, it can be assumed that failures in the sealing system and/or failures in the exhaust valve may have been the cause of the accident.

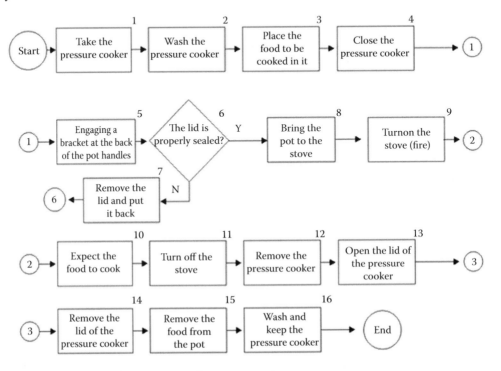

FIGURE 5.4 Flowchart of the task to use the pressure cooker.

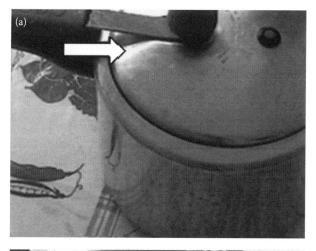

FIGURE 5.5   (a, b) Images of the pressure cooker—details of the gap in the cover.

The fault tree (Figure 5.6) was developed based on information gathered from the face-to-face interviews. These had been developed taking into account the steps that culminated in the accident.

It is perceived that not reading the instruction manual and inadequate closure of the pressure cooker lid are the main factors for the accident (steps 5 and 6). However, this latter fact is due to the inadequate design of this pressure cooker. Step 4 in the tree demonstrates the insecurity and reluctance of the user regarding the use of the pressure cooker.

In this fault tree, the steps flow from the bottom up. Solid lines represent the direct flow of activity to get to the accident. The dashed lines represent factors involved indirectly with the accident.

5.3.3.2   Accident Simulation with a Fan

The same procedure for selection of a volunteer was followed as in the previous case with the pressure cooker. During the video-recording process, the accident was replicated in as much detail as possible with regard to information reported by the user. Figures 5.7a through 5.8c demonstrate the artifacts used and a simulation of the steps that led to the accident.

The accident occurred when the user tried to clean the fan propeller. When trying to release the last lock that was holding the protection grid to another lock attached to the body of the fan, the user felt resistance and imposed more force with the thumb. The lock broke and the user bruised and cut her finger on the piece of hard plastic that was attached to what was the remaining part of the lock.

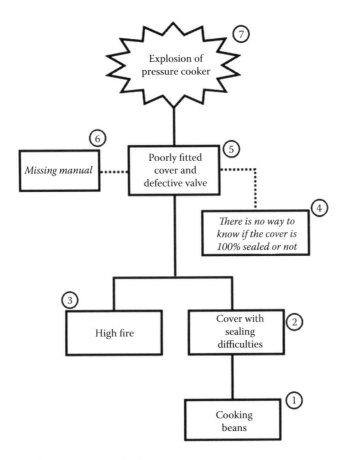

FIGURE 5.6   Fault tree for the accident with the pressure cooker.

The user suffered mild injury. The lock was broken beyond repair. However, for the simulation, the fan piece did not need to be broken.

The user reported that she read the instruction manual several times after she encountered a "huge" difficulty when removing the fan grid before she continued the process.

The flowchart of cleaning the fan task was developed based on the description and presentation of the activities of the user (Figure 5.9).

According to the flow chart, the user followed the correct procedures for cleaning the fan blades and did not deviate in any way. The user's response to consult the user manual when a problem was encountered only reinforces the difficulty of the task and the need to seek further information regarding the fan. The following steps preceding and culminating in the accident that occurred in the fan cleaning process were reported by the user:

- Attempting to open the grid
- Unlocking the safety locks of the grid (five locks)
- Breaking of lock five because of difficulty opening the lock
- Resulting in hurting and cutting the user

All steps described above can be seen more clearly in their logical sequence in the fault tree in Figure 5.10. According to the researcher, the use of the exact brand of fan involved in the accident is irrelevant because on inspection of other brands the same problem existed; some even had a greater degree of difficulty in opening and closing the locks.

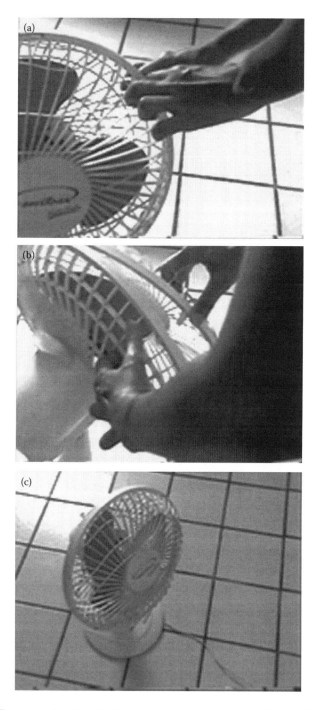

FIGURE 5.7 (a–c) Sequence of actions for the user when the fan is opened for cleaning.

Figure 5.11a and b show in detail the location of the latch found in the fan grid. It was confirmed during the simulations that there was great difficulty in removing the grid's lock. However, the manual reported the "ease of removal of the grid."

When designing products that require more safeguards, the use of stronger and more durable raw materials does not guarantee the safety of the user. It is necessary to include a safeguard subsystem to act in situations where there is a need to prevent physical injury to the user.

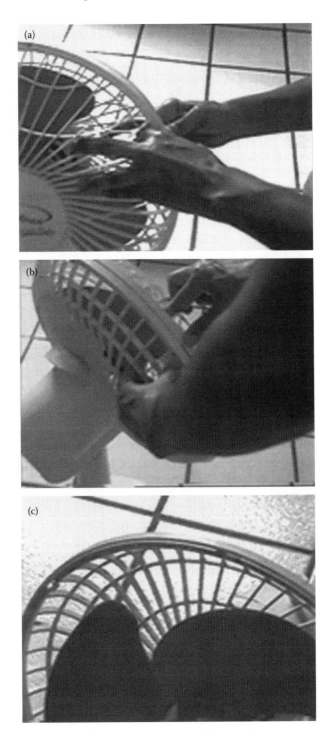

FIGURE 5.8   (a–c) Simulation of the accident with the fan.

5.3.3.3 Lessons Learned with the Accident Simulation

The main objective of this phase of the study was to examine more deeply the causes of the recorded accidents. Observe that, for each accident, it can be seen that a portion of the blame for the accident is either lack of attention or carelessness of the user.

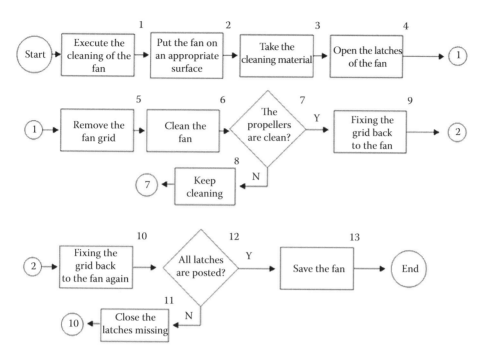

FIGURE 5.9   Flowchart for the cleaning activity of the fan.

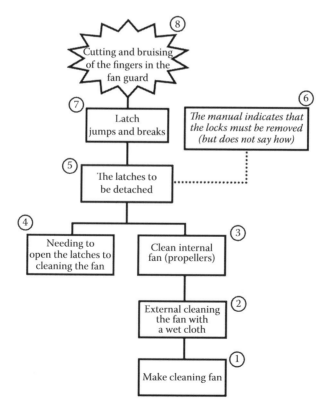

FIGURE 5.10   Fault tree for the accident during the cleaning activity of the fan (according to the user).

FIGURE 5.11   (a, b) Images of the fan with detail of the latches on the grid.

However, there are other factors that may have contributed to the accidents. One of these is the lack of emphasis on the elements of warning and safety items and another is the lack of barriers to avoid product misuse. This could have easily been prevented from the very beginning in the product development stage with the incorporation of ergonomic principles.

It can also be seen that the products analyzed did not present signs of being a "friendly" product. To be a "friendly" product, it is not only the product's usability that is considered, but also other aspects such as: (a) the composition of materials from which the product is made; (b) the components of the product (e.g., buttons, wires); (c) the quality of the instruction manual; and (d) the visibility of the product safety warnings. The products analyzed in this study have failed in at least one of these characteristics.

When designing products that require more safeguards, the use of stronger and more durable raw materials does not guarantee the safety of the user. It is necessary to include a safeguard subsystem to act in situations where there is a need to prevent physical injury to the user.

5.3.4   ANALYSIS OF USABILITY USING THE SYSTEM USABILITY SCALE

The SUS by Stanton and Young (1999) enables the user to follow a list of tasks to be performed while using the product.

5.3.4.1 Strategies of the Analysis of Usability

Stanton and Young (1999) propose the following procedures to the application of the SUS method that were adopted for this study. The steps are as follows:

- Give the SUS questionnaire to the respondents after the accident simulation.
- Instruct them to read the SUS questionnaire.
- Explain the questionnaire and its objectives.
- Clarify that questions can be asked of the researcher at any time during this process.
- Give the respondents plenty of time to complete the questionnaire.
- Tabulate the data based on the SUS guidelines after the questionnaire has been completed.

For the analysis of usability of the fan and pressure cooker, the same two volunteers were used. It is a requirement that these two volunteers have experience using the specified product and are knowledgeable of the products' capabilities.

The volunteers were instructed to answer the questions according to their level of satisfaction with the product using a Likert scale from 1 ("strongly disagree") to 5 ("strongly agree") for each questions, according to the SUS method. The degree of usability of the product was obtained after the questionnaire was completed and the responses were tabulated and calculated using a coefficient given by the SUS method.

These are the statements with which the Likert scale was used:

1. I would use this product frequently.
2. I think the product is unnecessarily complex.
3. I think the product is easy to use.
4. I think I would need technical support to learn how to use the product.
5. I find the functions in the product were very well integrated.
6. I think there was much inconsistency in the performance of the product.
7. I imagine that most people would learn to use this product quickly.
8. I find the product very uncomfortable to use.
9. I feel confident using the product.
10. I will need to learn many things before continuing to use the product.

5.3.4.2 Procedures for the Implementation of the System Usability Scale Score of Usability

It was possible to obtain the score of usability of the products after the completion of the question-naire using the calculations as follows:

a. For every odd question (1, 3, 5, 7, and 9), the scale score in the SUS method is obtained by subtracting 1 point. For example, for question 1, if the user responds "strongly agree," a final score equal to "4" ($5 - 1 = 4$) will be obtained.
b. For the even-numbered items, the score is obtained by subtracting the number on the scale by a constant value equal to 5. For example, if the user answers the second question with the value "2" ("almost strongly disagree"), a final score equal to "3" ($5 - 2 = 3$) will be obtained.
c. The sum of scores (even and odd questions) is multiplied by a constant 2.5 in order to obtain the score of usability (varying from 0 to 100).

5.3.4.2.1 Score of Usability for the Pressure Cooker

As can be seen in Table 5.5, the pressure cooker had an index of usability of 40 out of 100. This level of usability means that the product in question leaves much to be desired regarding

TABLE 5.5
Calculation of the Score of Usability for the Pressure Cooker

Score of Odd-Numbered Items = Position in the Scale minus 1 (−1)	Score of Even-Numbered Items = 5 Position in the Scale
Item 1: 2 − 1 = 1	Item 2: 5 − 3 = 2
Item 3: 4 − 1 = 3	Item 4: 5 − 2 = 3
Item 5: 1 − 1 = 0	Item 6: 5 − 5 = 0
Item 7: 4 − 1 = 3	Item 8: 5 − 5 = 0
Item 9: 2 − 1 = 1	Item 10: 5 − 2 = 3
Sum of odd numbers (NI)	= 8
Sum of even numbers (NP)	= 8
Total sum of the items = NI + NP	= 16
Total score of usability for SUS	= 16 × 2.5
Total of items × 2.5	= 40

performance and safety. In the opinion of the user of the product, this product should be taken off the market.

5.3.4.2.2   Score of Usability for the Fan

Using the same process to calculate the level of usability, the fan was found to have a level of usability of 62.5 out of 100, which is also considered to be unsatisfactory (Table 5.6).

5.3.4.3   Lessons Learned for the Implementation of the System Usability Scale Score of Usability

In general, it can be seen that the score for the pressure cooker (40) is low and that for the fan (62.5) may be considered low when one takes into account that the highest level of satisfaction of users is 100%. The score of usability for the pressure cooker and fan were not very different from the score of usability for other products researched in the complete study, which are not presented in this chapter. For comparison, one can see the score of usability for the electric shower (50), washing machine (50), and drill (50).

TABLE 5.6
Calculation of the Score of Usability for the Fan

Score of Odd-Numbered Items = Position in the Scale minus 1 (−1)	Score of Even-Numbered Items = 5 − Position in the Scale
Item 1: 5 − 1 = 4	Item 2: 5 − 3 = 2
Item 3: 3 − 1 = 2	Item 4: 5 − 1 = 4
Item 5: 2 − 1 = 1	Item 6: 5 − 3 = 2
Item 7: 4 − 1 = 3	Item 8: 5 − 5 = 0
Item 9: 5 − 1 = 4	Item 10: 5 − 2 = 3
Sum of odd numbers (NI)	= 14
Sum of even numbers (NP)	= 11
Total sum of the items = NI + NP	= 25
Total score of usability for SUS	= 25 × 2.5
Total of items × 2.5	= 62.5

5.4 CONCLUSIONS

In this study, it is clear that with many everyday products there is a lack of concern for user safety or ergonomic design, which make the product more consumer friendly and safe. A poorly designed product is a potential weapon in the hands of an inexperienced user. Although it is very easy to attribute accidents to user fault, this may not be the case in all circumstances, as seen in this study. Accidents in a particular situation can be initiated by several factors, including the environment, product design, faulty manufacturing, or even user misuse of the product. Accusing a user of being negligent is easy when you don't know which factors are intrinsically linked to the cause of the accident. According to Jordan et al. (1996), a bad design leads to misuse of the product.

It is important to mention that it was proved that most accidents in this study were a result of inadequate product design regarding safety and usability standards. Undoubtedly, there is a need for better governmental regulations regarding consumer safety. Specific safety technical standards and legislation on consumer products would avoid a number of accidents and save many lives.

REFERENCES

Correia, W.F.M. 2008. Product safety: An investigation on the usability of consumer products [In Portuguese]. MSc diss. Federal University of Pernambuco, Brazil.

Cushman, W.H., and Rosenberg, D.J. 1991. *Human Factors in Product Design*. Amsterdam: Elsevier.

Dillon, G.A. 2001. Quedas matam mais de 4 mil brasileiros por ano. Diário de Pernambuco, Recife, 15 de dezembro, Brasil, p. A8.

Frisoni, B.C. 2000. Ergonomics, usability and product quality: Comfort and users' safety; consumer defense [In Portuguese]. Research report. Rio de Janeiro. Catholic University, Brazil.

Guimaraes, L.B.M. 2000. *Process ergonomics. Serie Monography and Ergonomics* [In Portuguese]. v. 2. Porto Alegre, Brazil: PPGEP-UFRGS.

Iida, I. 1990. *Ergonomics: Project and Production* [In Portuguese]. Sao Paulo: Edgard Blucher.

Jordan, P.W. 1998. *An Introduction to Usability*. London: Taylor & Francis.

Jordan, P.W., Thomas, B., Weerdmeester, B.A., and McLelland, I.L. 1996. *Usability Evaluation in Industry*. London: Taylor & Francis.

Norris, B., and Wilson, J.R. 1999. Ergonomics and safety in consumer product design: development of a tool for encouraging ergonomics evaluation in the product development process. In *Human Factors in Product Design: Current Practice and Future Trends*, eds., Green, W. S. and Jordan, P. W., Chap. 8, 73–84. London: Taylor & Francis.

Salles, P., and Wolff, J. 1968. Hommes, besoins, activités. Paris, Dunod.

Soares, M.M. 1998. Translating user needs into product design for disabled people: A study of wheelchairs. PhD diss. Loughborough University, UK.

Spiegel, M. 1985. *Statistics* [In Portuguese], Chap. 2, 33–52. São Paulo: McGraw-Hill.

Stanton, N.A. 1998. *Human Factors in Consumer Products*. London: Taylor & Francis.

Stanton, N.A., and Barber, C. 2002. Error by design: Methods for predicting device usability. *Design Studies* 23:363–84.

Stanton, N.A., and Young, M.S.A. 1999. *Guide to Methodology in Ergonomics*. London: Taylor & Francis.

Weegles, M.F. 1996. Accidents involving consumer products. PhD diss., University of Delft, The Netherlands.

6 Product Design Issues Related to Safe Patient Handling Technology*

Thomas R. Waters

CONTENTS

6.1 INTRODUCTION

Workers in the healthcare industry who perform physically demanding patient handling tasks as part of their jobs, such as nurses, nurses' aides, physical therapists, and healthcare technicians, are at high risk for development of work-related musculoskeletal disorders (MSDs). These workers are exposed to significant risk factors for MSDs when lifting and moving heavy patients and equipment, pushing and pulling heavy equipment, working in extreme postures, and standing for long periods of time without adequate rest periods. When the physical demands of the job exceed the capabilities of the worker, the worker is at increased risk of developing an MSD, such as back pain. Moreover, the risk of developing a work-related MSD is even greater when the worker is exposed to more than one risk factor at the same time (NIOSH 1997; National Research Council and Institute of Medicine 2001).

Although healthcare workers report a high number of medical problems in the shoulders, neck, and legs, by far the most common occupational injuries in the healthcare industry involve back disorders (United States Department of Labor Bureau of Labor Statistics 2007). Often, these back

---

* The findings and conclusions in this chapter are those of the author(s) and do not necessarily represent the views of the National Institute for Occupational Safety and Health. This work was done by a U.S. Govt. employee and is not subject to copyright.

disorders prevent workers from doing their job and also cause many of them to change jobs because of back pain. In 2001, nurses working in the private sector reported 11,800 MSD cases, with the majority (nearly 9,000) of the reported injuries involving the back. More than one-third (36%) of the injuries resulted in lost time from work due to back disorders. Work-related MSDs are associated with excessive back and shoulder loading from manual patient handling, applying excessive forces during pushing and/or pulling objects, awkward posturing during patient care, and working long hours (Waters et al. 2006). Another study reported that 12% of nurses who planned to leave the profession cited back injuries as either a main or a contributing factor (Stubbs et al. 1986).

Work-related MSDs are very costly to the healthcare industry. Employees that suffer from musculoskeletal pain on the job are often less productive, more likely to make mistakes, and have more accidents at work. Workplaces with high levels of reported patient handling injuries report higher rates of lost/modified workdays, higher staff turnover, increased costs, and adverse patient outcomes (Collins et al. 2004). The extent of the problem is probably worse than reported in official injury reporting records due to likely widespread underreporting of injuries. A study by Cato, Olson, and Studer (1989) reported that 78% of nurses with back pain in the previous six months did not report it to management. In another study of nurses, Owen (1989) reported that 67% of nurses who reported low-back pain related to work did not report the incident in writing. Finally, work-related MSDs contribute to the critical nursing shortage in the healthcare industry, which also leads to more overtime for working nurses. In 2005, the U.S. Department of Health and Human Services reported 4,577 unfilled nursing jobs in Missouri, representing 8% of the total nursing workforce and the number is expected to grow to 17,024, including 25% of the nursing workforce, by 2020 (Elwood 2007).

In the past decade, tremendous strides have been made in identifying high-risk tasks and in developing and implementing solutions for reducing the risk. In the United States, effective programs for safe patient handling that rely on the use of ergonomic technology have been developed and implemented by many healthcare enterprises. Also, some states have passed legislation mandating implementation of safe patient handling programs. Nationally, ergonomic guidelines have been developed to provide useful information to the industry to help them efficiently implement effective programs that have been shown to reduce both the risks and costs associated with work-related MSDs. Most importantly, easy to use equipment that reduces the amount of physical demand required to perform these tasks has been developed for many of the high-risk tasks and is widely available. Studies have been conducted demonstrating that implementation of these programs and use of this equipment is cost effective, often paying for itself in less than three years (Collins et al. 2004; NIOSH 2006). Finally, recent findings have also suggested that implementation of a safe patient handling program can also increase the quality of care of patients while at the same time reducing the risk of MSDs for caregivers (Nelson et al. 2008).

It is clear that a large number of workers are routinely exposed to high levels of work-related risk factors for MSDs when performing patient handling tasks and these exposures lead to increased reports of work-related MSDs over their working lifetime. Nevertheless, it should be recognized that these hazards can be identified and controlled so that work-related MSDs can be prevented.

6.2 NEED FOR ERGONOMICALLY DESIGNED EQUIPMENT FOR PATIENT HANDLING

From about 1945, the healthcare industry began recommending reliance on "body mechanics" as a way to protect caregivers from risk of injury due to patient handling activities. Although there was little or no empirical evidence that this approach would actually reduce the risk of work-related MSDs, it was widely adopted because other solutions were not available. Recently, researchers have shown that there is no safe way to manually lift and transfer a fully dependent patient weighing as little as 110 lbs, even when two caregivers perform the task (Marras et al. 1999). Unfortunately, many schools of nursing and physical therapy continue to teach these outdated manual patient handling methods.

Adding to the problem is the fact that the majority of direct patient care workers are female employees who, on average, have a lower strength and lifting capacity than males. This means that females must work at a higher percentage of their maximum physical capabilities than males when performing the same strength demanding tasks. Thus, the risk to the worker would probably be greater for females than for males when performing most patient care tasks. Additionally, the healthcare industry workforce is getting older. The average age of registered nurses in 2004 was 46.8 years, up from 45.2 years in 2000 and 40.3 years in 1980. About 41.1% of the registered nurse population is aged 50 or older, compared to just 16.4% younger than 35 (Elwood 2007). It has been shown that, on average, physical capacity and strength decreases as workers age. Although older workers often have developed more knowledge and better skills, these improvements usually can't compensate for the decreased physical capacity, especially when the physical demands are increasing due to the obesity epidemic in the United States. The average body weight of both patients and caregivers is increasing over time and this increase in average body weight is likely to play a major role in increasing the risk of MSDs for healthcare workers. A person is defined as being overweight when their Body Mass Index (BMI) is 25 or higher and obese when their BMI is 30 or higher. The percentage of U.S. citizens who are classified as obese has increased from 15% in 1980 to 35.1% in 2006 (Flegal et al. 2010). The percentage of citizens who are reported to be both overweight and obese combined (BMI $\geq$ 25) was 68% in 2007–2008. The percentage of U.S. citizens who are classified as extremely obese (BMI of 40 or higher) has increased from 1.4% in 1980 to 6.2% in 2006 (National Center for Health Statistics 2008). Patients are becoming heavier, and it is not uncommon to see patients requiring hospitalization who weigh over 400 lbs. As the weight of the patient increases, the risk of injury to the patient and the healthcare worker who must transfer, move, and treat the patient also increases. Also, many tasks that may have been considered acceptable to perform manually in the past are no longer safe for the caregiver, such as lifting an arm or leg for treatment or reaching across a patient to perform a task. Adding to the risk for caregivers is the fact that, following surgery and other treatments, patients are not staying in hospital as long as they did in the past. In 1980, for example, the average length of hospital stay was 7.5 days compared with only 4.8 days in 2005 (National Center for Health Statistics 2007). This reduction in hospital stay time has resulted in two significant effects. First, there is a concentration of acute patient needs associated with patient transfers and movement while in the acute care environment. Secondly, there is now a need for a higher level of patient transfer assistance in the home care environment at an earlier stage of recovery than was previously required. Unfortunately, the home care environment is often lacking in the availability of assistive patient handling technology (Galinsky, Waters, and Malit 2001; NIOSH 2010). Both these factors have increased the risk of work-related MSDs for healthcare workers.

Recently, the healthcare industry has recognized the risks associated with the performance of physically demanding patient handling tasks, and to reduce costs and increase productivity, companies have begun implementing ergonomic programs or practices aimed at preventing these injuries. The core element of these programs is a reliance on the use of state-of-the-art ergonomically designed equipment to assist the worker in carrying out the prescribed task. As an added incentive to adopt technology-based patient handling practices, the Occupational Safety and Health Administration (OSHA) recently published an ergonomics guideline that provides an overview of the risks of work-related MSDs in nursing homes. The guideline provides information about the most effective approaches for mitigating or reducing those risks, and discusses training needs (OSHA 2009). One of the most important statements in the OSHA nursing home guideline is that, "manual lifting of residents be minimized in all cases and eliminated when feasible." This is best accomplished by implementing a technology-based safe patient handling program.

Finally, based on an assessment of typical patient handling tasks using the Revised NIOSH Lifting Equation, NIOSH researchers indicated that no caregiver should manually lift more than 35 lbs of a person's body weight for a vertical lifting task (Waters 2007). The NIOSH author recommended that when the weight to be lifted exceeded this limit, then assistive devices should be used. The Veterans

Health Administration (VHA) has adopted this recommendation and incorporated it into their current patient handling recommendations and patient handling algorithms. Moreover, other major interest groups, such as the American Nurses Association (ANA), the National Association of Orthopaedic Nurses (NAON), and the Association of periOperative Registered Nurses (AORN) have all adopted patient handling guidelines that recommend use of technology-based solutions for patient handling and movement (AORN 2007; de Castro, Hagan, and Nelson 2006; NAON 2009).

One of the most important issues to consider when choosing the proper technology-based solution is the level of dependency of the patient being moved. If the patient is fully dependent and cannot assist the caregiver in the patient handling task, then some type of powered technology will probably be most effective. On the other hand, if the patient is fully independent, then little assistance will be needed from the caregiver and some type of simple non-powered technology will probably be sufficient. For a partially dependent patient, however, decisions about which technology is needed must be determined based on an assessment of the patients' capabilities and the type of task being performed. The VHA, with the assistance of other groups, has developed a series of algorithms for helping caregivers decide what level of technology and assistance will be needed for a specific situation. The algorithms contain decision logic for each type of task, which recommends what technology is needed and how many caregivers would be needed to perform the task safely. The algorithms can be found at http://www.visn8.va.gov/patientsafetycenter/safePtHandling/default.asp.

6.3 TYPES OF ERGONOMIC PRODUCTS AVAILABLE

A wide range of ergonomic patient handling technology is available that can be applied to most types of patient handling activities (Baptiste 2005, 2007). The most common patient handling activities include: (1) lateral transfers of patients between two lateral surfaces, such as bed-to-bed or between a bed and an examining table; (2) vertical transfers of patients, such as from a bed to a chair or between chairs; (3) standing a patient from a sitting position; (4) ambulation; (5) transporting heavy equipment, such as pushing or pulling occupied or unoccupied beds, heavy treatment equipment, OR beds, etc.; (6) repositioning people in bed, side-to-side or up and down; and (7) working in extreme or awkward postures for long periods of time, including standing.

Each specific healthcare setting may have unique requirements for patient handling technology, such as operating rooms, critical care environments, orthopedic, home care, and rehabilitation settings.

The VHA has developed a technology resource guide that lists a wide range of technology for use in safe patient handling and movement. The technology resource guide classifies patient handling technology into the broad categories listed in Table 6.1. Brief descriptions of the most important categories of technology are provided below. The VHA technology resource guide can be found at http://www.visn8.va.gov/patientsafetycenter/safePtHandling/default.asp. Another resource for information about technology for use in home care environments can be found in two papers by Parsons, Galinsky, and Waters (2006a, 2006b). These papers provide information about solutions for lift and transfer assistance for partial weight-bearing and non-weight-bearing home care patients.

6.3.1 Air-Assisted Devices

Air-assisted devices typically consist of a flexible nylon mattress connected to an air pump that allows lateral transfers of patients from bed to stretcher by the release of low air pressure through openings on the underside of the mat. A pump maintains a cushion of air under the mattress, allowing the mattress to be moved laterally across the surface with little or no friction. Air-assisted devices have also been developed that allow a caregiver to raise a patient from the floor in a supine position and then laterally transfer them to a bed. An example of an air-assisted lateral transfer device is shown in Figure 6.1.

TABLE 6.1
Categories of Ergonomic Patient Handling Technology Available

Air-assisted lateral transfer aids	Bed systems (frames/surfaces/assist devices/other bed safety devices)
Car lifts/vehicle extraction	Ceiling lifts
Dependency/geri/specialty chairs/ wheelchairs	Floor-based lifts
Friction-reducing lateral transfer aids	Gait belts w/handles
Lifts (other)	Mechanical lateral transfer aids
Other mobility aids	Powered standing lifts
Repositioning devices	Sliding boards
Slings	Standing assist aids
Transport devices/powered beds	

Bariatric devices (patient BMI > 40)

Bariatric ambulatory/mobility aids	Bariatric bathing equipment (other)
Bariatric beds/mattresses/transportation	Bariatric ceiling lifts
Bariatric commodes/shower chairs	Bariatric lateral transfer systems
Bariatric repositioning systems	Bariatric powered lifts
Bariatric standing assist aids	Bariatric transfer/dependency chairs and cushions
Bariatric wheelchairs	

6.3.2 FRICTION-REDUCING LATERAL TRANSFER AIDS

This category of technology includes devices having ultra-low friction properties that will aid care-givers in moving or positioning patients during lateral transfers. These devices will also aid in maintaining safer postures during lateral transfers without causing undue stress to either the patient or the caregivers (Baptiste et al. 2006).

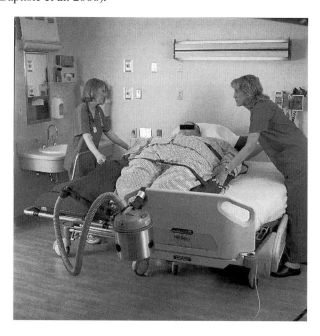

FIGURE 6.1 Air-assisted lateral transfer device. (Reprinted from Hill-Rom Services, Inc., Batesville, Indiana, © 2010. With permission.)

6.3.3   MECHANICAL LATERAL TRANSFER AIDS

This class of device includes systems that mechanically assist the user in performing a lateral transfer by actually applying lateral force to the sheet or pad on which the patient is resting. Some of these systems are driven by a controlled conveyor belt, or they may be devices that can be attached to a bed or procedure table to carry out lateral transfer or reposition the patient up in bed by mechanically pulling on the sheet or pad laterally. Some devices wrap the sheet around a powered spindle-type winch, and as it winds the sheet, the patient is transferred laterally with no manual force required.

6.3.4   FLOOR-BASED LIFTS

This category of handling equipment typically involves manual or battery-powered lifting assist devices that provide a vertical upward force to a sling that fits under the patient to perform the vertical lifting or lowering task. This class of device can accommodate a wide range of patient weights and patient characteristics. Manual floor-based lifts are operated by a hand or foot lever and may require a significant amount of manual force for operation. Battery-powered devices, on the other hand, are the easiest to use and require the least amount of manual effort to operate. In addition to vertical lifting tasks, these devices can also be used for other patient handling tasks, such as limb lifting or repositioning patients in bed. An example of a floor-based vertical patient lifting device is shown in Figure 6.2.

6.3.5   CEILING LIFTS

Similar to floor-based lifts, these devices provide a vertical upward force to a sling that fits under the patient to perform the vertical lifting or lowering task. These are battery-powered devices. They usually consist of a motor attached to a metal track that is mounted to the ceiling in the room where the lift will be performed, but they can also be attached to a wall with a swinging arm to support the lift motor. The track can consist of a single channel in which the motor can move linearly along the track or the track system can consist of an X-Y gantry system that can provide movement in multiple directions. These devices allow a single caregiver to perform vertical transfers of patients with no manual lifting. As with the floor-based models, these systems can be used for other patient handling

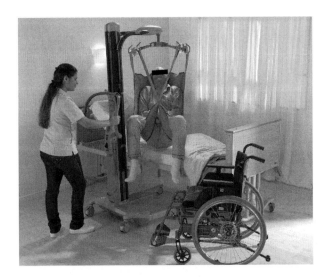

FIGURE 6.2 Battery-powered floor-based patient lifting device. (Reprinted from ArjoHuntleigh, Inc., Addison, Illinois, © 2010. With permission.)

tasks requiring vertical force application, such as limb lifting or repositioning in bed. Figure 6.3 shows an example of a ceiling-mounted, battery-powered vertical patient lifting device.

6.3.6 POWERED STANDING LIFTS (SIT-TO-STAND)

This class of device provides the capability to assist a patient in standing from a seated position. They typically consist of a shin pad and a base support on which the patient places their feet. A battery-powered motorized sling is applied around the back and arms of the patient. When the lift is initiated, the sling applies forward force to the patient and the resulting opposing force on the shins of the patient causes them to rise into a standing position. These devices require that the patient be weight bearing and able to stand without assistance. Some powered standing lifts can also be used as mobility aids and others are specifically designed for toileting and showering. Figure 6.4 shows an example of a powered sit-to-stand patient transfer device.

6.3.7 STANDING ASSIST AIDS

This miscellaneous category of technology consists of mounted handles, poles, and rails used for repositioning patients and providing assistance getting in and out of bed. It also includes specialty devices for assisting the transfer of patients from a wheelchair to a chair or bed. These devices typically require that the patient be able to transfer independently and have sufficient body balance and upper body strength. There are also standing lifts for children from 32" to 56" tall, which can also function as a desk and a standing aid, as well as fully motorized chairs that allow for the transfer of a patient from a sitting to standing position with power assistance and that allow mobility in either position. An example of a low-cost solution is a floor-to-ceiling pole that assists with standing, sitting, or transferring.

6.3.8 LIFTS (MISCELLANEOUS OTHER)

This category of handling technology consists of a range of products designed to assist in lifts and transfers in special environments. Some devices can provide a complete bedside treatment system

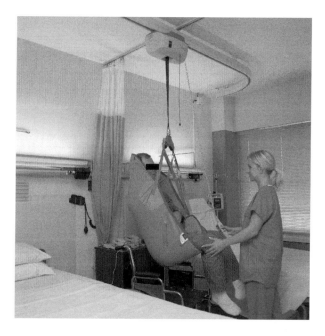

FIGURE 6.3 Battery-powered ceiling-mounted patient lifting device. (Reprinted from ArjoHuntleigh, Inc., Addison, Illinois, © 2010. With permission.)

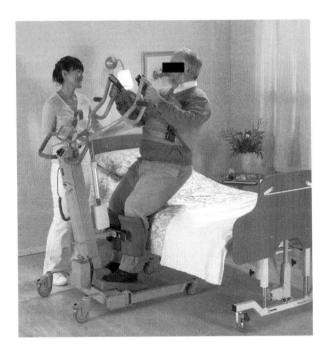

FIGURE 6.4 Battery-powered sit-to-stand patient lifting device. (Reprinted from ArjoHuntleigh, Inc., Addison, Illinois, © 2010. With permission.)

that lifts, transports, bathes, weighs, and turns the patient. Others provide a powered toilet lift that installs over the toilet in a bathroom and may include a totally integrated height adjustable bath system. This class of device includes an assortment of completely portable and deck-mounted, semi-portable, battery-operated pool lifts and lift systems powered by a 12 V motor with the power from the electrical system of a car to perform transfers (car, bath, home), as well as a wheelchair tilting device that allows patients to be examined by dentists, doctors, and other healthcare providers while the occupants remain in their wheelchair. An example of a specialized patient transfer device used to assist the patient in ambulation or during therapeutic therapy is shown in Figure 6.5.

6.3.9   TRANSPORT DEVICES/POWERED BEDS/POWERED TUGGERS

This category of technology consists of powered beds or powered tuggers that reduce pushing and pulling demands. Some devices can convert from a chair to a stretcher laterally and eliminate patient transfers. Another device in this category includes a wheelchair that is integrated with a patient examination table that eliminates the need to lift the patient onto the table. Other devices include attendant-driven motorized transport chairs with swivel seats and adjustable support arms that can move patients from a prone position to a seated one. Figure 6.6 shows an example of a powered wheelchair mover that reduces the pushing and pulling forces to near zero.

6.3.10   BARIATRIC PATIENT HANDLING DEVICES

There is a wide range of technology available for handling and moving bariatric patients (Baptiste, Meittunen, and Bertschinger 2004). Nearly all of the technology noted above has also been developed for use with heavier patients. Ambulatory devices, beds, and chairs are designed to handle as much as 1200 lbs. Heavy-duty lateral transfer aids can be used to move patients from bed to bed or bed to stretcher, and floor-based and ceiling lifts with extreme capacity are available. An example of a bariatric ceiling lift is shown in Figure 6.7.

FIGURE 6.5 Battery-powered ambulatory patient transfer device. (Reprinted from ArjoHuntleigh, Inc., Addison, Illinois, © 2010. With permission.)

6.4 GAPS IN TECHNOLOGY

Although a wide range of equipment is available, gaps in technology remain. For example, ceiling lift devices for vertical transfer of patients in operating rooms where the patient may need to be moved laterally but also may need to be turned from a supine to a prone position are lacking. Also, more devices specifically designed for use in rehabilitation settings where the handling equipment must provide dual function are needed. That is, the equipment must be designed so that it can provide adequate assistance for patient transfers when needed, but it must also be flexible enough so that it can

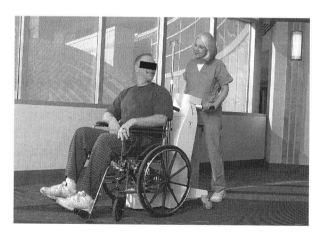

FIGURE 6.6 Battery-powered wheelchair mover. (Reprinted from ArjoHuntleigh, Inc., Addison, Illinois, © 2010. With permission.)

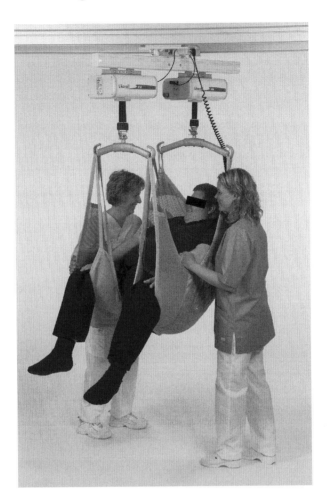

FIGURE 6.7 Battery-powered bariatric ceiling-mounted lift. (Reprinted from Hill-Rom Services, Inc., Batesville, Indiana, © 2010. With permission.)

quickly be adjusted for use during therapeutic procedures where the patient is required to use more and more of their own capabilities as they recover. Also, more research is needed to identify issues related to equipment usage. There are three areas of gaps in technology application. These include

1. Equipment that is needed, but has not been developed
2. Equipment that has been developed, but facilities have not obtained it
3. Equipment that has been developed and facilities have obtained it, but it is not being used

These issues need to be addressed in order to maximize the effectiveness of technology-based solutions in healthcare.

6.5   SUMMARY

There is strong evidence that: (1) manual patient handling presents a high risk for the development of work-related MSDs; (2) reliance on body mechanics alone is not effective in reducing the risk of patient handling injuries; (3) effective technological solutions are available to reduce or eliminate the risk of these health problems; and (3) implementation of a safe patient handling program that relies on use of state-of-the-art technology is cost effective and, on average, will provide a positive return on investment within three years of implementation. Unfortunately, many healthcare settings are still relying on unsafe manual lifting techniques rather than investing in newer, safer patient handling technology. Recent state legislation and possible U.S. federal legislation may force implementation of programs, but proactive ergonomics based on implementation of a safe patient handling program will provide a safe work environment for both workers and patients.

REFERENCES

AORN Workplace Safety Taskforce. 2007. *Safe Patient Handling & Movement in the Perioperative Setting.* Denver, CO: Association of periOperative Registered Nurses (AORN).

Baptiste, A. 2005. *New and Emerging Technologies for Safe Patient Handling and Movement.* London, England: Campden Publishing.

Baptiste, A. 2007. Technology solutions for high-risk tasks in critical care. *Critical Care Nursing Clinics of North America* 19 (2): 177–86.

Baptiste, A., Boda, S., Nelson, A., Lloyd, J., and Lee, W. 2006. Friction-reducing devices for lateral patient transfers: A clinical evaluation. *American Association of Occupational Health Nurses* 54 (4): 173–80.

Baptiste, A., Meittunen, E., and Bertschinger, G. 2004. Technology solutions for bariatric populations. *Journal of the Association of Occupational Health Professionals in Healthcare* XXIV (2): 18–22.

Cato, C., Olson, D.K., and Studer, M. 1989. Incidence, prevalence, and variables associated with low back pain in staff nurses. *American Association of Occupational Health Nurses Journal* 37 (8): 321–27.

Collins, J.W., Wolf, L., Bell, L.J., and Evanoff, B. 2004. An evaluation of a best practices musculoskeletal injury prevention program in nursing homes. *Injury Prevention* 10:206–11.

de Castro, A.B., Hagan, P., and Nelson, A.L. 2006. Prioritizing safe patient handling: The American Nurses Association's Handle with Care Campaign. *Journal of Nursing Administration* 36 (7–8): 363–69.

Elwood, J. 2007. The aging nursing population: Health systems and nursing schools face a confluence of issues related to aging baby boomers. *Springfield Business Journal Staff.* Originally published 10/8/2007. http://sbj.net/main.asp?Search=1&ArticleID=79035&SectionID=48&SubSectionID=108&S=1 (accessed January 13, 2011).

Flegal, K.M., Carroll, M.D., Ogden, C.L., and Curtin, L.R. 2010. Prevalence and trends in obesity among US adults, 1999–2008. *Journal of the American Medical Association* 303 (3): 235–41.

Galinsky, T., Waters, T., and Malit, B. 2001. Overexertion injuries in home health care workers and need for ergonomics. *Home Health Care Services Quarterly* 20 (3): 57–73.

Marras, W.S., Davis, K.G., Kirking, B.C., and Bertsche, P.K. 1999. A comprehensive analysis of low-back disorder risk and spinal loading during the transferring and repositioning of patients using different techniques. *Ergonomics* 42 (7): 904–26.

NAON. 2009. Safe patient handling. Special issue. *Orthopaedic Nursing* 28 (2S): 2–35.

National Center for Health Statistics. 2007. 2005 National hospital discharge survey. http://www.cdc.gov/nchs/data/ad/ad385.pdf (accessed February 2, 2010).

National Center for Health Statistics. 2008. Prevalence of overweight, obesity and extreme obesity among adults: United States, trends 1960–62 through 2005–2006. http://www.cdc.gov/nchs/data/hestat/overweight/overweight_adult.htm (accessed February 2, 2010).

National Research Council and Institute of Medicine. 2001. Musculoskeletal disorders and the workplace: Low back and upper extremities. Panel on Musculoskeletal Disorders and the Workplace Commission on Behavioral and Social Sciences and Education.

Nelson, A., Collins, J., Siddharthan, K., Matz, M., and Waters, T. 2008. Link between safe patient handling and patient outcomes in long-term care. *Rehabilitation Nursing* 33 (1): 33–43.

NIOSH. 1997. *Musculoskeletal Disorders (MSDs) and Workplace Factors.* DHHS (NIOSH) Publication Number 97-141. Cincinnati, OH: National Institute for Occupational Safety and Health.

NIOSH. 2006. *Safe Lifting and Movement of Nursing Home Residents.* DHHS (NIOSH) Publication No. 2006-117. Cincinnati, OH: National Institute for Occupational Safety and Health.

NIOSH. 2010. *NIOSH Hazard Review Occupational Hazards in Home Healthcare.* DHHS (NIOSH) Publication No. 2006-117. Cincinnati, OH: National Institute for Occupational Safety and Health.

OSHA. 2009. *Ergonomics for the Prevention of Musculoskeletal Disorders: Guidelines for Nursing Homes.* Document No. OSHA 3182-3R. Washington, DC: U.S. Department of Labor, Occupational Safety and Health Administration.

Owen, B.D. 1989. The magnitude of low back problems in nursing. *Western Journal of Nursing Research* 11 (2): 234–42.

Parsons, K.S., Galinsky, T.L., and Waters, T.R. 2006a. Suggestions for preventing musculoskeletal disorders in home healthcare workers, Part 1: Lift and transfer assistance for partially weight-bearing home care patients. *Home Healthcare Nurse* 24 (3): 158–64.

Parsons, K.S., Galinsky, T.L., and Waters, T.R. 2006b. Suggestions for preventing musculoskeletal disorders in home healthcare workers, Part 2: Lift and transfer assistance for non-weight-bearing home care patients. *Home Healthcare Nurse* 24 (4): 228–33.

Stubbs, D.A., Buckle, P.W., Hudson, M.P., and Baty, D. 1986. Backing out: Nurse wastage associated with back pain. *International Journal of Nursing Studies* 23 (4): 325–36.

United States Department of Labor Bureau of Labor Statistics. 2007. Nonfatal occupational injuries and illnesses requiring days away from work. http://www.bls.gov/iif/oshwc/osh/case/osnr0029.pdf2006 (accessed February 2, 2010).

Waters, T. 2007. When is it safe to manually lift a patient? *American Journal of Nursing* 107 (8): 53–59.

Waters, T., Collins, J., Galinsky, T., and Caruso, C. 2006. NIOSH research efforts to prevent musculoskeletal disorders in the healthcare industry. *Orthopaedic Nursing* 25 (6): 380–89.

7 Ergonomics of Packaging

Aleksandar Zunjic

CONTENTS

7.1 INTRODUCTION

Every day, the vast majority of consumers make contact with some packaging. With most products, the consumer realizes the first contact through interaction with the packaging. For this reason, the design of packaging is of great importance for consumers. There are a number of functions that packaging should meet, starting from the moment when the product leaves the production line. However, when the packaging comes in contact with the consumer, the ergonomic properties of packaging come to the forefront.

Manufacturers of packaging are often not aware of the dimensions of the problems that consumers have during interaction with the packaging. However, even when they are aware of the problem, manufacturers often do not know how to solve them. This chapter aims to indicate the dimension of the problem related to human–packaging interaction. Numerous examples will indicate the most common problems with packaging that consumers in practice encounter. The following text considers what should be avoided, what are possible solutions, as well as which aspects should also be considered in order to achieve ergonomic compatibility of packaging.

7.2 SAFETY ASPECT OF PACKAGING

Most consumers have had at least one bad experience with packaging, resulting in injury. The following sections will indicate the extent of this phenomenon. Data presented indicate an alarming situation. However, such data are rarely recorded systematically, and some organizations that have dealt with the problem for several years do not collect this type of data.

TABLE 7.1
Number of Emergency Room Treated Injuries Associated with Entire Class of Packaging and Containers for Household Products

No. ER-treated injuries	337.550	368.660	347.050	373.780	404.250	384.560	371.240
Year	1997	1998	1999	2000	2001	2002	2003

Source: Based on Spitler, V., Mills, A., Marcey, N. and O'Brien, C., Hazard screening report – Packaging and containers for household products, Consumer Product Safety Commission, Washington, 2005.

7.2.1 Injuries Caused by Packaging

According to the UK Department of Trade and Industry, about 67,000 people in the UK visit hospital every year because of an accident involving food or drink packaging (DTI 1997a, 1997b). Table 7.1 shows the number of injuries treated in the emergency room (ER), associated with packaging and containers for household products. Data presented in Table 7.1 are based on a U.S. hazard screening report (Spitler et al. 2005). Packaging and containers that contributed to the injuries are

- Bags (paper bags, plastic bags, and other kind of bags)
- Buckets or pails
- Glass containers (glass soft drink bottles, glass alcoholic beverage bottles, other glass bottles or jars, canning jars or lids, and glass tubing or test tubes)
- Pressurized containers (vacuum containers, aerosol containers, and other pressurized containers)
- Metal containers (self-contained openers, metal containers excluding aerosols, trash, and gasoline cans, and containers with key openers)
- Non-metal/non-glass containers (non-glass bottles or jars excluding baby bottles, plastic containers excluding bottles and jars, and wooden containers)
- Miscellaneous and other/not specified (bottles or jars not specified previously, containers not specified previously, and other containers, excluding vacuum or pressurized containers)

However, the number of medically treated injuries is greater than the number of ER-treated injuries. For example, the overall number of medically treated injuries in 2003 was 998,910. More troubling is the fact that 134 fatalities were recorded in 2001, as a consequence of human–package interaction.

Figure 7.1 shows the number of ER-treated injuries in 2003 in the United States, depending on the type of packaging. Figure 7.2 shows the number of all medically treated injuries in 2003 in the United States, depending on the type of packaging.

Winder et al. performed complex research on packaging-related injuries. Of the 200 subjects that participated in their research, 54.5% reported that they had injured themselves on food and drink packaging over the previous few years (Winder et al. 2002). Figure 7.3 shows the number of deaths in 2001 in the United States, depending on the type of packaging. Somewhat surprising are the data that the largest number of fatalities of all types of packaging is caused by bags.

It is interesting to consider the mechanism of appearance of injuries during interaction with packaging. The majority of accidents occurred at the early opening stage. The main causes of injuries that occur as a result of handling glass containers are explosion, foreign body in the eye, cut when opening a bottle, ingested glass, and sharp edges (Spitler et al. 2005). The greater number the injuries are sustained while disposing of broken glass (DTI 1999a, 1999b). The largest number of injuries from non-metal/non-glass containers occurs as a consequence of chemical burn, poisoning from the contents of a container, sharp edges, entrapment, or by getting fluid from a container into the eye. A cut is the main injury from buckets or pails and metal containers. In addition, swallowed

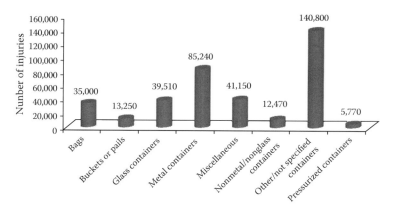

FIGURE 7.1 Number of emergency room treated injuries in 2003, depending on the type of packaging. (Based on Spitler, V., Mills, A., Marcey, N. and O'Brien, C., Hazard screening report – Packaging and containers for household products, Consumer Product Safety Commission, Washington, 2005.)

pull-tabs are often the cause of injuries from metal containers. Can burst, sprayed contents in children's eyes, and poisoning from the sprayed contents are the main sources of injuries from pressurized containers. Mouthing and allergy are also causes of injuries from miscellaneous (other) packaging products (Spitler et al. 2005). Individuals using a tool to open packaging (e.g., a knife) and using excessive force when trying to open packaging are additional sources of injuries. Especially frequent are injuries to consumers who use a knife when trying to open plastic packaging.

Although not as frequent, injuries from the caps on soft drink bottles can be very serious. Thus, injuries occur when, on opening a container, the cap leaves a bottle at great speed and hits the user in the face or eye. Using improper tools can also be a cause of accidents. If, over time, a tool for opening packaging becomes less usable (e.g., less sharp), this can be a reason for occurrences of injuries. The main cause for this is that the user then applies a greater force to open the packaging, which represents a potential risk for users. Lack of visible instructions for opening packaging

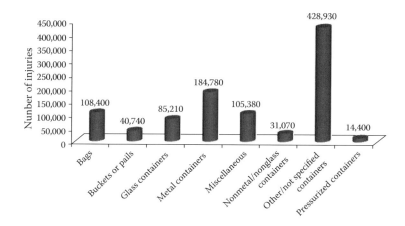

FIGURE 7.2 Number of all medically treated injuries in 2003, depending on the type of packaging. (Based on Spitler, V., Mills, A., Marcey, N. and O'Brien, C., Hazard screening report – Packaging and containers for household products, Consumer Product Safety Commission, Washington, 2005.)

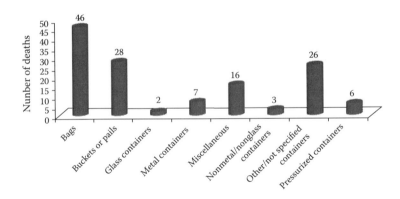

Type of packaging

FIGURE 7.3 Number of deaths in 2001, depending on the type of packaging. (Based on Spitler, V., Mills, A., Marcey, N. and O'Brien, C., Hazard screening report – Packaging and containers for household products, Consumer Product Safety Commission, Washington, 2005.)

often leads to the emergence of consumer frustration and the application of aggressive methods for opening, which are often not appropriate for the type of packaging. This approach increases the possibility of injury.

Gender is a factor that might have some influence on packaging accidents. Females report 40% more injuries caused by food and drink packaging (DTI 1997a, 1997b). Additionally, the UK Department of Trade and Industry suggests that the higher rate of injuries occur as women spend more time in the kitchen, where many injuries happen.

Children are a particularly vulnerable category of users, in relation to the human–packing interaction. A large number of accidents have occurred because children opened packaging when adults did not supervise them. Older people are also a vulnerable category of consumers in relation to the issue of packaging opening. They are aware of this problem, so that with greater or less success they develop a specific strategy for opening packages, in order to minimize the possibility of injury.

However, injury rates globally decline with age, from about the age of 35 onward. According to the UK Department of Trade and Industry, the reason for this result can probably be found in the assumption that the elderly spend less time opening food containers (DTI 1997a, 1997b). Additionally, these data may also reflect a cautious approach and developed strategies to opening packaging, applied by older people. However, it is necessary to bear in mind that a large number of accidents caused by opening packaging remain unrecorded, regardless of user categories. These consumers treat their injuries at home.

7.2.2 Packaging of Hazardous Material

There are several important properties that packaging should have in relation to the user, when it is necessary to pack a hazardous material (product). First of all, packaging should be a safe barrier between hazardous materials and the user. This feature primarily depends on the material used for packing, then on the technology used for closure, but it also depends on some other factors, such as the method of packing. Packaged hazardous material cannot come into contact with the user in any way, except in the intended or prescribed way. Packaging should also disable unwanted contact of other persons with dangerous materials. For example, packaging must prevent the direct contact of small children with certain medications intended for adults.

The Poison Prevention Packaging Act (PPPA) is administered by the U.S. Consumer Product Safety Commission (CPSC 2005). The PPPA demands special child-resistant and adult-friendly packaging of a wide range of hazardous household products, including most oral prescription drugs. Child-resistant packaging means packaging that is difficult for young children to open (usually below the age of eight), because the opening of such packaging requires the skills and abilities of older age. In addition, child-resistant packaging is often difficult to open by anyone, especially elderly and disabled patients with diminished manual strength or dexterity. For this reason, child-resistant packaging should have certain adult-friendly properties, which allow unrestricted opening by older users.

There have been considerable declines in reported deaths from ingestion by children of toxic household substances including medications, over the years that the regulations have been in effect. Regardless of this reduction in deaths, many children are poisoned or have potentially dangerous contact with medicines and household chemicals each year. Annually, there are about 30 fatalities of children under 5 years of age, who are unintentionally poisoned. Data from the National Electronic Injury Surveillance System (a CPSC database of ER visits) indicate that in 2003 about 78,000 children under 5 years of age were treated for poisonings, in hospital ERs in the United States. Furthermore, the American Association of Poison Control Centers reports over a million calls to poison control centers due to unintentional exposure to poisons of children under 5 years of age each year (CPSC 2005). Some of the reasons for the continuing ingestions are: availability of non-special packaging for prescription medication (on request), availability of non-special packaging size for medicaments that are sold directly to the customer without a doctor's prescription, inadequate quality control by manufacturers leading to defective closures, misuse of special packaging in the home (leaving the cap off or unsecured, transferring the contents to a non-special package), and violations of the law by the pharmacist.

A very large number of cases of poisoning in children are seen as a result of iron poisoning. A number of cases have had a fatal outcome. Generally speaking, iron is a safe substance if used according to the instructions of doctors. However, unsafe situations occur when packaging that contains iron ends in a child's hands. In order to prevent such situations, the Nonprescription Drug Manufacturers Association (NDMA) adopted a voluntary program in 1993, which includes a label warning statement for dietary supplements that contain iron. The CPSC and NDMA conducted a national consumer education program that advised adults to reclose child-resistant packages every time they are opened and to always store iron-containing products where toddlers cannot reach them. The U.S. Food & Drug Administration (FDA) has also proposed regulations that will make it harder for small children to gain access to high-potency iron products. The FDA requires that manufacturers wrap high-potency iron tablets and capsules individually, such as in blister packaging. The FDA supposes that the time and skill needed to remove tablets from unit-dose packaging would discourage toddlers to try it, or at least limit the number of tablets a child would swallow.

In order to reduce the risk of possible poisoning, human performance tests were developed to measure child-resistant and adult-friendly packaging effectiveness. Children aged 42–51 months were chosen as the test subjects in controlled experiments. The testing method was developed to simulate the situation that can be found at home. The experiment involved giving packages to pairs of children. The children were given 5 minutes for the package-opening task. If they did not open their package within that period, the children were given a single visual demonstration regarding package opening. Another 5 minutes were given to attempt to open the same package. The packaging was considered child resistant if no more than 20% of 200 children tested could open the package. The packaging was considered adult friendly if 90% of adults tested could open and close the same packaging (CPSC 2005).

A number of countries have stipulated which substances should have child-resistant packaging. For example, aspirin and paracetamol are substances that are regularly on the list of substances whose package must have child-resistant closures. However, in practice there is a significantly greater need for child-resistant packaging because this type of packaging should not apply only to those substances prescribed by law, but also for all substances that are potentially dangerous. Most of the products in the home chemistry domain belong to such substances.

Packaging of other hazardous chemical substances is largely governed by national standards and regulations. Most of these regulations refer to the guidelines related to the identification of hazardous chemical substances and give information to the user about these dangerous substances (usually in the form of a label), as well as to the obligation of safe packaging of these substances. For a long time, the European Union has made efforts to harmonize these regulations within the EU member states. It has already passed the new European Regulation on Classification, Labeling and Packaging of Substances and Mixtures (known as the CLP Regulation), which was published in 2008. The regulation provides a transition period to allow a gradual migration from the existing system to the new regime, which will end on June 1, 2015, when the CLP Regulation enters fully into force.

7.2.3   OTHER SAFETY ASPECTS OF PACKAGING

One of the aspects that manufacturers should certainly pay attention to when designing packaging is the mechanical contamination of a product by the packaging itself, during and after opening a package. Mechanical contamination can be especially dangerous to humans in the case of food products. An example of mechanical contamination of a product after opening the packaging are beverage cans with the stay-tab and pull-tab opening methods. These removable pull-tabs often dropped into cans, exposing consumers to the danger of accidentally swallowing the metal. To prevent this happening, stay-tabs were designed in the 1970s. However, recorded accidents indicate that stay-tabs may not have reduced the number of ingestions. Children often become victims as a result of this method of packaging opening.

Another example of contamination of products during the opening of a package is when a manufacturer uses a cork stopper. This type of stopper is frequently used to seal bottles containing alcoholic beverages. In this case, to open it, it is necessary to use a bottle opener. However, when the bottle opener penetrates the cork stopper, the cork often begins to fall apart. As a result, pieces of cork finish in the liquid, where they are often easily noticed by the naked eye. Sometimes the cork particles are not easily recognized, so the consumer can ingest them into the body during consumption of the contents from the bottle. In this case, we can act preventively, using a plastic stopper instead of a cork stopper.

A large number of products have a dual security system that should protect a product from contamination. The first is a lid. Below the lid, a diaphragm seal is located, which for most products provides hygienic protection and validity of the product over time (until it is opened). Although the main purpose of a diaphragm seal is to protect the product from contamination by the date of opening, in some cases it may cause contamination of the product during the opening phase. This phenomenon may represent a particular danger in the use of food products. Due to the application of a technological process, the diaphragm seal is too tightly coupled with the opening edge of a product, so that it cannot be easily removed when opening. An additional problem occurs when the dimensions of the diaphragm seal exactly matches the opening, so that there is no place for the customer to catch and remove the diaphragm seal by hand. As a result, the consumer is forced to use a tool for opening, usually a knife. Opening with tools is usually based on drilling the diaphragm seal, and removing it by hand when a hole is created. The diaphragm can be made of different materials that can decompose to some extent during the mechanical action of the tool and subsequent removal by hand. Small particles of the diaphragm can fall into the product, and with subsequent consumption of the contents of product, they may enter the body. From the rim of a packaging hole, the diaphragm in most cases cannot be eliminated completely, even by using tools. Regardless of the possibility of contamination of the products and users as described above (phases of opening are shown in Figure 7.4), the whole process of opening containers is often frustrating for consumers.

The solution is to provide controls for opening, e.g., a tab set in the middle of the diaphragm (Figure 7.5) or on the packaging rim (Figure 7.6). Dragging the tab should enable the consumer to remove the diaphragm seal completely. This means that the magnitude of the force required for

FIGURE 7.4 An example of a bad design solution for packaging, where the impossibility of adequate elimination of a protective closure system favors creating the condition for the mechanical contamination of products and consumers.

removing the diaphragm seal should be harmonized with the physical capabilities and limitations of humans.

It is well known that one of the basic functions of packaging is to prevent penetration of microorganisms into the product until the moment of opening, especially in the case of food and pharmaceutical products packaging. However, after opening, the packaging must not in any way cause contamination of products. This is an important property of packaging, which is often overlooked in some food products. Examples are cans with soft or alcoholic drinks, using the stay-tab opening mechanism. The main problem with this method of opening arises because the tab actually submerges into the can (and liquid) after opening (Figure 7.7). Thus, if microorganisms are on the tab, on opening they will pass to the fluid and cause its microbiological contamination. When a user consumes the contents of these cans, he/she will ingest the liquid and microorganisms from the packaging into the body to a greater or lesser extent, which can potentially lead to disease. From the aspect of microbiological contamination of products, cans with pull-tab mechanisms are the better solution.

FIGURE 7.5 An example of a good packaging design solution, where the tab located at the middle of the diaphragm seal allows easy opening, without the possibility of mechanical contamination of products and consumers.

FIGURE 7.6 An example of good packaging design solutions, where the tab located at the packaging rim allows easy opening, without the possibility of mechanical contamination of the products and customers.

When selecting materials for packaging, special attention should be paid to the selection of non-toxic materials. In the food industry, a special problem for consumers' health is the use of toxic materials for packaging production. A large number of packaging for frozen foods contains potentially carcinogenic substances. It is believed that the chemical perfluorooctanoic acid is present in the body of the vast majority of adult Americans and as an organism in all newborn babies. This substance accumulates in the body over time. Perfluorooctanoic acid is otherwise used for the

FIGURE 7.7 Stay-tab opening mechanism of packaging creates the possibility of microbiological contamination of products and customers.

production of Teflon pans. However, manufacturers of packaging for food products often use it when paper is utilized as packing material.

The use of potentially carcinogenic materials for the production of packaging is a big problem. The best solution is to substitute the carcinogenic materials with non-carcinogenic materials for packaging. Informing consumers about the potential dangers related to this issue is of particular importance, given that a large number of consumers, in general, are unaware of this problem. One possible solution is to pass legislation binding manufacturers of packaging to put information on their packaging that the product contains a potentially carcinogenic material (potentially carcinogenic substances are usually considered those materials that have been proven to be carcinogenic in tests on animals). The consumer then has to decide whether to buy that product.

7.3 INFORMATIONAL ASPECT OF PACKAGING

An important part of packaging design refers to information that is presented on the packaging. The design of this part of the interface between customers and products should be given special attention. Failure in this segment of the packaging design can have fatal consequences for consumers, as in the case with packaging of drugs and other medications.

The information contained on the packaging is sometimes the only form of communication between consumers and products. This especially applies to those products for which additional instructions for use are not designed. Most food products fall into this category. Also, a large number of cosmetic products contain the necessary information for consumers only on the packaging.

A number of accidents associated with inadequate use of the product due to the unsatisfactory presentation of information on the packaging had the impact on the state authorities in many countries to adopt regulations regarding the presentation of information on the packaging. These regulations primarily relate to the labels on the packaging of food products, drugs, cosmetic products, as well as some hazardous chemical substances. Although such regulations are extremely useful and largely based on ergonomic principles, the problem appears sometimes because such regulations are not uniform and may differ to a greater or lesser extent between countries. Such a situation may create confusion when a consumer buys an imported product. In addition, certain regulations in this domain are not fully explicit (e.g., location of specific information that is required on the packaging), which leaves the possibility that a manufacturer may make an error.

7.3.1 MOST FREQUENT MISTAKES RELATING TO THE PRESENTATION OF INFORMATION

Although efforts to introduce the order of presenting information on packaging have lasted several decades, in practice there are still many problems that can be confirmed by numerous examples. A large number of medication errors, for example, arise due to inappropriate packaging and its labeling. One study (Berman 2004) showed that a certain number of medication errors were created as a result of the following shortcomings related to the packaging:

* Small size and poor readability of presented information
* Similar appearing packages or labels on different products
* Poorly designed or disordered labels
* Poor use of color to differentiate products
* Inadequate warnings about drug use
* Poor legibility of presented information

Recognizing this problem, a number of national institutions introduced regulations on the information that should be present on certain types of packaging. For example, the Federal Food, Drug, and Cosmetic Act (FD & CA) is the basic food and drug law of the United States, which gives authority to the U.S. Food and Drug Administration (FDA) to oversee the safety of food, drugs, and cosmetics. On January 6, 1993, the FDA published 26 rules addressing various aspects of food labeling (Summers 2007). In the European Union, information that should be presented on packaging is also regulated. EU Directive 2000/13/EC (2000) sets the compulsory information that has to be included on all food labels on packaging. The presence of the following information on the packaging of food products is mandatory (Directive 2000/13/EC 2000):

- The name under which a product is sold
- The listing of ingredients
- The quantity of certain ingredients
- The net quantity in the case of prepackaged foodstuffs
- The date of minimum durability
- Any special conditions of use or storage conditions
- The name and address of the manufacturer or packager
- Particulars of the place of origin
- Instructions for product use when necessary
- The actual alcoholic strength by volume, with respect to beverages containing more than 1.2% by volume of alcohol

Without doubt, this directive and other similar directives for consumers are of great benefit. However, mistakes that occur in practice as a result of implementation of the directive will be pointed out, as well as the absence of specific information relevant to consumers that is not treated in the directive.

It can be noticed that the directive mandates that the package contains a minimum of ten required pieces of information. For larger sizes of packaging, presentation of such information on the packaging is not a problem. However, the problem appears with packaging of smaller dimensions, e.g., chocolate bars. As a consequence of the lack of surface area needed to accommodate the required information, packaging manufacturers resort to positioning information at locations that are less visible, while some of the information is not possible to see until the packaging is opened. An example of such packaging is shown in Figure 7.8.

Information should not be positioned in a location of the packaging that is not visible to the consumer, or only becomes visible when the package is opened. In addition, the information should not be located on parts of the packaging intended for opening (packaging controls). These parts of the packaging should be used only for positioning of the information that relates to the opening of the packaging (if it is not otherwise resolved).

Information regarding the opening of the packaging is of great importance for the user. However, directives that prescribe the information that should be present on the packaging typically do not prescribe the obligation of specifying information concerning the method of opening the package. The presence of this kind of information on the packaging is important, especially when a specific type of packaging has not been used previously, or when the method of opening is opposite to the

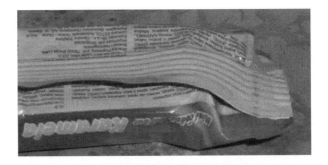

FIGURE 7.8 An example of packaging where some information is not visible to the consumer, until the packaging is opened.

FIGURE 7.9 An example of packaging that does not possess an instruction for opening, while the method for opening is contrary to the expectation of the consumer.

user expectations. In practice, packages that contain this kind of information are very rare, and are largely related to food products.

The following example illustrates the emergence of a problematic situation when trying to open packaging that does not contain instructions for opening. Figure 7.9 shows a package that contains a ring for opening (controls for opening). This package does not contain any information relating to opening the packaging, so consumers must try to open it relying on their experience and intuition. The user most often tries to open this package by pulling the ring toward himself/herself. However, this approach leads to a rapid separation of the ring from the can. In that case, finding a new tool is the only solution. The better solution is to put an instruction on the packaging, relating to the opening. The message "push the ring to open the tin and then pull it toward yourself" would probably be appropriate in this case.

7.4 PACKAGING OPENABILITY

The biggest problems for the consumer arises when he/she tries to open individual packages. In such situations, the possibility of injury increases significantly. Particularly vulnerable are the following consumer categories: children, the elderly, people with disabilities, and females. In relation to this packaging feature, there are also a number of complaints from left-handed persons. However, occasionally, healthy adults also have a problem with opening packaging.

7.4.1 MOST FREQUENT MISTAKES CONCERNING OPENING PACKAGING

The following is an overview of the most common problems that consumers have in relation to package opening. One of the most common problems that consumers have relates to their inability to open the package manually. Occasionally or constantly, older people, children, invalids, and even adults are unable to open a package because opening it requires the use of great force. In the following section, the forces and movements necessary to open packaging will be analyzed in more detail. Figure 7.10 shows the percentage of respondents (Winder et al. 2002) who occasionally or constantly have problems related to a lack of power when opening certain types of packaging.

Users often do not know how to open packaging (opening is not obvious), which is one of the main problems related to packaging opening. Figure 7.11 shows the percentage of respondents (Winder et al. 2002) who occasionally or constantly do not know how to open certain types of packaging.

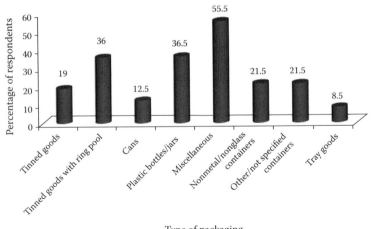

FIGURE 7.10   The percentage of respondents who always or sometimes have problems with lack of strength needed to open certain types of packaging. (Based on Winder, B., Ridgway, K., Nelson, A. and Baldwin, J., *Applied Ergonomics, 33*, 433–38, 2002.)

The solution to this problem is in providing controls intended for packaging opening. These controls must be easily distinguishable. The way to use controls should be obvious and in accordance with user expectations and previous experience.

Often, users are unable to determine where to start opening, which is a very common problem when opening a package. The main reason for this is that there are no visible controls on the packaging, which could be used for opening. In addition, there are packages that do not possess visible controls for opening, but it is enabled their relatively easy opening on the basis of application of some specific design solutions. Figure 7.12 shows such a design solution that has no visible control.

Most users try to open packaging from the picture placed at the junction of the cardboard and plastic part of packaging. However, given that for most users this package fails to open by hand,

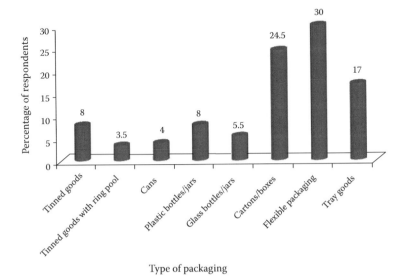

FIGURE 7.11   The percentage of respondents who always or sometimes do not know how to open certain types of packaging. (Based on Winder, B., Ridgway, K., Nelson, A. and Baldwin, J., *Applied Ergonomics, 33*, 433–38, 2002.)

FIGURE 7.12   An example of packaging that does not possess an opening place that is easy to locate and that is obvious for the majority of users.

they look for a tool (a scissors or knife) to open it. However, this package possesses a more simple method of opening. The place to start opening the package is located in the upper right corner of the packaging.

A symbol indicating the location of the place where it is necessary to start opening the package is not easily visible, which presents a problem for users. Additionally, the meaning of this symbol isn't obvious to most users.

The best solution for this problem is shown in Figure 7.12. The solution is to redesign the packaging. An enhanced design solution could be based on the old solution, but it would be necessary to add the visible control on the packaging. This control could use an identical mechanism of opening. However, if the company making the product has no material resources for investing in a new packaging design solution, from the aspect of consumers it is also acceptable that the solution could be based on a redesign of the existing symbol, which indicates where to start opening. Figure 7.13 shows the percentage of respondents (Winder et al. 2002) who occasionally or constantly cannot see where or how to start opening certain types of packaging.

A large number of packages are not very easy to open. We need to invest a lot of work and effort to open certain packages. An example is a clamshell plastic packaging. Many consumers at first try to open this type of packaging by hand. They look for a place to start opening, and attempt to determine how the package opens. Sometimes, after failing to open the package, they try to meet this goal by using a tool. In this case, the whole process of opening is disappointing and unpleasant for the consumer. Figure 7.14 shows the percentage of respondents (Winder et al. 2002) who consider that some packagings are "too fiddly to open easily."

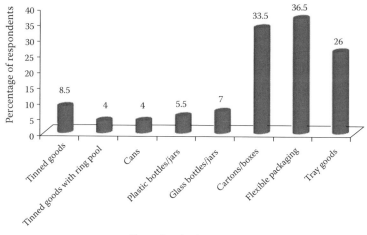

FIGURE 7.13   The percentage of respondents that always or sometimes cannot see where or how to start opening certain types of packaging. (Based on Winder, B., Ridgway, K., Nelson, A. and Baldwin, J., *Applied Ergonomics*, 33, 433–38, 2002.)

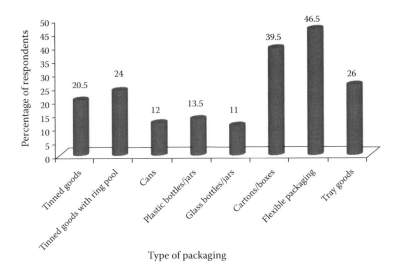

FIGURE 7.14   The percentage of respondents that consider some packaging too fiddly to open easily. (Based on Winder, B., Ridgway, K., Nelson, A. and Baldwin, J., *Applied Ergonomics,* 33, 433–38, 2002.)

A large number of users, especially the elderly, have previously had at least one negative experience related to injuries when opening packaging. For this reason, the concern of a number of consumers related to the occurrence of injuries when opening packaging is understandable. These kinds of injuries are particularly common when sharp and inadequate tools are used for opening. Figure 7.15 shows the percentage of respondents (Winder et al. 2002) who are occasionally or constantly worried in relation to being injured when trying to open a package.

For the purpose of opening and handling certain packages, it is necessary to use great force or use tools. In such circumstances, damage to the product itself may occur, especially when the consumer is not aware of the location of the product or its volume within the packaging. Shedding of

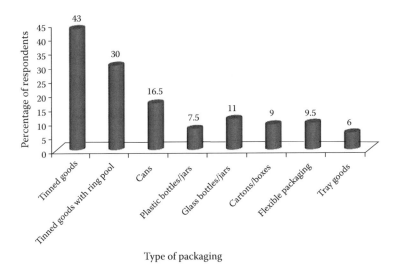

FIGURE 7.15   The percentage of respondents who are always or sometimes worried in relation to being injured when trying to open certain types of packages. (Based on Winder, B., Ridgway, K., Nelson, A. and Baldwin, J., *Applied Ergonomics,* 33, 433–38, 2002.)

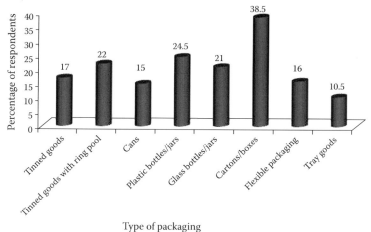

FIGURE 7.16 The percentage of respondents who are always or sometimes worried in relation to spilling or wasting the products because of opening. (Based on Winder, B., Ridgway, K., Nelson, A. and Baldwin, J., *Applied Ergonomics,* 33, 433–38, 2002.)

the contents of packaging is not a rare phenomenon, in cases where the packaging is hard to open. Figure 7.16 shows the percentage of respondents (Winder et al. 2002) who are occasionally or constantly worried about spilling or wasting the products because of opening.

The packaging designers should bear in mind that consumers will take various, sometimes even extreme, steps to open a particular package. In such situations, damage to products may occur, while consumers in some cases of damages will not be responsible.

Inadequate opening of the packaging and a contrary way of opening it in relation to the way that is foreseen are frequent in practice. During the user trials, 10 of the 68 subjects were seen to turn the caps the wrong way (Norris et al. 2000). Children and the elderly most likely turn the cap the wrong way. In addition, it may happen that consumers create a too big or too small hole when opening a packaging, so the product cannot leave the package in an appropriate manner. This type of error is also a consequence of poorly designed packaging. In the interaction with the packaging, the user is often not able to estimate how much power will be necessary to open it. Figure 7.17 shows the percentage of respondents (Winder et al. 2002) who are occasionally or constantly opening the packaging incorrectly, so the product does not leave the packaging properly.

Instructions on the packaging associated with the use of products or opening of the packaging can sometimes be very confusing. Due to such omissions, the consumer may remain deprived of information that is vital to use the products. An example of such a confusing instruction on packaging is shown in Figure 7.18. The instruction refers to the shelf life of a product.

On the package is literally written:
Date of production: 36 months before the date specified on the packaging.

At the bottom of the packaging, the date is specified in the form of the following information, which is imprinted on the packaging (the information is poorly visible): *22F301 09 03 12*. It can be assumed that the last six numbers relate to the date. Detailed examination of the packaging indicates that there is no further information pertaining to the date. If we connect previously mentioned information from the packaging, we get the resulting information: the product was produced 36 months before 09 03 12. However, this futuristic information means little to the consumer. On the basis of this confusing information, the consumer cannot determine the exact date of production or the shelf life of a product. Therefore, it is probably better not to open or use such a product. Figure 7.19 shows

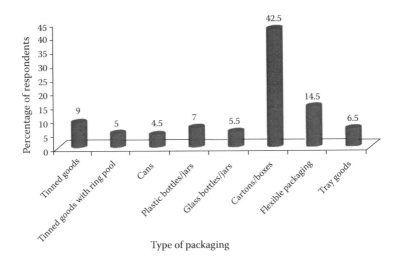

FIGURE 7.17   The percentage of respondents who always or sometimes open the packaging incorrectly, so the product does not leave the packaging properly. (Based on Winder, B., Ridgway, K., Nelson, A. and Baldwin, J., *Applied Ergonomics,* 33, 433–38, 2002.)

the percentage of respondents (Winder et al. 2002) who occasionally or constantly meet confusing instructions on packaging.

In practice, there are a large number of child-resistant mechanisms (controls) for opening packages. In the description that follows, an overview of ways of activating controls on child-resistant packaging is given. The most frequent instructions regarding the opening of packaging that is not intended for children are: random push down while turning; localized push in while turning (force must be applied to designated place on closure); localized squeeze force while turning; turn top cap until stops and then push down and turn; align two points then push up on tab or lip; press to release and then lift hinged tab (dispensing cap); remove a portion (tab), rotate the blister to orient, push through; hold fitment while turning; turn closure until stops, then lift and continue trying to open; push up to release; pull up to release and lift hinged lid (dispensing cap); squeeze two points

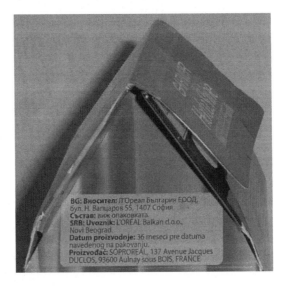

FIGURE 7.18   An example of a confusing instruction on the packaging, which refers to the shelf life of a product.

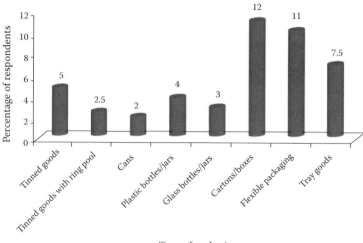

FIGURE 7.19 The percentage of respondents who always or sometimes meet confusing instructions on packaging. (Based on Winder, B., Ridgway, K., Nelson, A. and Baldwin, J., *Applied Ergonomics*, 33, 433–38, 2002.)

simultaneously to open; requires key device or fingernail or coin or other tool to open; localized squeeze while lifting removes overcap (actuates normally); localized press down then pull up at arrow; line-up arrows on the overcap and ring to remove; push out; remove a portion (tab) and push out; push tab while rotating directional pump to spray position, then pump with finger; press down on a point to release lock, rotate orifice to spray position, and squeeze trigger; localized squeeze while lifting up, then pressing two tabs while lifting lid to open; squeeze two specific points simultaneously, lift zipper tab and pull to open; localized push up to remove; press hold, pull out (parts remain together), push out; pull trigger, lift flap, push out; press then flex and lift to open; push in, squeeze and hold, hold and pull.

For a number of users, especially elderly persons, the opening of child-resistant packaging can sometimes be a tremendous process. The reason for this can be that the instructions for packaging opening are sometimes unclear for the user. An additional reason may also be the existence of a large number of methods for opening child-resistant packaging, which can be contrary to the methods that the consumer previously used in practice. Certainly, the decline of the physical and mental ability of humans with age may have a negative impact on this phenomenon. Figure 7.20 shows the percentage of respondents (Winder et al. 2002) who consider that tamper/child-proof mechanisms makes opening difficult.

7.4.2 Forces for Opening Packaging

Almost every day we can see packaging that is difficult to open. An example that is probably well known to all is the problem with opening a jam jar. It is important that the force necessary for packaging opening be in accordance with the physical capabilities of the consumer. In addition to the use of great force for opening packaging, an additional problem is that the material from which the packaging is made is often smooth. Thus, although a consumer can provide sufficient force, he/she cannot transmit it to the packaging.

In order to design a packaging that is customized to the consumer from the aspect of ergonomics, designers should be aware of the forces that people can develop when opening the packaging. Several experiments have been conducted in order to determine the forces that people develop when opening the packaging. Daams performed such an experiment in 1994. Twelve male and ten female

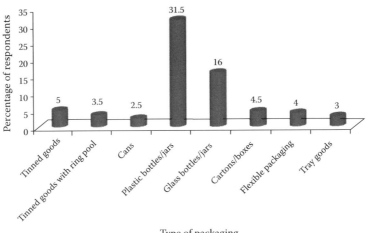

FIGURE 7.20 The percentage of respondents who consider that a tamper/child-proof mechanism makes opening difficult. (Based on Winder, B., Ridgway, K., Nelson, A. and Baldwin, J., *Applied Ergonomics*, 33, 433–38, 2002.)

subjects participated in this research. The average age of the subjects was 22 years. Subjects were in a standing position, while torque (maximal static force) was measured on a fixed jar (at 0.95 m from the floor) and on a freely movable jar.

When the jar (lid ⌀ 66 mm, jar ⌀ 75 mm, jar height 113.5 mm) was at 0.95 m, average torque was 11.7 Nm for males, 6.36 Nm for females, and 9.27 Nm for both (males and females). When the position of the jar was not fixed, average torque was 9.67 Nm for males, 5.91 Nm for females, and 7.96 Nm for both (Daams 2001).

The torque that humans can achieve when opening a package depends on the shape and dimensions of the packaging. Crawford, Wanibe, and Nayak (2002) investigated the maximum wrist torque strength of a group of 20 young adults (average age for males and females was 26.2 and 25.9 years, respectively) and 20 older adults (average age for males and females was 73.8 and 68.7 years, respectively) during packaging opening. Height (first number in parenthesis) and diameter (second number in parenthesis) of the 12 test pieces (lids) in this research were as follows: A (10,20); B (20,20); C (30,20); D (10,50); E (20,50); F (30,50): G (10,50); H (20,50); I (30,50); J (10,80); K (20,80); L (30,80). All dimensions are in millimeters. A, B, C, D, E, F, J, K, and L denote circular test pieces. G, H, and I denote square test pieces.

Figure 7.21 shows the maximum torque that the younger subjects (male and female) achieved in relation to each test piece. The data shown in the figure were obtained on the basis of calculation, based on the data from the study of Crawford, Wanibe, and Nayak (2002).

Figure 7.22 shows the maximum torque that the older subjects (male and female) achieved in relation to each test piece. The data shown in the figure were obtained on the basis of calculation, based on the data from the study of Crawford, Wanibe, and Nayak (2002).

Besides knowing the maximum torque that a man can achieve when opening packaging, for the purpose of design it is necessary to know the size of a comfortable torque for opening packaging. It is not difficult to conclude that consumers will not be satisfied if every time they open a package they must use maximum force. The change of the comfortable opening torque depending on age is shown in Figure 7.23. Comfortable opening torque is measured on the caps of bottles.

Crawford, Wanibe, and Nayak (2002) measured the torque necessary for opening different food products. The torque was measured on lids with diameters ranging from 28 to 73 mm. The height of the caps ranged from 9 to 23 mm. Figure 7.24 shows the measured values of the torque during opening of nine food products.

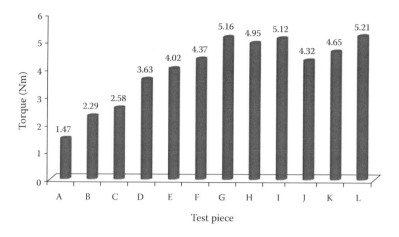

FIGURE 7.21 Mean torque strength for younger men and women in relation to each test piece. (Based on Crawford, O.J., Wanibe, E. and Nayak, L., *Ergonomics*, 45, 13, 2002.)

If we compare the data from Figure 7.24 with the data from Figure 7.23, depending on the age group of users, it can be noticed that a certain number of products do not meet the conditions for comfortable opening of the packaging. For example, in the case of consumers older than 60 years, all tested packages, except for coffee, are uncomfortable to open. A comparison of the data in Figures 7.21, 7.22, and 7.24 reveals that the torques required to open the packaging for marmalade and beetroot are greater than the maximum torque that users can achieve, regardless of the age of the consumer. This means that these packagings can only be opened by using additional tools. This analysis shows the necessity for redesigning packagings from the viewpoint of reducing the torque, which is required to open the majority of food products whose packagings possess the lid.

7.4.3 Tools for Opening Packaging

A *Yours* magazine survey stated that three out of five people over the age of 50 have purchased tools to open packaging (Yoxall et al. 2009). Customers are offered a number of different tools

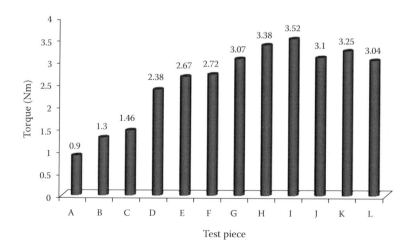

FIGURE 7.22 Mean torque strength for older men and women in relation to each test piece. (Based on Crawford, O.J., Wanibe, E. and Nayak, L., *Ergonomics*, 45, 13, 2002.)

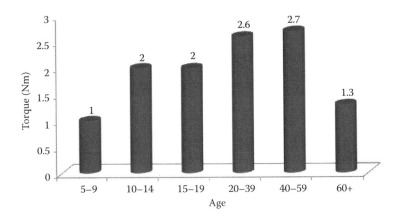

FIGURE 7.23 Mean comfortable opening torque for men and women in relation to age. (Based on Norris, B., Hopkinson, N., Cobb, R. and Wilson, J., *Injury Control & Safety Promotion*, 7, 4, 2000.)

intended to open packaging. Some consumers recognize these tools as useful, but they are often considered expensive. In addition, the use of these tools sometimes requires certain skills, or even a short training. For these reasons, a certain number of consumers continue to open packaging using knives, pliers, scissors, or other household tools. This consumer' behavior is risky and often leads to injuries.

Despite the use of tools, many consumers still have problems with opening packaging. The problem of the efficiency of individual tools for packaging opening has been recognized. In order to determine the efficiency of certain tools, Norris et al. (2000) measured the maximum torque that can be achieved for opening packaging using tools such as a rubber mat, a rubber cone, and nutcrackers. The subjects in this research were five males and five females, aged 22–32 years. Torque was determined for opening and tightening. Figure 7.25 shows the maximum torque (mean for 10 subjects) accomplished by using the tools for opening the packaging.

If we compare the data in Figures 7.24 and 7.25, we can see that the rubber cone is not an appropriate tool for opening the following packages: pickled onions, lemon juice, jam, beetroot, and marmalade. Similarly, the rubber mat is not an appropriate tool for opening the following packaging of food products: pickled onions, jam, beetroot, and marmalade. Given that the performance of individual tools for packaging opening is limited, the solution may be to design packages that are

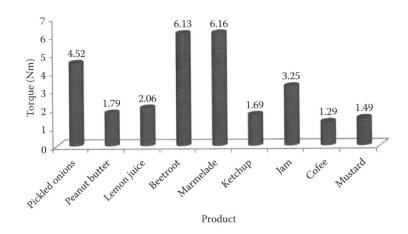

FIGURE 7.24 Torque required to open food products. (Based on Crawford, O.J., Wanibe, E. and Nayak, L., *Ergonomics*, 45, 13, 2002.)

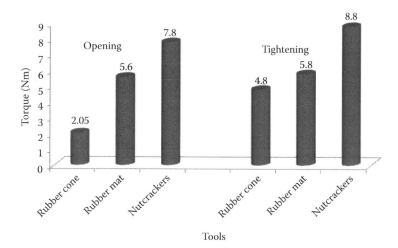

FIGURE 7.25 Maximum torque strength using tools. (Based on Norris, B., Hopkinson, N., Cobb, R. and Wilson, J., *Injury Control & Safety Promotion*, 7, 4, 2000.)

easy to open, i.e., packages that do not require additional tools. Design improvement of tools for opening packaging could also contribute to the solution of this problem.

7.5 DESIGNING ERGONOMIC PACKAGING

In addition to the various functions that packaging should meet, it should also have ergonomic features. This means that the packaging should be customized to the consumer from psychological, physiological, and anatomical aspects. From the ergonomic aspect, the package should be comfortable (easy to open, convenient to carry, and ease of use of the product), safe to handle (without causing injury), and should allow efficient use (should not cause errors when using the product, without prolonged duration of activity in relation to anticipated or expected time). In order to accomplish the previously mentioned ergonomic functions, the size, shape, weight, material, texture, and interface of the packaging should be customized to human needs. In addition, packaging must meet certain aesthetic properties for pleasant usage.

The dimensions of packaging must comply with the anthropometric characteristics of the male and female. Given that the human being primarily interacts with packing by using the hands, in the first place packaging designers need to know the anthropometric characteristics and dimensions of the hand. When the packaging is of small dimensions, for the purpose of simple opening, packaging has to be designed so that the 95th percentile of consumers can successfully open the packaging. When the packaging is opened, easy usage of products we can also ensure by creating such operative hole in the packaging, which will match the hand with its size. If the packaging has large dimensions, designers should take into account the 5th percentile of the user population. The user should be able to encompass the packaging. Easy but clumsy packaging will not be comfortable to handle.

Packaging shape should also be harmonized with the anthropometric characteristics of humans. If this is not the case, in some circumstances the packaging can be dropped, causing injury to the consumer or damage to a product. In addition, the shape of packaging can be successfully used as a method of coding. Coding of packaging by shape can prevent errors when using products that have different purposes, but a similar label and other signs. An example is the shape of packaging for shampoo and hair conditioner made by the same company.

The weight of a package containing a product largely depends on the product itself. If a product is heavy, comfort can be increased by installing a handle or handrail, which will facilitate handling and transport. In the case of products over 25 kg, the packaging should be envisaged with at least two handles, i.e., adjusting the package to the possibility of manipulation by two or more persons.

Material used for packaging primarily should be safe for users. Avoiding sharp edges on the packaging is very important. Light materials are suitable for handling, so they have priority if they also meet the other functions required of the packaging. It is important that the material used for making packaging does not contain carcinogenic substances.

The texture of the packaging should also support the feature of easy handling. In some cases, a matted or roughed surface can increase friction during contact between the hand and packaging, which facilitates the opening of some packaging (e.g., packaging containing a lid). As a shape, texture can also be used successfully as a method of coding of packagings, for the purpose of their mutual distinguishing. This feature is particularly important for coding of packaging for blind and partially sighted people.

The interface of packaging, almost certainly, is the most important part of the packaging for the user. Similarly, as the machine has its own interface, packaging also has its own interface, which allows successful interaction in the system man–packaging. Although one can accomplish interaction with packaging in different ways, controls and displays on packaging are essential and very important segments, which participate in the interaction with man (as is the case with machines). Further explanation is necessary, since this approach in the domain of interaction of man and packaging is, to some extent, innovative.

As is known, controls are the parts of a machine on which man acts, so that machine performs a certain action. Analogously to this, the controls on the packaging are parts of the packaging on which the user acts, in order to reach the product. For example, all the previously mentioned types of child-resistant closures are different types of controls on packaging. Every quality-designed packaging possesses controls, which allows easy opening of packages. In addition, the ergonomically designed controls on packaging allow the rest of the packaging to remain undamaged during opening, as well as to retain its function.

A machine using a display shows the specific information in the system. Similarly, the display on a packaging shows the necessary information related to the packaging and product. Each surface of packaging that can be used to accommodate information is its display. By optimizing the interface elements, as well as other mentioned elements that affect the interaction with the user, we can influence the improvement of the properties of ergonomic packaging.

7.6   CONCLUSION

Packaging is all around us. Each product has packaging. Therefore, it is difficult to imagine how many different types of packaging exist. However, relatively little packaging possesses ergonomic features. As a result, there is a whole range of problems that burden the consumer in relation to packaging. They manifest themselves through dissatisfaction, frustration, injury, and abandonment of the use of the product.

Earlier, when designing the packaging, producers primarily insisted that the package be visually attractive, with the aim of attracting the consumer to purchase the product. Such practice still exists. One gets the impression that manufacturers and distributors are not aware of the dimensions of the problem, and the benefits that can be obtained if packaging is designed from ergonomic aspects.

Designing packaging requires a multidisciplinary approach, which involves the participation of experts in the field of ergonomics. Designing packaging that has ergonomic features is not a simple task that can be solved quickly. A number of ergonomic factors have to be considered when designing packaging. The common case in practice is that designing of packaging only starts when the product is already made. In this way, designers of packaging have very little time and opportunity to achieve good quality. Designing of packaging should be initiated in the early stages of product development, especially when it is necessary to design packaging for a new product. Such an approach will ensure that all the functions of packaging are represented, including the function of the ergonomic compatibility of packaging.

REFERENCES

Berman, A. 2004. Reducing medication errors through naming, labeling, and packaging. *Journal of Medical Systems* 28 (1): 9–29.

CPSC. 2005. *Poison Prevention Packaging: A Guide for Healthcare Professionals*. Washington: Consumer Product Safety Commission.

Crawford, O.J., Wanibe, E., and Nayak, L. 2002. The interaction between lid diameter, height and shape on wrist torque exertion in younger and older adults. *Ergonomics* 45 (13): 922–33.

Daams, J.B. 2001. Torque data. In *International Encyclopedia of Ergonomics and Human Factors*, ed. W. Karwowski, 334–42. London and New York: Taylor & Francis.

Directive 2000/13/EC. 2000. Directive of the European Parliament and of the Council on the approximation of the laws of the member states relating to the labelling, presentation and advertising of foodstuffs. The European Parliament and the Council of the European Union.

DTI. 1997a. Consumer Safety Research. *Domestic Accidents Related to Packaging*, Vol. I. London: Department of Trade and Industry.

DTI. 1997b. Consumer Safety Research. *Domestic Accidents Related to Packaging: Analysis and Tabulation of Data*, Vol. II. London: Department of Trade and Industry.

DTI. 1999a. Government Consumer Safety Research. *Assessment of Broad Age-Related Issues for Package Opening*. London: Department of Trade and Industry.

DTI. 1999b. Government Consumer Safety Research. *Use and Misuse of Packaging Opening Tools*. London: Department of Trade and Industry.

Norris, B., Hopkinson, N., Cobb, R., and Wilson, J. 2000. Investigating a potential hazard of carbonated drinks bottles. *Injury Control & Safety Promotion* 7 (4): 245–59.

Spitler, V., Mills, A., Marcey, N., and O'Brien, C. 2005. Hazard screening report – Packaging and containers for household products. Consumer Product Safety Commission, Washington.

Summers, J.L. 2007. *Food Labeling Compliance Review*. Iowa, IL: Blackwell.

Winder, B., Ridgway, K., Nelson, A., and Baldwin, J. 2002. Food and drink packaging: Who is complaining and who should be complaining. *Applied Ergonomics* 33 (5): 433–38.

Yoxall, A., Langley, J., Luxmoore, J., Janson, R., Taylor, C.J., and Rowsan, J. 2009. Understanding the use of tools for opening packaging. *Universal Access in the Information Society* (October 27): 1–9. http://www.springerlink.com

Section II

Design for Usability

8 User Research by Designers

Stephen J. Ward

CONTENTS

8.1 INTRODUCTION

In daily life, we are surrounded by and make use of environments, buildings, products, and information systems that have been designed by others. The design disciplines that contribute to our built environment and material culture each have domains of knowledge and methods that enable practitioners to understand relevant issues, make judgments, communicate ideas, and ultimately, finalize designs for the products, buildings, environments, and systems that will be made. The act of designing involves making predictions, or at least conjecturing, about how future users will respond to and interact with something that does not yet exist (Redstrom 2006). It follows that designers, and specifically product designers, should have methods for developing their knowledge of users and testing their assumptions about the interaction of users and the products being designed.

In the ergonomics literature, the discussion and proposed guidelines on how to do this are commonly referred to as part of a "user-centered" approach to design. Most of the published research on ergonomics in product design has the product–user relationship as its focus and is available as a body of knowledge to those involved in designing future products. There is, however, relatively little study of how industrial designers think about users and how they go about finding out what they need to know about potential users of the things they are designing. It is important that designers' perspectives and methods are reviewed from time to time so that there is an opportunity to reflect on why particular methods may be preferred in the practice of design and how ergonomics expertise can best be integrated with the design process.

In this chapter, the results of interviews and a survey of industrial designers undertaken by the author are discussed together with published work by others on relating user research methods to design. Reflections and recommendations arising from the discussion are intended to have practical application for designers and also for human factors specialists who provide advice to designers or are involved in other ways in the design of products for everyday users.

The underlying motivation for this study, from the author's point of view, is the hope that it will help to raise the profile of user research in the practice of design and encourage deeper understanding of the user experience. It is hoped that this will contribute to design activity that will result in everyday products and systems that more comprehensively address user needs and provide lasting satisfaction.

8.2   DEFINITIONS

It may be helpful, at this point, to define some of the terms used in this chapter.

"Users" refers mainly to those who could ordinarily be expected to be among the owners or end-users of particular types of consumer products. Designers sometimes describe an intended user population as the "target market" for the product being designed. For some types of products, the people involved in installation, and maintenance could also be counted as users—if their needs are identified as impacting on the design process and outcomes.

"User research" may be taken to refer broadly to methods by which information about users and the product–user relationship is gathered and applied by a design team. This could include, but is not limited to

- Finding and applying published data and descriptions of users, e.g., anthropometry, market demographics
- Collecting customer or user feedback in relation to current and past products
- Observation of users with existing products or prototypes of new products
- Interviews with actual or potential users of a product
- Focus group discussions
- User trials of products, mock-ups, and prototypes
- Seeking advice from people with expert knowledge of intended users
- Constructing imagined scenarios of user interaction, for reflection and discussion

Thus, the definition of user research includes activities that are commonly associated with market research or applied ergonomics.

Finally, the terms "ergonomics" and "human factors" may be taken in this chapter to be interchangeable and to include the study of physical, cognitive, and social aspects of the interactions between people and the products and systems they use.

8.3   USER RESEARCH METHODS IN DESIGN

In addition to the traditional application of data and guidelines derived from human factors research, newer methods are well established and widely discussed as design methods. The following is a brief description of some of the commonly discussed "new" human factors methods.

8.3.1 FOCUS GROUPS

Focus groups are widely used as a qualitative research method for market research (Ives 2003). A key characteristic of the focus group method is the interaction between group members and the synergistic effect where participants respond to and build on contributions by others. In product design, focus groups can be combined with other methods. For example, Caplan (1990) describes a study in which qualitative and quantitative results were achieved by asking the participants in a focus group to discuss and select assessment criteria for a product and then individually rate alternative concepts according to the selected criteria. Bruseberg and McDonagh (2003a) make the comment that focus groups provide a good common ground for collaboration between designers and researchers.

8.3.2 ETHNOGRAPHY

Ethnography is a tool used in anthropology. Its purpose is to record and describe human cultures using techniques such as observation, listening, and (sensitively) joining in with activities taking place in their natural setting (Taylor, Bontoft, and Flyte 2002). There is growing interest in ethnography as a research tool in design, particularly for its potential to assist designers in identifying opportunities for new types of products. The method can be used not only for studying foreign cultures, but also for subcultures such as teenagers and shoppers (Sanders 2002). It is a qualitative research method and seeks depth and richness of information by immersion in, but minimal interference with, the culture being studied. Ethnographic studies of foreign cultures typically involve observations over months or even years. In adapting the technique for use in design, the time frames are reduced and video is commonly used as a means of communicating observations to the design team.

Techniques similar to ethnography, but with a product-focused market research approach are described as "observational research" by Abrams (2000). Abrams describes techniques for video recording behaviors and conversations in users' own homes, sometimes using a prototype of a new product.

8.3.3 SCENARIO BUILDING

Scenario building is a modeling method in which the designer creates imagined situations of a user interacting with a product in a context of use (Hasdogan 1997). Scenarios may be explored in the designers' imagination, but they can also be made explicit and communicated to others by means of stories, illustrations, and by acting out a simulation. A designer may think through many scenarios in which different phases of use, different users, and different contexts are played out. Suri and Marsh (2000) describe a structured approach to building scenarios and offer a definition of scenario building as, "the development of a series of alternative fictional portrayals – stories – involving specific characters, events, products and environments, which allow us to explore product ideas or issues in the context of a realistic future."

Scenarios are likely to be more reliable when they take into account information that has a basis in observation and empirical studies rather than being wholly dependent on the personal experience of the designer. Thus, several people can contribute their experience and professional perspective to building a scenario (Joe 1997).

8.3.4 IMMERSION: EXPERIENCING A USER PERSPECTIVE

There are occasions when designers choose to put themselves in a scenario to gain an understanding of what it would be like to be a particular type of user. For example, Mueller (2000) describes the design of a car in which members of the design team wore special suits that restricted movement and added weight to simulate the feeling of being an older or less mobile user. Experiencing a user perspective through "immersion" in a task or environment is described by Bruseberg and McDonagh (2003b) as a method by which designers may seek to expand their "empathic horizon."

8.3.5 EXPERT SOURCES

Apart from the users themselves, other people may be regarded as expert sources of information about users or the things users might wish to do. Ulrich and Eppinger (2000) recommend interviewing "lead users," people who have made innovative use of products or technology ahead of other users. They may, therefore, be able to help identify emerging user needs and market trends (Meadows 2002). Jordan (2000) described an expert, in this context, as someone who, by virtue of their education, professional training, and experience, is able to make an informed judgment about likely user response.

8.3.6 USER TESTING

User testing methods generally employ a mock-up or prototype of a product to be evaluated in some kind of simulated use, preferably by people outside the design team. Testing methods have been widely discussed in the ergonomics and human factors literature in connection with product design and human–computer interface design. While test methods can be rigorous, involving realistic prototypes tested by a representative sample of users and following a well-considered test protocol, there is also evidence that "quick and dirty" test methods with low fidelity prototypes and a small number of test subjects can also provide useful guidance in the product development cycle and this has the benefit of enabling testing to be started earlier and/or repeated more often in an iterative design process (Thomas 1996; Virzi 1989, 1992). The purpose of user testing in the early stages of a design process can be to evaluate usability, provide an early warning of potential problems, generate new insights, or explore user preferences. The method can make use of the mock-ups and prototypes throughout the design process and, as outlined above, can be used in conjunction with other methods like focus groups and adapted ethnographic studies.

8.4 REVIEW OF DESIGN PRACTICE

The brief discussion of selected research methods above has been drawn mainly from contributions by others to the human factors and design management literature. Many of these contributions have been made by researchers and specialists in human factors disciplines and some refer to projects undertaken in organizations large enough to support multi-disciplinary design teams. The purpose of the study described in this chapter was to find out if and how designers themselves, and especially those in small design teams, integrate user research in typical product design projects. This study was done in the expectation that it would provide insights into how well different methods are understood by designers, how rigorously they are applied, and the perceived usefulness of different methods in design projects.

The study, undertaken in Australia by the author (Ward 2006), involved group interviews with nine design teams, followed by a questionnaire distributed to individual designers working in design consultancy companies that were identified from business directories on the internet. Group interviews used open questions and participants in each group (all members of the same team) were encouraged to support their responses with a discussion of their experience in recent design projects. Topics and questions used to guide the group discussions were

- Describing users—What words or descriptions are used within the team to identify user groups and characteristics?
- Published information—What "facts and figures" about users are employed? (e.g., anthropometric data, market research statistics)

- Expert input—Have you consulted human factors specialists or others with expert knowledge of relevant user groups?
- Further investigation—How do designers in the team use (for example), observation, interviews, and focus groups to gain a better understanding of the user?
- User testing—How have you involved people in testing of design ideas and mock-ups?
- Product safety—How do you assess the potential for unsafe user interaction with things you are designing?
- Product–user scenarios—How do you use imagined scenarios as a way of understanding likely user interactions with things you are designing?

The number of participants in each group interview ranged from three to eight. In total, there were 45 participants in the discussions, nearly all held a qualification in industrial design, and over half had more than 5 years experience in product design. Discussions were recorded and transcribed for analysis and subsequent development of an online individual questionnaire for further exploration of issues that emerged in the group discussions. Respondents to the questionnaire included some participants of the group interviews as well as designers in similar product design groups throughout Australia. Nearly all of the 36 respondents to the questionnaire, representing at least 15 different design companies, identified themselves as either industrial designers or design managers and more than half had over 10 years experience in product design.

8.4.1   What Was Learnt from the Study

In this section, the results of the group interviews and questionnaire are compared and discussed. Quotations are from the interview transcripts. The figures with graphical information are based on questionnaire results.

8.4.2   Describing Users

In the group interviews, designers sometimes identified user groups of the products they designed with general cultural stereotype labels like "soccer mums" and "Joe six-pack," but narrowed down their descriptions to specific occupations or different levels of expertise where the distinction was significant to the design project.

> I would often describe them in terms of how sophisticated they are, or how educated ... or try to put some descriptor on where they might be coming from in relation to that particular product...

Demographic data were sometimes part of the description, perhaps reflecting input from market research, and designers often spoke of "consumers" or "customers" in preference to "users."

The questionnaire asked respondents to rate the usefulness of different types of information, if it had been available, about expected users in recent design projects. When these information types are ranked in order of usefulness (Figure 8.1), it is apparent that, with the exception of anthropometric data, the qualitative information types were generally considered more useful than the quantitative or demographic descriptors. The reason for the high ranking of anthropometric data may be that, if available, this information can be directly related to decisions to be made about the dimensions and details of the design. However, responses in the group discussions suggested that the specific anthropometric data sought by designers were often not found.

Customer feedback reports identify likes and dislikes with existing designs and help to show what qualities or features should be either retained or changed in a new design. Stories and anecdotes are a way of describing user experience and drawing attention to issues that should be considered. They also provide a basis for creating scenarios of future user interaction. On the other hand, it is less likely that information, such as customer location, income, and ethnicity of users (the lowest

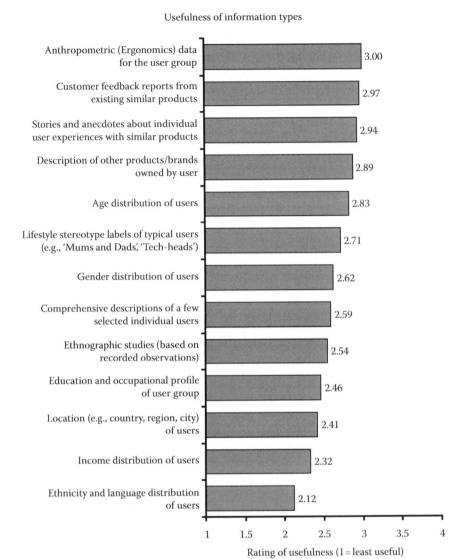

FIGURE 8.1   Information types ranked in descending order of usefulness.

ranking information types), can be confidently translated into product requirements, although this type of information is relatively easy to derive from population surveys and past sales statistics.

> ...we get that information from our marketing team. We know how far away the typical customer is from where they buy their products. Pretty intimate detail.. but sometimes you choose not to use that information.

8.4.3 GETTING INFORMATION ABOUT USERS

From a list, the questionnaire respondents were asked to select all the methods that they had used to get information about users during the previous year. The list included the methods that had emerged from the group discussions. The results are shown in Figure 8.2.

The two most commonly used methods of getting information about users were, "talking to someone familiar with the user group" (used by 31 of 35 respondents) and "observation of users with

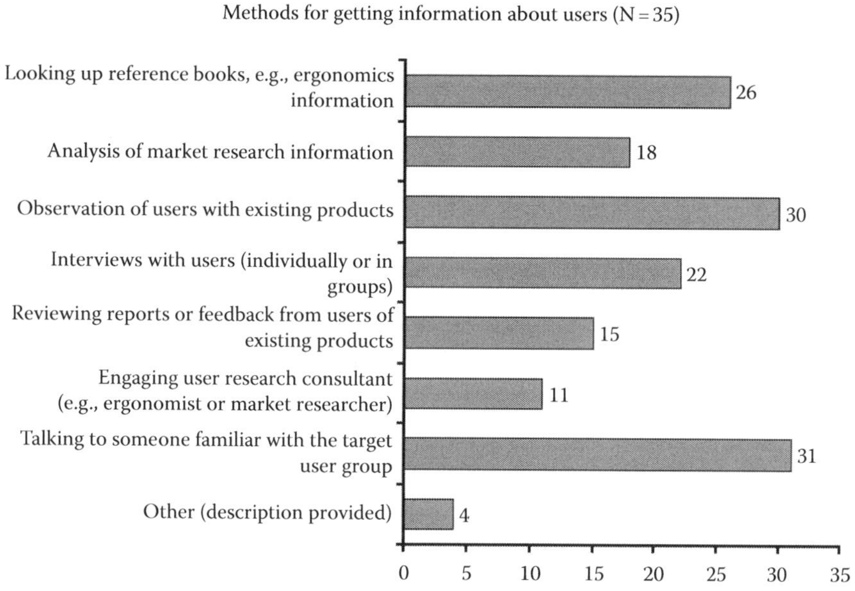

Methods for getting information about users (N = 35)

FIGURE 8.2 Methods used by respondents during the previous year to get information about users.

existing products" (30 of 35). Each of these methods had been used by more than 85% of respondents in the last year. The least commonly used of the listed methods was "engaging a consultant with user research expertise," although the discussions indicated that advice from consultants sometimes came via their clients.

8.4.3.1 Using Published Data

Looking up reference books (e.g., ergonomics information) was ranked the third most common method of getting information about users by respondents to the questionnaire. About 75% of respondents claimed to have looked up "reference books, e.g., ergonomics information" in the past year. It is possible that some respondents claiming to have used reference books may have been referring to sources other than ergonomics references. However, it seems reasonable to take the response to this question as being predominantly about ergonomics sources given that the question prompted for ergonomics information and that anthropometric or ergonomics information was the most commonly mentioned type of reference material in the group interviews.

Questionnaire results showed anthropometric data to be perceived as useful and commonly used by designers. However, this seems to conflict with some of the opinions expressed in the group interviews. Comments recorded in the group interviews revealed concerns over the limitations of anthropometric or ergonomics data, with some participants expressing the view that the data were often not as useful as simple testing of mock-ups, and that the information sought for specific design projects could not be found, at least in the most commonly mentioned reference publications, such as Tilley and Henry Dreyfuss Associates (2002) and Pheasant and Haslegrave (2006).

A way of reconciling this apparent discrepancy might be to conclude that designers were generally familiar with anthropometric information and found it useful in situations where it was applicable. Much of the frustration expressed in the interviews related to situations where designers had looked for anthropometric data but found there was a mismatch between the data available and the design problem to be resolved.

There's very little information available. We've done a few [head mounted products] and we've found the information very thin. So we've had to develop our own ... to create that kind of very specific detail.

The idea that the attributes of existing products could be regarded as a form of published information was put forward in some of the group discussions. The shapes and dimensions of many types of products are the outcome of a history of design refinement and market acceptance. A number of group interview participants mentioned using existing designs of apparently successful products as an alternative or supplement to anthropometric data in determining appropriate sizes. The use of existing products as a reference is not limited to matters of shape and size. Products also informed designers about styles, operating conventions, and features that are apparently acceptable to users.

> ...we're not introducing something that's never been in the market before, and those small refinements have been made along the course of it, so yes, they work pretty well. Any handle details or things that weren't functioning, we would have flagged that by now...

There are some obvious problems with using existing products as a guide to new designs. It is the nature of product design to continually innovate and differentiate from what already exists. In cases where the product being designed is a new category of product or will be a significant departure from past practice, then existing products may not provide the guidance required.

8.4.3.2   Customer Records and Feedback

The group interview participants from in-house design teams described systems within their companies that provided designers with feedback collected from users of the company's existing products. Comments indicated that feedback came from sources such as warranty registration cards and records of repairs. In some cases, consultant designers had access to similar information via their client. Overall, only 15 of the 35 questionnaire respondents (43%) indicated that they had obtained information by "reviewing reports or feedback from users of existing products" in the previous year. However, "customer feedback reports from existing or similar products" was considered the second most useful type of information, according to responses shown in Figure 8.1.

8.4.3.3   Expert Sources of Information

The group interview participants were asked if they had consulted someone with particular expertise to get more information about users. It was intended that this question would identify professionals with specific research capabilities, such as ergonomists, market researchers, or social researchers. The comments indicated that it was relatively uncommon for designers to engage consultants, although sometimes the designers' clients had engaged a consultant independently of the designer. This is consistent with the questionnaire results, shown in Figure 8.2, that engaging a consultant was the least used of the listed methods for getting information about users. However, about half of the respondents to the questionnaire indicated that they had provided 2D or 3D models for "expert evaluation by a consultant or specialist" (Figure 8.3).

An interesting and common practice to emerge from the group interviews was the way that designers consulted people who were not professional researchers, but who were considered competent to provide information about users of particular product types because of their experience with the intended user group. Retail sales people, for example, were mentioned in some of the discussion groups. This method was perceived to be quick and cost effective compared with methods of seeking information directly from a sample of individual users.

> ...one of the first things we do is, we go out and have a look and interact with the salespeople and talk to people...

For consultant designers, the client was often regarded as an expert source of information about users. Clients also provided other kinds of information and facilitated direct contact with users.

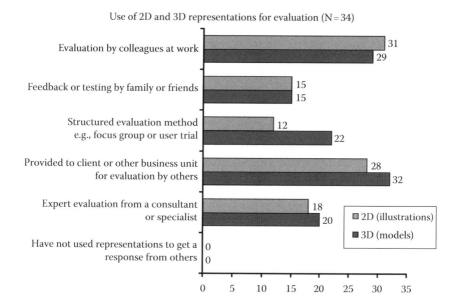

Use of 2D and 3D representations for evaluation (N = 34)

FIGURE 8.3 Use of 2D and 3D representations of designs to evaluate user response.

> ...we also use the clients themselves for feedback ... and the clients' understanding of their customers as people.
>
> ...clients will tell us what people do to their products and we just go, 'you're kidding!', I never thought that would happen...

For in-house designers, other business units were expert sources of information about users. Marketing and customer service divisions were sources of expert knowledge that were mentioned in the group interviews.

> ...we have [product] demonstrators ... all over the country, and through those demonstrators we can get a lot of knowledge as well...

To some degree, designers may regard themselves as experts on a user group in relation to a particular product category. A number of comments from the group interviews suggested this, particularly from the in-house designers who worked mainly in the same product category, had developed a familiarity with "the customers," and had become increasingly confident to make decisions with less external input.

8.4.3.4 Experiencing Use

Designers learn about the user experience by doing what the users have to do. For example, with a project to design a tool used in bone surgery, designers in one of the group interviews had spent some time using an existing tool with meat bones. Other similar examples were provided, especially where the product being designed needs some level of skill on the part of the user.

> There is another way of learning - when we were doing the espresso machines we went and had lessons, so we got a professional to teach almost everyone in the company how to make coffee.

The relative importance and frequency of use of this method, which was detected in the group interviews, suggests further study might be useful in developing guidelines for this type of research in design.

8.4.4   GETTING INFORMATION DIRECTLY FROM USERS

The group interviews and study of the literature revealed a great variety of ways of getting information directly from users, or people who could be considered representative of a user group. Some approaches involved a mixture of methods, such as user trials built into a focus group setting, or combining observation with interview methods. The themes that emerged from the interviews, and used to develop the questions in the questionnaire, served as a framework for this discussion, but many of the practices found were not easily separated into categories.

8.4.4.1   Interviews and Focus Groups

About two-thirds of the questionnaire respondents claimed to have used "interviews with users (individually or in groups)" during the previous year. This made interviews the fourth most commonly used of the eight methods shown in Figure 8.2. Focus groups were the type of interview most commonly mentioned in the group interview responses to questions in this part of the study. In some groups, the view was expressed that facilitating a focus group was best done by a specialist, not a designer. However, there seemed to be general agreement on the benefit of designers observing the focus group (either by being there or via video) if any products or prototypes were shown in the focus group meeting.

> ...as soon as we saw the video, the way they picked it up, it was great for us because we saw the distance between the fingers and whether they picked it up left or right-handed...

8.4.4.2   Observation

The group interview discussions produced a number of comments and examples about the mostly informal approaches used for observation. Examples were given of observation without interaction with those being observed, but in other examples, observation was combined with talking to people.

"Observation of users with existing products" had the second highest frequency of use out of the methods shown in Figure 8.2, with about 90% claiming to have used this method in the previous year. A number of comments in the group interviews suggested that seeing a user doing something was often the moment of insight for a designer. In these situations, observation provided a situation in which the designer could recognize actions that would be relevant to the design, rather than a means of collecting data for analysis. Overall, observation would seem to be both a frequently used and effective method of getting information.

> I don't think there would be many projects where we wouldn't actually go out and observe people...
>   I've watched people using parking meters ... So – anything. I watch people doing stuff, so it's not just related to the job at hand...
>   You go and observe and see something quite different happening. That's when the opportunity comes out of it.

8.4.4.3   Using Design Models in User Research

Asked about how they had used either 2D or 3D representations of designs for feedback and evaluation during the last year, questionnaire respondents provided the information shown in Figure 8.3. For the purpose of this question, 2D representations were defined as "any form of illustration whether hand drawn or computer generated" and 3D representations were "all types of physical models from simple representations of size and shape to realistic prototypes." Respondents could select any number of the options offered for each of the 2D and 3D representations. The relative use of 2D and 3D representations was similar in most situations except for "structured evaluation methods, e.g., focus group or user trial," where 3D representations had been used by nearly twice as many respondents. The most common situations where either type of representation had been used were "evaluation by colleagues at work" and provision of the representation to a "client or other business unit for evaluation by others."

8.4.4.4 Testing with Users

The wide spectrum of approaches to testing was revealed in comments from the group interviews. Examples were provided of quick approaches, such as getting feedback from people near at hand, e.g. work colleagues and family members. Comments from the group interviews confirmed, as would be expected, that informal "quick and dirty" testing was done much more frequently than testing with formal procedures.

> When you've got a prototype ... we may give it to a number of people ... to have a look at and use, ...our families, including young children, an older grandma and grandpa...
>
> ...quite often I'll give someone a prototype and show them how to use it, or not show them how to use it, depending on what I'm trying to ascertain from it, and sometimes I'll take photographs and things like that.

The responses to the questionnaire, shown in Figure 8.3, confirm that evaluation of 2D and 3D representations of designs by colleagues at work was more common than evaluation through a structured method, such as a focus group or user trial. However, the responses indicate that 2D and 3D models were subjected to a "structured evaluation" more often than they were used for feedback and testing by family or friends. It is not clear how the respondents interpreted the expression "structured evaluation," but as this question also had a category of "provided to client or other business unit for evaluation by others," it seems likely that structured evaluations would have generally involved the designer.

8.4.4.5 Scenarios

The questions put to designers in both the group interviews and the questionnaire, on the topic of how knowledge of people is incorporated into the design process or in design thinking, were based on how scenarios are used. Scenarios were defined in the group interviews as imagined situations of a person using a product in a particular context or environment. In the group interviews, the discussion was stimulated by asking participants who they imagined when thinking about the use of a product. This approach was also used in the questionnaire by asking, "who do you most often picture in your mind as the user?"

In the group interviews, participants spoke most often about imagining themselves or someone they knew using the product. They also talked about discussing and acting out scenarios with the design team, with the aid of a mock-up or simple props.

> I definitely use my mother [in imagined scenarios] because I can picture her frustration at putting things that I think are very simple together...
>
> ...we've been doing it for years, and role-playing and setting things up on phone books and whatever you've got laying around to go through the whole activity, whatever the product is...

Also mentioned, but less frequently, were examples of "constructing" fictional users with characteristics based on evidence and consensus from a group.

> ...we had a theme board were we had different characters ... this was generated for our own purpose, in discussion with the client, a board that had images of each of those characters.

A nearly opposite picture emerged in the questionnaire responses (Figure 8.4) with only 3 of the 35 respondents stating that they most often imagined themselves using the product. None said they most often imagined someone they knew, although several group interview participants had mentioned imagining relatives in product usage scenarios. About half of the questionnaire respondents claimed that they most often tried to bring a range of people into their imagined scenarios, while about one-third claimed they most often imagined a typical user.

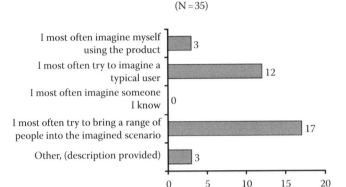

FIGURE 8.4   Who is the user in imagined scenarios?

8.5   DISCUSSION

8.5.1   THE ROLE OF THE CLIENT

One of the issues that emerged consistently from the interviews with the design consultancy groups was the relationship between the design team and their client. This was mentioned as part of the response to many of the questions and clearly impacted on the way most forms of user research were done. This is a significant finding from this study and one that has practical implications for anyone involved in a commercial design practice.

The relationship between designer and client was sometimes characterized in the group interviews as a partnership. At other times it was described more as a purchaser–supplier relationship, where the client purchased specific services and skills. In all cases, the client exercised control, through the project budget, over what services and activities were provided by the designer. Project management entails an agreed schedule of activities and outcomes.

Both parties, designers and their clients, will typically contribute knowledge and research results to the project. The balance of responsibilities will vary according to the nature of the project and the client. In some cases, clients provided information about users and used their contacts and resources to conduct user trials and other evaluations of design concepts. In other cases, designers had to compensate for a lack of client capability in this area and undertake their own research or, sometimes, alert the client to the risks of proceeding without the investigation considered necessary by the designer.

The general scope of design activity in Australian consultancies and the corresponding capability acquired by designers may have been shaped over time by the requirements of clients. However, the client's expectation of what a designer does could also have been shaped by the services and benefits that designers have traditionally marketed to their clients. Comments from this study indicated designers wanting to be more involved in user research, but finding the client unwilling to fund the activity. The finding by Bohemia (2004), that manufacturers had given relatively low importance to "conduct user trials" as a function expected of consultant designers is consistent with this view.

The overall picture that emerges from this study shows designers using many strategies and methods to find out what they need to know about users of the things they are designing. The results are broadly in line with the findings of a study by Hasdogan (1996), who interviewed designers and design teams in the UK and identified a number of ways in which designers developed their understanding of the user. Hasdogan makes reference to a "user model," which she described as, "any representation of the potential user, created by or available to the designer, to assist him in making predictions about the actual user." This is compared by Hasdogan with the more familiar idea of "design models," which are representations of the object being designed, including sketches, scale drawings, mock-ups, and prototypes. In adopting models of both the user and the product, the designer makes assumptions or plays out imagined scenarios about the product–user interaction.

8.6 CONCLUDING COMMENTS

Much can be learned from a review of design practice. The Australian groups interviewed revealed the need for designers to be opportunistic, to get information quickly and cheaply. Designers' focus was on getting information that would allow the design to progress rather than on refining and documenting research methodology. Different projects have different needs for information and offer different opportunities for getting it. Clients usually provided some information and avenues for collecting more information, but they also left gaps that the designers had to fill with their own user research (or get by without it). One of the strengths that designers contribute to user research is the ability to produce sketches and models, which are integral to the development of a design, and which are integral to many user research methods.

The questionnaire produced a roughly even split between agreement and disagreement on a question about whether product evaluation involving users should be done by someone other than the designer. Comments from the interviews suggested that some participants did not feel it was commercially justifiable for designers to dedicate time to research that could be done by others, although some indicated they would like to do more research if this could be supported by the client. At least some interview participants stated, from their experience, that there were distinct benefits in being involved face-to-face in research activities, even if the research was organized and conducted by another specialist. The main benefit of being directly involved was that the information received directly from a discussion with or observation of users was richer and more illuminating than it could have been if it was another person's interpretation in a report of what had taken place. Depending on the nature of the research, the designers' involvement could provide the benefit of interaction as the designer can seek clarification and engage in a discussion about hypothetical alternatives. These benefits were expressed in the group interviews and are found similarly expressed in studies by others, e.g., Bruseberg and McDonagh (2003).

The questionnaire responses in this study revealed general agreement from the designers surveyed that:

- User research is a necessary part of product design
- Designers are expected to carry out user research
- In the future, designers will need a stronger capability in user research

However, there seems to be some development needed if more or better user research is to be done by designers. While responses to the questionnaire indicated that designers felt they were expected to carry out research, interview comments suggested that some clients currently tended to separate user research from design. Arguably, clients will be more likely to fund user research within the designers' scope of work if designers and design teams can demonstrate the unique benefits of the kinds of research they could do from within the design process. The published *IDEO Method Cards* (IDEO 2003), provide examples of methods that have been adapted to the design process.

It has been common for product design consultancies in Australia to provide "back-end" services of production engineering and detailing as part of their service to the client. By contrast, it appears from the literature that "front-end" capability in exploratory user research is a potential growth area in design practice. The relevant literature shows that benefits can be derived from user research methods being more responsive to the needs of designers, both in terms of the types of information required and the time at which information is obtained in order to mesh with the development of ideas in the design process. Potential commercial benefits for designers could include being able to expand the services offered to their clients and arguably produce superior design outcomes.

However, the enhancement of user research capacity in design teams will depend on a shift in the way industrial designers are perceived by clients, as well as by designers themselves. Thus, the change would have to be driven on several fronts and could involve

- Designers developing their research skills, with a focus on methods that make use of design representations (drawings, mock-ups, etc).
- Ensuring that students are exposed to the theory and practice of user research methods in their studio projects.
- Continuing development of multi-disciplinary design teams—integration of human factors expertise with the design process.

REFERENCES

Abrams, B. 2000. *The Observational Research Handbook: Understanding How Consumers Live with your Product.* Chicago: NTC Business Books.

Bohemia, E. 2004. The difference between in-house and contracted industrial designers. Paper read at Futureground, Design Research Society International Conference, Melbourne, Australia.

———. 2003b. Organising and conducting a focus group: The logistics. In *Focus Groups: Supporting Effective Product Development*, eds. J. Langford and D. McDonagh, 63–72. London: Taylor & Francis.

Bruseberg, A., and McDonagh, D. 2003a. Focus groups in new product development: Designers perspectives. In *Focus Groups: Supporting Effective Product Development*, eds. J. Langford and D. McDonagh, 21–45. London: Taylor & Francis.

Caplan, S. 1990. Using focus group methodology for ergonomic design. *Ergonomics* 33 (5): 527–33.

Suri, J.F., and Marsh, M. 2000. Scenario building as an ergonomics method in consumer product design. *Applied Ergonomics* 31:151–57.

Hasdogan, G. 1996. The role of user models in product design for assessment of user needs. *Design Studies* 17 (1): 19–33.

———. 1997. Scenario-building in the product design process. Paper read at The Challenge of Complexity: 3rd International Conference on Design Management, at University of Art and Design, Helsinki.

IDEO. 2003. *IDEO Method Cards*. Palo Alto, CA: IDEO.

Ives, W. 2003. Focus groups in market research. In *Focus Groups: Supporting Effective Product Development*, eds. J. Langford and D. McDonagh, 51–62. London: Taylor & Francis.

Joe, P. 1997. Stories for success. *Innovation* 16 (3): 20–23.

Jordan, P. 2000. *Designing Pleasurable Products*. London: Taylor & Francis.

Meadows, L. 2002. Lead user research and trend mapping. In *The PDMA ToolBook for New Product Development*, eds. P. Belliveau, A. Griffin, and S. Somermeyer, 243–66. New York: John Wiley.

Mueller, J.L. 2000. Growing up without growing old. *Innovation* 19 (4): 46–51.

Pheasant, S., and Haslegrave, C.M. 2006. *Bodyspace: Anthropometry, Ergonomics, and the Design of Work*, 3rd ed. Boca Raton, FL: Taylor & Francis.

Redstrom, J. 2006. Towards user design? On the shift from object to user as the subject of design. *Design Studies* 27 (2): 123–39.

Sanders, E. 2002. How "applied ethnography" can improve your NPD research process. *Visions Magazine* April 2002. Product Development and Management Association. http://www.pdma.org/visions/print.php?doc=apr02/applied.html (accessed February 11, 2005).

Taylor, K., Bontoft, M., and Flyte, M.G. 2002. Using video ethnography to inform and inspire user-centred design. In *Pleasure with Products: Beyond Usability*, eds. W.S. Green and P. Jordan, 175–87. London: Taylor & Francis.

Thomas, B. 1996. 'Quick and dirty' usability tests. In *Usability Evaluation in Industry*, eds. P.W. Jordan, B. Thomas, B.A. Weerdmeester, and I.L. McLelland, 107–44. London: Taylor & Francis.

Tilley, A.R., and Henry Dreyfuss Associates. 2002. *The Measure of Man and Woman – Human Factors in Design*. New York: John Wiley.

Ulrich, K.T., and Eppinger, S.D. 2000. *Product Design and Development*, 2nd ed. New York: McGraw-Hill.

Virzi, R.A. 1989. What can you learn from a low fidelity prototype? In Proceedings of Human Factors 33rd Annual Meeting 224–28.

———. 1992. Refining the test phase of usability evaluation: How many subjects is enough. *Human Factors* 34 (4): 457–68.

Ward, S.J. 2006. Designers and users: A survey of user research methods employed by Australian industrial designers. Unpublished MSc thesis, University of New South Wales.

9 Near Field Communication: Usability of a Technology

Bruce Thomas and Ramon van de Ven

CONTENTS

9.1 INTRODUCTION

9.1.1 NEAR FIELD COMMUNICATION AS CONSUMER TECHNOLOGY

At a major trade fair such as the Consumer Electronics Show (CES) in Las Vegas, the visitor can expect to see a wide variety of cool gadgets designed to excite the consumer electronics enthusiast. But at such a fair, it is not only the consumer products themselves that are on display, but also some of the technologies that make them work. At CES 2006, Philips Semiconductors (now NXP) presented near field communication (NFC) (Philips 2005), a technology that enables short-range communication between electronic devices, such as mobile phones, televisions, PDAs, computers, and payments terminals via a fast and easy wireless connection. This chapter describes some of the human factors involved in bringing NFC to the CES, and also discusses some of the implications of spending human factors efforts on developing technology as much as on developing products.

NFC is a wireless technology that has evolved from a combination of existing contactless identification and interconnection technologies (NFC Forum 2006a). NFC is both a "read" and "write" technology, so that communicating devices can both send and receive information. The communication between two NFC compatible devices takes place only when they are brought very close to each other—less than 4 cm. This has two advantages:

- To establish a connection the users need only to bring the devices close together
- There is no need for extensive dialogues to determine which devices should communicate, since this is established through the context of being close together

Once a connection is established, the devices need no longer remain together, that is the touch need only be long enough for a "handshake" between the devices. Once the handshake has taken place, further interaction between devices, if required, can continue using other wireless technologies with greater bandwidth, such as Bluetooth or WiFi.

The potential advantage for the consumer is that a mobile device can be used to store and access all kinds of personal data both at home or on the move (NXP 2006a). When two NFC-enabled devices are placed close together, they automatically initiate network communications without requiring the user to configure the setup. In this way, NFC-enhanced consumer devices can easily exchange and store personal data, such as messages, pictures, and MP3 files. So, for example, digital cameras can send their photos to a television set with just a touch. When it is used in combination with contactless payment technology, NFC can also enable the user to make secure and convenient purchases with the mobile device.

9.1.2   NEAR FIELD COMMUNICATION FORUM

For a technology such as NFC to be successful, a variety of organizations must cooperate. The communication between devices must take place irrespective of the manufacturer, and links to service providers must be reliably and securely established. In order to ensure this, in 2004 the NFC Forum was formed "to advance the use of Near Field Communication technology by developing specifications, ensuring interoperability among devices and services, and educating the market about NFC technology" (NFC Forum 2006b). The forum has over 100 members, including manufacturers, applications developers, financial services institutions, and others who work together "to promote the use of NFC technology in consumer electronics, mobile devices, and PCs."

The goals of the NFC Forum (NFC 2006b) are to

- Develop standards-based Near Field Communication specifications that define a modular architecture and interoperability parameters for NFC devices and protocols
- Encourage the development of products using NFC Forum specifications
- Work to ensure that products claiming NFC capabilities comply with NFC Forum specifications
- Educate consumers and enterprises globally about NFC

The NFC Forum and the organizations within the forum are working to establish NFC as a unique and intuitive interaction paradigm. The members of the forum have the ambition to make NFC the interaction paradigm of choice for many applications, such as payment, ticketing, etc. However, the success and broad acceptance of this "touch and connect" paradigm depends on people's response to it. It is potentially much more than "another way of interacting with technology."

9.2 BACKGROUND FOR THE CASE STUDY

The case study described here was part of the effort to evaluate people's response to the "touch and connect" paradigm and to ensure acceptance of it. The study primarily concerns a project carried out between Philips Semiconductors (NXP) and Visa, both members of the NFC Forum (NXP chairs the forum). Philips and Visa conducted the study in December 2005 in North America. The study was part of the companies' efforts to develop programs that bring convenience, ease of use, and security to mobile transactions. The purpose of the study was to take an in-depth look at usability and learn about consumer behavior when interacting with the technology.

The study concerned the development of scenarios of use for NFC and the testing of those scenarios. The study was intended to test not only the usability of the NFC technology, but also to evaluate the acceptance of the technology on the basis of actual use and to provide recommendations to minimize dissatisfaction and maximize user acceptance. This work was carried out as part of the build up to presenting NFC at the CES in 2006.

Although the bulk of the work reported here concerns the development and testing of scenarios in preparation for the CES, it represents part of an ongoing process to ensure the usability and acceptance of NFC. A technological development such as NFC has no clear start and end; there is a continual development as each participating organization contributes to the maturation process. Consumer products launched with the technology will incorporate the state of the art, but the technology will continue to develop beyond the release of individual products or systems.

The study illustrates that human factors effort is not solely restricted to specific product development, and highlights where human factors specialists can still have an impact in a business where very short timescales and tight budgets limit the scope for human factors activity.

9.3 OBJECTIVES

Philips Design became involved with the development of NFC at a time when the number of strategic partners in the NFC Forum was growing rapidly and approaching the desired critical mass. The various member of the NFC Forum were rapidly entering a phase of NFC pilots and early implementations. Nokia began to ship NFC-enabled phones in 2005.

The task for Philips Design was to look at the usability and acceptance issues regarding NFC. The objective was to get a better insight into consumer acceptance of the technology and to test the attractiveness of potential applications. Visa, another partner in the NFC Forum, was willing to join forces with Philips in this.

At the time that Philips Design became involved, Philips Semiconductors was ramping up the introduction of NFC and wanted to maximize its potential uptake. This assignment had therefore to be seen in light of certain business drivers:

- The new interaction paradigm was a unique opportunity, but it also increased the responsibility of Philips Semiconductors to perform and lead in the development of the interaction paradigm. Therefore, it was important to ensure that this paradigm was fully compatible with the users' experience, desires, and expectations.
- Because it was a new interaction paradigm, the number of potential applications of NFC was enormous, including critical interactions such as payment. The broad range of potential applications made it necessary to provide users with a consistent experience. Business success would rely on users perceiving NFC as a consistently intuitive solution.
- Philips Semiconductors was leading the technological development and the standardization of NFC. They were also leading the application rollout by means of the NFC Forum.

This meant that Philips Semiconductors and the forum needed to acquire knowledge of end-user acceptance and usability to stimulate the successful implementation by others.

- NFC is technically related to radio frequency identification (RFID), which suffers some resistance owing to privacy issues, mainly in the United States. NFC was not positioned as an RFID variation, but could suffer some resistance as well. This situation created a need for Philips Semiconductors to efficiently test and acquire knowledge with end-user input to know better the points of resistance and how to maximize acceptance.

On the basis of these needs, it was decided to create a number of scenarios of use and test them with potential users. Since interaction and usability were key factors to be investigated, the scenarios would have to be developed to an extent where hands-on experience could be evaluated. It was then planned to carry out the evaluation in a usability test, where the main NFC applications and variations could be tested concurrently, including some variations and user interface alternatives. Further, in order to address the issues relating to the acceptance of the technology, it was decided to include interviews in the test to explore the test participants' experience of the technology, its perceived usability, and their attitudes to it. The test was to be conducted on an individual basis to ensure that the participants had a full experience of what NFC could do, as well as having the scope to express their opinions fully.

The goals agreed for the test were

- To test the usability and pleasure of NFC
- To identify specific acceptance issues during the study
- To test the consistency of the NFC interaction paradigm
- To identify opportunities to maximize the implementation potential from an end user perspective

In addition to the specific objectives of the test, both Philips Semiconductors and Visa wanted to make maximum use of the scenario material and the results of the study to demonstrate the attractiveness of NFC to a wider audience, particularly at the CES and other trade fairs.

9.4 SCENARIO DEVELOPMENT

9.4.1 REQUIREMENTS

A preliminary approach was to base the study on four core applications that had already been identified within the NFC Forum (touch and go, touch and confirm, touch and connect, and touch and explore). There were also a number of potential variations in these concepts covering different application areas and user interface alternatives (e.g., with or without buttons to activate communication). A particular requirement for Visa was that payment scenarios should have the highest priority.

With an extremely versatile technology like NFC, there was no problem identifying potential applications. Indeed, a wealth of ideas already existed, ranging from exchanging contact details by touching two phones together, to buying music via a smart poster and having the files downloaded to a phone through the mobile network. However, it would not be practical to include every possible idea in a test, so the challenge was to develop a limited set of scenarios that would illustrate the widely different areas of application, highlight issues that might be considered problematic, and explore not only the current but also the future potential of the technology.

Another consideration was the availability of existing demonstrators. A variety of these had been created in various locations by different people to test the functionality of the technology and to illustrate the commercial benefits. However, it was not clear how many simulations or demonstrators already existed, what degree of functionality they incorporated nor whether they could be used in the test situation, e.g., owing to potential inconsistencies between them.

9.4.2   Idea Generation

In order to explore ideas and to create realistic scenarios to be tested, workshops were held with representatives from Philips Design, Philips Semiconductors, and Visa.

The first step in the creation process was to create an extensive list of possible NFC applications. This list was started with cases gathered from existing papers and from the outcome of earlier discussions. In the first creative workshop, this list was extended. The workshop generated a list of over 40 applications, including payment, information exchange, ticketing, interactive gaming, programming domestic appliances, etc. For each application, the user interaction was described, as well as the immediate result of the touch contact and whether it involved payment. Some examples of these applications are described in Table 9.1.

This list of applications was used as input for a second workshop with a small team of people from Philips Semiconductors, Philips Design, and Visa. In this workshop, a selection was made of scenario concepts based on four main application areas that were defined in discussions during the workshop:

- Retail payment
- Home shopping
- Sharing
- Smart media

A further session in this workshop investigated the touch metaphor in more depth: What does "touch" mean? What does it communicate? Is it intuitive? etc. Following this, another creative session was held to develop concepts for scenarios that involved the most relevant use cases and to define how the touch should be applied.

The outcome of the second workshop was a range of scenario concepts sorted according to the four main application areas.

TABLE 9.1
Examples of NFC Application Scenarios

Application	Obtain article purchased via internet or interactive television
Source	Creative workshop
Type of contact	Touch = pay
Payment	Yes
Interaction at contact	Read advertisement. Forward request to operator. Confirmation
Setup	Probably none for user
Other interaction	Purchased and to place order
Application	Order (advertised) tickets with phone
Source	Philips Semiconductors
Type of contact	Touch = order
Payment	Yes
Interaction at contact	Forward request to operator. Confirmation/security required
Setup	Probably none for user if Smart Ticketing provided as standard feature on phone
Other interaction	Feedback on phone needs to be provided when ticket received and available for use
Application	Obtain information from non-interactive poster or other medium
Source	Philips Semiconductors
Type of contact	Touch = grab
Payment	No
Interaction at contact	Information transferred from poster to mobile phone or other device
Setup	Probably none for user
Other interaction	Navigate new content after transfer. Content needs to be configured for the device onto which it has been transferred.

9.4.3 Scenarios

Following the workshops, Philips Design further refined the concepts using feedback from various stakeholders within the partner organizations. After a number of iterations, three scenario concepts were selected for further development:

- Coffee corner
- Home shopping
- Ticket purchase

The "coffee corner" scenario included the interactions (Figure 9.1):

- Order coffee using a profile
- Add an additional element to the order
- Pay for the order
- Transfer access rights for the local Wi-Fi to a personal laptop

The issues addressed in this scenario involved the retail experience and included:

- How to set up a customer profile
- The level and mechanism of security required (need for security vs. convenience of use)
- Competing possibilities for transfer between phone and laptop (selection of "sending" and "receiving" device when both have capability)

The "home shopping" scenario included the interactions (Figure 9.2):

- Obtain information from smart packages (download a movie trailer from a DVD package in a store)

FIGURE 9.1   "Coffee corner" scenario sketch.

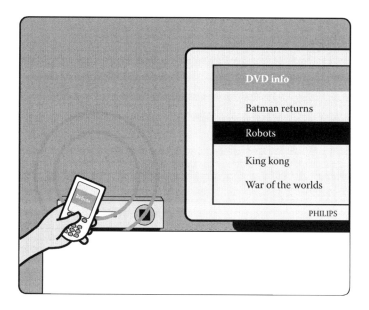

FIGURE 9.2 "Home shopping" scenario sketch.

- Match the mobile phone to a television and play information stored on the phone via the television
- Order movie-related items via the television and pay for the items ordered

The issues addressed in this scenario involved the transfer of information outside and inside the home and included:

- Alternative selection mechanisms between smart poster and phone menu.
- Competing possibilities for information transfer. Selection on "sending" device vs. selection on "receiving" device.
- Position of reader on devices.
- Location and nature of interaction.
- Use of card vs. other NFC devices (phone, remote control, etc.).

The "ticket purchase" scenario included the interactions (Figure 9.3):

- Obtain and pay for tickets from a smart poster
- Give a ticket to a friend (pass from one mobile phone to another)

The issues addressed in this scenario involved purchasing from smart media and sharing information with others. They included

- Presentation of options separately on poster vs. menu in phone
- Level and mechanism of security required (need for security vs. convenience of use)
- Required interaction to set up and transfer specific information between two equivalent devices
- Presentation of feedback

9.4.4 INTERACTION DESIGN

With the content defined, the next step was to work out the concepts into complete realistic scenarios. Every step of the three scenarios was described in detail and visualized in a storyboard. The

FIGURE 9.3 "Ticket purchase" scenario sketch.

main purpose of creating these storyboards was to identify the points of interaction: What possible interaction do we want to test? What actions do we expect from the user and what is the result of these actions?

Then, the scenarios were again discussed by the core team and adjusted if necessary. The outcome of this session was a complete storyboard per scenario. Figure 9.4 shows an example of part of the coffee corner scenario.

Once the main steps of the test were described in the scenarios, the next step was to describe the communication between user and devices in detail. The user interface dialogue was developed for every interaction in the scenarios. This involved specifying in detail what the user would see on the display of his/her own NFC device (which was a mobile phone) or another supporting device, and what user input would be required. Every element of communication to be included in the demo

1.1

1.2

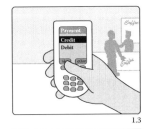
1.3

| Rachel walks into a Starbucks coffee bar. On entering she touches the 'menu card' by the entrance with her phone. Because she has more or less the same routine every time she goes to Starbucks her phone has stored a 'Starbucks routine'. | She confirms that she wants the 'usual' by pressing ok. Her favorite coffee and access to the WiFi is selected. She adds an extra chocolate muffin to her order. | Rachel selects the payment application from her phone and selects her credit account. |

FIGURE 9.4 Extract from "coffee corner" scenario.

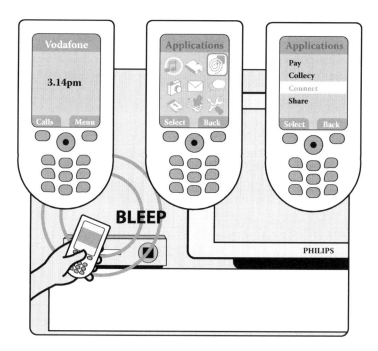

FIGURE 9.5   Example of interaction design.

had to be worked out. An example of interaction and screen design is shown in Figure 9.5. This was also used as communication to Philips Research, who were responsible for building the working demonstrators, so they would know exactly what had to be implemented.

9.4.5   CONSTRUCTION OF DEMONSTRATORS

To build realistic demonstrators, cooperation was needed between Philips Design and Philips Research. At Philips Design the content of the simulations was defined. The front end was designed consisting of physical mock-ups (e.g., kiosk, bar, DVD rack) and simulations in flash that were played on a touch display, a television set, and a laptop PC. The front end in flash was linked to the back-end software developed by Research. The NFC-enabled phones were delivered by Nokia and prepared by Research with input from Design.

The front-end user interface was developed in such a way that the users would be able to interact with a realistic user interface. This was done so the "touch contact" was tested in a realistic context rather than performing each interaction in isolation. The same NFC-enabled mobile phone was used as a central element for all scenarios, which each test participant could identify as his or her "own" phone for the duration of the test.

9.4.6   RE-USE OF DEMONSTRATORS

Some of the demos, especially the "coffee corner" kiosk, were later used as demonstration material at the CES in Las Vegas. The kiosk was also adapted and used on the 3GSM 2006 in Barcelona. Here, business partners of Philips Semiconductors and Siemens were given an NFC-enabled phone at registration, which could be used to exchange business contacts by touching phones with others. At the kiosk, they could order a CD or DVD and pay for this by touching the kiosk with their mobile phones.

The reaction of the visitors to the trade fairs was generally positive. They gave positive feedback about the intuitive touch of the demo, especially that not many clicks were necessary on the phone.

Those people who were already familiar with the demo were evidently not as impressed (which suggests that the interaction becomes familiar very quickly). Those who hadn't seen the demos before were quite impressed with what was shown.

9.5 USABILITY TEST

9.5.1 METHOD

The test was based on the methodology described by Thomas (2001). This methodology is based on the definition of usability given by ISO DIS 9241-11 ergonomics requirements for office work with visual display terminals (VDTs): Part 11: Guidance on usability:

> The extent to which a product can be used by specified users to achieve specified goals with effectiveness, efficiency and satisfaction in a specified context of use.

Effectiveness concerns whether users are able to achieve their goals. Efficiency concerns the resources a user must invest in order to achieve these goals (time, steps, errors, etc.). Satisfaction concerns whether users have a good feeling that their goals have been achieved successfully.

In usability studies carried out by Philips Design, users and goals (tasks) are usually specified in cooperation with the client with reference to the target market and the key tasks a user will need to perform with the device, equipment, or system being studied. The context of the usability testing is normally that of a usability laboratory, where salient aspects of the environment are reproduced in which the device, equipment, or system being tested will normally be used (e.g., lighting conditions, working posture of user, and relevant ancillary equipment).

The specific measures normally used are

- Effectiveness—whether participants were able to complete the selected tasks, evaluated on a 4-point scale:
  - First attempt: Success with minimum deviation from a recognized path without assistance, using their first strategy or approach. For example, they may press the wrong key and then recover before going on to succeed.
  - Later attempt: Success, but only after they have tried different strategies or approaches, without assistance.
  - Help/Direction for Use (DFU): Success, but needed to look at manual or have a "hint" (minor help) from tester.
  - Failure: Participant gives up, or requires significant help from tester. Also, if participant takes an unacceptable time to complete the task (defined by tester).
- Efficiency—time to complete the task or number of steps taken to complete the task. If realistic measures cannot be taken due to the test setup, subjective opinions of test participants regarding task organization are sometimes used.
- Satisfaction—subjective opinions of test participants on whether the user interface matches their needs and expectations.

The advantage of the ISO definition and this test methodology is that they make it possible to quantify usability. Despite this, however, satisfaction remains subjective and therefore somewhat difficult to measure, although some good attempts have been made; e.g., Travis (2003) gives a list of measures including subjective rating scales and the ratio of positive to negative remarks. A problem with measuring satisfaction is defining precisely what is being measured, and how this might differ from another subjective attribute of a product, such as pleasure (Jordan 2000) or perceived usability (Tractinsky, Katz, and Ikar 2000). Indeed Lindgaard and Dudek (2003) discovered that satisfaction is "a complex construct comprising several concepts," and they also suggested that rating scales and interview statements tap different interface qualities.

Jordan (2001) argues that usability is only one layer in a hierarchy of values:

1. Utility
2. Usability
3. Pleasure

Utility concerns the basic functionality of a device. The primary requirement is that it should work. For example, a watch should keep time. Usability can only be achieved if utility is provided. A device can satisfy usability principles, but if it serves no purpose, it will not be used. Going beyond usability, pleasure is achieved when additional emotional or hedonistic criteria are met. In order to examine attitudes, it is necessary to go beyond the simple evaluation of usability and explore some of the wider issues relating to its utility and also to its attractiveness.

Studies at Philips Design focus less on satisfaction per se, but use in-depth interviews to explore the test participants' opinions, preferences, and experience with the device, equipment, or system. The interview is designed to actively probe and prompt users to comment and give feedback in order to gain further insights on how information should be presented and how the interface can be improved. In the NFC study, it was particularly important to address these subjective issues relating to the acceptance or rejection of the technology, so in this study extensive time was given to the interviews in order to discuss these topics.

9.5.2 SET UP

The usability test was conducted in Atlanta in the first weeks of December 2005. Atlanta was identified as a key location in the United States for the introduction of contactless technology because of a strong technology focus in the city as well as it being the site of another NFC trial taking place at the Philips Stadium.

The demonstrators were set up in two rooms. The first room was furnished to resemble a living room, in which the home shopping scenario was set up. The other room was divided into three areas: the coffee corner (Figures 9.6 and 9.7), a public space in which smart posters were placed, and a DVD store. Both test rooms were equipped with cameras and microphones. The test facility also included an observation room with a video feed from the cameras in the test rooms. The test sessions were recorded onto DVD (Figure 9.7).

Each test session took from 1.5 to 2 hours. The sessions were conducted on an individual basis so that every test participant had an equal chance to obtain hands-on experience with the demonstrators, but without being influenced by the experience or the opinions of others. The participants

FIGURE 9.6   The "coffee corner" demonstrator.

FIGURE 9.7   The "coffee corner" kiosk.

generally went through the scenarios in the same order (coffee corner, TV shopping, buy and sell), although some participants were shown the TV shopping scenario first. This order was chosen because it represented a progression from the familiar to the unfamiliar in the type of interaction involved. It is certain that participants learned from one scenario to the next, which could have been controlled through balancing the order of presentation, but it was felt that the progression of presentation better reflected the real-world probability of exposure and would necessitate less explanation from the tester (especially for the third scenario). Interim interviews were held after the presentation of each scenario, and an extensive closing interview was held after all three scenarios.

The NFC applications were tested concurrently, including user interface alternatives. Where alternative scenarios were tested (e.g., providing multiple touch points on a poster when several options are possible vs. providing a single touch point with a subsequent menu in the mobile device), each participant experienced one of the options and was shown the other option during the subsequent interview. By doing this, the experience of each scenario was not disrupted, but the test participants were still able to give their opinions and preferences for each variation. Balanced numbers of participants were shown each variation. Thus, objective usability measures were obtained for each such alternative from half the participants and subjective opinions were obtained from all participants.

9.5.3 Participants

Twenty men and women with ages ranging from 18 to 40 were recruited from the Atlanta population to participate in the study. All participants were regular users of credit or debit cards and all were frequent users of cell phones.

There has been considerable debate in the usability community about the numbers of participants necessary to conduct a usability test. Virzi (1992) and Nielsen (2000) have argued that a usability test can be conducted with as few as four or five participants. The reasons given for this are primarily that it is cheaper and that most of the serious usability problems can be detected. However, it should be pointed out that Nielsen argues for repetitive testing, in that it is better to fix the problems you find and then test again to find further problems and so on; "The best results come from testing no more than 5 users and running as many small tests as you can afford" (Nielsen 2000).

Nielsen's approach is qualitative and focuses on solving problems. Nielsen (2004) is very critical of a quantitative approach to usability testing, arguing that, "number fetishism leads usability studies astray by focusing on statistical analyses that are often false, biased, misleading, or overly narrow. Better to emphasize insights and qualitative research." In this, he echoes the sentiments attributed to Benjamin Disraeli that "there are three kinds of lies: lies, damned lies, and statistics."

However, the purely qualitative approach also has its problems. Sauro (2004) advocates a quantitative approach arguing that, "the risks of relying heavily on a qualitative approach can lead to a severe misdiagnosis especially when usability problems are difficult to detect."

Travis (2004) concludes that Nielsen and Sauro differ only in emphasis. Nielsen's approach is suitable where usability problems are found and fixed and another "throwaway" prototype can be quickly tested; while Sauro's approach is necessary to set usability acceptance targets, compare a product's usability to the competition or a predecessor, to correlate a product's usability with sales or customer returns, or to answer the question: "How usable is this product?"

The NFC study was intended to discover rather than quantify usability problems, which would suggest that the more cost-effective approach advocated by Nielsen would be appropriate. Nevertheless, there was still a requirement to obtain some indication of how usable the technology was. This meant that a larger sample would be necessary. Further, it was also intended to explore attitudes as much as usability. Even if the most important usability problems might be discovered with a small sample, it was by no means certain that this would be sufficient to discover the potential variety of opinions that might be expressed when using a larger sample.

In market research studies, in which attitudes are important, it is common to interview large numbers of people. Since the NFC study was addressing such issues, irrespective of the objective requirements concerning participant numbers, there was also a need to appear credible at face value for audiences more familiar with market research studies, and ultimately for a wider audience in the publication of findings from the study in presentations at trade fairs and in the media.

One possible way of incorporating large participant numbers cost effectively into a study is to interview them in groups. Focus groups have the advantage that they not only enable the opinions of larger numbers to be sampled, but they can also reveal and prioritize issues that are important to users, particularly concerning issues that the investigator had not considered (van Vianen, Thomas, and Nieuwkasteele 1996). However, focus groups also have a number of disadvantages, particularly the problems of group dynamics; participants can influence each other and the opinions of dominant individuals can influence the outcome. More importantly, however, focus groups do not provide sufficient opportunity to directly observe the hands-on experience of using the products being tested. Because the direct experience was important to study the usability of NFC, as well as for participants to understand its capabilities and benefits, a group approach was inappropriate. Further, with individual participants a much deeper questioning was possible during the interviews than would have been possible when addressing the opinions of each member of a group in turn.

Ultimately, it was decided to test with 20 participants because this number provided sufficient credibility and the numbers for reasonable confidence, while being cost effective in obtaining in-depth qualitative input. This became apparent in practice, since clear trends emerged in the responses and by the twentieth participant the testers and observers were agreed that no new insights were forthcoming.

9.5.4 Results

The study showed that the participants liked the convenience, ease of use, and "coolness" of making transactions with mobile phones. They accepted and appreciated the concept of incorporating information transfer and secure payment functionality into mobile phones. Retail purchases with the mobile phone were particularly well received, as the participants found the NFC technology and contactless payment easy to understand, convenient, and fast.

Some of the main findings of the study were

- Mobile transactions are cool. The participants enjoyed downloading content from NFC "smart" posters and responded favorably to the idea of purchasing tickets through posters. They described the technology as "cool" and "awesome" and liked the idea of then being able to use the phone to gain entry into an event.

- Mobile payment is easy to use. The participants found it easy to make contactless payment using the mobile phone. Learning curves were very short, with all participants interacting confidently with the mobile phone and payment terminals. They found it intuitive to initiate a transaction by holding up the phone to a terminal, as an alternative to presenting a payment card.
- Mobile payment is convenient and fast. The test participants appreciated the ease of use, convenience, and the speed of contactless payments on an NFC-enabled phone. They also liked the idea of not always having to carry a wallet or purse.
- "Receiving" transactions should be automatic. When using the phone to make a purchase or download information, the transaction should be automatic; participants liked the simplicity of transactions that were initiated just by holding the mobile phone to an NFC-enabled reader. However, for "sending" applications, such as giving a ticket to a friend, users may prefer to initiate the transaction with a command.
- There is a need for a clear, consistent mark. The test participants generally looked to find a mark to indicate exactly where an NFC transaction could take place. The most intuitive place for a mark was directly over the communication point. Users did not want to guess where and how to orient their mobile phones to complete a transaction.

The study obtained a positive reaction from Philips Semiconductors and Visa with parallel press releases on both their web sites (NXP 2006b; Visa 2006a). These included the following statements from each organization:

- "The usability study clearly demonstrates that consumers like the simplicity of using NFC to access and securely pay for entertainment, information and services while on the move," said Christophe Duverne, vice president and general manager, Identification, Philips Semiconductors and chairman, NFC Forum. "Now it's up to us – the industry – to cooperate effectively and deliver on the promise of the technology by driving standardization and building the ecosystems that will ensure commercial success. And of course, we must keep the end-user experience first at all times."
- "Visa was very pleased with the results of the study, which demonstrated not only that the participants reacted positively to contactless payments with mobile devices, but also that there is real excitement among consumers," said Gaylon Howe, executive vice president, Consumer Product Platforms, Visa International. "The study provides a strong validation for Visa as we continue to drive acceptance for contactless technology, which will be critical for widespread uptake of mobile payments."

The results of the study provide NXP, Visa and, ultimately, all members of the NFC Forum with some interaction guidelines for the development and implementation of NFC technology. Such guidelines provide the basis for a consistent approach to the implementation of the technology across multiple organizations, which is focused on user expectations and needs. Ultimately, it is hoped that the guidelines from the current study, along with the results of future research, will provide a foundation for the industry-wide standardization of the key user interactions using this technology.

9.6   WIDER CONTEXT

The study reported here was not an isolated exploration of the use of NFC technology; it is part of a continuing effort to understand and improve the user experience. Several major trials have been carried out around the world to understand the benefits that NFC technology can bring to people's everyday lives. Three examples are described below.

9.6.1 PHILIPS ARENA ATLANTA, USA

From December 2005 to September 2006, Visa USA, Philips, Nokia, Cingular, Atlanta Spirit, Chase, and ViVOtech worked together on a major NFC trial at the Philips Arena stadium in Atlanta, Georgia, which allowed 150 Atlanta Thrashers and Atlanta Hawks season ticket holders to buy goods at concession stands. Additionally, they were able to access and download mobile content such as ringtones, wallpapers, screensavers, and clips from favorite players and artists by holding an NFC-enabled phone in front of a poster embedded with an NFC tag.

Steinmeier (2006) and Visa (2006b) report a positive outcome of this study:

- The trial participants overwhelmingly embraced the technology and expressed that the mobile device and applications significantly improved their arena experience.
- Data usage increased during game days; trial participants used their NFC-enabled mobile device more frequently to search for and purchase digital content.
- Participants indicated that they would like to use their mobile devices for payment at a variety of merchant locations, for any purchase size, and saw value in accessing multiple payment accounts on their mobile devices in the future.
- Trial participants also expressed interest in having multiple applications on a single mobile device, including transit, loyalty, and digital content applications.

9.6.2 CITY OF CAEN, FRANCE

From October 2005 to May 2006, Philips, in collaboration with France Telecom, Orange, Samsung, retailer Group LaSer, and Vinci Park, conducted a major multi-application NFC trial in Caen in Normandy, France. During the six-month trial, 200 Caen residents used Samsung D500 mobile phones with an embedded Philips NFC chip as a means of secure payment in selected retail stores, parking facilities, and to download information about famous tourist sites, movie trailers, and bus schedules.

Again, Steinmeier (2006) reports a positive outcome:

- There was a strong adoption and very positive feedback from users about the ease of use of services.
- The value of the phone was perceived as higher.
- The reliability and security of the application was well received.
- The credit card company, Cofinoga, saw about a 30% increase in transactions among the trial group of consumers. Some users made purchases as high as €1500 in the 20 stores where they could use NFC.
- Users were not concerned about the security of paying with credit from their handsets. The ability to block the credit card via the network operator was perceived as an additional security feature.

9.6.3 RHEIN-MAIN-VERKEHRSVERBUND, GERMANY

On April 19, 2006, Philips, Nokia, Vodafone, and the Rhein-Main-Verkehrsverbund (RMV)—the regional public transport authority for the Region Frankfurt Rhine-Main in Germany—announced the world's first commercial launch of NFC (NXP 2006c). This followed a 10-month field trial in which Nokia 3220 mobile phones with integrated NFC technology could be used as electronic bus tickets and act as loyalty cards for discounts at local retail outlets and attractions.

The field trial was conducted to evaluate the technical feasibility of NFC for mobile ticketing as well as to evaluate its acceptance in comparison to smart cards. Preuss (2006) reports that 146 end users and 15 conductors participated in the trial. They were interviewed by phone and participated in focus group discussions after the trial had been running for 6 months. As in the other trials, there was a positive outcome:

- The trial showed a high acceptance with high scores on total satisfaction.
- The NFC mobile ticketing was described as "practical," "comfortable," and "easy to use," as well as being "fast and more trouble-free log-in/log-out."
- Participants who had previously used a smart card before the mobile phone stated that they would prefer to use the mobile phone in future.
- However, activation of the phone for ticketing purposes was seen as complicated.
- The phone was still regarded as a phone so there was a need for different phones for various target groups and it is important for users to be able to receive calls while using the phone for ticketing.

Overall, Preuss concludes that the NFC phone was seen as a more convenient media, since "it is always with you anyway."

9.7   DISCUSSION

9.7.1   Human Factors in Technical Development

The role of the human factors specialist is not limited to the development of consumer products. This case study has shown that human factors methods can also be applied in the design and development of other technologies to be used in consumer products.

The focus of this case study was the development and testing of technology demonstrators. Such demonstrators provide the opportunity to create a tangible experience for potential users of the technology and, in this respect, they do not fundamentally differ from prototypes for products. The value of building a demonstrator is to illustrate the capability of the technology, which would otherwise be invisible. It is not necessarily the purpose to create a commercial product, but to show the potential applications for which the technology can be used, and also to show that the technology actually works. Such demonstrators, therefore, do not require fully worked out user interfaces, as is the case for a consumer product, but just enough to expose the user to the utility of the technology. Essentially, however, the role of the human factors specialist is the same, that is to bring a user-oriented focus to the development team and as much as possible to ensure usability and to involve users in the process.

NFC has been conceived and developed over a number of years. This is an ongoing process with no clear start and finish, and is largely concerned with the production of technical standards. It is by no means self-evident that users should be involved in such a process, but the study presented here, even though it concerns only a small part of the overall process, shows that even in technical development the feedback from potential users can be valuable in shaping the vision of what the technology can provide and how it should do this.

9.7.2   Communication

The various NFC trials carried out demonstrate a commitment by the various organizations involved to ensure that the technology functions and that the potential users obtain a clear benefit (utility). All these organizations are looking at users' response to the technology and they are all contributing to its development.

There are many players in the development of NFC, located in the member organizations of the NFC Forum. Like other roles in the process, the human factors role is divided between people in those organizations who don't always know each other or even know of each other. This means that knowledge concerning the technology is distributed and there is a danger that important findings may not be communicated throughout the development community. However, in this case, the NFC

Forum plays a vital role in coordinating the effort between the different organizations. The involvement of the NFC Forum and its member organizations at events like the CES further enhance such communication.

9.7.3 PUBLICITY

This case study not only contributed to the development of the NFC technology demonstrators, but also to its presentation at the CES. The market story was strongly enhanced through the possibility of providing evidence that it is acceptable to users as well as being user friendly. This in itself is a strong argument for user involvement in development, though of course it is not possible in advance to guarantee such a positive outcome! On the other hand, it does provide an incentive to make the required improvements and to test again.

However, this study also highlighted a need for face value in methodology when results are publicized. Even if it can be argued that a discount method using only five participants is methodologically sound, this does not convey a strong message to the news media, who are more used to hearing about user studies involving numerous participants reported by market research organizations. The problem here is to find a balance between relatively superficial but quantitative feedback from large numbers of people, obtained through survey methods or focus groups, and the depth of information, obtained for example through interview methods, required to drive development further.

9.7.4 RELATIONSHIP WITH MARKETING

In dealing with issues of attitude and acceptance rather than usability, there is an overlap with consumer studies carried out by marketing specialists. However, there need be no conflict of interest here as long as the goals of the study are clear. The NFC study reported here was concerned with identifying issues that might be of concern in developing the technology so that at an early stage, steps could be taken to overcome them. It was not intended to answer the marketing question of the commercial viability of the technology, although the positive results of the study provide some indication that this might be the case.

There is even a potential overlap in the methods used by human factors specialists and market research, and the two disciplines have a strong potential to cooperate. Van Vianen, Thomas, and Nieuwkasteele (1996) show that this can be achieved. They describe some of the advantages and disadvantages of the methods available, e.g., using focus groups to explore and discuss new ideas as opposed to individual interviews to obtain in-depth information. The methods used in any study should be driven by requirements rather than who actually carries it out.

9.7.5 CONSUMER PRODUCT DEVELOPMENT

The consumer electronics business of today is dominated by extremely short development cycles and outsourcing of much of the development work (see Philips 2004). In such an environment, it is difficult to provide extensive user input and multiple iterative cycles of design and development are no longer feasible (see Jansen and Thomas, Chapter 17 this volume). One solution to this problem is to focus human factors efforts on applications and components. This case study has shown that human factors input can, indeed, play a valuable role in the development of technologies to be incorporated in consumer products.

ACKNOWLEDGMENTS

This work would not have been possible without the close cooperation of our colleagues at Visa, NXP, and Philips. In particular, the authors would like to thank Sandy Thaw, Francesco Prato, Steffen Reymann, Robert Blake, and Dolf Wittkamper.

REFERENCES

ISO-DIS 9241-11: Ergonomic requirements for office work with visual display terminals (VDTs): Part 11: Guidance on usability.

Jordan P.W. 2000. *Designing Pleasurable Products: An Introduction to the New Human Factors.* London: Taylor & Francis.

———. 2001. Pleasure with products. In *User Interface Design for Electronic Appliances*, eds. K. Baumann and B. Thomas, 303–28. London: Taylor & Francis.

Lindgaard G.L., and Dudek C. 2003. What is this evasive beast called satisfaction? *Interacting with Computers* 15:429–52.

NFC Forum. 2006a. *Near Field Communication and the NFC Forum: The Keys to Truly Interoperable Communications.* Wakefield MA: NFC Forum.

———. 2006b. The NFC Forum. www.nfc-forum.org/aboutus/

Nielsen J. 2000. Why you only need to test with 5 users. www.useit.com/alertbox/20000319.html

———. 2004. Risks of quantitative studies. www.useit.com/alertbox/20040301.html

NXP. 2006a. Near Field Communication. www.nxp.com/products/identification/nfc

———. 2006b. "How would you like to pay for that? Cash, card or phone?" www.nxp.com/news/content/file_1231.html

———. 2006c. Philips and SKT join forces to simplify NFC development around the world. www.nxp.com/news/content/file_1237.html

Philips. 2004. Supply markets. www.philips.com/About/businessesandsuppliers/Suppliers/Section-14746/Index.html

———. 2005. Come closer, go further. Philips document order number 9397 750 15334.

Preuss P. 2006. Case study: Mobile ticketing in RMV – German transportation. Paper presented at the NFC Europe 2006 Conference, Frankfurt am Main, October 2006.

Sauro J. 2004. The risks of discounted qualitative studies. www.measuringusability.com/qualitative_risks.htm

Steinmeier S. 2006. Near Field Communication (NFC) in use. Paper presented at Het 9e e-Nederland Congres [The 9th e-Netherlands Congress], Zeist, November 2006.

Thomas B. 2001. Usability evaluation. In *User Interface Design for Electronic Appliances*, eds. K. Baumann and B. Thomas, 295–302. London: Taylor & Francis.

Tractinsky N., Katz A.S., and Ikar D. 2000. What is beautiful is usable. *Interacting with Computers* 13:127–45.

Travis D. 2003. Discount usability: Time to push back the pendulum? www.userfocus.co.uk/articles/discount.html

———. 2004. Comment: Getting the right measure of usability. www.usabilitynews.com/news/article1567.asp

Vianen E. van, Thomas, B., and Nieuwkasteele M. van. 1996. A combined effort in the standardization of user interface testing. In *Usability Evaluation in Industry*, eds. P.W. Jordan, B. Thomas, B.A. Weerdmeester and I.L. McClelland, 7–17. London: Taylor & Francis.

Virzi R.A. 1992. Refining the test phase of usability evaluation: How many subjects is enough? *Human Factors* 34: 457–68.

Visa. 2006a. "How would you like to pay for that? Cash, card or phone?" www.corporate.visa.com/md/nr/press291.jsp?src=home

———. 2006b. Consumers give two thumbs up to first North American mobile phone payment and content trial. www.usa.visa.com/about_visa/press_resources/news/press_releases/nr341.html

10 Empathy Meets Engineering: Implanting the User's Perspective into a Systematic Design Process

Matthias Göbel

CONTENTS

10.1 INTRODUCTION

10.1.1 THE ROLE OF ERGONOMICS IN PRODUCT DESIGN

In the translation of Grandjean's (1989) *Fitting the Task to the Man*, ergonomics in product design deals with the fitting of the product to the user, or, more precisely, to the needs of the user. This not only includes the person, but also the circumstances of usage, such as the purpose of use and the environmental conditions. Considering ergonomics with a focus on handling attributes, the term *usability* (and, respectively, *usability engineering*, for the design process) has been established synonymously with ergonomics.

A first general design challenge can be identified for meeting the different needs of a large target group. As consumer products are not manufactured for a specific user, but aim for a most widespread group of users (at best a global market), a large variety of users, tasks, and circumstances of use must be considered. Furthermore, not only rational aspects of use play a role in consumer products. Customers may enjoy aesthetic and prestige aspects as well, and even accept a restricted practical value in favor of such emotional aspects.

Another particularity for consumer product design arises from the fact that the buying decisions are made on the product information that is available within a very limited amount of time, and often with very restricted testing opportunities, e.g., during a product demonstration by a sales agent. Furthermore, future users often do not have the knowledge to overlook all product functions and attributes, e.g., when purchasing a new type of product that they haven't owned (and thus experienced) before. For those reasons, not only the objective quality, but also the immediate experience is an important success factor.

In fact, the relevance of ergonomics varies for different products. It equally varies with the cost segment and with the expectations of the users. Some users may consider ergonomic aspects as distinct attributes for decision making, while other customers might not consider usability criteria. However, ergonomics criteria are receiving more and more attention, as products differ less and less in technical performance and technical features, a large number of features are seen as critical for ease of use, and customers pay more attention to the sustainability of use in terms of efficiency in operation (see Figure 10.1). Furthermore, the experience of low effort in use attracts more and more attraction, as products are becoming increasingly complex. However, an association of specific brands with high ergonomic standards has been established only in a few cases (such as Apple for user-friendly home computers and computer-controlled communication and entertainment devices).

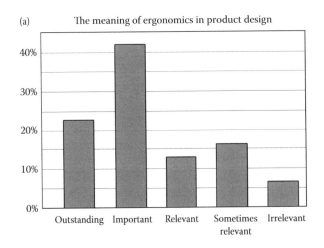

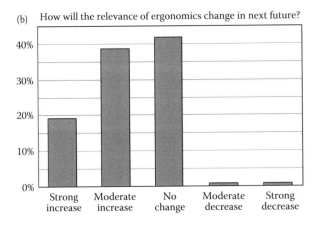

FIGURE 10.1 The relevance of ergonomics estimated by product developers and product designers. ($n=31$ professionals in different domains.) (Data from Fietz, A., Ergonomie im Produktentwicklungsprozess. Diplomarbeit am Lehrstuhl für Arbeitswissenschaft und Produktergonomie. Technische Universität Berlin, 2004.)

10.1.2 ERGONOMICS CHALLENGES IN PRODUCT DESIGN

Focusing on the design level, what are the ergonomic challenges in the design process?

Observing the sheer amount of literature in ergonomics, one would expect a vast number of guidelines to be available for proper design. In fact, hardly any guidelines can be applied directly for design purposes.

At a very elementary level, such as the physical interaction between user and product, many considerations are available for sizing buttons and labels. Common recommendations for key size follow the shape of the human hand, thereby suggesting keys of 13–15 mm in length (Schmidtke 1989a). Empirical tests support this recommendation. Such a key size is, however, not practical for many devices. Particularly for small and portable devices, keys of 3–4 mm are common and still usable. This is not in conflict with ergonomic recommendations, but just a different prioritization of size against performance, as small keys are usable, but at the cost of speed and error rate. From a design perspective, this is a balancing task between size on the one hand and typing speed and error rate on the other hand. For this reason, it does not make sense to create miniaturized keyboards if there is no serious space restriction. This balancing is further complicated if other variables have to be considered that induce even more degrees of freedom in design, such as the detrimental sensory-motor performance of elderly users, or the operation and environmental conditions that will affect the required size of labels. Finally, all those variables do not have constant settings, but vary for different users. A technocratic design process thus would have to make use of methods of fuzzy logic. Similar considerations are given for almost any application of human measures, such as forces, movement ranges, perceptions, and motor control.

On a higher design level, interaction procedures that control information flow have to be designed. Typical examples of this are knobs and dials or menu controls for various functions in a computer-controlled device. For design, no strict rules may exist that prescribe the one best design. Only two abstract rules must be applied: (1) consistency and (2) compatibility (Strasser 1995). The latter criterion considers public conventions (e.g., red as an alerting color) as well as individually shaped references (e.g., used with the Microsoft Windows dialog scheme). For design, this requires consideration of widely used standards, common experiences of specific user groups (e.g., senior customers), and, ideally, individual customs. Apart from likely inconsistencies between those references (which may conflict with the consistency criterion as well), there is, once again, no applicable rule to make design decisions, but only a fuzzy compromise may be sought after consideration of the check criteria.

On a higher level of functionality, and the selection of functions and features being considered for a product, this is again conflicting as many functions and features enhance the versatility of use, whereas the ease of use is improved by "keeping it simple," eliminating unnecessary functions. A proper design may indeed allow joining both requirements to a certain degree, however, the basic conflict will remain (apart from cost and market considerations, etc.). This criterion is mostly affected by the aim of the producer to cover a broad range of users and use cases (which require a varied functionality) on the one hand and the wish of users to get a product that does exactly what they need, and thus does not need a long-winded setup on the other hand.

Summing up the aforementioned aspects, ergonomic design is a complex problem-solving task, as straight facts that do exist cannot be directly implemented for design. This is further complicated by varied user attributes and corresponding demands, e.g., elderly persons as a target group, as well as by economic pressure for the design process, the production process, and the design outcome ("better, quicker, cheaper").

10.1.3 THE NEED FOR THE USERS' PERSPECTIVE

For a long time, probably until the 1980s, users had not been greatly involved in the design process, as product design was mostly critical from a technological perspective and thus engineering and production technologies were the most crucial factors for product performance. Furthermore, most products were simple enough to estimate the needs of the users from a common sense point of view. With more saturated markets and decreasing technological differentiation, aspects of aesthetics and usability became more relevant. Concurrently, this trend was emphasized by the increasing complexity of products, requiring a more careful adjustment to the users' needs in order to allow the technical benefits to be exploited. During this phase, starting in the 1990s, usability criteria received increasing attention, initially focusing on marketing studies and then involving users more and more in the design process. Today, almost any quality product has been designed with end users involved. However, the hype around a user-centered design has significantly slowed down, as the downfalls of user participation have been experienced as well.

This raises the pragmatic question of how to make this contribution most effective in terms of improving quality and facilitating the design process. This again raises the more academic question about the nature of a user's contribution.

From a very simplistic perspective, a user's participation in the design process only makes sense if the user can add something that the designers do not have at hand. This is very unlikely to be technical expertise or insight, but rather the perspective of an ideally typical user. This may include experience about the operational conditions and users' expectations, what is typically part of a marketing survey, but user participation may also provide insight into the perception and logic of an individual, thus helping to identify incompatibilities between designers' and users' concepts. Of course, each user is an individual, so that a representation of the broad range of users would require a larger variety of participating users.

Göbel (2007) described the different perspectives of designers and users mainly by their different experiences and insight. Users have difficulties in imagining new ideas and have little understanding about feasibility and compromise due to a lack of technical understanding and a lack of reference experiences. As a consequence, users tend to change preferences when being asked the same thing repeatedly or when experiencing them in reality or in a prototype. This inconsistency in expressed requirements appears equally for features that were initially required, but later considered as redundant, as well as for features that were, initially, not considered relevant, but turned out to be useful in a later stage of experience.

Designers, however, face the complementary problem of understanding all the details of the concept and its technical realization, thereby lacking imagination of the obstacles to get there (Darses and Wolff 2006). This compares to the gap between the understanding of a clever and a simple-minded person: it is equally difficult or even impossible to raise the understanding of a simple-minded person to a clever person as it is vice versa to lower the understanding of a clever person to that of a simple-minded one. However, in the design process, the differences are not (only) caused by different levels of cleverness, but mainly different degrees of experience. This adds to some differences in intellectual capacities, as designers are mostly academics, whereas users cover the whole range from very simple-minded to very clever persons. In particular, the consideration of "stupid" users remains a challenge for most designers.

The consideration of the users' perspective is thus a most useful contribution in product design, although one must expect inconsistent behavior during the design and test process due to lack of deeper knowledge.

10.2   THE DESIGN PROCESS

10.2.1   Basic Strategies

In any case, the design process has to start at the highest, most abstract level and process down to the detail design aspects: the details follow the concept. However, for designing the higher-level aspects, an anticipation of the consequences on the lower-level design aspects is required. So, any design process is implicitly a type of chicken-egg (catch-22) problem (Luczak 1995).

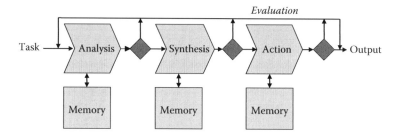

FIGURE 10.2 Archetype of systematic problem-solving cycle. (Rhombs describe comparison with targets and corresponding decision making.)

Any optimization processes in design are hence iterative in nature, and cannot be split into segments without compromising the outcome. This is particularly important for more complex products (such as cars) that cannot be designed as a whole, but only in many concurrent processes for reasons of resource organization and time to market. Many basic ergonomics deficits that are still present in even highly optimized products have to be accounted for in this problem, as they become obvious only when assembling the entire product, but cannot be rectified in the available time frames.

Technically, the design process cannot be broken down into an algorithm, but is a creative process that can be described as a problem-solving cycle. In order to achieve most optimal output (and not random results), systematic problem-solving algorithms (as shown in Figure 10.2) are widely applied.

The minimal three steps, "analysis," "synthesis" (elaboration of problem solutions and selection of preferred option), and "action" (implementation), are part of an internal feedback cycle. Depending on the outcome of each step, the process proceeds with the next step or the present or even previous steps are repeated until a satisfactory result is found.

Memory is relevant in each of the steps for referencing and in order to introduce experiences. The memory differences between different designers, and also between designers and users, explain the likely different outcome for identical tasks given to different individuals.

This very basic scheme is, however, not practical for designing an entire product at once. Even rather simple products are thus designed in a segmented process. Therefore, two basic strategies have been established: The more engineering type of process (left side in Figure 10.3) tries to follow a straightforward process by breaking up into sequences that are ideally processed one after the other. For the design process, this is typically performed by concept development, rough design, detail design, functional prototype, etc. Although feedback loops are part of the process, they have a rather secondary role for evaluation purposes. Ideally, the design process passes straight through.

Whereas this works well for structured units that can be widely anticipated, more holistic design tasks, like ergonomics design, mostly require a more iterative procedure, as highlighted on the right side of Figure 10.3. Although the structures of both procedures in Figure 10.3 similarly correspond to the problem-solving cycle shown in Figure 10.2, the user-centered design process emphasizes more the iterative character of the process that is required to check and to refine the product step by step.

In fact, iterations are time consuming, and, depending on the level of finish of the tested prototype, costly. Most important is the lack of anticipation of when an iteration cycle will come to an end. Thus, from a managerial point of view, one should try to minimize or avoid iteration cycles wherever possible, particularly in the later phases. In order to reduce time and cost of trials, simulation environments have become increasingly popular.

10.2.2 Different Stakeholders in the Development Process

Different experts have to be involved in the product design process, and ergonomics aspects are part of different design activities. This raises questions about the coordination of ergonomics aspects

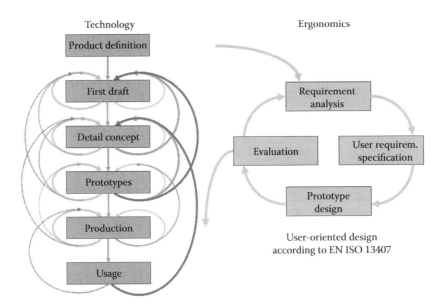

FIGURE 10.3 Core structures of an engineering design process (left side, VDI 2221) and an iterative process using the example of a user-centred design cycle according to ISO 13407 (right side).

among the different experts, and, again, about the coordination of the different development strategies that have to be processed concurrently to a large extent.

Ergonomics is the object of marketing, industrial design, mechanical and electrical engineering, and computer science, to mention the most important. Thus, ergonomics is represented by different experts with different roles, aims, and ways of thinking, causing a significant risk of ending in an inconsistent design.

In the few cases in which explicit ergonomic experts are involved, this may end up even more conflicting: because other experts are responsible for defined parts of the total object, an ergonomist is necessarily interfering with their areas of responsibility (see Figure 10.4). Thus, ergonomists must cooperate closely with the other experts, adapting to their work style while still facing a risk of conflicts. This is equally the case if user participation is considered. Who does control this interaction and how is user feedback channeled to the different fields of expertise? The problems associated with those interferences often cause designers to have reservations about users participating in the process.

Possibly the most important challenge for the integration of ergonomics into the product development process are the very different strategies of engineering, design, and ergonomics. Designers usually work very intuitively, but use a step-by-step approach to refine a general concept found in the early phases to a detailed physical or virtual model in the later phases. Engineers have highly developed straightforward algorithms available, which enable an almost systematically organized development process (e.g., VDI 2221; Pahl and Beitz 1993). In contrast to both these approaches, ergonomic product development is organized analytically, but highly iteratively (e.g., according to ISO 13407).

As a consequence, for example, engineers often need to have structural aspects been defined and approved, while ergonomists may still identify the need for changes on a conceptual level. The idea to start work on the ergonomics process earlier than the engineering process brings the problem that ergonomists need technological decisions coordinated to target a realizable product concept.

10.2.3 THE LANGUAGE OF ERGONOMICS

Any consideration of ergonomics aspects needs specifications in order to communicate the user needs to the designers and engineers and to coordinate the different stakeholders. This raises the question of how ergonomics requirements, qualities, or attributes can be formally expressed.

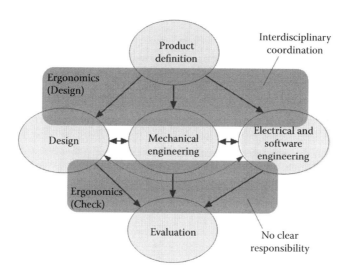

FIGURE 10.4 Example for the interdisciplinary roles in the development process and the involvement of ergonomics. (From Göbel, M. and Friesdorf, W., *Ergonomics in the Digital Age*, Proceedings of the XVth Congress of the International Ergonomics Association, 24–29, August 2003, Seoul, Korea, Santa Monica: IEA Press, 2003.)

The first problem already occurs when ergonomic approaches are expressed during the early conceptual phases of the product development. Requirements such as "easy to use," "easy to learn," and "low workload" are a common part of almost any specification sheet. Neither engineers nor designers nor product managers will deny the necessity of such attributes, but they cannot be specified without having a more precise concept or even a design at hand. Only with the main structure defined, a more concrete specification may be expressed (e.g., sizes of buttons and display units), though ergonomic quality is mostly being created during the conceptual phase.

Furthermore, ergonomic quality can only be evaluated on a whole product and can only be expressed by ergonomics deficits. As there is no positive description for ergonomic qualities, the ideal (however fictitious) product design would be the one that has no deficits. This again requires a comprehensive/holistic design proposal for evaluation.

This might explain why ergonomics considerations for practitioners and in practice are mostly based on checklists (e.g., Schmidtke 1989b), and not on design guidelines.

10.3   USER PARTICIPATION IN PRODUCT DESIGN

10.3.1   FORMS AND ATTRIBUTES OF USER PARTICIPATION

Considering a user's contribution to the design process, what benefits and risks may occur that could worsen the outcome or lengthen the design process?

The most basic form of user participation is a survey of user experiences and expectations in general or with only a vague description of the intended concept. This is usually done in the form of questionnaires or interviews. Such a simple, rather marketing-like form of survey is easy to perform and not problematic in terms of uncovering still secret design or technology highlights to outsiders. It is, however, very much limited to well known and already experienced concepts, as users are mostly unable to express feedback on innovative ideas or design details that they have yet to experience. Even more critical, this type of assessment is very insensitive to discover overloading of the users: If users cannot imagine an idea or a feature correctly, most of them tend to judge their appreciation anyway, either because they have not been aware that they did not understand or because they do not want to admit a limited understanding. Hence, this method might be suitable for exploring general wishes and views, but delivers increasingly unreliable results if focused on the product details.

Another popular method of user participation is prototype testing. In this late phase of product development, users can clearly explore a new product as it appears almost like a consumer product.

Hence, this type of assessment is very useful as a final check and for detail adjustments (e.g., labeling, instruction manual, and packaging). Many companies do practice such a type of assessment as an effective quality control measure before launching any new product. If the testing is performed with early prototypes, final adjustments, if necessary, can be induced during production setup without delaying the time to its introduction to the market. This type of assessment is very effective in terms of the quality of the feedback, but very impractical if more substantial usability problems are encountered owing to its late application in the development process. In most cases, such products are launched with known deficits anyway, as corrections would be too time consuming and costly when identified.

For this reason, the approach of user-centered design, e.g., as suggested by ISO 13407 (Figure 10.3, right side), considers user participation within the design process. It follows a cyclic loop of design, testing, and redesign, which can be applied to a certain extent as an iteration loop from a rough design concept to a detailed prototype.

The basic concept is indeed convincing, however the practical experiences were not as positive as expected. Although any exit is (theoretically) feasible only after having met the targets, the development process cannot be controlled in terms of time and effort required to succeed, and further tends to divert or to turn around in circles, apart from the challenge to involve a group of non-experts more or less frequently. This is particularly the case for more complex products with interaction effects between different design features.

Bönisch (2005) explains such a development cycle for the design of an anesthetic respirator, requiring 14 loops with a tendency to oscillate between different options, as it was not possible to meet all requirements with one single design. In an industrial setup, it is mostly impossible to leave the time frame open until a satisfactory design is found. More realistically, the iteration cycles have to be stopped after a set time has elapsed, regardless of the extent to which the design is still compromised at that point in time. In addition, a significant effort might be required to provide sufficiently functional demonstrators for evaluation. If this is compromised, user evaluation is again compromised with the consequence of even more iteration cycles being required. Furthermore, users mutate to experts with time, but a replacement with novice users during the optimization process will cause confusion among the new users due to the lack of previous knowledge.

Göbel and Yoo (2006) demonstrated the effects of an extreme form of participation to study the differences between users' and designers' perspectives. In order to design a mobile phone for elderly persons, a group of typical users, seniors aged between 60 and 75 without a specific technological background or education were asked to design a mobile phone for elderly persons on their own. Eight seniors worked in a team with alternating team meetings and individual working periods. No other expert support was provided to ensure an unbiased design outcome. The documentation of the meetings kept by the senior team members was then analyzed. The requirements listed at the initial meeting are very typical for technologically non-affine users: minimal functions, large keys and display signs, and easy to use and emergency functions were considered the most relevant attributes. After eight meetings, the group worked out a more specific design. Over 24 pages, the seniors reported about 80 main functions. Although no mobile phone was available on the market that had exactly this specific functionality, the final design by the seniors appeared as a compilation of available functions and interaction principles. Particularly, the product complexity in terms of number of functions was within the range found on popular commercial mobile phones. At first glance, one could argue that, in fact, the novice users became experts with time. However, in this study there was no induction of technological knowledge, and the team did not have any opportunity to set up prototypes or to do testing. Thus, with time, only the insight into the options of this technology increased (comparable to the effect of counseling).

The conclusions that may be drawn from this study indicate that the requirements expressed by users change very much with experience and insight. This raises the question, which level of experience participating users should have to contribute in a beneficial way. On the one hand, focusing on first-time buyers (considering the initial requirements in the above-mentioned example) will likely cause frustration if users get more experienced with a product. On the other hand, focusing on the requirements of more mature users (considering the final specification in the above-mentioned example) will frustrate naïve users—who would possibly not decide to purchase such a product.

Further, the more that users get involved in the details of the product design, the more their requests will correspond with those of professional designers (and marketing experts). Three important consequences can be drawn from this experience:

1. From a rational point of view, many products seem to be designed in a technocratic way, ignoring user needs to a certain extent. But, in fact, they often meet the needs of more mature users and cover the variety of different users.
2. In practice, product design needs to consider both perspectives—the needs of the mature users as well as the needs of the naïve users—although the requirements of both groups may conflict.
3. Naïve users must not be experienced, not even with other products design, as they learn to abstract from their own perspective.

Although the above-mentioned example focused mostly on the functional aspects of a product, one may assume a similar effect for the interaction level, however with few interactions between both levels.

Another aspect of user participation is the identification of unexpected handling problems or misconceptions about user experience. An example from Göbel and Neth (2004) might illustrate this phenomenon. During a laboratory study, seniors were observed when exploring the menu structure of a commercial mobile phone in search of a specific function. An error occurred as one facilitator covered the display by mistake. However, the test person continued to enter the navigation keys, obviously without display control. It turned out that this person was not aware of the principle of a dialog control, the selection of a function by navigation through a list of options displayed (in contrast to static functions that were directly selected by pushing the according key). Follow-up tests provoking this type of "mistake" showed that 63% of the seniors ($n=60$; age range: 55–90 years) had not properly understood the dialog interaction principle, but only 19% had difficulties reading information from the display, and 25% reported problems operating the keys. Interestingly, the seniors were not aware of this misconception with the menu interaction. When observing young people using their mobile phone, they expected the users had learned the steps of interaction by heart. One of the seniors commented accordingly: "Those young people can memorize everything so easy. In my age now, the brain does not work that well anymore and I have to write down everything step by step." As a consequence, the subjects did not realize that the task could not be properly performed and for the evaluation setup this type of misconception was not considered either. Although very beneficial in its outcome, user participation for those types of misconceptions is somewhat random, as it does not specifically check for deficits. A more comprehensive study would have to be structured accordingly, for example, in the form of a fault tree analysis. However, this type of testing may be extremely extensive in terms of the numbers of factors to be tested with varied subjects that must not get to an experienced status.

Apart from the fact that users can reliably express needs only when being aware of functions and interaction principles ("Only when I could test the prototype I got an idea what I should have wished"), the communication between users and designers might cause misconceptions as the referencing experiences are mostly different.

10.3.2 FROM DIRECT TO INDIRECT USER PARTICIPATION

User participation, as mentioned above, turns out to be essential to ensure ergonomic quality on the one hand, but provides only some limited information for the design process and may cause significant delays on the other hand.

This raises the question of how to design the most effective participation in terms of inducing user input in an appropriate form and at an appropriate point of time in the course of the design process. Further, the quality of user feedback is a serious concern, as it may cause extensive rework if misconceptions occur, either in terms of information transfer or in terms of revised opinions.

In an idealistic approach, the process control would remain with professionals, however the user input would be elicited and fed accordingly to the process. Playing this concept through, a systematic acquisition of user requirements and user capabilities could build a basis for design.

One example of such a systematic approach was developed for the creation of suitable product ideas during the early product definition, using a resource perspective: The functionality and the features of a product are derived from linking available resources to required resources. This resource transformation approach (RTA) considers all types of human, technical, and environmental resources, and further applies the available human resources to a maximum extent. The reason for this is the applied paradigm of maximizing control of a product by reducing its functionality to the level that is required for operating it and avoiding unnecessary human resource exploitation while applying available skills, which is particularly relevant for elderly persons.

In detail, a three-dimensional matrix is created that lists lacking or scarce resources (requested functions and services of a product that disable or at least impede the fulfilment of a task) against available (not yet exhausted) resources (see Figure 10.5). This is performed separately for the different tasks to be considered for the designed product. Those resource categories encompass personal resources, the social environment, and the technical environment, thus aiming for a holistic approach even if not all types of resources need to be considered in any case. The resource categories are defined mostly by user input, albeit a more common wording is used to elicit user information. In order to focus on user needs (and to avoid technological overkill), first the lacking resources should be elicited, and, in a second step only, the available resources. A more hierarchical structure for the resources may be applied where deemed useful.

For the different fields built in this matrix, a creative process is then performed with design professionals striving for ideas or concepts that link available resources to lacking resources. Not all matrix fields have to be filled, one field per column is sufficient. If there are more available, the most appropriate one may be selected, or a combined solution may be worked out.

The basic concept behind this approach is to translate the user's perspective to an engineering design scheme. However, this is still a creativity tool. This concept is basically intended to generate

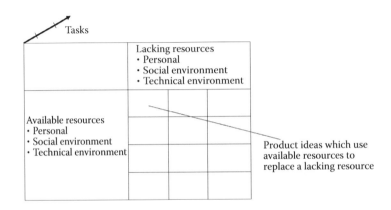

FIGURE 10.5 Basic feature matrix of the resource transformation approach (RTA) creativity method.

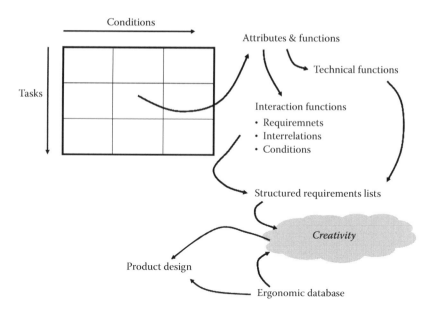

FIGURE 10.6 Pathway for the detail and interaction design phase.

product ideas, helping elderly persons to stay independent while making use of available capacities as much as possible.

For the more detailed interaction design, a second, rather pragmatic approach was developed that re-uses the user input and the task specification (see Figure 10.6). A matrix is set up listing all combinations of tasks (including sub-tasks) and conditions (in which the user will handle the product and in which the product will be operated). In the matrix cells, attributes and functions that emerge from each task and each condition are noted.

The attribute and function list is then split into such aspects as touching technical functions (e.g., storing measured data) and ones considering interaction functions. The latter group is then specified for functional requirements, interrelations (between different interaction functions), and specific conditions to be operated. A first check of possible conflicts (e.g., space) can be performed at that time. From this information base, structured requirements lists for the subsequent creative phase can be generated, e.g., sorted by groups using similar types of interactions, or sorted by simultaneous or subsequent actions. During a creative phase, the product is designed with the additional help of an ergonomic database to specify physical attributes (sizes, forces, etc.) or, at least, indicating the necessity to consider specific factors (human perception, human motor action characteristics).

Although the development process is drawn as a fixed straightforward sequence, real development processes often need some iteration loops to match product requirements, in particular during the later development steps. Furthermore, the most important design decisions are made during the creative phase, but no unique optimum pathway may be outlined for every type of product and every type of development team. Thus, such an approach has to be seen as a toolbox to structure and facilitate the development process. Figure 10.7 shows software that implements this strategy in the form of an electronic toolbox.

Figure 10.8 shows, as an example, a mobile phone designed for elder users. It comes in a popular style as seniors dislike a special appearance because this is often perceived as stigmatizing. However, it applies very simple interaction principles, e.g., a menu control that operates like a mechanical selector, and it is equipped with a large speaker area and lenses in the keys that allow reading without reading glasses. The uncertainty of elderly people to manipulate the setting by mistake is met by a hardware reset button to undo all recent changes in settings. Emergency functions are realized in a way that no specific keys need to be pressed, but only the shell needs to be pressed rhythmically.

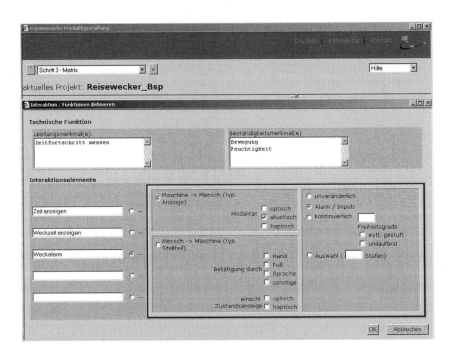

FIGURE 10.7   Computer-supported interaction design tool.

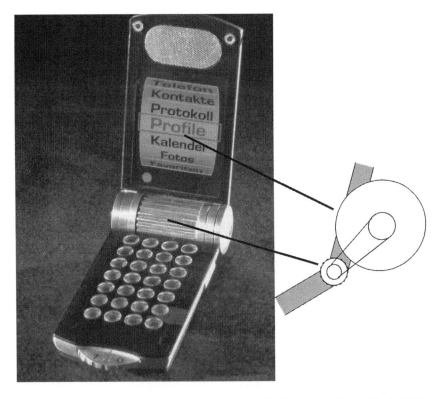

FIGURE 10.8   Example of a mobile phone designed for elderly person. (From Goebel, M. and Neth, K.-U., *Gesellschaft für Arbeitswissenschaft e.V, Personalmanagement und Arbeitsgestaltung, Bericht zum 51. Kongress der Gesellschaft für Arbeitswissenschaft v.,* GfA-Press, Dortmund, 2005.)

10.4   DOES A SYSTEMATIC APPROACH HELP TO IMPROVE DESIGN RESULTS?

Systematic approaches mostly aim for improved efficiency of the design process, allowing shorter time to market and clearer budget and capacity planning. However, such approaches are considered as technocratic in nature, restricting creativity and resulting in functional but tedious products that do not have originality that would inspire potential customers.

However, can a systematic consideration of user information, e.g., as suggested in the aforementioned RTA approach, improve design quality as well?

In order to study such effects, a comparative study applying both the aforementioned approaches versus the popular creativity methods "brainstorming" and "morphology" were performed.

Twenty-seven student groups of three students each, designed one product (within a period of 14 weeks, and 4 hours per week). The students were equally skilled with a basic education in ergonomic product design (20 hours duration) and received training for either one of the applied sets of methods (4 hours). The 27 groups were assigned in a $3 \times 3 \times 3$ factorial design to three different types of products, three different disciplinary compositions (considering students with specialization in engineering, economics, and social sciences), and three different arrangements of systematic vs. creative procedures in the concept and detail design phase.

Results were evaluated by an expert panel ($n=5$) and the students ($n=78$), rating total quality and six product attributes. The order of evaluation was randomized, and the referees were not aware which strategy was applied for each design.

As a result, the evaluation of the design quality demonstrated significantly better design results when the systematic approach was applied (Figure 10.9). This is equally true for the product concept as well as for detail design. The effect, however, depends on the disciplinary structure of the design team as well as on the type of product.

For the different attributes (Figure 10.10), the addressed target groups are more specific on the conceptual side and broader on the detail design side when applying the systematic approaches. The product concepts are more rational, but detail design is evaluated as more emotional if the systematic methods are used. The variety of use is considered more specific for both design aspects, whereas the design allows a more disperse use if the systematic approach is applied. The systematic approach, however, brings out products that were characterized more to adorn than those being designed with creativity methods.

It is shown that the application of a systematic approach results in more specific products that are, however, more disperse in use and more adorned and emotional in detail design. In total, the products designed with this approach were rated significantly higher than those designed using traditional creativity methods.

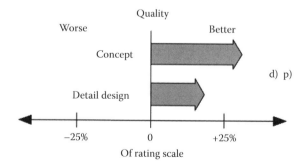

FIGURE 10.9   Evaluation of design quality (gray arrows denote a significant difference between both tested design strategies $p<0.05$, white arrows denote a non-significant value; d) denotes a significant effect of disciplinary team composition on result $p<0.05$; p) denotes a significant effect of product type on result $p<0.05$).

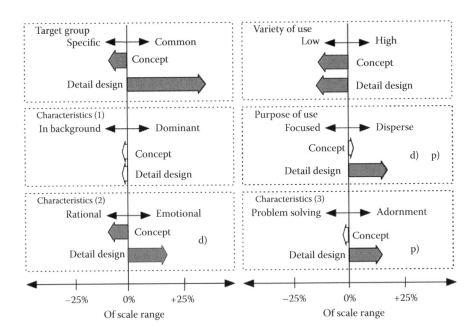

FIGURE 10.10 Evaluation of product attributes (gray arrows denote a significant difference between both tested design strategies $p < 0.05$; d) denotes a significant effect of disciplinary team composition on result $p < 0.05$; p) denotes a significant effect of product type on result $p < 0.05$).

10.5 CONCLUSIONS

User participation has to be considered as an essential contribution to the product design process. However, the lack of insight and hence inconsistent responses of the users make an introduction of the users' perspective difficult and user participation will likely lengthen the design process.

The basic challenge of matching the users' perspective with the needs of a mostly straight-forward design and engineering process can be met by the more systematic approach of integrating user needs indirectly instead of considering user feedback on as yet incomplete design proposals during the development process. It can be shown that products designed by applying this strategy provide a higher design quality and can better meet emotional and other soft criteria.

REFERENCES

Bönisch, B. 2005. *Partizipatives Interface Design am Beispiel Anästhesierespirator*. Aachen: Shaker Publ.

Darses, F., and Wolff, M. 2006. How do designers represent to themselves the users' needs. *Applied Ergonomics* 37 (1): 757–64.

Fietz, A. 2004. Ergonomie im Produktentwicklungsprozess. Diplomarbeit am Lehrstuhl für Arbeitswissenschaft und Produktergonomie. Technische Universität Berlin.

Göbel, M. 2007. Ergonomic design of computerized devices for elderly persons – the challenge of matching antagonistic requirements. In *Universal Access in Human Computer Interaction*, ed. C. Stephanidis, 894–903. Heidelberg: Springer.

Göbel, M., and Friesdorf, W. 2003. Optimization strategies of ergonomic product development processes with regard to ergonomics. In *Ergonomics in the Digital Age*, ed. E.S. Jung. Santa Monica: IEA Press. Proceedings of the XVth Congress of the International Ergonomics Association, 24–29 August 2003, Seoul, Korea.

Göbel, M., and Neth, K.U. 2004. Junge Alte – der demografische Faktor bei der Produktgestaltung. In *Ergonomie und Design*, ed. R. Bruder, 111–22. Stuttgart: Ergonomia.

Goebel, M., and Neth, K.-U. 2005. Partizipative Entwicklung und Bewertung seniorengerechter Produkte. In *Gesellschaft für Arbeitswissenschaft e.V, Personalmanagement und Arbeitsgestaltung, Bericht zum 51. Kongress der Gesellschaft für Arbeitswissenschaft v. 22–24 März 2005*, 507–10. Dortmund: GfA-Press.

Göbel, M., and Yoo, J.W. 2006. Ergonomic product design for elderly users. In *Proceedings of the IEA2006 Congress of the International Ergonomics Association*, eds. R.N. Pikaar, E.A.P. Koningsveld and P.J.M. Settels. Elsevier Ltd.: Oxford.

Grandjean, E. 1989. *Fitting the Task to the Man.* London, New York: Taylor & Francis.

ISO 13407 1999. Human-centred design processes for interactive systems. Brüssel: CEN.

Luczak, H. 1995. Macroergonomic anticipatory evaluation of work organization in production systems. *Ergonomics* 38:1571–99.

Pahl, G., and Beitz, W. 1993. *Konstruktionslehre – Methoden und Anwendung.* Berlin, Heidelberg, New York: Springer.

Schmidtke, H. 1989a. *Handbuch der Ergonomie.* Koblenz: Bundesamt für Wehrtechnik und Beschaffung.

———. 1989b. *Ergonomische Prüfung von Technischen Komponenten, Umweltfaktoren und Arbeitsaufgaben- Daten und Methoden.* München, Wien: Carl Hanser Verlag.

Strasser, H. 1995. Ergonomics efforts aiming at compatibility in work design for realizing preventive occupational health and safety. *International Journal on Industrial Ergonomics* 16:211–35.

VDI 2221. 1993. *Methodik zum Entwickeln und Konstruieren technischer Systeme.* Düsseldorf: VDI Verlag.

11 Building Empathy with the User

Louise Moody, Elaine Mackie, and Sarah Davies

CONTENTS

11.1 INTRODUCTION

This chapter discusses the importance of empathy to the understanding and application of ergonomic principles. The potential challenges of developing ergonomic and empathic skills in design students are discussed. Teaching methods that are employed within the Industrial Design department at Coventry University, and which are more broadly applicable to the design industry as pragmatic research and evaluation techniques, are described.

11.1.1 PHEASANT'S FALLACIES

A user-centered design approach requires an understanding of user characteristics, expectations, desires, and needs within the context of tasks and situations. This should be translated and communicated through sensitive design solutions. Often, designers show a preoccupation with the visual experience of the user (Pheasant 1992), neglecting the wider physical, psychological, social, and emotional requirements. Pheasant (1992, 8) explains the misuse or non-use of ergonomics by designers in terms of "five fundamental fallacies":

1. This design is satisfactory for me—it will, therefore, be satisfactory for everybody else.
2. This design is satisfactory for the average person—it will, therefore, be satisfactory for everybody else.
3. The variability of human beings is so great that it cannot possibly be catered for in any design—but since people are wonderfully adaptable it doesn't matter anyway.
4. Ergonomics is expensive and since products are actually purchased on appearance and styling, ergonomic considerations may conveniently be ignored.
5. Ergonomics is an excellent idea. I always design things with ergonomics in mind—but I do it intuitively and rely on my common sense so I don't need tables of data.

Pheasant links the first, second, and last of his fallacies with the concept of empathy.

11.1.2 EMPATHY

Consideration of empathy in the literature is often related to the training of health professionals rather than designers (Wiseman 1996; Blatner 2002). Blatner (2002) defines empathy as

> an ability to imagine with some degree of accuracy what it's like to be in the predicament of the other person; … empathy entails the ability to communicate that awareness so the other person feels understood.

Carkhuff (1969, 58) describes empathy as involving "crawling inside another person's skin," seeing the world through their eyes and experiencing it in the same way. While Wiseman (1996) defines the key attributes of empathy as: seeing the world as others see it, being non-judgmental, understanding another's feelings, and communicating that understanding. These definitions lend themselves to the insight and understanding of users that is required in ergonomics and that should be communicated through design.

Pheasant (1992) differentiates intuition and empathy in design: both provide personal understanding, but intuition refers to "immediate insight or understanding without conscious understanding" (*Pocket Oxford Dictionary* 1996). By contrast, empathy refers to a process of gathering information about a user population in order to consciously understand it. This involves testing intuitions more objectively through reasoning and problem solving, rather than blindly relying on them. An appropriate application of research methodology should provide an empathic understanding, which can then be communicated through designs to reflect the desires, expectations, and needs of the user.

Fulton Suri (2001) advocates empathy as a means of broadening an understanding of people and situations, and as a legitimate and useful part of the ergonomist's tool kit. She argues that empathy provides a sensitive tool for understanding how people interact with products. An empathic design approach, therefore, involves designers trying to gain insight into the lives and experiences of users, and applying this knowledge to the design process to ensure that a product is usable and meets the needs of the consumer. It can provide a commercial advantage by enabling designers to realize needs that the users themselves are perhaps unaware or unable to articulate (Leonard and Rayport 1997). It typically involves study within the world of end users to recognize and understand users as real people and not just as subjects (Mattelmaki and Battarbee 2002).

Traditionally, psychological theorists regarded empathy as a stable and measurable personality trait (an innate disposition) that could be measured through psychometric tests (Hogan 1969; Smither 1977). More recently, empathy has been related to a range of individual differences, including age, gender, personality, life experiences, and socialization (Blakemore and Choudhury 2006;

Zahn-Waxler 1991). Research suggests higher levels of empathy among females than males (Zahn-Waxler 1991) and, as a result of pre-frontal cortex development during puberty and into the early twenties, teenagers can have difficulty analyzing emotions and thoughts, and displaying empathy (Blakemore and Choudhury 2006).

Empathy is likely to have both "trait" and "state" components and is now more widely recognized as a skill (Wiseman 1996). Consequently, although individuals have a natural disposition to be empathic, the level of empathy displayed will vary depending on the situation, and can be developed and therefore taught. This is important to the caring professions and we would argue, to the design profession too. However, based on our experience, we would suggest that empathy can be particularly challenging for students to acquire and, as such, represents troublesome knowledge (Perkins 1999).

11.1.3 EMPATHY AS A THRESHOLD CONCEPT

There is increasing educational interest in the notion of "threshold concepts" (Meyer and Land 2003; Clouder 2005; Davies and Brant 2006; Davies and Mangan 2005). Within all subject areas there are thought to be particular concepts that represent a new and previously inaccessible way of thinking about something that is particularly challenging to acquire. A threshold concept represents a transformed way of understanding, or "thinking" within a discipline that without it the learner cannot progress effectively. The transformed view represents how people think or practice within the discipline and can reveal the previously hidden interrelatedness of the subject that defines its boundaries.

Threshold concepts represent, or lead to, troublesome knowledge that is conceptually difficult, counter-intuitive, or "alien" (Perkins 1999). Difficulty in grasping the threshold concept may result in the learner getting "stuck" and holding an understanding that lacks authenticity and depth. In attaining the threshold concept, the learner moves from a common sense understanding, and from previously held and apparently obvious beliefs, to a transformed view of the subject matter. This allows further progression. Meyer and Land (2003) identify the transition from the old understanding, to acquisition of the threshold concept as a "conceptual gateway" or portal. Movement through the portal is transformative, as the perception of the subject matter is shifted irreversibly: as a result, the learner does not return to viewing it in a more primitive way.

Here, we argue that empathy is a threshold concept for students of design ergonomics. A user-centered approach that recognizes individual diversity and complexity can be conceptually difficult or alien to students who have previously focused on aesthetics, their own requirements, or reducing others to stereotypes. For them, the application of empathy is often a new way to approach design that is based on research and a recognition of the importance of design for all. Those who find it difficult to place themselves in the position of other people are unlikely to be able to translate user needs into effective user-centered products.

Without true empathy, methodology will be applied in an artificial sense. This represents the troublesome period during which existing beliefs and ways of working are challenged by new knowledge, leading students to lack authenticity or engage in mimicry (Clouder 2005). Often, we see confusion in students as to what embodies an ergonomic issue. For example, students will consider product reliability without reference to the impact on the user. Similarly, they undertake the steps of research without gaining deep insight into user needs, or accessing the true meaning of the information. With the acquisition of the threshold concept, the subject area becomes more broadly recognizable and defined. Once empathy is developed, the transformed view of design ergonomics should represent how designers (and certainly ergonomists!) practice within the discipline.

11.1.4 LEARNING STYLES OF DESIGNERS

The application of rigorous research methods and empathy can be counter-intuitive to a designer's natural and learned style of thinking and working. Learning styles are stable indicators of how learners perceive and interact within learning environments (Schmeck 1985). They are the preferred manner in which information is processed and although they are subject to variability due to student abilities, personal preferences, and field of study (Kolb 1981), many of the influences on study behavior are stable (Nulty and Barrett 1996).

Subject disciplines traditionally adopt a particular educational process, which influences a student's choice of study behaviors; a mismatch between the learning style of the student and the discipline can lead to difficulties (Nulty and Barrett 1996). It is likely be beneficial if students learn to converge on the learning style that most closely meets the knowledge and methodology associated with the discipline.

Carl Jung's theory of psychological types illustrates preferred ways of adapting and learning. The Myers-Briggs Type Indicator (MBTI) based on Jung's theory is a psychometric tool for assessing 16 "types" and their associated learning styles (Myers and McCaulley 1985). It has been used to assess the learning styles of various occupational groups. Durling, Cross, and Johnson (1996) identify through extrapolation from similar groups (architects and artists) and a study of design students, that designers are quite different to the general population and to other disciplines.

This is illustrated in Figure 11.1, which shows where occupational groups tend to cluster in relation to learning preferences. As a result, Durling, Cross, and Johnson (1996) summarize that designers broadly prefer teaching that

1. Starts with the big picture and then explains details
2. Focuses on future possibilities and gives alternative viewpoints
3. Has a lightweight structure, allowing for guided exploration
4. In the main, shows objective data, is logical and analytical, and is based on exemplars showing things (although one-third of designers are happier with more subjectivity, a person-centered approach, and the utilization of value judgments)

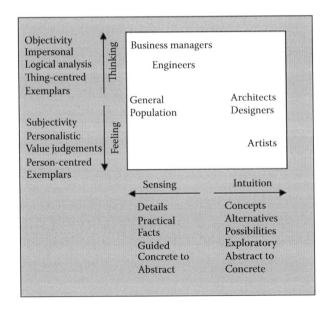

FIGURE 11.1 Learning preferences based on Myers Briggs types. (Adapted from Durling, D., Cross, N. and Johnson, J., *Proceedings of International Conference on Design and Technology Educational Research '96.* Loughborough University, UK, 1996. With permission.)

The strong reliance on intuition is apparent in Figure 11.1 and supports Pheasant's fifth fallacy as well as other research suggesting a strong link between designers' creativity and intuition (MacKinnon 1962; Davies and Talbot 1987). Practicing designers report that when choosing between the various ideas generated, they know when the right one presents itself, though this process is difficult to verbalize (Davies and Talbot 1987). The reliance on intuition is reinforced by the complexity, time, and expense involved in empirical approaches and the belief that ergonomics is common sense (Helander 2000; Scott 2005). As a result, Kanis and Wendel (1990) report that industrial designers tend to rely on five foundations for their design approaches:

1. Literature searches
2. Experimentation with existing products
3. Trials with new designs by the designer
4. Generalized private experiences
5. Presuppositions

In Figure 11.1 the disparity between engineers and designers is highlighted, and Durling, Cross, and Johnson (1996) point out the difficulties of teaching designers "other subjects," such as engineering and computer skills. The natural leaning toward intuition and away from facts and a guided approach, points to the challenge of teaching designers the systematic research approach required of ergonomics. Design education is largely studio based and experiential (Lawson 2006). Designers' learning is exploratory and flexible. This is well matched to the adaptable, project-based methods of teaching involving large measures of personal tuition. For subjects such as engineering and ergonomics (especially as student numbers increase), there is a tendency to rely on lecture-based teaching and the provision of learning resources to encourage student-directed learning, which is less likely to be suited to design students.

Furthermore, Durling, Cross, and Johnson (1996) point out that "other" subjects (e.g., ergonomics, engineering) are often taught to designers by a subject specialist (i.e., a non-designer) whose style is likely to differ. Given the varied backgrounds of ergonomists, it would be interesting to see where they cluster in terms of Figure 11.1, but we would assume that for those coming from an engineering or psychology background it is likely to differ from that of designers. Where there is a resulting style mismatch, the student may experience psychological discomfort, and knowledge transfer may be impeded. This is not only relevant to design within higher education, but to the professional development of designers, and more broadly to how ergonomic principles are communicated to colleagues and clients.

11.1.5 THINKING STYLES

There is a popular tendency to regard divergent thinking as the core skill in art and design (Lawson 2006). Divergent thinking is characterized by the production of multiple alternative answers from the available information, and a departure from the known and predictable (Durling, Cross, and Johnson 1996). A problem may be solved in a variety of different ways that are all valid, and often unique or novel. This serves as an effective approach where difference and originality is valued, and is often linked to designers' creativity. However, there is now increasing recognition that creative design requires convergent, as well as divergent thinking (Cropley 2006; Lawson 2006). Convergent thinking is directed toward finding the best or correct solution to a problem and is traditionally associated with the study of science and engineering. It is characterized by speed, logic, accuracy, accumulating knowledge and information, and applying set techniques (Durling, Cross, and Johnson 1996; Cropley 2006).

We would argue that one of the reasons ergonomics can pose a challenge to designers is due to the natural tendency to focus on creativity, intuition, and divergent thinking. Ergonomics requires the ability to accumulate and apply factual knowledge, observe closely, draw "correct" conclusions, focus in on, and recognize a solution, and then weigh up the feasibility of solutions. These are characteristics of convergent thinking (Cropley 2006) and are more aligned to a scientific or empirical approach. If design students are naturally less inclined to problem solve in this way, a user-centered approach may pose a challenge for the teaching and learning of ergonomic methodology.

The risk of adopting a more convergent approach is that new ideas are blocked and thinking "leads only to production of tried and trusted, 'correct' answers" (Cropley 2006, 402). Lawson (2006) believes that designers need convergent and divergent thinking in near equal proportions, in order to solve problems, to meet the needs of others, and to create aesthetically pleasing products. Finke, Ward, and Smith (1992) similarly identify creativity as requiring first the generation of effective novelty and then the exploration or evaluation of this novelty to ensure the creativity is genuine, workable, and acceptable (Csikszentmihalyi 1999). Convergent thinking provides the knowledge base and ability to explore variability and identify effective aspects on which creativity can be built.

As teaching staff and professional ergonomists, it is important for us to foster both styles through our teaching, rather than expecting the impossible from creative students. By effectively teaching a convergent approach, we are enhancing the skill set that students can apply to ergonomics and creative design. The means by which potentially "troublesome" ergonomics knowledge is communicated to designers to suit their natural styles is key.

11.1.6 Developing Empathy through Role-Play

Empathy as an interpersonal skill is arguably best developed through experiential learning (Blatner 2002). Blatner (2002) argues that this builds a deeper type of understanding and more flexible thinking than traditional teaching methods. It also suits the learning styles of design students where we see a preference for exemplars, guided exploration, and studio-based learning (Durling, Cross, and Johnson 1996). Simulation is "an attempt to represent reality in various forms" and provides an experiential learning experience (Meister 1990, 203). Meister highlights that simulation is important to ergonomists to help find answers that may not be found using other methods. Role-play is a form of simulation that can support the understanding and application of ergonomics among professionals and students.

Role-play is used in many professions, including the health and social sciences, to develop empathy toward patient's feelings and emotions (Nestel and Tierney 2007), and in economics and business to teach a range of subjects including ethics (Brown 1994). IDEO, the innovative design company, makes use of role-play throughout the design process to "get under the skin" of people. They define it as

> the practice of group physical and spatial pretend where individuals deliberately assume a character role in a constructed scene with, or without props. (Simsarian 2003, 1012)

They recognize the process as "experiential and creatively generative" (Simsarian 2003, 1012). Seland (2006) advocates the use of role-play in design to

1. Understand users and context of use
2. Explore, test, and communicate ideas
3. Involve users
4. Enhance the design process
5. Work with mobile technology

Fulton Suri (2001) also recognizes its value in developing working relationships and communicating important ergonomic issues to implementers by facilitating empathy with users.

Role-play is a useful technique to inform inclusive design (Richardson et al. 1996) and has been demonstrated by Pastalan (1982), who simulated age-related visual changes in order to help architects determine design requirements for the visually impaired in different environments. As Poulson, Ashby & Richardson (1996) highlight, the technique is equally applicable in other design contexts. The use of props and scenarios to simulate elements of a disability and place the designer in the position of a disabled user and improve the ability to empathize has been described (Nicolle and Maguire 2003; Poulson et al. 1996). Role-play has also been demonstrated commercially by the Ford Motor Company through use of the Third Age Suit designed by Loughborough University. The suit simulates reduced joint mobility and declining sensory acuity (reduced tactile and blue light sensitivity), which can occur with age (Hitchcock et al. 2001). It has been used to raise awareness of the physical capabilities of older users, and allow designers and engineers to recognize elements of a product design that may present issues to older users (Hitchcock and Taylor 2003). Nicolle and Maguire (2003) describe the use of empathic modeling, or role-play, through workshops in their teaching of inclusive design at Loughborough University. They found that students were encouraged to think more about how senses and abilities are taken for granted, as well as the coping strategies and adaptation techniques employed by older and disabled users.

In order to address the threshold concept of empathy, and help designers develop their understanding of the field of ergonomics and its relevance, we are making use of role-play in our teaching of industrial design students at Coventry University. As a teaching method, this is particularly suited to the learning and thinking styles of design students.

11.2 ERGONOMICS AT COVENTRY UNIVERSITY

The Department of Industrial Design at Coventry University has been running undergraduate courses in transport design with an ergonomics input since 1972. As well as our transport courses, we run Industrial Product Design and Consumer Product Design courses. These are typically four-year master of design (MDes) programs, including a professional experience year.

As a fundamental part of all the courses, students receive taught ergonomics modules in their first and second year of study. The acquired knowledge and understanding is integrated with design and engineering through projects undertaken during the third and final years. These enhance the students' design capability by encouraging due consideration of aspects that impinge on and direct decision making.

The first and second year modules focus on introducing the principles of human factors/ergonomics and the role of the subject in the design of products or transportation. In the first year this concentrates on developing an understanding of the broad spectrum of users, their experience/interactions with products, and learning how to evaluate these interactions as part of the design process. Students are expected to develop a basic understanding of relevant physical and cognitive ergonomics; understand where to find sources of ergonomic data; and gain practical experience of research and evaluation methods. In the second year, this is built on through practical application to a design project that is required to show evidence of a user-centered approach through the research, design, and evaluation phases.

The focus here is on the teaching of ergonomics during the first year of the degree program. The starting point for the introduction to ergonomics is empathy, and this theme continues throughout the degree. The emphasis is on students understanding different users, thinking beyond themselves and their peer group, and being able to translate this into designs for a range of different people. Empathy, and research methods to develop it, are introduced through lectures; a series of workshops to develop and reinforce the concept; and then application and effectiveness is assessed through coursework, supported by tutorials.

We aim to encourage active learning through materials that provide variety and relate to topics that may interest the students. The activities are not necessarily novel, but their combination and application seek to embed the identified threshold concept of empathy and to follow sound educational principles. While some of the specific activities we describe, e.g., a workshop session, focus purely on the development of empathy, others require integration of various ergonomic concepts and methodology, e.g., coursework activities. Frequent revisiting of the concept of empathy helps its development by reinforcing it through different applications and thought processes.

11.2.1   USER SCENARIOS AND PERSONAS

A systems approach to design is taught to ensure a broader perspective is taken than just individual user interaction. This is achieved through consideration of the context in which the study of user behavior is presented. For example, first year group coursework might require the examination of user requirements within the context of a wider activity or scenario. A broad user scenario is provided such as shopping at the supermarket, which is then broken down into task areas, e.g., making a shopping list, getting to the supermarket, undertaking the shopping task, unloading the trolley, packing the goods at the checkout, and getting the goods home. Through exploration of the scenario, students identify the key tasks, which they break down through task analysis. These are then analyzed from the perspective of various users or personas.

Personas are created to represent a range of different user types. They are given characteristics, which the students research and further develop. They are useful in providing a focus for developing empathy and in evaluating how well resultant design solutions meet the needs of the persona. By working in groups of 8–10, the students explore one of 4–5 personas in depth, first by working in pairs, and then, by sharing the conclusions, they compile the evaluations as a group.

The assessed output is a visual presentation showing detailed insights into the different personas through role-play activities, use of a range of research skills and user testing techniques, a task analysis, product comparison, and an anthropometry study. Finally, a set of user requirements is generated for each persona, together with a prioritized list for the activity context they have explored. The approach is not purely empathic, and the adoption of a participatory design approach and empirical methods are also encouraged.

11.2.2   RESEARCH METHODS

Understanding scenarios and personas gives a context to the use of various research methods. These are formally taught through lectures, and practiced through workshops and coursework. Library and internet research is expected to promote an understanding of user characteristics and specific conditions, whether through books and journal papers or discussion groups. Understanding the psychosocial issues associated with living with a particular characteristic is essential to gaining empathy and preventing simple reduction of a condition/trait to more tangible physical characteristics.

Observing and interviewing representative users to gain detailed insights into how tasks are performed, and individual differences in behavior is encouraged. The importance of end-user input to design is emphasized. These are recorded using cameras, video-taping, and annotated notes. Given the visual focus of the students, picture recording is usually preferred. This is encouraged as long as pictures can be adequately annotated to demonstrate the findings, interpretations, and reflections.

Task analysis is taught as an important tool for use in understanding, designing, and evaluating. Formal methodologies such as the hierarchical task analysis (Shepherd 2001) are taught, but in application this may take the form of a written model or a photographic breakdown. Teaching focuses on core methodology which is practiced during workshop sessions; other methods such as diaries, cognitive walkthroughs, focus groups, questionnaires, and checklists are introduced and often applied during project work. Role-play is one of the research methods we have found most useful in teaching empathy and supporting students in the acquisition of this troublesome knowledge.

11.3   THE USE OF ROLE-PLAY

Ergonomics teaching is often delivered through lectures and student-directed learning, which is unlikely to be the most appropriate approach for many design students. For students who do not have good empathic skills, the relevance and applicability of methodology and theory (as evidenced by Pheasants fallacies and the earlier discussion) may be far from apparent. This may lead them to believe that they are required to apply a set of formal models and methods routinely within a design process without necessarily understanding and applying user needs in their designs. They may also fail to see the interlinking of concepts given that ergonomics is taught as a discrete subject within a much broader design curriculum.

Role-play can be used to instill ergonomic understanding by tapping into the thinking and learning styles that are preferred by design students. Through practical workshop teaching, it is essential to ensure empathy can be applied and related to design thinking. Following introduction of an impairment, application in a workshop helps bring to life and reinforce concepts, e.g., the specific features of a visual condition, and allows more effective connection with the associated day-to-day difficulties.

Learning is known to be more effective when the learner is motivated, able to take responsibility for their own learning, and to make use of their own experience and critically appraise evidence (Knowles 1990). These requirements can be met through role-play workshops and coursework. We aim to make the research methods as accessible and relevant as possible and to demonstrate that ergonomics, as well as sketching, can be enjoyable!

Taught as a research method, role-play introduces a more convergent approach to solving problems, by observing and recognizing the needs of others, analyzing and drawing conclusions, and using this knowledge in order to produce alternative solutions. As an activity, it encourages experiential learning by offering an experience that can be reflected on, integrated with existing knowledge, and then applied to subsequent design work.

While IDEO relate role-play to improvisational theatre, Jordan (2004) links it to the process of method acting used by actors such as Marlon Brando and Robert De Niro. "The Method" involves actors immersing themselves into a character in order to sense and experience their life and emotional conditions. The actor aims to create a realistic performance by drawing on his/her own emotions, memories, and experiences. This is more in line with the use of role-play described here. It is used in quite an unstructured way, in contrast to the use of actors and controlled scenarios that is often seen in medical training. Students are encouraged to experience an activity from the perspective of a user or persona, relate this to how they might feel, and create scenarios that meet the goals of the session or coursework outputs. The experience is quite loosely defined; teaching staff provide direction on the desired outcomes of the activity and follow up with a feedback and reflection session. We aim to build on students own experiences and encourage a more active, self-directed approach to learning.

Students are asked to try and place themselves in the position of another user by taking on single or multiple props, thereby dealing with unfamiliar circumstances. The experience should increase understanding of the user role that has been adopted and the circumstances they face, hence increasing the empathic understanding that can be applied to designing for that user. The sessions make use of various props that simulate a variety of the physical elements of aging, disabilities, and individual differences. Role-play takes place in a real-world context, which has been shown to be beneficial for learning and motivation (Lunce 2006). The learning experience is intended to present psychosocial aspects, as well as physical changes to the students. For example how an individual feels emotionally in social situations with a screaming baby, or frustration at performing a fine motor task with arthritic fingers.

Simulation of third age and disability issues through role-play has been well reported, but we would argue that often it does not describe the range of issues that designers need to consider. Empathy is not just about understanding "old people" and those with disabilities, but appreciating that other users may have a completely different experience due to their gender, physical, or psychological characteristics. The approach at Coventry University goes beyond the empathic modeling of disability and third age issues and considers wider user characteristics.

11.3.1 Female Issues

Over the past five years, on average over 93% of Coventry University Industrial Design students have been male: as design professionals they will need to be able identify with certain female characteristics.

Students are encouraged to role-play using false fingernails (see Figure 11.2) to explore physical ergonomic issues, such as the postural changes of the hand, e.g., when using a keyboard, alternative use of controls, and the emotional sense of protection of one's nails.

The impact of the female anatomy, clothing such as straight skirts, and accessories such as high heels and handbags on behavior, needs, desires, and expectations is important. Some male students have been willing to engage in role-play to consider the influence of female attire on daily tasks, such as the associated postural changes and psychological feelings (e.g., vulnerability) that can accompany using a handbag. Figure 11.2c and d show students wearing restrictive clothing and false breasts for research purposes. As well as enhancing empathy, this serves as a reminder of the sorts of allowances that may be made for the female anatomy, clothing, and nails when designing for physical fit.

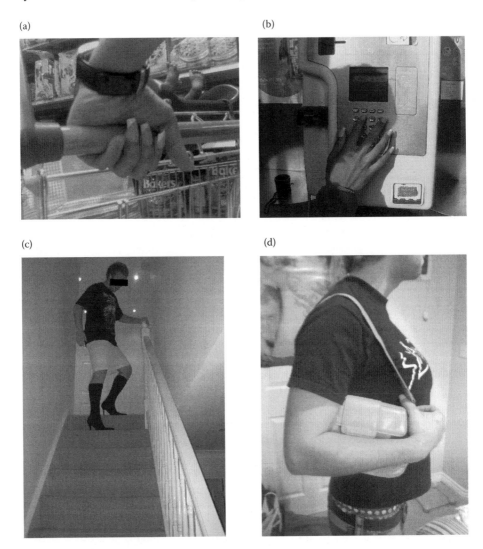

FIGURE 11.2 Male students role-playing female characteristics (a, b: false nails; c: high heels and a skirt; d: handbag and breasts).

11.3.2 PARENTING ISSUES

We have two pregnancy suits—an Empathy Belly® (Birthways, Inc., USA) and The Bump (Life Choice Ltd., UK). Both garments enable the user to experience some of the symptoms and changes experienced during pregnancy, e.g., weight gain, backache, shifting of the center of gravity, and postural changes. The suits are weighted ("Empathy Belly" 30 lb/13.6 kg, "The Bump" 26.5–27.5 lbs) to provide enlarged breasts and a pregnant belly within a maternity smock. They were designed for educating expectant fathers, school children, medical and nursing students, and managers about their workforce.

Our students are taught about the physical and psychological effects of pregnancy and hormones. They use the belly (see Figure 11.3) to help reinforce this understanding and to increase their sensitivity toward pregnant women and their needs. The suits have also been used to simulate obesity and provide insight into the experience of being overweight. The suits are especially important in the consideration of design for physical fit issues, comfort and postural changes.

The department has also acquired "an infant simulator", christened Polly-Esther. She is a "Baby Think it Over" RealCare® Baby II (Realityworks, Inc., WI, USA), originally designed to allow teenagers to appreciate the pressures and responsibilities associated with parenthood. She is a life-sized infant simulator that weighs 6.5–7 lbs and has a flexible neck that requires the head to be supported at all times. An internal computer simulates the baby crying at realistic intervals that can be adjusted to increase the amount of care time required and reduce time between episodes. Polly-Esther senses that she is held, moved, and the type of care received—whether she is fed, burped, rocked, or has her nappy changed. She is equipped with a change of nappy, clothes, and bottle, and is used in conjunction with a car seat and push chair.

Polly-Esther allows consideration of the issues associated with caring for a child during the performance of everyday activities. It has been particularly enlightening for many of the students to experience operating products one handed while looking after a child, negotiating spaces with the pushchair, and managing the amount of kit associated with a small child (see Figure 11.4). Dealing with a screaming baby in public and prioritizing a baby's immediate needs have drawn attention to the emotional impact of children on operating products and systems, and the need for good usability. For example, both Polly-Esther's baby seat and push chair are difficult to assemble and fit. "Working it out" often presents a challenge to a design student, but with a crying baby in their arms, in a public car park they experience frustration and embarrassment.

FIGURE 11.3 A male student using "The Bump" pregnancy belly.

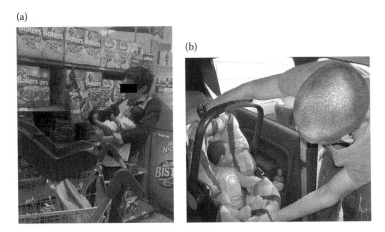

FIGURE 11.4   (a, b) Polly-Esther role-play outings.

11.3.3 PHYSICAL DISABILITIES

To explore the impact of reduced mobility, a number of props are available to simulate physical disabilities. Crutches, wheelchairs, and the use of sticks and knee braces provide some experience of the physical and mental aspects of having a disability and the effect on product interaction. Neck collars are used to allow consideration of visibility and the changed sense of awareness in densely populated areas. They have a direct impact on the ability to visually scan the immediate environment quickly and to collect visual information.

It is estimated that up to 80% of people will experience back pain lasting more than a day at some time during their life (Maniadakis and Gray 2000). The widespread nature of the condition and the increasing prevalence with age makes it an important design consideration. Taken from the experience of one of the author's training as a physiotherapist, taping methods are used to simulate the postural changes that occur in acute back pain. Using micropore and zinc oxide tape, the back is taped to restrict mobility, which causes discomfort when bending (Figure 11.5). This has been found especially useful for considering seating posture in vehicles and the positioning of controls and displays in public spaces.

11.3.4 SENSORY IMPAIRMENTS

The students have the opportunity to perform tasks with impairments to their tactile, visual, and auditory senses. Ten pairs of glasses that simulate various visual impairments have been acquired

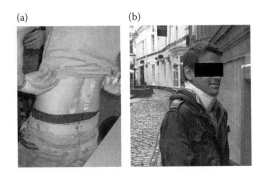

FIGURE 11.5   Simulation of back stiffness (a) and use of a neck collar (b).

from a charity that supports blind and partially sighted people—Visual Impairment North-East, England. They simulate conditions such as loss of peripheral vision, loss of half visual field in each eye, severe loss of vision giving light projection only, and reduced visual acuity leaving approximately 3/60 to 1/60 vision (registered as blind with this level of visual acuity) (see Figure 11.6a). Glasses smeared with Vaseline have also been used. The glasses help identify some of the problems visually impaired people experience when navigating around spaces and handling everyday products. They draw attention to control and display issues, such as text size, color contrast, and the use of shape and tactile cues.

Gloves of varying thicknesses can be used to simulate the loss of tactile sensitivity associated with aging and conditions such as Reynaud's phenomenon. Figure 11.6b shows the use of surgical gloves for this purpose. Earplugs are also used to reduce auditory stimulation.

11.3.5 THIRD AGE MODELING

The effects of aging are experienced through simultaneous use of a number of the available props as it is important to consider the combined effect on life experiences given multiple impairments. For example, the use of neck braces and tape to restrict joint mobility (see Figure 11.7), glasses and gloves to mimic changes in vision and loss of tactile discrimination that occur with age. Students are encouraged to remember that they will be designing for users of all ages, their future selves, and the market share this represents.

11.3.6 SKILL PROGRESSION

The empathy research workshops and coursework are undertaken in year one of the degree program. It is anticipated, and reflected in the course design, that as a result of increased empathy, sensitivity to others, and an ability to identify key users and their characteristics, that the students are able to develop their own personas from the second year onward. The use of props continues throughout their time at Coventry for research, testing mock-ups, and evaluating existing designs.

11.3.6.1 Scale Models

Scale models are used to understand the experience of a smaller or larger user, or how products are perceived and used from a child's perspective. Application of a scaling process to design models allows students to gain first-hand experience of the usability of the products from the perspective of users at the extremes of the population in terms of anthropometric dimensions. For example, Figure 11.8 shows a project involving the scaling of a mobile phone to fit very large adult (97.5th percentile male) and small female (2.5th percentile female) hands. This was achieved by calculating

(a)   (b)

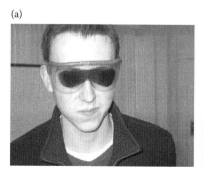 

FIGURE 11.6 Simulation of sensory impairments including Visual Impairment North-East simulation glasses (a) and surgical gloves (b).

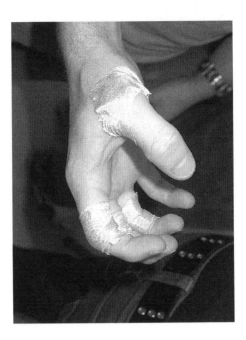

FIGURE 11.7   Simulation of physical restrictions with tape.

a scale factor between the user's hand dimensions and those of the student and making representative models of the mobile phone based on the scaled dimensions.

11.3.6.2   Product Testing and Design Evaluation

Role-play is encouraged in the assessment of existing products. Workshops have focused on the comparison of different brands and models of potato peelers, tin openers, and cars in terms of their usability by individuals with different physical problems, such as arthritis or a broken arm (see Figure 11.9).

The use of props in the testing of mock-ups provides a "first pass" evaluation to highlight potential problems for specific user groups. This is introduced during the second year of the degree and is particularly important in the later stages when students have the opportunity to work through the full user-centered design process, balancing out aesthetic, ergonomic, and engineering requirements. Subsequent evaluation with representative users is then expected.

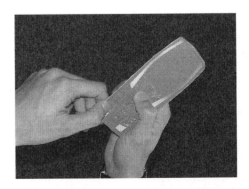

FIGURE 11.8   Simulating the size of a mobile phone in a child's hand.

FIGURE 11.9   Kitchen utensil evaluation workshop.

11.4   EVALUATION OF THE APPROACH

As a teaching and research method, we have found role-play to be efficient in terms of the time required to prepare, carry out, and reflect on activities. It encourages students to empathize with the position and feelings of others, and to look beyond their immediate assumptions and expectations. Used in the first year of a four-year degree program, it helps build group interactivity, communication, and team-working skills when run as a group exercise. Post-task reflection allows practice of presentation skills and analysis of the information collected. Sharing the experience of role-playing allows comparison, reflection, and discussion around the experiences of others.

11.4.1   Student Feedback

A survey was given to a first-year student cohort following a role-play workshop to inform our teaching practice. The workshop required the students to undertake a short journey by bus, train, taxi, or car in pairs, with one observing and the other performing the activity with a role-play prop. During the activity the students were asked to consider the tasks involved at each stage of the journey, how the task differs for them given the characteristics being simulated, the difficulties experienced, and then to make suggestions of design features that would make life easier for users with the given characteristics.

The survey requested responses to a number of questions on 5-point Likert scales as well as free text feedback regarding the perceived value of the role-play exercise. The survey was administered during a lecture and was returned by 54 out of 74 students (73%), 53 of whom were male (mean age 19.5 years). The results are summarized in the graphs in Figure 11.10.

Ninety-six percent of respondents felt that they had learned something useful in the session. A mean response of 4.26 was given when asked "How useful did you find role-playing as a way to understand other people their characteristics and needs?" (median and mode = 4). Some of the specific feedback comments given are listed in Table 11.1. The feedback suggests that the experience had begun to challenge some of their assumptions. Comments such as "tasks which I thought were very simple became a struggle for people with disabilities" and "helped you notice points I did not believe" suggest the experience has helped them reflect on their pre-existing beliefs (and Pheasants fallacies 1 and 5).

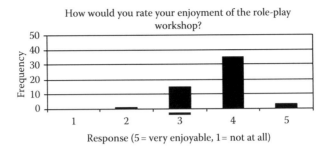

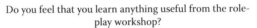

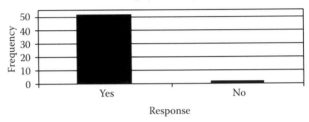

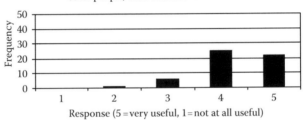

FIGURE 11.10   Student feedback survey results.

The session was not as successful for all the students: "ergonomic problems are obvious, role-play is not needed" and "most of it is common sense though." Other comments perhaps suggested poor depth of understanding, e.g., "I got to understand how it feels to be partially blind." These views reflect in terms of the threshold concept, a lack of authenticity and true application of empathy. This is unsurprising as the workshop was held during the second week of their first ergonomics module and understanding is expected to develop as the module progresses. The session ended with a facilitated discussion, which aimed to focus on more subtle observations and findings, and ensure students were not relying on generalizations and assumptions.

Enjoyment of the session was also rated highly. Table 11.2 represents some of the comments received when asked "what are your feelings towards undertaking role-play as an activity?" Some students found the experience embarrassing, but for most this was short-lived. The majority seemed to recognize it as a valuable experience, and perhaps preferable to two hours in the lecture theatre!

The students' subjective assessment of role-play as a valuable and enjoyable learning experience is positive and indicates student engagement; however, it does not indicate whether the knowledge and skills acquired will be retained and transferred to later research and design scenarios.

Going forward, it is important to demonstrate the value of the method both in terms of improved empathy and more broadly in the application of ergonomics. In the long term, we hope to be able to demonstrate the development of empathy throughout the degree course and to see

TABLE 11.1
Student Feedback on Learning from the Role-Play Workshop

What did you learn from role-playing /observing the role-play activity?

- To hold a child, and how difficult it was to cater for the child
- How difficult it is to interact with a car while wearing fingernails
- That being pregnant is very tiring
- The implications involved with having a baby—much heavier than I thought
- Wearing a neck brace significantly reduced movement and was awkward for controls and visibility
- Public transport does not accommodate everyone, especially pregnant and fat people would struggle with the seating
- Watching others showed the different postures and speeds they travelled at
- How variable different types of people's movement and comfort thresholds are
- Tasks which I thought were very simple became a struggle for people with disabilities, e.g., crossing the road
- Really helps to see how much of a disadvantage back problems are
- Helped you notice points I did not believe
- How it is harder to do very simple tasks because nails get in the way
- How hard it really can be for older people and disabled people, especially things as simple as getting in and out, and getting in the back (of a car)
- The way that certain conditions affect a task may not go as you'd think
- Neck movement with the brace restricts certain activities, i.e., picking up dropped items, drinking from a glass

this reflected in final year design work. Long-term evaluation is planned to determine whether the threshold concept of empathy is acquired over the duration of the teaching program and to further determine the nature of the concept. Evaluation will need to be sufficiently discriminatory to differentiate those students who can effectively apply empathy and those who have only a superficial understanding.

11.4.2 MAKING IT WORK

Seland (2006) points out that role-play as a method is fresh, open, and entertaining, but is not suited to everyone. The approach taken here does not require, or test, acting skills as can be the case with the way it is employed by other disciplines, and in other scenarios. This is felt to be important to reduce the potential discomfort and embarrassment felt by students and to remove any pressure to perform a role, thereby focusing on the learning experience. The fear of looking foolish might initially inhibit some students, but this is minimized by operating it as a group task, and feedback suggests most students do engage with the activity.

Given the fluidity of the activity, it is important to provide clear guidelines and indications of the required outputs. This typically involves identifying the key questions or issues that should be addressed during the task by both the role-player and observer. The students leave university premises to carry out the activities, and there is little direct control over the learning experience.

To ensure the safety of the students while undertaking role-playing activities, which often take place in public spaces, they work in pairs. One participant undertakes the role-play activity while the other acts as a guide and observer. The guide ensures the safety of the role-player and practices their observation and recording skills by capturing key interactions on camera. In the consideration of driving tasks, props that may affect driving performance, and therefore safety, are only to be used in undertaking tasks associated with a stationary vehicle. Students are also required to consider the risks involved in the activity and take appropriate steps to reduce them.

TABLE 11.2
Student Feedback on Their Feelings during the Role-Play Workshop

What are your feelings toward undertaking role-play as an activity?

- Fun
- Necessary
- Slightly embarrassing, useful, and enjoyable
- Insightful
- A good way of understanding the activities
- Useful
- It was quire comical
- Fun, slightly embarrassing at times
- I'm very positive, it's good to experience new situations
- It's essential to empathize in a worthwhile way with others
- Uncomfortable physically (re taped back)
- Different, hard work
- Interesting
- Enjoyable method of understanding the abilities of others
- Useful—a quick way to empathize with user type
- Difficult, but useful
- Important/beneficial
- Good hands-on experience
- It is a bit embarrassing to begin with but after a while it was fun and very useful
- A bit strange, but helpful. It's probably better to make someone with a disability test the design and give feedback
- Pointless
- It was fun, but I'd prefer there were no observers

The activity is recorded through notes and photographs; these are brought back to a debriefing session. This debrief, which we use to end the workshop, allows group reflection on the exercise and ensures that the activity is not just regarded as a game. We have found it essential to ensure that findings and experiences within the groups are shared with those using different props to facilitate in-depth thinking around user issues and to reinforce learning. Expert facilitation can draw out some of the more subtle product interactions and emotional issues that may otherwise be ignored, and to ensure that generalizations and simplifications of user issues are not made. It also provides an opportunity to link the experience back to the learning or research objectives of the session.

When using role-play for assessment purposes, coursework submission has required photographic evidence and reflective annotation of the experience. Assessment has been based on the level of insight gained into a user's characteristics, behavior, and difficulties, and the student's reflections on these issues.

11.4.3 LIMITATIONS OF ROLE-PLAY

Despite the benefits of role-play, it is essential to be aware of its limitations. It is important to emphasize that the characteristics simulated through the use of props are only being adopted for a limited time frame. This is very different to experiencing a real condition in the longer term. A real characteristic or disability results in adaptation and behavioral compensation over a period of time, which cannot be fully appreciated by temporary use of a prop. It is not possible to simulate all elements of conditions or, for example, all the effects of aging, especially the full cognitive and emotional implications. The methods used are only a crude representation. Individual differences are equally significant within aging and disabled populations, and misrepresentation is inevitable if the understanding of a user population is entirely reduced to the use of props. Cognitive impairments are especially hard to represent reliably and to relate to, and the interlinking with physical difficulties is an important consideration (Hitchcock and Taylor 2003).

Thus, there is the risk of oversimplification, and that designers will fail to recognize the complexity of people and their situations. Not recognizing the limitations of the data available may provide a false sense of security in the findings and, as a consequence, poor design implementation. Role-play is not adequate as the only research or evaluation tool and designers should not become reliant on their own assessment of a situation, perhaps even using it to reinforce pre-existing beliefs and intuition. Certainly, in the context of teaching, the use of observers and facilitators to draw out the more subtle findings is very useful to stimulate reflection and refine the application of the data.

As a method, role-play does not, and should not replace the involvement of representative end users in product research, testing, and evaluation. The success of the technique is dependent on the extent to which it is embraced as a research method within the context of a wider empirical methodological approach.

11.5 CONCLUSIONS

This chapter has focused on the importance of empathy for designers to effectively apply ergonomic principles within a user-centered design process. Ergonomics is often perceived to conflict with aesthetics, but is, in fact, about juggling constraints and different user needs through a creative problem-solving approach. A consideration of people's needs can also provide creative, ergonomics-led design opportunities. Empathy binds the understanding and application of ergonomic principles, without which designers will struggle to effectively form and integrate user requirements into innovative and credible design solutions.

While variability exists within all occupational groups, the natural styles and dispositions of designers have been discussed to elucidate the potential challenges in learning and applying ergonomics. This is in line with the belief that research and teaching techniques are likely to be more successful if tailored to designers' characteristics and strengths. This has directed our teaching of ergonomics through experiential learning, a relatively lightweight structure (with low theoretical emphasis), allowing for guided, practical application. The focus on empathy as a threshold concept for design students is important; failure to grasp it will block the learner from accessing the real meaning of content that is subsequently taught (Davies and Brant 2006).

Although not fully robust as a research method, role-play has proved to be a great introduction to design research, to help build an understanding of how life is experienced by various users. We believe it can be effective in aiding the students to think beyond their own gender, age group, and abilities and to provide highly memorable experiences that will later be reflected on. It provides students with a "feel" for situations and characters, increases insight, and encourages a problem-solving approach to address the difficulties they experience. It offers a high level of involvement and engagement in the learning experience and has a potential impact on wider professional skills such as team working.

Role-play provides a way to introduce more empirical research methods, such as field observation. Through the adoption of both empathic and participatory design approaches, it is hoped that students will be able to blend a consideration of needs, feelings, and emotions through empathic experiences with more objective data from interviewing, focus groups, and task analysis methods. Recognition of the limitations of role-play and the need for user participation to verify and add to findings is essential to ensure objective design and user-centered products.

Our description of role-play and its theoretical justification in relation to developing empathy is largely informed by its use as a teaching tool. However, the ideas and props used are more broadly applicable to the product design industry and in continuing professional development as pragmatic techniques for improved user-centered design. Role-play as a simulation method builds on empathic design principles, and provides useful information in situations where it is difficult to access or involve real users.

REFERENCES

Birthways, Inc. USA. www.empathybelly.org/home.html (accessed March 27, 2007).

Blakemore, S.J., and Choudhry, S. 2006. Development of the adolescent brain: Implications for executive function and social cognition. *Journal of Child Psychology and Psychiatry* 47 (3–4): 296–312.

Blatner, A. 2002. Using role-playing in teaching empathy. http://www.blatner.com/adam/pdntbk/tchempathy.htm (accessed March 15, 2007)

Brown, K. 1994. Using role-play to integrate ethics into the business curriculum a financial management example. *Journal of Business Ethics* 13 (2): 105–10.

Carkhuff, R. 1969. *Helping and Human Relationships.* New York: Holt, Rinehart and Winston.

Clouder, L. 2005. Caring as a 'threshold concept': Transforming students in higher education into health (care) professionals. *Teaching in Higher Education* 10 (4): 505–17.

Cropley, A. 2006. In praise of convergent thinking. *Creativity Research Journal* 18 (3): 391–404.

Csikszentmihalyi, M. 1999. Implications of a systems perspective for the study of creativity. In *Handbook of Creativity*, eds. R.J. Sternberg, 313–35. Cambridge: Cambridge University Press.

Davies, P., and Brant, J. 2006. *Teaching School Subjects: Business and Enterprise.* London: Routledge.

Davies, P., and Mangan, J. 2005. Recognising threshold concepts: An exploration of different approaches. European Association in Learning and Instruction Conference, 23–27 August 2005, Nicosia, Cyprus. http://www.staffs.ac.uk/schools/business/iepr/docs/working-paper19.doc (accessed March 29, 2007).

Davies, R., and Talbot, R. 1987. Experiencing ideas; identity, insight and the imago. *Design Studies* 8 (1): 17–25.

Durling, D., Cross, N., and Johnson, J. 1996. Personality and learning preferences of students in design and design-related disciplines. In *Proceedings of International Conference on Design and Technology Educational Research '96*, ed. J.S. Smith, 88–94. UK: Loughborough University.

Finke, R.A., Ward, T.B., and Smith, S.M. 1992. *Creative Cognition.* Boston: MIT Press.

Fulton Suri, J. 2001 The next 50 years: Future challenges and opportunities for empathy in our science. *Ergonomics* 44 (14): 1278–89.

Helander, M.G. 2000. Seven common reasons to not implement ergonomics. *International Journal of Industrial Ergonomics* 25 (1): 97–101

Hitchcock, D., Lockyer, S., Cook, S., and Quigley, C. 2001. Third age usability and safety – an ergonomics contribution to design. *International Journal of Human-Computer Studies* 55:635–43.

Hitchcock, D., and Taylor, A. 2003. Simulation for inclusion…true user-centred design? *Include 2003, Royal College of Art London*, eds. R. Coleman and A. MacDonald, 105–10. The Helen Hamlyn Research Centre, Royal College of Art.

Hogan, R. 1969. Development of an empathy scale. *Counselling Psychologist* 5:14–18.

Jordan, P. 2004. Keynote speech at the Fourth International Conference on Design and Emotion, 12–14 July 2004, Ankara, Turkey.

Kanis, H., and Wendel, I.E.M. 1990. Redesigned use, a designers dilemma. *Ergonomics* 33 (4): 459–64.

Knowles, M. 1990. *The Adult Learner: A Neglected Species.* Houston, TX: Gulf Publishing.

Kolb, D.A. 1981. Learning styles and disciplinary differences. In *The Modern American College*, eds. A.W. Chickering & Associates, 232–55. San Francisco: Jossey Bass.

Lawson, B. 1990. *How Designers Think.* 2nd ed. London: Butterworth Architecture.

Leonard, D.A., and Rayport. J. 1997. Spark innovation through empathic design. *Harvard Business Review* 75 (6): 102–13. (HBS Working Paper No. 97-606.)

Life Choice Ltd., UK. www.thebump.co.uk/ (accessed March 27, 2007).

Lunce, L.M. 2006. Simulations: Bringing the benefits of situated learning to the traditional classroom. *Journal of Applied Educational Technology* 3 (1): 37–45.

Mackinnon, D.W. 1962. The personality correlates of creativity: A study of American architects. In *Personality and Learning Preferences of Students in Design and Design-related Disciplines* (International Conference on Design and Technology Educational Research 2–4 September 1996), eds. D. Durling, N. Cross, and J. Johnson, 11–39. Loughborough University, Copenhagen: Munksgaard.

Maniadakis A., and Gray, A. 2000. The economic burden of back pain in the UK. *Pain* 84:95–103.

Mattelmaki, T., and Battarbee, K. 2002. Empathy probes. In *Proceedings of the 7th Biennial Participatory Design Conference*, Malmo 23–25 June 2002. 266–71.

Meister, D. 1990. Simulation and modelling. In J.R. Wilson and E.N. Corlett (eds) *Evaluation of Human Work: A Practical Ergonomics Methodology*, 180–99. London: Taylor & Francis.

Meyer, J.F.K., and Land, R. 2003. Threshold concepts and troublesome knowledge: Linkages to ways of thinking and practising. Occasional Report 4: ETL Project, Universities of Edinburgh, Coventry and Durham, May 2003. http://www.tla.ed.ac.uk/etl/docs/ETLreport4.pdf (accessed March 30, 2007).

Myers, I.B., and McCaulley, M.H. 1985. *Manual: A Guide to the Development and Use of the Myers-Briggs Type Indicator.* Palo Alto, CA: Consulting Psychologists Press.

Nestel, D., and Tierney, T. 2007. Role-play for medical students learning about communication: Guidelines for maximising benefits. *BMC Medical Education* 7: 3. http://www.biomedcentral.com/1472-6920/7/3 (accessed March 23, 2007).

Nicolle, C., and Maguire, M. 2003. Empathic modelling in teaching design for all. In *Universal Access in HCI, Inclusive Design in the Information Society* ed. C. Stephanidis (Vol. 4, Proceedings of the 2nd International Conference on Universal Access in Human–Computer Interaction, 22–27 June, Crete, Greece), 143–47. Mahwah, NJ: Lawrence Erlbaum Associates.

Nulty, D.D., and Barrett, M.A. 1996. Transitions in students' learning styles. *Studies in Higher Education* 21 (3): 333–45.

Pastalan, L.A. 1982. Environmental design and adaptation to the visual environment of the elderly. In *Aging and Visual Functions*, eds. R. Sekuler, D. Kline, and K. Dismukes, 323–33. New York: Alan R. Liss.

Perkins, D. 1999. The many faces of constructivism. *Educational Leadership* 57 (3): 6–11.

Pheasant, S. 1992. *Bodyspace: Anthropometry, Ergonomics, and the Design of Work.* London: Taylor & Francis.

Poulson, D., Ashby, M. and Richardson, S. (1996). *USERfit, A Practical Handbook on User-Centred Design for Assistive Technology.* Brussels: ESC-EC-EAEC.

Realityworks. http://www.realityworks.com/ (accessed March 20, 2006).

Schmeck, R. 1985. Learning styles of college students. In *Individual Differences in Cognition,* eds. R. Dillon and R. Schmeck, Vol. 1, 233–79. New York: Academic Press.

Scott, A. 2005. *The Black Art of Ergonomics.* In *D for Design*, 91–94. QUT School of Design, Brisbane. http://eprints.qut.edu.au/2915/ (accessed March 20, 2007).

Seland, G. 2006. System designer assessments of role-play as a design method: A qualitative study. In *Proceedings of the 4th Nordic Conference on Human-Computer Interaction: Changing Roles*, eds. A. Mørch, K. Morgan, T. Bratteteig, G. Ghosh, and D. Svanaes, 222–31. NY, USA: ACM New York.

Shepherd, A. 2001. *Hierarchical Task Analysis.* London: Taylor & Francis.

Simsarian, K.T. 2003. Take it to the next stage: The roles of role-playing in the design process. In *CHI '03 Human Factors in Computing Systems*, eds. G. Cockton and P. Korhonen, 1012–13. ACM New York: NY.

Smither, S. 1977. A reconsideration of the development study of empathy. *Human Development* 20:253–76.

Visual Impairment North-East. http://www.vine-simspecs.org.uk/simspecs.htm (accessed March 27, 2007).

Wiseman, T. 1996. A concept analysis of empathy. *Journal of Advanced Nursing* 23:1162–67.

Zahn-Waxler, C. 1991. Empathy: A developmental perspective. *Psychological Inquiry* 2 (2): 155–58.

12 Usability Testing of Three Prototype In-Vehicle Information Systems

Margaret J. Trotter, Eve Mitsopoulos-Rubens, and Michael G. Lenné

CONTENTS

12.1 INTRODUCTION

The sophistication of in-vehicle technologies is rapidly increasing. Vehicles now come equipped with driver assistance systems, such as collision avoidance and electronic stability control, as well as complex in-vehicle information systems (IVIS), such as navigation, telecommunication, and entertainment systems. As IVIS complexity has increased, consumer tolerance for poor design has decreased (Stanton and Young 2003), and where consumers once considered usability a bonus, they are now likely to expect that the products they purchase are usable (Jordan, McClelland, and Thomas 1996). Therefore, designing usable IVIS can provide vehicle manufacturers with an important competitive advantage.

Also, while IVIS have the potential to enhance the driving experience and driver performance of consumers, poorly designed IVIS have the potential to detrimentally affect driver performance and safety by increasing driver workload and the potential for driver distraction (Green 2009). Therefore, it is vital that IVIS are designed with the expectations, requirements, and capabilities of potential end users in mind. Traditionally, the technology-driven approach has predominated in the automotive industry (Brace et al. 2007). In this approach, system design is guided by goals of technological achievement (Marmaras and Pavard 1999). In contrast to the technology-driven approach, a human-centered design approach provides an opportunity through which IVIS developers can address user needs, capabilities, and limitations as part of the overall design process. The extent to which user requirements are being met through the human-centered design framework can be explored through usability testing. Deficiencies in IVIS designs, identified through usability testing, can be addressed in the next design iteration prior to system manufacture and deployment.

Drawing on the framework for assessing usability that was adapted by Regan and colleagues for the in-vehicle context (Regan et al. 2007; Mitsopoulos et al. 2003; Mitsopoulos, Regan, and Tierney 2001), the current study proposed to assess the usability of three prototype IVIS design concepts for music selection using a suite of questionnaires, including the System Usability Scale (SUS) and two custom-designed questionnaires. As a result of this approach, a secondary aim of the study was to further explore the suitability of the SUS for the IVIS context. The three IVIS systems examined in this study were developed by and tested for an Australian vehicle manufacturer. Usability feedback from this study was intended to inform the future IVIS designs of this manufacturer.

12.1.1 Usability

Usability is a multi-dimensional construct (Alshamari and Mayhew 2009; Nielsen 1993) that applies to all aspects of a system with which users might interact (Nielsen 1993). As defined by ISO 92411-11 (1998), the usability of a system refers to the extent to which the system can be used to achieve the goals of effectiveness, efficiency, and satisfaction. Effectiveness refers to the degree of accuracy and completeness with which users can achieve specific goals within the system; efficiency refers to the level of effort required by users in order to achieve the degree of accuracy and completeness necessary to achieve specific goals; and satisfaction refers to the extent to which users have positive attitudes toward system use.

12.1.2 Usability Evaluation

In industrial settings, such as vehicle manufacturing, design concepts often need to be evaluated within strict time frames and budgets (Thomas 1996). This has led to the development of a number of informal usability tests that are more suited to the industrial context, requiring fewer participants over a shorter time period, while still providing important usability information. One such test is the SUS (Brooke 1996), a 10-item scale designed to provide subjective ratings of system usability. The SUS is an easy to use tool requiring few resources to apply, and has been used to evaluate the usability of various systems across different domains. Other usability questionnaires do exist, but the SUS has proved a popular choice for industry. Bangor, Kortum, and Miller (2008, 2009) identified over 200 studies using the SUS over the last decade with the most commonly studied user interface designs being internet-based Web pages and interactive voice response phone systems. The SUS has also been used to evaluate the usability of in-vehicle entertainment systems (e.g., Stanton and Young 1999, 2003). However, while the SUS is quick and easy to administer and score, it does not offer detail on the specific factors that shape individuals' usability scores. This limitation has prompted some researchers and practitioners to develop their own questionnaires for use instead of, or to supplement, scales such as the SUS, that allow for the collection of data on specific usability barriers for a given system.

Three usability approaches were used in the current study: the SUS questionnaire, custom-designed questionnaires, and IVIS task performance data. This chapter focuses only on the former two approaches. Data from the latter approach are described elsewhere (see Mitsopoulos-Rubens, Trotter, and Lenné in press). Custom-designed questionnaires have been used to evaluate the usability of in-vehicle technology (e.g., Adell and Varhelyi 2007; Regan et al. 2007; Mitsopoulos et al. 2001, 2003). This study builds on the approach adopted by Regan and colleagues, employing custom-designed usability questionnaires.

12.2 METHODS

12.2.1 PARTICIPANTS

Thirty participants (16 males and 14 females) aged between 24 and 55 years (M 31.5 years; SD 8.1 years) took part in the study. Participants were recruited from Monash University staff and students using advertisements in the university's weekly newsletter and online career gateway. All held a current full (unrestricted) car driver's license and had been licensed for between 5 and 32 years (M 11.5 years; SD 7.7 years). Across participants, there was a wide spread in terms of the number of kilometers driven per year, but over half the participants reported driving between 5,001 and 15,000 km in a given year. The majority of participants (86%) reported that they listen to music "every time" or "most of the time" they drive.

12.2.2 MATERIALS

Studying usability in an automotive setting has many challenges. While it is important to immerse the participants in an environment that replicates the key elements of the intended context of use (Bevan and Macleod 1994), there are also important legal and ethical concerns. For these reasons, the use of a simulated driving environment holds much appeal.

This study used a simulated driving environment presented on a desktop PC with a 17" monitor, a Logitech MOMO Racing Force Feedback steering wheel with accelerator and brake foot pedals, and IVIS interface. The steering wheel was mounted on a table in line with the monitor, and the foot pedals were located on the floor. A schematic of the experimental set up is provided in Figure 12.1. The IVIS interface (dimensions: 14.5 × 7.5 cm width × height) replicated the radio fascia of a current Australian vehicle, and was able to interpret button presses into serial data sent to a PC. The PC could then convey this message by playing a sound file, making a screen change or both. Participants interacted with the interface via the manual manipulation of several buttons and a rotary control. The buttons served a similar function across the three designs, although the effect of the rotary control on scrolling varied. The IVIS interface was positioned to the left of the monitor table, in the same relative position to the steering wheel as in an actual Australian vehicle. A laptop computer running DirectRT software (v. 2006; Empirisoft) with connecting headphones was used to administer the IVIS tasks.

12.2.3 TASKS

Several IVIS task lists were constructed, with each task list comprising seven items that asked participants to play an artist, album, or song (e.g., "Play the song 'Take a Bow' by Madonna"). The goal of each item was to select a song by navigating through the menu structure using the main rotary control and available button options.

The tasks were constructed so as not to be overly prescriptive, allowing participants to perform the task via a number of routes. For example, to complete the item "Play the song "Victim of Love" by The Eagles", participants could either search through the song menu or select The Eagles from the artist menu first and then look for the specific song among the available songs by that artist. However, regardless of route choice, successful completion of each item required multiple steps. Six task lists, and two versions of each task list, were prepared (Table 12.1). The task lists were

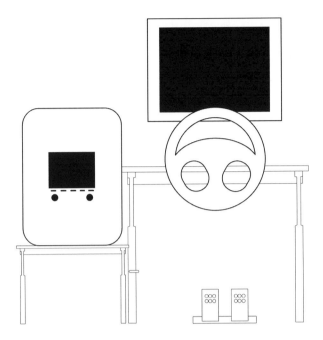

FIGURE 12.1   Schematic representation of the experimental set up.

specifically constructed so that each list required approximately the same number of button presses in total. The two versions differed only in the order in which the tasks in each list were presented. The task lists for interface Design C differed from those used for the other two interfaces as, at the time of testing, only the search by artist function was active in this system. Tasks were administered aurally via a headset in a self-paced manner.

Participants were required to perform the IVIS task while engaged in a simulated driving task. This study used the Lane Change Test (LCT; Mattes and Halén 2008), which has been used previously as an effective surrogate driving test in a large number of IVIS studies (e.g., Engström and Markkula 2007; Harbluk et al. 2007). The LCT simulates driving on a 3 km straight three-lane road (see Figure 12.2). Participants begin in the center lane, then follow signs positioned on either side of the road that direct them to move into different lanes. The lane change information appears on the approaching sign when the distance between the driver and the sign is 40 m. Once the information on the sign appears, the participants' task is to move into the lane indicated on the sign as quickly as possible. Participants must drive at a constant 60 km/h—the maximum speed achievable in the system.

12.2.4   DESIGNS

The three IVIS designs differed in how information was displayed on the screen, although a hierarchical menu structure was used in each case to structure the information. The music information in each design was categorized alphabetically by artist, album, and song and participants could navigate directly to these three lists. From the artist menu, the next menu level below a selected artist was the albums of that artist, and below a selected album, the songs specific to that album. Likewise, for the Album menu, the next level below a selected album was the songs specific to that album.

Interface Design A (Simple List) comprised a list-style menu where the first item in the list was highlighted for selection. This design concept is based on a standard, linear list format (Figure 12.3, top). Items in Design A were displayed in a red font against a black background, with the exception of the selected item, which was displayed in white. Items on interface Design B (Modified Fisheye) were fanned out around a point on the left of the screen in a fisheye menu

TABLE 12.1
Version 1 Tasks Lists

Version 1/List 1 (VE Modified and Fisheye)

Start playing the album "The Day" by Babyface

Play the song "Showtime"

Play the song "Take a Bow" by Madonna

Play the song "Forbidden Love" by Madonna

Start playing all songs by Chris Brown

Start playing the album "Flag"

Play the song "Victim of Love" by the Eagles

Version 1/List 3 (VE Modified and Fisheye)

Play the song "Guiltiness" by Bob Marley

Start playing all songs by Michael Jackson

Start playing the album "Best of Van Halen" by Van Halen

Play the song "Stupid Girls"

Start playing the album "Exclusive"

Play the song "Killer Queen" by Queen

Play the song "Crazy Little Thing Called Love" by Queen

Version 1/List 5 (Cover Art)

Play the album "Loose" by Nelly Furtado

Play all songs by Queen

Start playing the album "Shock Value" by Timberland

Start playing all songs by Yello

Play all songs by the Eagles

Start playing the album "JackInABox" by Turin Brakes

Play the song "When your Body Gets Weak" by Babyface

Version 1/List 2 (VE Modified and Fisheye)

Play the song "Leave Me Alone (I'm Lonely)" by Pink

Play the song "Fingers" by Pink

Play the song "Seven Bridges Road"

Start playing all songs by Babyface

Start playing the album "Off the Wall"

Start playing the album "Diary of Alicia Keys" by Alicia Keys

Play the song "My Love" by Justin Timberlake

Version 1/List 4 (VE Modified and Fisheye)

Start playing the album "Chant Down Babylon" by Bob Marley

Play the song "Simple Days"

Play the song "Promiscuous" by Nelly Furtado

Play the song "Do It" by Nelly Furtado

Start playing all songs by Pink

Start playing the album "Future Sex Love Sounds"

Play the song "With You" by Chris Brown

Version 1/List 6 (Cover Art)

Play the song "Heartburn" by Alicia Keys

Start playing all songs by Van Halen

Start playing the album "Something to Remember" by Madonna

Play all songs by Bob Marley

Start playing the album Greatest Hits (Parlophone) by Queen

Play all songs by Justin Timberlake

Play the song "My Love" by Justin Timberlake

style where the font size of the items increases toward the center, and the largest, central item can be selected (see middle of Figure 12.3). Items on interface Design B were displayed in a red font against a blue background, with the exception of the selected item, which was displayed in white. Interface Design C (Cover Wheel) was a variation on a fliptych-style menu, in which the letters of the alphabet were depicted on fliptychs and positioned around a rotating wheel (see the bottom panel of Figure 12.3). The letter turned to the center front of the wheel could be selected, bringing up a list of items under that particular letter. The album covers of the items under each letter were displayed above the letters on the wheel. Album covers were displayed in full color, while the wheel was displayed in green and the letters of the alphabet in white. These were all displayed against a blue background.

12.2.5 Usability Questionnaires

12.2.5.1 System Usability Scale

The SUS (Brooke 1996) is a 10-item scale that produces a single numerical measure of system usability. Participants respond to each item on a 5-point Likert scale. Responses are then scored to produce a combined score out of a possible 100. The higher the composite score, the greater the usability of the system. Brooke (1996) does not provide any details as to the interpretation of the SUS scores; however, Bangor, Kortum, and Miller (2008) offer an interpretation of the SUS scores.

FIGURE 12.2   Screen shot taken of a participant's experience of the Lane Change Test.

Systems with SUS scores of 70 or above are considered acceptable, those with scores between 50 and 70 are considered marginally acceptable systems, while those systems with SUS scores below 50 are judged as being unacceptable and almost guaranteed to be associated with usability problems in the field. These scores are used as a guide for the interpretation of SUS results in the current study.

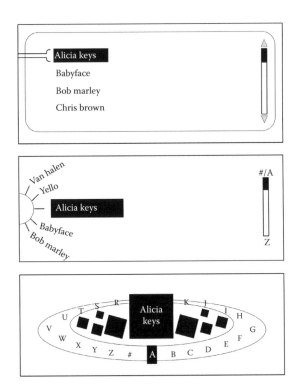

FIGURE 12.3   Schematic representation of the design concepts for music selection: Design A: Simple list (top), Design B: Modified fisheye (center), and Design C: Cover wheel (bottom).

12.2.5.2 Custom Questionnaires

One general and one system questionnaire were developed to address the usability attributes of effectiveness, efficiency, and satisfaction (ISO 9241-11 1998). The majority of the questions in the system questionnaire contained two parts. The first part required participants to respond "yes" or "no", or to provide a rating on a 5-point Likert scale. The second part of the question required participants to justify their responses in an open text field. For example, "Was it easy to get lost in the menu levels (i.e., forget where you were)? Yes or No. If you responded "Yes," what was it about the system that made it easy for you to lose your place?" As part of the general questionnaire, participants were asked to rank the three designs on each of effectiveness, efficiency, and satisfaction.

12.2.6 Procedure

Participants took part in individual sessions of approximately 2 hours in duration. On arrival, participants were first given instructions on the performance of the LCT and then completed two practice drives of approximately 3 minutes each. Next, participants were instructed in the use of, and given practice with, each of the IVIS interface designs. Participants performed a baseline drive before beginning the experimental tasks. While driving performance data were collected as part of the broader study, this section outlines only the aspects of the procedure that are relevant to the collection of the usability questionnaire data. For the experimental tasks, participants first carried out the tasks in a given task list without driving (IVIS baseline), and then carried out the tasks in another list while also driving (dual task). This pattern was followed for each of the three IVIS, with the order that participants experienced the IVIS counterbalanced to ensure against any practice effects. The task lists were administered via headphones. Participants would hear a task such as "Play the song Hotel California by The Eagles" and would then interact with the IVIS in order to complete the task. Each task was judged complete when the correct song was playing in the system. Following completion of each dual task scenario, the participants completed the SUS and the system questionnaire, meaning participants completed the system questionnaire three times, once for each IVIS design. To conclude their session, participants completed the general questionnaire.

12.3 RESULTS

When completing both the SUS and the custom questionnaires, participants were asked to consider their combined experience from the baseline and dual task conditions. Thus, the usability questionnaire data that were collected are for this collective assessment.

SUS questionnaire data were analyzed using a one-way repeated measures analysis of variance (ANOVA) where design was the independent variable. The data from the custom questionnaires were analyzed descriptively and, where appropriate, with a relevant non-parametric statistical test.

12.3.1 System Usability Scale

SUS questionnaires were scored according to Brooke (1996). The SUS is designed to produce a single composite score and the scores of individual items are not meaningful. As can be seen in Figure 12.4, mean SUS scores ranged from approximately 52 to 66, with Design A (Simple List) receiving the highest score, Design C (Cover Wheel) the next highest score, and Design B (Modified Fisheye) receiving the lowest score. A one-way repeated measures ANOVA revealed a significant effect of IVIS design (F (2, 58) = 5.35, $p < 0.05$). A series of three paired sample t-tests with Bonferroni adjustment ($\alpha = 0.017$) revealed that Design A scored significantly higher than Design B in overall usability. Using Bangor, Kortum, and Miller's (2008) guide to SUS score interpretation, the scores of all three IVIS can be considered marginally acceptable and could all benefit from further design improvement to enhance their usability. The SUS scores for Design B are bordering on unacceptable, so this design in particular could benefit from improvement. The reasons underlying this pattern of results were explored further using the information collected as part of the custom usability questionnaires.

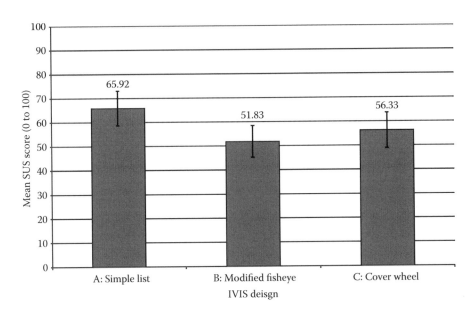

FIGURE 12.4 Mean SUS scores for designs A (Simple list), B (Modified fisheye), and C (Cover wheel). Error bars represent the standard error of the mean (SEM).

12.3.2 CUSTOM QUESTIONNAIRES

12.3.2.1 Effectiveness

Effectiveness here refers to the ease at which the music selection goal could be achieved. As part of a set of questions designed to address effectiveness, participants were asked to indicate on a scale from "very difficult" to "very easy" how easily they were able to navigate through the menus of each system. Approximately one-third of participants found it "easy" to navigate through the menus. Analysis of the data using the Friedman test revealed no significant difference between the designs (Fr=5.37, $p>0.05$).

Table 12.2 summarizes the main difficulties experienced by participants under the three usability attributes. As indicated in Table 12.2, the main effectiveness issue reported by participants for Design A (Simple List) was a difficulty seeing information on the screen due to the red font and small font size. However, only a minority of participants reported these difficulties. A small number of participants also reported a difficulty seeing information on the screen in Design B (Modified Fisheye) due to the color and orientation of the font. However, by far the main effectiveness issue experienced by participants in relation to Design B was the lack of correspondence between the direction in which the rotary control was turned and the direction that turning the rotary control caused the list to move. The main effectiveness issue for Design C (Cover Wheel) again related to the ability to see information on the screen. In this case, participants attributed this to the lack of clarity of the graphics.

As an overall indicator of effectiveness, participants were asked to rank the designs from the most easy to use (Rank 1) to the least easy to use (Rank 3). The designs were given points for each ranking they received: Rank 1, three points; Rank 2, two points; and Rank 3, one point. The summation of these ranking points indicated Design A was considered the most effective system to use (indicated by the rating "1" in Table 12.2).

12.3.2.2 Efficiency

Efficiency here relates to the speed with which participants felt that the music selection tasks could be accomplished accurately. Participants were asked to respond to the question "How efficient was the system in allowing you to select music?" A Friedman test revealed that the pattern of responses varied significantly across the designs (Fr=6.69, $p<0.05$). A series of three Wilcoxon tests with

TABLE 12.2
Main Issues and Ratings of the Interface Designs

Usability Attribute		Interface Design		
		Design A (Simple List)	Design B (Modified Fisheye)	Design C (Cover Wheel)
Effectiveness	Rating:	1	3	2
	Main difficulty:	Red font and small font size	Direction of rotary dial movement not corresponding to screen movement; font orientation	Cover graphics unclear
Efficiency	Rating:	1	3	2
	Main difficulty:	Speed of scrolling too slow	Speed of scrolling too slow	Speed of scrolling too slow; return to "A" too slow
Satisfaction	Rating:	2	3	1
	Main difficulty:	Issues with effectiveness and efficiency	Issues with effectiveness and efficiency	Issues with effectiveness and efficiency
Liked most		• Simplicity • Clear format, familiarity	• Font size of selected item • Items visible on either side of selected item	• Ability to search by alphabet • Album cover graphics
Liked least		• Slow scrolling • "Dull" design	• Discrepancy between expected screen movement and actual screen movement when using rotary dial • Slow scrolling	• Slow return to "A" when going back to previous menu

Bonferroni adjustment ($\alpha = 0.017$) revealed the source of this significant difference to be between Designs A and B, with Design A being perceived as more efficient than Design B.

The main efficiency issue identified by participants for all designs was the slow speed of scrolling. To improve the efficiency of Designs A and B, participants suggested incorporating search by alphabet as the first step in searching by artist, album, and song in order to reduce the time spent scrolling through long lists. Participants also suggested increasing the speed of scrolling by making scrolling speed correlate with the speed at which the rotary control is turned. Design C already incorporated searching by alphabet as the first step in the music selection process. Nonetheless, to improve the efficiency of Design C, participants suggested having the screen return directly to its previous letter position when a menu is selected rather than first to the letter "A."

As with effectiveness, the designs were ranked according to their relative efficiency. The summation of these ranking points confirmed that Design A was considered the most efficient (Table 12.2). The efficiency rankings reflected a similar pattern to those obtained for effectiveness.

12.3.2.3   Satisfaction

In order to be usable, a system must be satisfying to use, meaning that users must like the way the system looks and feels and it must serve a purpose. Participants were asked to indicate to what extent they were satisfied, overall, with each of the IVIS designs. A Friedman test revealed that satisfaction levels differed significantly between designs ($Fr = 6.42$, $p < 0.05$), with the majority of responses for both Design A and Design C falling within the range of "unsatisfied" to "satisfied," while responses for Design B were spread across "very unsatisfied" to "satisfied."

In general, the satisfaction issues identified reflected the issues discussed above regarding effectiveness and efficiency. In addition, some participants indicated that they would like the option of being able to personalize the display with their own color schemes and font sizes. Summation of ranking points indicated that Design C was considered the most satisfying to use of the three designs (Table 12.2).

12.3.2.4 Overall

The responses collected for the final questions of the system questionnaire provided some insight into the overall usability of the designs. Participants were asked what they liked most and also what they liked least about each interface design. As shown in Table 12.2, aspects of the visual display dominated what participants liked in the interfaces, although the ability to search by alphabet at the top level of a menu was also appreciated. The participants considered the length of time it took to scroll the menus and discrepancies between their expectations of system function and what was actually experienced, as particularly dislikeable in the interfaces.

12.4 DISCUSSION

The current study explored the usability of three prototype IVIS music selection design concepts by asking participants to perform representative tasks using prototype versions of the designs. The SUS questionnaire was used to compare usability across designs, while custom usability questionnaires were designed specifically for this study in order to provide the qualitative information necessary for a more in-depth understanding of any usability issues associated with the current designs.

Design A (Simple List) was rated the most effective and efficient. This can be attributed largely to its simpler and more familiar display interface, as evidenced by participants indicating that the familiarity and simplicity were two aspects that they liked the most about this design. This design was similar to that used in devices that participants were likely to have had some experience with, such as iPods and MP3 players, explaining its familiarity. Compared to the other two interfaces, Design A also contained minimal use of graphics, a possible explanation for its considered simplicity. Design C was rated the next most effective and efficient, but was rated the most satisfying to use. Design B was rated the lowest in effectiveness, efficiency, and satisfaction. The main usability issue with Design B was the lack of correspondence between expectation and reality in terms of the direction that turning the rotary control caused the list to move. That is, the direction in which the list would scroll when the rotary control was turned in a particular direction was the reverse of these users' expectations. Participants in this study felt that a clockwise turn of the rotary control should move items *down* a list of artists/albums/songs, when in fact it moved the items *up*. Issues such as scrolling speed and font color were also identified for Design B; however, these were not emphasized to anywhere near the extent of the scrolling direction issue.

The pattern of results found in the custom questionnaires reflected the global usability scores found using the SUS, supporting the use of the SUS as a global usability assessment tool for the IVIS context. Nonetheless, the interface scores were reasonably close and the SUS is of limited value in terms of identifying specific usability barriers. Therefore, it is recommended, where practicable, that additional questions be asked of participants in order to gain some insight into the reasons underlying the SUS scores. The SUS scores for all three interfaces were of marginal acceptability according the Bangor, Kortum, and Miller (2008) assessment guide, with Design B bordering on unacceptable. The low scores may be due to the very early stage of the design process at which the interfaces in the current study were evaluated. Even the design considered most efficient and effective (Design A), still requires a great deal of improvement in terms of its usability in order to get its scores to a level at which its manufacturers could consider the system usable.

One of the major concerns identified by participants across all designs was the rate at which it was possible to scroll through long lists. It is important to note, however, that the rate of scrolling was constrained by the technology used at the time of testing. While this is a limitation of the study, it does highlight the issue that a slow rate of scrolling impacts adversely on the usability of an IVIS. Participants indicated that more efficient scrolling would reduce the amount of time looking at the visual display while driving. Findings from the 100 Car Naturalistic Driving study (Klauer et al.

2006) indicate that total eyes-off-road durations of greater than 2 seconds can significantly increase drivers' crash risk. Therefore, future studies could incorporate eye-tracking measures in order to establish the extent to which IVIS designs affect participants' eyes-off-road time.

12.5 RECOMMENDATIONS

A number of the findings in this study could aid the development of usability guidelines or checklists for the development of IVIS music selection designs. While guidelines for the development of in-vehicle systems do exist, such as the European Statement of Principles (ESoP 2008) and the Alliance of Automobile Manufacturers (AAM) guidelines (2006), these are high-level guiding principles, made deliberately broad so as to cover the vast range of current and future in-vehicle technologies (Stevens 2009; Green 2009). Examples of these broader guidelines include "The allocation of driver attention while interacting with system displays and controls remains compatible with the attentional demand of the driving situation" (Design goal II, ESoP 2008); and "Systems with visual displays should be designed such that the driver can complete the desired task with sequential glances that are brief enough not to adversely affect driving" (Principle 2.1, AAM 2006). Checklists, such as the Transportation Research Laboratory's (TRL) checklist (Stevens et al. 1999) also exist, and provide system designers and engineers with specific guidance for ensuring that IVIS are designed in accordance with key human factors design principles; for example, the TRL checklist asks "Does the information presented on the IVIS display appear legible?", then, more specifically, the checklist asks whether legibility is compromised by size of image, color, contrast, and other image-related features, and, among other questions, asks whether red/green and blue/yellow color combinations are avoided.

Although this was not the principal aim of this study, the following findings from this research support recommendations made in several existing guidelines and checklists, as well as suggesting several additional issues for consideration by IVIS interface designers and manufacturers as part of the design process:

- The SUS provides useful quantitative data; however, using a suite of questions, including those specifically designed to obtain relevant qualitative data, is likely to provide great assistance to designers in the development of the next version of the IVIS.
- Difficulty seeing information on the IVIS screen due to the color and size of the font used in some of the design concepts was an issue for participants in this study. Testing the font and background colors and contrast in an early phase of development is likely to ensure prototypes are constructed using colors that will be easily seen by drivers.
- Participants in this study found it difficult to use IVIS designs when the use of controls did not correspond to their expectations. By maximizing the level of correspondence between the manual manipulation of controls (such as the turning of a rotary control) and the effect drivers expect such movement to have on the system (such as the direction a movement causes an onscreen list to move), designers are likely to enhance the usability of their IVIS designs.
- Usability is likely to be enhanced if IVIS designers keep scrolling to a minimum and maximize the speed at which scrolling can be accomplished. The speed of scrolling was of major concern to participant in this study across all the IVIS designs evaluated.

12.6 CONCLUSION

Taking a human-centered approach to IVIS design is important for the development of systems that minimize any negative impacts on driver performance. The assessment of IVIS usability is an important part of the human-centered design process. While there are numerous ways to measure system usability, questionnaires provide a method that is not highly resource intensive. As part of this study, data from custom-designed usability questionnaires supplemented data from the SUS in order to assess the usability of three prototype IVIS music selection designs. The system with the most familiar operation mechanisms and the simplest visual display was thought by participants to be the most effective and most efficient, and was preferred overall. Across all three designs, the design issues that most affected participants' assessments of effectiveness were font size and color and the lack of correspondence between expectations and how certain controls functioned in actuality. The issue most affecting participants' assessments of efficiency was the speed of scrolling. The results of the custom questionnaires reflected the pattern found by the SUS, indicating that the SUS may be useful as a quick and inexpensive means of comparison between different IVIS designs in the in-vehicle context. However, the SUS scores for all three designs were comparatively low. Further use of the SUS to evaluate IVIS in future studies would better enable researchers to determine its validity in this context. The qualitative data from the custom-designed questionnaires allowed the formation of a number of recommendations for future IVIS design improvement. In an extension to this research, we are currently examining the impacts on measures of driving performance associated with the use of these prototype designs. This will provide for a more in-depth assessment of usability issues and safety implications of prototype design features.

The following represent the main findings of this study:

- Design A was preferred because of its ease of use and familiarity. Designs B and C used unfamiliar scrolling concepts and graphical representations and this appeared to negatively affect participants' impressions of the systems' effectiveness and efficiency. When designing future IVIS, designers should focus on simple menu systems that build on familiar constructs.
- Minimizing the scrolling required, even to the extent of adding extra levels to the menu hierarchy in order to reduce list length, is likely to have a positive impact on drivers' impressions of the usability of systems in the driving environment. Allowing any necessary scrolling in a system to be accomplished quickly is also likely to enhance a system's usability.
- IVIS using large, clear fonts in colors that contrast highly with the background are preferred. Simple, uncluttered screen layouts will also enhance drivers' perception of the system's usability.
- The SUS can be successfully used to provide a global measure of the usability of systems in the in-vehicle environment; however, in order to derive the information necessary to make recommendations for design improvement, designers need to administer questionnaires designed to extract relevant qualitative information on the aspects of the tested systems that participants like and dislike.

ACKNOWLEDGMENTS

The research described in this chapter was carried out under contract for the Cooperative Research Centre for Advanced Automotive Technology (AutoCRC) in Australia. The views expressed in this chapter are those of the authors, and do not necessarily represent the views of the AutoCRC. Many thanks to Dr. Paul Salmon for his comments on earlier versions of this chapter.

REFERENCES

Adell, E., and Várhelyi, A. 2008. Driver comprehension and acceptance of the active accelerator pedal after long-term use. *Transportation Research Part F* 11:37–51.

Alliance of Automobile Manufacturers. 2006. Statement of principles, criteria and verification procedures on driver interactions with advanced in-vehicle information and communication systems. Washington, DC.

Alshamari, M., and Mayhew, P. 2009. Technical review: Current issues of usability testing. *IETE Technical Review* 26:402–6.

Bangor, A., Kortum, P.T., and Miller, J.T. 2008. An empirical evaluation of the System Usability Scale. *International Journal of Human-Computer Interaction* 24:574–94.

———. 2009. Determining what individual SUS scores mean: Adding an adjective rating scale. *Journal of Usability Studies* 4 (3): 114–23.

Bevan, N., and MacLeod, M. 1994. Usability measurement in context. *Behaviour & Information Technology* 13:132–45.

Brace, C., Young K.L., Lenne, M.G., and Archer, J. 2007. Human factors research to improve safety through enhanced design of in-vehicle systems. Presented at *2007 Intermodal Conference on Safety Management and Human Factors*, Swinburne University of Technology, Hawthorn, Melbourne.

Brooke, J. 1996. SUS: A 'quick and dirty' usability scale. In *Usability Evaluation in Industry*, eds. P.W. Jordan, B. Thomas, B.A. Weerdmeester, and I.L. McClelland, 189–94. London: Taylor & Francis.

Commission of the European Communities. 2008. Recommendation on safe and efficient in-vehicle information and communication systems: Update of the European Statement of Principles on human machine interface. Brussels.

Engström, J., and Markkula, G. 2007. Effects of visual and cognitive distraction on lane change task performance. In *Proceedings of the Fourth International Driving Symposium on Human Factors in Driver Assessment, Training and Vehicle Design* (July 9–12, 2007), 199–205. Stevenson, WA, USA.

Green, P. 2009. Driver interface safety and usability standards: An overview. In *Driver Distraction: Theory, Effects and Mitigation*, eds. M.A. Regan, J.D. Lee, and K.L. Young, 445–61. Boca Raton, FL: CRC Press.

Harbluk, J.L., Burns, P.C., Lochner, M., and Trbovich, P.L. 2007. Using the Lane Change Test to assess distraction: Tests of visual-manual and speech-based operation of navigation system interfaces. In *Proceedings of the Fourth International Driving Symposium on Human Factors in Driver Assessment, Training and Vehicle Design* (July 9–12, 2007), 16–22. Stevenson, WA, USA.

ISO 92411-11. 1998. Ergonomic requirements for office work with visual display terminals (VDTs) – Part 11: Guidance on usability (International Standard). International Organisation for Standardisation.

Jordan, P.W., McClelland, I.L., and Thomas, B. 1996. Introduction. In *Usability Evaluation in Industry*, eds. P.W. Jordan, B. Thomas, B.A. Weerdmeester, and I.L. McClelland, 1–3. London: Taylor & Francis.

Klauer, S.G., Dingus, T.A., Neales, V.L., Sudweeks, J.D., and Ramsey, D.J. 2006. The impact of driver inattention on near-crash/crash risk: An analysis using the 100-car naturalistic driving study. Report No: DOT HS 810 594. National Highway Traffic Safety Administration (NHTSA), Washington.

Marmaras, N., and Pavard, B. 1999. Problem-driven approach to the design of information technology systems supporting complex cognitive tasks. *Cognition, Technology and Work* 1 (4): 222–36.

Mattes, S. and Halén, A. 2008. Surrogate distraction measurement techniques: The lane change test. In *Driver Distraction: Theory, Effects and Mitigation*, eds. M.A. Regan, J.D. Lee, and K.L. Young, 107–22. Boca Raton, FL: CRC Press.

Mitsopoulos, E., Regan, M.A., and Tierney, P. 2001. Human factors and in-vehicle intelligent transport systems: An Australian case study in usability testing. In *Proceedings of the 8th World Congress on Intelligent Transport Systems*. Sydney, Australia. [CD ROM]

Mitsopoulos, E., Regan, M.A., Triggs, T.J., and Tierney, P. 2003. Evaluating multiple in-vehicle intelligent transport systems: The measurement of driver acceptability, workload, and attitudes in the TAC SafeCar on-road study. In *Proceedings of the 2003 Road Safety, Education and Policing Conference*. Sydney, Australia. [CD ROM]

Mitsopoulos-Rubens, E., Trotter, M.J., and Lenné, M.G. In press. Effects on driving performance of interacting with and in-vehicle music player: A comparison of three interface layout concepts for information presentation. *Applied Ergonomics*.

Nielsen, J. 1993. *Usability Engineering*. San Diego, CA: Morgan Kaufmann.

Regan, M., Stephan, K., Mitsopoulos, E., Young, K., Triggs, T., Tomasevic, N., Tierney, P., and Healy, D. 2007. The effect on driver workload, attitudes and acceptability of in-vehicle intelligent transport systems: Selected final results from TAC SafeCar project. *Journal of Australasian College of Road Safety* 18 (1): 30–36.

Stanton, N.A., and Young, M.S. 1999. *A Guide to Methodology in Ergonomics*. London: Taylor & Francis.

———. 2003. Giving ergonomics away? The application of ergonomics methods by novices. *Applied Ergonomics* 34:479–90.

Stevens, A. 2009. European approaches to principles, codes, guidelines, and checklists for in-vehicle HMI. In *Driver Distraction: Theory, Effects and Mitigation*, eds. M.A. Regan, J.D. Lee, and K.L. Young, 395–410. Boca Raton, FL: CRC Press.

Stevens, A., Board, A., Allen, P., and Quimby, A. 1999. A safety checklist for the assessment of in-vehicle information systems: Scoring proforma. Report No: PA3536-A/99. Transport Research Laboratory, London.

Thomas, B. 1996. 'Quick and dirty' usability tests. In *Usability Evaluation in Industry*, eds. P.W. Jordan, B. Thomas, B.A. Weerdmeester, and I.L. McClelland, 107–14. London: Taylor & Francis.

13 Ergonomics and Usability in an International Context

Sinja Röbig, Muriel Didier, and Ralph Bruder

CONTENTS

13.1 INTRODUCTION

The scientific discipline of ergonomics has been strongly influenced by its historical and geographical development. Interest in working conditions and occupational health had already started in the eighteenth century and can be considered as the premise of ergonomics. The onset of such human preoccupations is strongly related to industrial development. Thus, countries that were part of the industrial revolution were first to consider the human as a component of the production system.

The UK, Germany, the United States, Italy, and France are the countries that started to promote the need for better working conditions to protect workers' health. In some countries, the recommendations have been transformed into laws that aim to protect human beings from the negative effects of work. This growing interest in working conditions, human factors, and related topics is reflected in the emergence of structured scientific communities, e.g., the Ergonomics Research Society in 1949 (UK), the Gesellschaft für Arbeitswissenschaft (GfA) in 1953 (Germany), the Human Factors and Ergonomics Society in 1957 (USA), the Société d'Ergonomie de Langue Française in 1963 (France), and the Società Italiana di Ergonomia in 1961 (Italy). At the international level, the International Ergonomics Association (IEA) was founded in 1959 with the mission to elaborate and advance ergonomics science and practice, and to improve the quality of life by expanding its scope of application and contribution to society.

Thus, the economic development was the starting point for scientific research related to the human being in working environments. This evolution resulted in the founding of national and international ergonomics organizations to offer scientists an opportunity to exchange scientific information. Western countries were first to establish ergonomics organizations, but today a large number of countries in South America, Africa, and Asia are founding their own ergonomics associations (see IEA triennial report 2006–2009).

Although the IEA definition of ergonomics[*] is widely spread, the historical and geographical context in which ergonomics research was and is conducted influences the focus of the research and the fields of applications. For example, two broad fields of ergonomics can be differentiated. The first field is industrial ergonomics, which is related to the physical aspects of human work

---

[*] Ergonomics (or human factors) is the scientific discipline concerned with the understanding of the interactions among humans and other elements of a system, and the profession that applies theoretical principles, data, and methods to design in order to optimize human well being and overall system performance. (Definition of the International Ergonomics Association 2000.)

213

and the working environment. The second field is cognitive ergonomics, which deals with human behavior, decision-making processes, organizational design, and product design. Further distinctions can be made, for example, based on the field of application. In Germany, a well-accepted definition of ergonomics ("Arbeitswissenschaft") can be translated as: "Analysis, systematization and design of organizational and social conditions of work processes with the goal that workers can work productively and efficiently without any physical or psychological damages, with social fairness and the possibilities to develop further competencies" (Luczak and Volpert 1987; Schlick, Bruder, and Luczak 2010). This definition partially covers the IEA definition but differs, for example, in the field of application. The German definition focuses on the work processes while the IEA definition covers 'the interaction among humans and other elements of a system" without precisely defining the context of application. Actually, the wide scope of the IEA definition of ergonomics covers a lot of the national specificities.

In the same manner as the beginning of ergonomics, the current technological development influences the focus of the research. For example, the rapid development of computer science has greatly modified and continues to modify the working environment, with the emergence of software tools and machines that are controlled by software. Computer science modifies daily life as well, with the introduction of electronics in consumer products. As observed in the initial emergence of ergonomics, countries that are involved in these new developments have started to develop new fields of research and new analysis tools to adapt these new technical systems to human capabilities, needs, and wishes. One of these fields of application, usability, places a very strong emphasis on the user-centered design.

The first international organization to assemble scientists concerned with usability was created in 1991, under the name the Usability Professional Association (UPA). The next section will give an overview of the usability research field with a special focus on the role of the internationalization and globalization of markets on usability studies.

13.2   USABILITY IN AN INTERNATIONAL CONTEXT

Scientific literature and web pages are currently an almost endless source of information on usability studies. There are books available in different languages, e.g., *Methoden der Usability Evaluation* (Sarodnick and Brau 2006), *Designing Web Usability* (Nielsen 2001) and *Mesure de l'utilisabilité des Interfaces* (Baccino, Bellino, and Colombi 2005). Websites are hosted by research organizations; e.g., UPA Germany, Usability Net; or by companies conducting usability studies (Sirvaluse in Germany, Userfocus in the UK, and Nielsen Norman Group in the USA). Some experts in the field of usability publish almost exclusively on their websites, like Jacob Nielsen (http://www.useit.com, last access: 30/04/2010).

Since usability is a growing market, there is also the possibility to attend seminars on usability (English, American, French, and German) to acquire some competences in the field of usability. In some countries, usability has been translated into the local language, like "utilisabilité" in French or "Gebrauchstauglichkeit" in German.

But do all users of the term usability, translated or not, have a common understanding of it? Are the publications describing the same concept? Does the country of application influence the usability methods?

A widespread definition can be found in the standard ISO 9241-110 (2006). It defines usability as "the extent to which a product can be used by specified users to achieve specified goals with effectiveness, efficiency and satisfaction in a specified context of use." Although this standard was originally designed for human–computer interaction, it is also often used as a general definition of usability, independent of the product type.

Some countries developed their own additional standards, especially concerning usability testing. A well-known standard is the published ANSI INCITS 354 (2001), named Common Industry Format for Usability Test Reports. This national standard became an ISO standard, ISO/IEC 25062 (2006) Common Industry Format (CIF) for Usability Test Reports.

There are more examples of standards related to usability, like ISO 20282 regarding the "Ease of operation of everyday products" or ISO 16982 "Ergonomics of human-system interaction." These mainly focus on usability methods, selection, and usage. The target groups of the existing standards differ. The target group of ISO 9241 "Ergonomics of human-system interaction" is usability experts, whereas ISO 16982 "Ergonomics of human-system interaction – Usability methods supporting human-centered design" is written for project managers to support them in planning usability tests and in choosing methods.

Most of those documents are based on the usability definition of the standard ISO 9241, which refers to effectiveness, efficiency, and satisfaction. But, as shown in some studies, there are differences concerning the relevance of effectiveness, efficiency, and satisfaction in different countries due to different cultural approaches. Evers and Day (1997) already showed that effectiveness and ease of use are perceived differently in their importance for Chinese, Indonesians, and Australians. Further studies show that for Chinese users, the satisfaction, joy of use, and design of a product are the most important factors, whereas Danish users are more likely to focus on the effectiveness and efficiency of a product and the lack of frustration (Frandsen-Thorlacius, Hertzum, and Clemmensen 2009). The observed differences are not due to different interpretations of the definition, but to the different roles of these factors in different cultures.

Since market internationalization is gaining importance, there is a need for new terms to describe the cultural influence on different technologies and information. One such term is "cultural usability." There are some studies on how cultures can be described in cultural models to explain their differences. The most well known are the cultural models by Hofstede (1997) and Nisbett and Norenzayan (2002), but there are others.

Hofstede defines five dimensions to describe a culture: the power-distance, collectivism versus individualism, femininity versus masculinity, uncertainty avoidance, and long- versus short-term orientation. The five dimensions are described as follows:

- Power distance: This is the extent to which a person in a lower position in a company, organization, etc., accepts and expects that the power is disposed unequally.
- Collectivism versus individualism: In cultures with more individualism, relationships between people are rather loose. People are concentrated on themselves. Cultures with more collectivism are those where people are integrated in strong groups. People take care of others in their group and they are very loyal.
- Femininity versus masculinity: This dimension is usually measured by the roles that the genders are participating in the culture. Cultures are more masculine, where the roles are very traditionally separated, which means that men are tough and work and women are tender and take care of the home and children. Cultures in which gender roles are more overlapping, are more feminine.
- Uncertainty avoidance: In uncertain countries, people feel uncomfortable and threatened by the unknown and uncertainty. They prefer strict, well-known structures, and they often hide under strict laws and religions.
- Long- versus short-term orientation: Long-term oriented societies usually have stable social requirements. They are usually oriented toward skill development, working hard, patience, etc. (Marcus and Gould 2000; Hoft 1996).

The cultural model by Trompenaars is also one that is frequently discussed. He describes culture in different layers— artifacts and products, norms and values, and basic assumptions.

Artifacts and products contain all the explicit artifacts. It is the layer that is visible to the outside. It includes the heroes that represent a culture, or regularly participated activities, for expample typical sports like soccer in Germany or Baseball in the United States, or holidays (e.g. Independence Day in the United States, Day of Reunion in Germany), symbols, etc. (Trompenaars and Hampden-Turner 1997).

Norms and values present the emotional conception of a culture. They represent the values that are desirable for a culture, e.g., good or evil, exciting or boring. They also represent the code of behavior of a culture and tell people how to behave in different situations (norms).

The inner layer, the basic assumptions, cannot be recognized by people from the outside; it is the implicit part of the culture and consists of the basic rules for human survival (Trompenaars and Hampden-Turner 1997).

Trompenaars describes, as Hofsteede does, different dimensions of culture. The seven dimensions for Trompenaars are universalism versus particularism, individualism versus collectivism, neutral versus emotional, specific versus diffuse, achievement versus ascription, sequential versus synchronic, and internal versus external control (Trompenaars and Hampden-Turner 1997).

These dimensions are described as follows:

- Universal versus particularism: This dimensions shows if it is possible to describe rules within a culture and if those rules are demanded and enforced.
- Individualism versus collectivism: This dimension expresses if the people of this culture function better in groups or as individuals.
- Neutral versus emotional: This dimension describes if it is common for a culture to show emotions in public.
- Specific versus diffuse: This dimension questions if private life and work life are kept separate from each other.
- Achievement versus ascription: This dimension questions if need to prove ourselves to achieve a status or if status in the culture is given to us by birth.
- Sequential versus synchronic: This dimension describes if people of a culture usually do one thing after the other or if they are used to doing different tasks at a time.
- Internal versus external control: This dimension describes if we control our environment or if we are controlled by the environment.

Nisbett differentiates between Easterners and Westerners in his cultural model. The general difference in his opinion is in the cognition and perception of people. He says: "Cultural practices and cognitive processes constitute one another. Cultural practices encourage and sustain certain kinds of cognitive processes, which then perpetuate the cultural practices" (Nisbett and Norenzayan 2002). In his opinion, Eastern and Western cultures are different in categorizing objects, functions, and concepts (Nawaz and Clemmensen 2007). In his theory, he names two different kinds of cultural orientations (Sanchez-Burks, Nisbett, and Ybarra 2000). He differentiates between task-focused people and socio-emotional people. Task-focused means that during a task, people are concentrated on the goals of the task, which they want to complete as accurately as possible. Socio-emotional people are more sensitive to the interpersonal climate of the situation. Social harmony is most important to them.

Hall, in his cultural model, sees culture as a program for behavior and something learned by a person. He says that the context is the main factor of communication, "context is the information that surrounds an event" (Hall and Hall 2002). The information itself is connected to the event itself. The local value of the context for the communication varies in different cultures. Hall defines a scale on which he measures the context of a culture between "high-context culture" and "low-context culture." For him it is important that these two forms of communication are not two dimensions that exclude each other. Most cultures have aspects of "high-context" that are not communicated directly but through gestures and mimic. But they also have criteria of "low-context" where they communicate mainly with spoken language and many details. The other dimension he separates is monochronic versus polychronic culture.

In a high-context culture, non-verbal communication is characteristic for contextual communication. By body language, gestures and mimics are playing an important role. Less is spoken out loud or written down because information is already present in the physical environment (Hall 1976). A characteristic of high-context cultures is "reading between the lines."

In a low-context culture, the communication is more explicit and direct aspects of private life and professional life are clearly separated. The focus of communication is on the content of the communication (Le Mont Schmid 2001).

In a monochronic culture, time is very important (Le Mont Schmid 2001). Time is seen as linear and can be divided into segments. Members of monochronic cultures strongly concentrate on a specific task or activity at a specific time. Appointments are exactly timed and the timetable is strictly followed (Hall 1983).

In polychronic cultures, time is a sub-item. Working processes are not gradual, but they are executed unordered. Different tasks are done at the same time; the focus is on the task executed at that time, less on the plans or timetables made before (Hall 1983). Not being on time is not seen as impolite.

Cultural models show that there are many differences between people of different cultures. These differences can also greatly affect usability studies. Often, the cultural models are used to explain those effects and give advice on the choice of a method in an intercultural context. If a culture in the dimensions of Trompenaars, for example, is defined more collectivistic and less individualistic, methods like focus groups might work better than tests with a single person.

Cultural dimensions are not only used to give support in choosing the right method for a purpose, but they are also used to give advice on how a specific method is to be applied in different cultures. The Power Distance Index, which was defined by Hofstede, is used to explain cultural influences on methods like thinking aloud, interviews, or questionnaires. In Malaysia, for example, the test results of a usability study are much better if the conductor is in a higher position than the subject, they will not give any bad critics about the product and when asked they will say that the product is just fine (Clemmensen and Plocher 2007). This shows, especially in countries with a high power distance, that the hierarchical relation between the test conductor and the subject has to be carefully chosen to limit the influence on the results. This is especially important if the results of different studies have to be compared.

How big this influence is, also depends on the method used. For example, thinking aloud is a frequently used method in usability studies, and many studies describe it as quite sensitive to cultural differences.

For example, in Eastern Asian countries, a quiet testing environment and self-monitoring during a task positively affects the level of thinking. If the subject has to think aloud during a test, then it might affect the accomplishment of the task (Shi 2008). In China, thinking aloud is often more a combination of thinking aloud and interviews. Since the subjects are not used to thinking aloud, they do not talk while doing a task. They first do the task without talking, and then the evaluator asks questions. The evaluator requires very good communication skills to keep the user talking (Shi 2008). If the usability study is conducted by a local evaluator, thinking aloud should not be used for the test since the evaluator will not interrupt the subject during the test, especially if they are fulfilling the task without any problems (Clemmensen et al. 2007).

Thus, using the thinking aloud method in Eastern Asian countries can be difficult to apply because the people are not used to verbalizing their thoughts due to their holistic way of thinking. They are often more concentrated on the interpersonal climate of a situation, because of their socio-emotional orientation. Therefore, the relation between the subject and the evaluator is of special interest in Eastern countries, while this is less important in Western countries (Shi 2008).

To improve the reliability of the results, subjects and evaluators must be chosen very carefully, especially in Eastern countries. This led to the idea of modifying this technique in order to make people talk more openly about the problems they are having with the product. The Bollywood method was developed for this purpose. Scenarios are created, which are very dramatic and emotional, like scenes from a Bollywood movie. In India, these movies are one of the known possibilities to openly criticize social or political topics. By creating these emotional and dramatic scenes in a usability study, the subjects are motivated to give qualitative feedback on a product (Clemmensen et al. 2007, 2008; Oyugi, Dunckley, and Smith 2008). Another possibility is to try to give guidelines for applying the method (see Table 13.1)

TABLE 13.1

Advice for Practical Thinking Aloud Tests in Cross-Cultural Settings

Advice	Explanation and Corresponding Psychological Principle
Explain the background of the test	Easterners want to know the broader context and background of the test; Westerners are less likely to focus on it. This is related to differences in what people attend to (sections "Cultural cognition" and "Instructions and tasks").
Allow for more pauses when Easterners think aloud	Easterners have more difficulty in thinking aloud (section "The user's verbalization").
Thinking aloud might adversely affect Easterners' task performance	Thinking aloud might impair the performance of Easterners and enhance the performance of Westerners. Relates to the principles about expressing holistic and analytic thinking verbally (section "The user's verbalization").
Rely less on expressions of surprise when Easterners are test participants	The extent to which people express surprise differs between cultures (section "Reading the user"). Using surprise as a main marker of usability problems is thus problematic.
Be aware of and mitigate cross-cultural biases in analysis TA results	The attribution of causes to behavior differs among cultures (section "Reading the user" and "The overall relationship between user and evaluator"). Further, the grouping and perception of similarities among behaviors and usability problems may differ depending on the evaluators' cultural background.
Critique of interfaces is likely to seek a compromise and be indirect when users are Easterners	Easterners use conversational indirectness and often attempt to find a middle path (section "The overall relationship between user and evaluator").
Use evaluators and users with similar cultural backgrounds, if possible	Difference in culture may impact the number of identified problems and redesign proposals. Familiarity between evaluator and user may also impact results (section "The overall relationship between user and evaluator").
TA tests also concern non-task issues	Easterners are more likely to have a socio-emotional orientation (section "The overall relationship between user and evaluator"). Thus, they may perceive the relationship with the evaluator as more than solving tasks or thinking out loud.

Source: Clemmensen, T., Herzum, M., Hornbaek, K., Shi, Q., and Yammiyavar, P., *Cultural Cognition in the Thinking-Aloud Method for Usability Evaluation*. ICIS 2008 Proceedings, Paper 189, 2008.

Another example of the potential influence of cultural differences on usability studies can be found in the choice of colors. Colors have different meanings in different countries. Red and green seem to be the most stable colors, meaning "danger, wrong" or "true" (Zühlke 2004). Others are not so clearly understood. The color yellow, for example, in European countries and in the United States is related to "caution, cowardice" and is therefore often used as a color for warning. In Asian countries, yellow stands for "sacred, imperial, royalty or honor" or in the Middle East for "happiness, prosperity" (Aykin and Milewski 2005). In an Asian country, if a yellow light is used as a warning color, e.g., in a control center, this might not be understood without an explanation.

A further example of the potential influence of cultural differences can be observed with the meaning of symbols. In Japan, for example, "x" is the symbol for something negative. This might lead to confusion if a Japanese participant is asked to answer with an "x" when he/she wants to give a positive answer to a question. The test results might then be falsified (Nielsen 1993).

These examples show that cultural differences can have some effect on the results of a usability study if they are not taken into account during the preparation, conductance, analysis, or interpretation of usability tests. The importance of the effect of cultural differences on the usability test results not only depends on the cultural differences themselves, but also on the way they have been considered during the preparation, conductance, and interpretation of the test.

Due to the internationalization of the markets, more and more studies on product usability are also conducted internationally. Additionally, usability, in a similar way to ergonomics, develops itself further according to technological development. One of the latest developments in the domain of usability concerns the higher significance given to feelings when interacting with a product. This tendency is generally covered by the term user experience (UX). According to ISO 9241-210, 2010 user experience is defined as "A person's perceptions and responses that result from the use or anticipated use of a product, system or service." It focuses on the user's feelings while using a product, system, or service, his/her perceptions and physical and psychological responses. It is influenced by the expectations and experiences of the user. A product that is usable might not have a good user experience. For example, an MP3 player may have good usability because the structure of the menu is clear and it can be used intuitively, but it has no UX because the design does not please the user. It is also possible the other way around. A nice looking MP3 player might have a good UX, but the usability may be poor because of a poor menu structure.

Culture is, like usability, one of the components influencing the user experience of a product (Arhippainen and Tähti 2003). Some studies concerning the influence of culture on user experience have been conducted in the field of user interfaces or websites. They are also based on cultural models like Hofstede's (1997) (compare Marcus [2006], Lee et al. [2008] and Eune and Lee [2009]). This is why many criteria that affect usability are also found in the field of user experience. For example, the color of a product might affect the feelings of the participants toward that product and the perceived user experience. Since the perception of the meaning of a color is often culture dependent, the color of a product might be important for the user experience and might lead to different results in different countries.

As can be seen, there are many facets of a culture that influence usability studies. Many scientists have already tried to define cultural dimensions that describe a culture. These dimensions are also used to describe why some usability methods might not work in the same way in different countries or why they need to be applied to each country. Some of the dimensions, like the Power Distance Index defined by Hofstede, are more obvious. Others, as the basic assumptions defined by Trompenaars, are not so obvious. They show that usability studies are not as easy to plan and conduct as people often think, especially if results are supposed to be comparable, the same methods have to be used in the same way. This is not easy since some methods have to be modified so they can be used in different countries or special basic conditions have to be considered (for example in questionnaires: which questions are allowed in which country). They need to be planned and conducted carefully to get reliable results.

13.3 CURRENT UNDERSTANDING OF ERGONOMICS, USABILITY, AND USER EXPERIENCE

As shown in the first two sections, definitions for ergonomics, usability, and user experience can be found in different international documentation. However, only a few official documents define the relationship between those terms and their potential overlaps. For example, the compendium "Ergonomics practice and usability testing of products" (Adler et al. 2010) refers to the definitions of ergonomics* und usability† as defined in the respective standards ISO 6385 (2004) and ISO 9241-11:1999 (equivalent to ISO 9241-110 2006) because they are considered the most widespread definitions. This compendium underlines the fact that the two terms are sometimes defined differently, including within standards, and that it is not unusual to find some identical issues that are designated

---

* The scientific discipline concerned with the understanding of interactions among human and other elements of a system, and the profession that applies theory, principles, data, and methods to design in order to optimize human well-being and overall system performance.
† The extent to which a product can be used by specified users to achieve specified goals with effectiveness, efficiency, and satisfaction in a specified context of use.

differently. To overcome the difficulty of clearly identifying the boundaries and potential overlaps of the terms ergonomics and usability, the content of the compendium combines them under one entity.

The difficulty in distinguishing the terms has been observed among usability experts. In the framework of a study conducted by the Institute of Ergonomics of the Darmstadt University of Technology on this topic, 20 experts from 8 European countries were invited to present their own understanding of these terms and to describe how they are related to each other. They were asked to give their own opinion based on their daily practice and not the "academic" version. The analysis of the collected answers shows that it is not easy to differentiate ergonomics, usability, and user experience, even for experts. The definitions are so closely related in daily practice that experts are not always able to formalize their own conception without hesitation and long discussions. As a consequence, most of the experts refer to the definitions that can be found in standards with possibly some slight variations (see Section 13.2). Concerning the relationship between ergonomics, usability, and user experience, the experts provided their answers in the form of graphical descriptions. The results were amazingly different: none of the graphics were identical!

Three different approaches can be described:

- *Hierarchical model.* Experts consider the three research fields as different layers of a pyramid (Figure 13.1). At the base of the pyramid is "ergonomics," which corresponds to the basic requirements that have to be fulfilled for any type of product. If the requirements at the ergonomics level are not fulfilled, it does not make sense to consider the level usability. The same remark applies for the higher-level user experience: before developing the user experience, the ergonomics and usability requirements have to be fulfilled. Each level is seen as an independent entity without overlapping the other dimensions. However, the higher level cannot be reached if the foundation is not completed.

  Figure 13.2 shows another representation of the overlapping of the disciplines. The participant sees ergonomics in the center of the three fields because they consider that products cannot be developed without ergonomics. The next step is usability, which is analyzed when ergonomic requirements are fulfilled. The last circle is the UX, which is evaluated last, but all three disciplines have to be considered in designing a product that corresponds to the user.

  Flowcharts were also used to demonstrate the hierarchical order of the three research fields. An example is shown in Figure 13.3. All the flowcharts drawn started with ergonomics, followed by usability and user experience. Differences in the flowcharts could be found in the feedback loops. In the flowchart in Figure 13.3, there are feedback loops for each research field. Other flowcharts did not have feedback loops or they had an additional back loop between user experience and ergonomics.

FIGURE 13.1 Relationship between usability, ergonomics, and user experience (UX) seen as a hierarchical order.

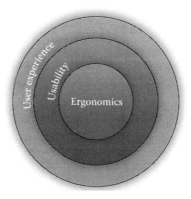

FIGURE 13.2   Relationship between usability, ergonomics, and user experience (Version D).

- *Hierarchical and parallel mixed model.* Other experts believe that usability and ergonomics are two independent dimensions that have to be considered parallel to each other. Both types of requirements have to be fulfilled before user experience can be considered; they are the basis (Figure 13.4) while user experience is situated at a higher level. All three dimensions—ergonomics, usability, and user experience—must be fulfilled in order to have a user-oriented product. However, all these aspects need not be considered simultaneously.

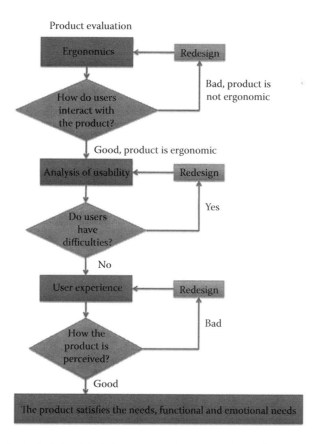

FIGURE 13.3   Flowchart of relationship between usability, ergonomics, and user experience.

FIGURE 13.4   Relationship between usability, ergonomics, and user experience.

- *Overlapping model.* There were also experts who thought that the three disciplines are overlapping. Figure 13.5 shows the result. Ergonomics and user experience were not seen as absolutely separate from each other. They are overlapping in different concerns. Usability is seen in the intersection of those two research fields, included in their common part. Usability consists solely of the area within the intersection between ergonomics and user experience, without additional points.
- *Further models.* Beyond the models described above, there were also more creative descriptions of the relationship between the three research fields. One of the experts linkened the relationship of the three research fields to a soccer game. By his definition, the game itself is the user experience. The rules of the game are the ergonomics and the soccer player is the usability professional. The goal is the usability.

      This is supposed to show that the user experience is the global goal behind all the other disciplines. The ergonomics as the rules of the game are the basics for a product. The usability professional as the player has to consider the rules of the game, symbolically representing ergonomics, in order to reach his/her goal, the usability of the product.

      The differences in the understanding of the relationship between ergonomics, usability, and user experience not to be only intercultural problem. When comparing the results of the experts from one country, it became clear that it is also an intracultural problem. Although there are definitions for each of the three research fields, it is still not clear how they are related to each other. This shows that there is still a need for a model to clearly describe the relationship between these three research fields.

13.4   CONCLUSION

Ergonomics started early in the twentieth century. There have been many achievements in this field to improve working conditions and, recently, the ergonomics of products. Ergonomics is widespread and is known in most countries around the world.

FIGURE 13.5   Relationship between usability, ergonomics, and user experience, overlapping model.

Usability came up in the 1980s when computers started to become widely available. It is a relatively new research field that is developing rapidly, but it is still not present in some countries. Compared to ergonomics, the term usability can be understood differently depending on the country; sometimes it even differs within a country. This shows that further intracultural and intercultural exchanges are necessary to develop a common understanding of usability. Furthermore, the relationship between ergonomics, usability, and user experience has to be clarified in order to avoid confusion, particularly at the crossing between disciplines. An intercultural common definition and understanding of usability is necessary.

Due to the potential influence of cultural differences on usability testing, there is a need for further research to clearly identify those differences, thereby offering the possibility to better control intercultural artifacts when performing international usability studies. This can be done by creating new guidelines for usability studies, especially when conducted in an international context. Such guidelines should contain information about what method can be used and how it needs to be applied to ensure the comparability of results collected in different countries.

Another possibility is to design new usability methods that can be applied in the same way in each country without modification or without having to meet special basic conditions. Those methods could then provide results that are internationally comparable without adaption in the countries.

Since most of these propositions cannot be realized within a short time period, the best way to overcome the potential cultural differences at this time is to cooperate with local experts, if available. They can give the best support for usability studies because they know their country best and have experience in how the methods can be applied there.

Further research should include the development of new methods that are less sensitive to cultural differences, of tools that support professionals in their choice of usability methods and show them how to ensure the comparability of the results achieved by their analysis.

REFERENCES

Adler, M., Herrmann, H.-J., Koldehoff, M., Meuser, V., Scheuer, S., Müller-Arnecke, H., Windel, A., and Bleyer, T. 2010. *Ergonomiekompendium – Anwendung Ergonomischer Regeln und Prüfung der Gebrauchstauglichkeit von Produkten*. 1. Auflage. Projektnummer: F 2116. Dortmund: Bundesanstalt für Arbeitsschutz und Arbeitsmedizin.

ANSI INCITS 354. 2001. Common Industry Format for Usability Test Reports. Standard.

Arhippainen, L., and Tähti, M. 2003. Empirical evaluation of user experience in two adaptive mobile application prototypes. In *Proceedings of the 2nd International Conference on Mobile and Ubiquitous Multimedia*. Norrköping, Sweden.

Aykin, N., and Milewski, A.E. 2005. Practical issues and guidelines for international information display. In *Usability and Internationalization of Information Technology*, ed. N. Aykin, 21–50. Mahwah, NJ: Lawrence Erlbaum Associates.

Baccino, T., Bellino, C., and Colombi, T. 2005. *Mesure de l'utilisabilité des interfaces*. Paris: Hermes Science.

Clemmensen, T., Herzum, M., Hornbaek, K., Shi, Q., and Yammiyavar, P. 2008. *Cultural Cognition in the Thinking-Aloud Method for Usability Evaluation*. ICIS 2008 Proceedings, Paper 189. AIS Electronic Library (AISeL).

Clemmensen, T., and Plocher, T. 2007. The Cultural Usability (CULTUSAB) Project: Studies of cultural models in psychological usability evaluation methods. In *Usability and Internationalization: Second International Conference on Usability and Internationalization*, ed. N. Aykin, S.274–80. UI-HCII 2007, held as part of HCI International 2007, Beijing, China, July 22–27, 2007. Berlin, Heidelberg: Springer.

Clemmensen, T., Shi, Q., Kumar, J., Li, H., Sun, X., and Yammiyavar, P. 2007. Cultural usability tests – how usability tests are not the same all over the world. In *Usability and Internationalization*, Part I, HCII 2007, LNCS 4559, ed. N. Aykin, 281–90. Berlin, Heidelberg: Springer Verlag.

Eune, J., and Lee, K.P. 2009. Cultural dimensions in user preferences and behaviors of mobile phones and interpretation of national cultural differences. In *DGD '09 Proceedings of the 3rd International Conference on Internationalization, Design and Global Development: Held as Part of HCI International 2009*, ed. N. Aykin, 29–38, San Diego, CA. Berlin, Heidelberg: Springer-verlag.

Evers, V., and Day, D. 1997. The role of culture in interface acceptance. In *INTERACT '97 Proceedings of the IFIP TC13 Interantional Conference on Human-Computer Interaction*, 260–67. London, UK: Chapman & Hall, Ltd.

Frandsen-Thorlacius, O., Hertzum, M., and Clemmensen, T. 2009. Non-universal usability? A survey of how usability is understood by Chinese and Danish users. In *Proceedings of the 27th International Conference on Human Factors in Computing Systems*, eds. S. Grennberg, S.E. Hudson, K. Hinckley, M.R. Morris, and D.R. Olsen Jr. CHI 2009, Boston, MA, April 4–9, 41–50. New York: ACM.

Hall, E.T., 1976. *Beyond Culture*. Oxford: Anchor Books.

———. 1983. *The Dance of Life: The Other Dimension of Time*. New York: Anchor Books.

Hall, E.T., and Hall, M.R. 2002. Key concepts: Underlying structures of culture. In *Readings in Cultural Communication: Experiences and Contexts*, 2nd ed., eds. J.N. Martin, T. Nakayama, and L.A. Flores, 199–206. Boston: McGraw-Hill.

Hofstede, G. 1997. *Cultures and Organizations: Software of the Mind, Intercultural Cooperation and its Importance for Survival*. New York: McGraw-Hill.

Hoft, N.L. 1996. Developing a cultural model. In *International User Interfaces*, eds. E.M. del Galdo and J. Nielsen, 41–73. New York: John Wiley.

IEA triennial report. 2006–2009. http://www.iea.cc/upload/Triennial%20Report%20FINAL.pdf (accessed April 30, 2010).

ISO 9241-11. 1999. Ergonomic requirements for office work with visual display terminals (VDTs) – Part 11: Guidance on usability.

ISO 6385. 2004. Ergonomic principles in the design of work systems. Standard.

ISO 9241-110. 2006. Ergonomics of human-system interaction – Part 110: Dialogue principals. Standard.

ISO 9241-210. 2010. Ergonomics of human-system interaction – Part 210: Human-centred design for interactive systems. Standard.

ISO 16982. 2002. Ergonomics of human-system interaction – Usability methods supporting human-centered design. Standard.

ISO 20282. 2006. Ease of operation of everyday products. Standard.

ISO/IEC 25062. 2006. Common Industry Format (CIF) for Usability Test Reports. Standard.

Lee, I., Choi, G.W., Kim, J., Kim, S., Lee, K., Kim, D., Han, M., Park, S.Y., and An, Y. 2008. Cultural dimensions for user experience: Cross-country and cross-product analysis of users' cultural characteristics. British Computer Society Conference on Human-Computer Interaction. In *Proceedings of the 22nd British HCI Group Annual Conference on HCI 2008: People and Computers XXII: Culture, Creativity, Interaction* – Volume 1, 3–12. Swinton, UK: British Computer Society.

Le Mont Schmid, P. 2001. *Die amerikanische und die deutsche Wirtschaftskultur im Vergleich. Ein Praxishandbuch für Manager*. 3., neu-bearbeitete Auflage. Göttingen: Hainholz Verlag GmbH.

Luczak, H., and Volpert, W. 1987. *Arbeitswissenschaft. Kerndefinition – Gegenstandskatalog – Forschnugsgebiete*. Eschborn: RKW-Verlag.

Marcus, A. 2006. Cross-cultural user-experience design. In *Proceedings of the 4th International Conference*, eds. D. Barker-Plummer, R. Cox, and N. Swoboda, Diagrams 2006, Stanford, CA, June 28–30, 2006, 16–25.

Marcus, A., and Gould, E.W. 2000. Crosscurrents: Cultural dimensions and global user-interface design. *Interactions* 7 (4): 32–46.

Nawaz, A., and Clemmensen, T. 2007. Cultural differences in the structure of categories among users of clipart in Denmark and China. In *The Seventh Danish HCI Research Symposium* (DHRS 2007), ed. A.H. Jørgensen. Copenhagen: IT University.

Nielsen, J. 1993. *Usability Engineering*. Boston: Academic Press.

———. 2001. *Designing Web Usability*. 2nd edn. Munich: Markt+Technik-Verlag.

———. 2010. http://www.useit.com (accessed April 8, 2010).

Nielsen Norman Group. 2010. http://www.nngroup.com/ (accessed April 8, 2010).

Nisbett, R.E., and Norenzayan, A. 2002. Cultural and cognition. In *Stevens' Handbook of Experimental Psychology*, ed. D.L. Medin and H. Pashler. 3rd ed. Vol. 2, 561–97, New York: John Wiley and Sons.

Oyugi, C., Dunckley, L., and Smith, A. 2008. Evaluation methods and cultural differences: Studies across three continents. ACM International Conference Proceeding Series; Vol. 358, In *Proceedings of the 5th Nordic Conference on Human-Computer Interaction: Building Bridges*, 318–25. New York, NY: ACM.

Sanchez-Burks, J., Nisbett, R.E., and Ybarra, O. 2000. *Cultural Styles, Relational Schemas and Prejudice Against Outgroups*, 318–325. Ann Arbor, MI: University of Michigan.

Sarodnick, F., and Brau, H. 2006. *Methoden der Usability Evaluation: Wissenschaftliche Grundlagen und praktische Anwendung*. Bern: Huber.

Schlick, C., Bruder, R., and Luczak, H. 2010. *Arbeitswissenschaft*. 3. Auflage. Heidelberg: Springer.

Shi, Q.A. 2008. Field study of the relationship and communication between Chinese evaluators and users in thinking aloud usability tests. ACM International Conference Proceeding Series; Vol. 358, In *Proceedings of the 5th Nordic conference on Human-computer interaction: building bridges*, 344–52. New York, NY: ACM.

Sirvaluse. 2010. http://www.sirvaluse.de/leistungen/methoden/index.html (accessed April 8, 2010).

Trompenaars, F., and Hampden-Turner, C. 1997. *Riding the Waves of Culture, Understanding Cultural Diversity in Business*, 2nd edn. London: Nicholas Brealey.

UPA. 2010. http://www.upassoc.org/usability_resources/about_usability/index.html (accessed April 8, 2010).

UPA Germany. 2010. http://germanupa.de/ (accessed April 8, 2010).

Usability Net. 2010. http://usabilitynet.org/home.htm (accessed April 8, 2010).

Userfocus. 2010. http://www.userfocus.co.uk (accessed April 8, 2010).

Zühlke, D. 2004. *Useware-Engineering für technische Systeme*. Berlin, Heidelberg, New York: Springer-Verlag.

14 Consumer eHealth

Christopher P. Nemeth and Richard I. Cook

CONTENTS

14.1 INTRODUCTION

Healthcare seeks to improve an individual's physical and mental well-being through a well-considered, consistent program of diagnosis and treatment. However, a recent study of consumer healthcare by McGlynn et al. (2003) found that adults in 12 U.S. metropolitan areas received about half (54.9%) of the recommended care processes. Receiving half of the healthcare treatment that has been prescribed has significant implications for patient health, particularly when treating chronic conditions. In the case of diabetes, for example, routine blood sugar monitoring is essential to assess treatment effectiveness, in order to ensure appropriate responses to poor glycemic control and to identify complications of the disease early enough to prevent serious consequences. Yet, only 24% of participants in the study who had diabetes received three or more glycosylated hemoglobin tests over a two-year period. In McGlynn et al.'s study, there was little difference among the proportion of recommended preventive care (54.9%), acute care (53.5%), and care for chronic conditions (56.1%).

Three years later, Asch et al. (2006) found similar results when they examined medical records and conducted telephone interviews with 6712 randomly selected patients who visited a medical office within a two-year period in 12 metropolitan areas from Boston to Miami to Seattle. The survey examined whether people received the highest standard of treatment for 439 measures ranging across common chronic and acute conditions and disease prevention. The study looked at whether the patients got the right tests, drugs, and treatments. Overall, patients received only 55% of the recommended steps for top-quality care, with no group doing any better or worse. Health experts blame the overall poor care in the United States on an overburdened, fragmented system that fails to keep close track of patients with an increasing number of multiple conditions (Donn 2006).

Chronic conditions dominate healthcare in most parts of the world, including the United States (Clark 2003). Those who suffer from chronic conditions such as diabetes must follow a care regimen in order to manage their illness or suffer dire consequences. For example, failure to manage diabetes can result in the condition worsening with further complications such as neuropathy (nerve damage) and sight loss. The treatment of diabetes and other chronic diseases is also costly, absorbing a disproportionate amount of healthcare funds. Improvement in care for chronic illness promises to both improve the health of those who are afflicted, and to mitigate the cost of their care.

The provision of healthcare for consumers, particularly those who suffer from chronic conditions, needs to be improved. Recent initiatives, such as self-management, seek to improve patient continuity and quality of care, particularly among populations in greatest need, such as those who have chronic conditions. Day-to-day responsibilities for chronic illness care fall most heavily on patients and their families and benefit from the development of an effective collaborative relationship with care providers, which includes collaborative problem definition, goal setting and planning, creation of a continuum of self-management and support training, and active and sustained follow-up (Von Korff et al. 1997). Self-management support systematically provides education and assistance to increase patients' skills and confidence in the management of their health problems. Such support encourages daily decisions that improve health-related behaviors and clinical outcomes, and makes it possible for patients and their families (the primary care providers in the case of chronic illness) to improve care (California Healthcare Foundation 2005).

Making it possible for consumers to participate in the care process promises to improve self-care as well as mitigate costs, and use of communications systems has made this possible. For example, Noel et al. (2004) report that the treatment of complex heart failure, chronic lung disease, and/or diabetes mellitus in homebound elderly patients using home telehealth units significantly reduced resource use, improved cognitive status and treatment compliance, and stabilized chronic disease. Finkelstein, Speedie, and Potthoff (2006) found that video conference "virtual visits" between a skilled home health nurse and chronically ill homebound patients can improve patient outcome at a lower cost than traditional skilled face-to-face home healthcare visits.

McGlynn et al. (2005, 2644) concluded that one key to rectifying the failure to receive recommended care is "the routine availability of information on performance at all levels. Making such information available will require a major overhaul of our current health information systems, with a focus on automating the entry and retrieval of key data for clinical decision making and for the measurement and reporting of quality." Better data collection, such as monitoring information from glucose meters, and better communication between patients and care providers can improve care for chronic conditions (Saporito 2005). The cognitive work that is necessary to perform such care is a shared activity requiring well-qualified action based on accurate, timely information. High-bandwidth information technology (IT) provides a means to present, find, and use that information. The manner in which that is done must reflect the needs of all participants in the care process, particularly the patient–physician relationship.

14.2   THE EVOLUTION OF EHEALTH

While telemedicine began with remote image analysis in 1959 (Perednia and Allen 1995) and accelerated with the use of video for therapeutic consultation (Wittson and Benschoter 1972), it has evolved most rapidly since the inception of the Internet. Much of the current discussion related to IT and consumer health revolves around technological capabilities. Features of technology attract attention to possibilities that have not previously been available. For example, "telemedicine" is the use of telecommunications (as simple as a telephone or sophisticated as a satellite) to provide medical information and services (ATSP 2007) and, increasingly, is shorthand for remote clinical consultation. These include interactive video consultations and "store and forward" radiology evaluations, and telerobotic surgery (Perednia and Allen 1995). These are typically conducted between and among physicians, who share a base of professional training and protocols that guide their interactions.

As IT in healthcare quickly evolves, there is no clear consensus on the meaning of "eHealth" beyond two universal themes of health and technology (Oh et al. 2005). This chapter uses the term eHealth to refer to cognitive work that is related to an individual's healthcare that may be conducted in person or over electronic media from the telephone to other means such as the Internet.

Figure 14.1 shows the current state of support for clinical healthcare cognitive work. Patients and care providers exist in an information ecology that includes the patient, other clinicians, devices, information systems, and physical artifacts. Care providers attend to individual patients (shown as direct contact by black lines) using their own observation as well as the views of consultants and patient self-reports. Patients rely on themselves, friends, and family members as well as the resources that are available via the worldwide web. Care providers direct and monitor (through indirect contact shown by dotted lines) diagnostic and communications equipment, which is shown in the lower portion of Figure 14.1. Such devices have recently been connected to communication networks to "push" significant information to clinicians and related systems. They consult physical cognitive artifacts

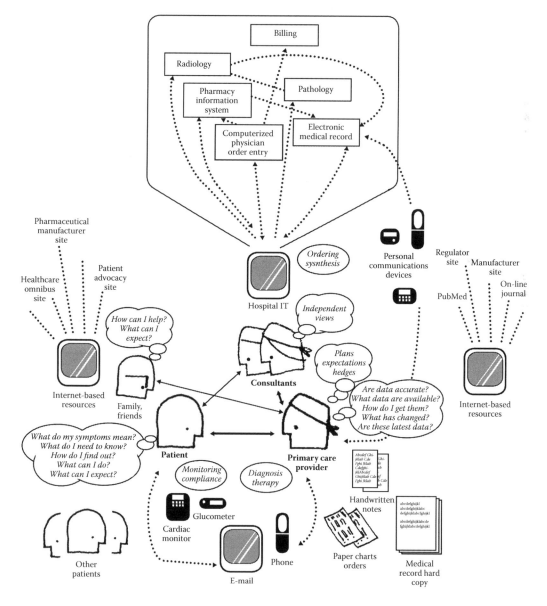

FIGURE 14.1 Consumer healthcare information ecology. (Reprinted from Cognitive Technologies Laboratory, Chicago, Illinois, © 2007. With permission.)

(Hutchins 2000), such as paper charts, orders, hard copy medical records, and status boards. They request and synthesize data from a variety of information systems and departments (shown in the upper portion of Figure 14.1). They also have their own set of information resources that are available via the worldwide web, although the professional resources are different from those that patients use. All this happens in the context of caring for multiple patients (at lower left), who each have unique diagnoses, needs, and care trajectories that must be planned and coordinated.

The role of IT in healthcare is likely to continue and increase. In regions with limited infrastructure or in developing countries, healthcare IT will link care providers who are based at health centers, referral hospitals, and tertiary centers (Heinzelmann, Lugn, and Kvedar 2005). Industrialized countries that are coping with personnel shortages and decreasing third-party reimbursement are likely to use the Internet and the worldwide web to move healthcare delivery from the hospital or clinic into the home (Heinzelmann, Lugn, and Kvedar 2005) as an adjunct to cost-effective healthcare and delivery (Doyle, Ruskin, and Engel 1996). For example, healthcare organizations have opened virtual intensive care units (VICUs) in which a single intensivist and/or ICU nurse attend patients in several geographically dispersed ICUs. Radiographic images at U.S. facilities are being sent to India and Australia for interpretation, and these interpretations are sent back to the United States via computers. More and more specialists are using advanced communications technology to look at computerized tomography scans (CTs) of the head, electrocardiographs (EKGs), ultrasounds, and more at night or on weekends. Before this, specialists would have had to come into the hospital to obtain this information. This gives greater access to expertise for some of these images (typically the less important ones). It also means that for many specialists the context and other cues available from the physical environment, such as actually touching the patient, have been removed. These changes are likely to lead to huge changes in performance, communication, and coordination (Robert Wears, personal communication, March 7, 2007).

14.3   CLINICAN–PATIENT RELATIONSHIP

In addition to economics and technology, the success of eHealth will depend on human factors (Heinzelmann, Lugn, and Kvedar 2005) because the clinical problems that are most amenable to eHealth solutions are largely cognitive (Perednia and Allen 1995). This requires an understanding of cognition, the nature of the healthcare process, and the quality of information that participants find and use.

Cognitive activity includes attention, perception, memory, decision making, and decision selection (Nemeth 2004, 33–66) and each participant in the care process performs it to improve their health or the health of others. *Patients* need to know where to look for information related to their care, and evaluate the quality of that information. They assess and report on their own condition subjectively, may perform diagnostic data collection through the use of devices, such as a glucometer, and monitor trends over time. Their ability to comply with guidance from care providers, such as prescriptions, relies on understanding information that is provided to them. *Family members and friends* seek ways to help the patient. This includes understanding what to expect, collaboration to figure out aspects of care such as taking medications, performing therapy, and assisting with activities of daily living. *Primary care providers* listen to and examine the patient to build a mental model (Lipschitz and Shaul 1997) of the patient's past, current, and potential course. They rely on their own clinical expertise, results of tests, consultation with other clinicians, and the professional literature to refine their understanding and intentions. *Consulting clinicians* provide expertise that the primary care provider may need, which can include a report of prior care that the patient has received, or specialized diagnostic or therapeutic knowledge.

The nature of healthcare issues and the cognitive work that is required to engage them is variable and complex. For example, the decision to proceed with a certain treatment for a patient relies in part on the trade-offs between what is known about certain courses of treatment and their anticipated harms and benefits. The majority of these activities do not occur in what could be described as familiar territory, where the data are sufficient and the patient's recovery is certain. Instead, patient condition and prognosis often exist in the kind of circumstances in which the available evidence on what to do is weak (Sharpe and Faden 1998, 214–20). Some practitioners contend that much of medical practice takes place where there is little proven knowledge and anticipated harms/benefits are equivocal (Nemeth 2005). Making trade-off decisions that are related to anticipated harms and benefits of courses of treatment are directly linked to the clinician interaction with the patient.

Healthcare necessarily relies on a cognition that is distributed (Hutchins 1995) among patients, clinicians, and various systems to collect, share, retrieve, and use information that changes rapidly and can be difficult for consumers to understand and apply. The information ecology in Figure 14.1 illustrates the distribution of cognition among individuals, units, information and communications systems, and physical artifacts. The potential exists for eHealth to enfranchise large segments of the population that have so far been at, or outside, the margins of adequate treatment. Access to the Internet should make a large number of resources and capabilities available to both clinicians and patients. The distances that separate caregivers from patients and each other suggest that a focus on communications using computers is likely to be particularly beneficial to these populations. Consumer populations, such as the elderly, poor, and those with debilitating conditions, are typically disadvantaged in terms of access to medical resources. Rural populations are typically late to receive new medical diagnosis and therapy.

While interaction between patients as consumers of healthcare and care providers is not new, growing reliance on high-bandwidth communications, such as the Internet, creates a qualitatively new ground for consumer–physician interaction in a technology-mediated relationship; a new diagnostic or therapeutic modality in itself (Perednia and Allen 1995). The use of both written and verbal communication has been shown to increase patient understanding and compliance (American Medical Association 2002). Schillinger, Wilson, and Bindman (2003) suggest two influences on success in chronic disease care: the system of care itself may influence the degree to which guideline-based care is delivered, and the extent to which physicians and patients truly engage in a partnership that allows for a mutual exchange of opinions, ideas, and solutions. *Medication concordance* is one example of the kind of cognitive work that needs to occur between clinician and patient that eHealth can assist. Medication concordance is the notion of agreement between a physician and patient; specifically, that a patient understands the information that a care provider communicates (MedicineNet 2007). Decision making is another example of cognitive work. Recently, increased patient participation in making choices related to healthcare is a popular theme. Levinson et al. (2005) found that while nearly all patients in their study wanted to have the opportunity to make decisions, nearly half (52%) preferred to leave the final decisions to physicians and 44% preferred to rely on physicians for medical knowledge rather than seek it for themselves. This is a delicate and sophisticated negotiated relationship between patient and care provider, requiring information tools that are equally sophisticated but are not currently available.

Support for this essential link between patient and physician, such as written instructions and simple web sites, are inadequate for the level of understanding that medication concordance implies. While cognitive aids, such as hard copies of instructions, can provide insights into the kind of cognitive work that people do (Nemeth et al. 2006), information system development requires a well-grounded understanding of the work domain that it is intended to support (Nemeth et al. 2005).

The quality of information that participants find and use also affects their ability to understand and make well-informed decisions. Consumers currently use the Internet to seek out as many as ten sources for health-related information (Brennan 2007). While over 110 million Americans have used the Internet to obtain healthcare information (Smith et al. 2006), little has been done to learn how citizens actually use the Internet for healthcare (Powell et al. 2005). Adolescents who have grown up with the Internet as a primary source of information lack the skills to filter credible information that varies in quality and reliability (Smith et al. 2006). Physicians also need critical appraisal skills to determine whether information that a patient finds is relevant to the patient's condition and based on the best available evidence (Schwartz et al. 2006). The current offering of web sites makes information available to a broad population, whether it is accurate or erroneous, current or outdated, biased or impartial. Recent studies (Akerkar, Kanikar, and Bichile 2005) of web search engine use in developing countries shows that consumers can easily find and embrace health information using web search engines. However, that information may be influenced by agendas such as commercial interests. By contrast, clinicians rely on health information that has been subject to peer review and is far more likely to be reliable. This discrepancy can put patient perceptions at odds with clinicians. The trend may also evolve beyond developing countries.

Successful use of the Internet to obtain healthcare information depends on several different types and levels of literacy, including basic literacy, computer literacy, information literacy, health information literacy, and health literacy (Anton and Nelson 2006). *Health literacy* is the ability to read, understand, and act on healthcare information. The Institute of Medicine (Adams and Corrigan 2003) cited both as a priority area for national action. The challenge becomes clearer in light of data that show nearly half (46%) the U.S. population in 1999 was functionally illiterate when dealing with the healthcare system. Indeed, those who are 60 and over and those who subsist on low incomes need health information most but are least able to read and understand it (American Medical Association 2002). This presents significant implications for patient ability to understand, use, and remember health-related information.

14.4 DISCUSSION

The telehealth literature has enumerated the benefits of technology-mediated relationships, and the popular press has shown how the use of telehealth can serve dire needs such as the Haiti earthquake relief (Freudenheim 2010). However, the benefits of technology also include shortcomings that can impoverish the care process if they are not addressed (see Elford [2008] for a review of recent and prospective practical issues in telehealth). The promise of improved quality and efficiency of healthcare through IT applications has been blocked by a variety of obstacles to the development, adoption, and use of this type of automation. Both narrow technical factors, such as the lack of common standards and poor integration of technology platforms, and wider issues of whether technological offerings are appropriate have dogged the introduction of new IT in healthcare.

IT projects in healthcare and other domains tend to focus narrowly on the technology itself (what a system can do), rather than on the work that the system is intended to support (what users need to accomplish). This has more to do with the nature of IT than the domain in which it is applied. Routinely, a basic information system (platform) is installed with the intention to develop application interfaces to serve specific needs in an organization. However, most of the automation that is built using IT platforms ends up being *technology centered* rather than *user centered* (Billings 1999). Technology-centered IT tends to be less useful than its designers intended and is the source of new forms of failure and new types of workload for its operators. Avoiding these pitfalls requires adoption of a user-centered design approach to IT. That is, it requires stepping away from a narrow focus on technology and adopting a broader perspective in which IT becomes one of many ways in which agents can cooperate in a workplace. This latter approach is the paradigm of computer-supported cooperative work (CSCW).

14.4.1 eHealth Issues

Communication among participants who are remote from each other has potential side effects. Patient care is intended to be continuous: a seamless connection among care providers, facilities, and programs that are intended to restore a patient to good physical and/or mental condition. Gaps in the continuity of care (Cook, Render, and Woods 2000) are the underlying condition that threatens care through problems such as poor handoffs of care, difficulties with medication reconciliation, insufficient communications between caregivers, and deficient medical records. Although gaps are ubiquitous in healthcare, there are a variety of patients, locations, and illnesses that increase the likelihood and consequences of gaps. Patients who are at increased risk for gaps in care continuity include those who have multiple illnesses, cognitive deficits, complex treatment regimens, or who are physically and organizationally separated from their caregivers.

Shifting the role of data collection to the patient implies a role shift. Patients must now take on the traits of a technician that were once reserved for those who had completed a program of training. What do the data tell us about patient success in the performance of such tasks? The accuracy of blood glucose self-monitoring depends on technique and the user (Tomky and Clark 1990). Rumley (1997) determined that even use of glucometers by hospital staff members was unacceptable without training, quality control procedures, a quality assurance scheme, and improved meter technology that was backed by laboratory expertise. The role of the technician requires an aptitude for such tasks that not every patient will be able to perform well, particularly if enervated by illness. Incorrect actions can occur when information and interfaces do not fit the abilities, their past experiences, or the demands of their daily lives among patients, family members, and friends (Woods 2000).

The equipment that patients use and the way they use it also affects their ability to perform such work. Obradovich and Woods (1996) found that medical device interface deficiencies induce users to develop their own tailoring strategies in order to cope with the shortcomings. Deficiencies in medical devices include poor feedback about device state and behavior, complex and ambiguous sequences of operation, multiple poorly distinguished modes of operation, and ambiguous alarms. In the case of self-monitoring, Bastanhagh et al. (2007) found that none of the glucometers they studied met the American Diabetes Association (ADA) criteria of reading within 5% of laboratory reference results. Others (e.g., Tate, Clemens, and Walters 1992; Dean et al. 1981) have found the performance of various glucometer models varies widely. Bamberg et al. (2005) found that the test strips used in blood sugar monitoring are vulnerable to the effects of exposure to the elements.

While Dunn et al. (1977) found little difference in diagnostic accuracy, tests required, patient management practices, efficiency and referral rates, or patient attitudes when using four different means of communication (audio only, audio plus still images, interactive video, and face-to-face). However they did find evidence that when using different communication modes, physicians use different methods to arrive at a diagnosis and different patient management practices. Understanding the nuances of physical and mental health requires sensitivity to a large array of phenomena. Making decisions about patient health require as rich a set of information as possible. In-person experiences rely on sensitivity and observation to make use of as many phenomena as are available. Electronic media, though, necessarily convey a subset of information compared with face-to-face encounters. Less information from physical cues, direct observation of the patient, and the ability to clarify and adjust meanings can erode the ability of patient and care provider to communicate.

The highly personal nature of healthcare may also be affected by the use of electronic communication to mediate sensitive information. Matters related to personal care may not lend themselves to full, open description via electronic means. This may be due to modesty, fear, or uncertainty. For example, nurse coordinators in an anesthesia department wrote notes on scraps of paper on items they were not sure about, hiding them for later verification rather entering them into an electronic scheduling system for others to see (Nemeth 2003, 7).

14.4.2 Representations Aid Cognitive Work

As two of their strategies to improve patient safety, the Institute of Medicine (Kohn, Corrigan, and Donaldson 1999, 177, 183) recommended improving access to accurate, timely information, and making relevant information available at point of patient care. Healthcare relies on the description of process and condition. For each patient, multiple diagnostic and therapeutic processes are underway, about to be started, or being concluded. Each patient's condition can be accounted for by a spectrum of variables that are interrelated. The interactions among those variables exceed the ability of clinicians to perceive them. Multiple individuals care for multiple patients and frequently start, stop, and resume care tasks. The presentation of information directly affects clinicians' ability to develop an effective mental representation of past, current, and prospective states of the patients who are under their care. Even under the best circumstances, there is an irreducible uncertainty that makes it difficult for clinicians to fully grasp the phenomena for which they are accountable. Recent increases in coordination demands due to staff resource limits place an even greater need for the reliable exchange of information in instances such as between-shift hand-offs.

Patients and care providers operate in what has been termed a *joint cognitive system* (Woods and Hollnagel 2006), including individuals as well as information systems that are intended to accomplish a shared goal. *Joint cognitive activity* is an extended set of actions that are carried out by an ensemble of people who coordinate with each other. In order to succeed as a team player in the conduct of joint cognitive activity (Christofferson and Woods 2002), participants must enter into an agreement to work together, be mutually predictable in their actions, be mutually directable, and maintain a common ground (the pertinent knowledge, beliefs, and assumptions that the involved parties share) (Klein et al. 2004). This includes information systems that are intended to aid cognitive work.

Existing representations (especially paper artifacts) combine pertinent data to allow clinicians to create, update, and evaluate the implications of their mental models. The paper artifacts that clinicians create, help them to handle the uncertainty and contingency inherent in the care setting. The complexity and urgency of the cognitive work that clinicians perform is demonstrated by the varied artifacts that already exist in the workplace. Artifacts include hard copies of notes, check-lists, and status boards, as well as control/display interfaces on individual pieces of electronic equipment, assignment schedules that are shown throughout a facility on computer monitors, and more.

Information displays serve as cognitive artifacts (Hutchins 2002) and directly influence the delivery of patient care and the coordination of care across and among patients. The way that a problem is presented can improve or degrade the cognitive work of clinical care providers and patients, because artifacts shape cognition and collaboration (Woods 1998). Representations can make a task easier (Zhang and Norman 1994). They can draw together essential elements of information (domain semantics) for participants to consider, and compactly integrate multiple kinds of information (Heiser and Tversky 2006). Rule-based programs have tried to transfer decision making to a machine agent. By contrast, skillfully crafted representations facilitate and empower human expert judgment. They do so by portraying the domain semantics that describe the current state, constraints, goals, and opportunities for action.

Representations that simultaneously show both constraints and possible solutions are crucial to orient and re-orient clinicians and patients. In the case of consumer health, representations can include crucial information such as composite profiles that depict the interactions of different important laboratory values, or depictions of trends in laboratory values. Representations that support cognitive work can improve the reliability and efficiency of clinical work that, in turn, promote patient safety and minimize gaps in the continuity of care. More effective representations would *spare both patients and clinicians some of the effort of data synthesis* by accounting for a wealth of discrete elements through summary, or abstraction (Rasmussen and Pjetersen 1995). Such representations to aid data synthesis would also *enrich their ability* to contemplate problems and to envision opportunities. Because of this, effective representations offer substantial potential to benefit clinical care and improve patient safety.

Healthcare IT (particularly in the consumer sector) is in an early stage of development. As a result, few of the information tools that are intended to aid cognition reflect consumer healthcare domain semantics. The means to make that possible are available through the study of cognitive work.

14.5   STUDYING COGNITIVE WORK

Studies of clinical decision support systems (see Garg et al. 2005; Ramnarayan and Britto 2002) suggest that prior attention has largely focused on the clinician. However, information tools to aid the patient–clinician relationship must necessarily be based on an understanding of the work that *all* participants perform. This requires the ability to understand *work as done*, rather than *work as imagined* (Nemeth 2007) through the study of the way that people actually perform in the real world.

The field of human factors studies cognitive work in the real world in order to reveal context, underlying mechanisms, and behavioral patterns through the use of research methods. We mentioned earlier in this chapter that patients and care providers operate in a joint cognitive system. The study of such a system in the real world discovers how people behave and develop strategies to accomplish what they need to do within the constraints, the domain semantics, they face (Woods and Hollnagel 2006). For example, patients who manage chronic conditions must routinely follow a cycle of taking samples such as blood specimens, having samples evaluated, reviewing sample values, and making decisions based on those values in the context of prior values. Decisions can take the form of changes to current medication therapy, deletion or addition of medications, or the choice to make no change. The development of representations to support such decisions must be based on understanding the cognitive work that is involved.

The following human factors methods are an example of the resources that can be used to develop such an understanding and translate it into useful results.

14.5.1 COLLECT DATA

Ethnographic methods that are performed as part of field studies make it possible for researchers to understand the cognitive work of healthcare (Aarts, Doorewaard, and Berg 2004). Such contact with the details of work as it exists in the real work domain is essential to understand it. This is because the regularities of cognitive work can be discovered only through examination of the details of the specific settings to see how these patterns play out (Mackay 1999; Norman 1991).

Field studies such as *direct observation* and *informal interviews* supplemented by audio and video recording look underneath current practice to see what domain semantics has adapted to and how changes reverberate to transform roles, judgments, difficulties, strategies, and vulnerabilities (Larkin 1989; Hutchins 1995). Identification and description of *cognitive artifacts*, including hard copy print-outs, status boards, and computer-generated and equipment-generated displays that represent problems and constraints, assist the solution of problems, represent solutions, and provide solutions to individuals or groups as part of their *distributed cognition*.

14.5.2 ANALYZE DATA

Constraint-based work domain analysis (Vicente 2000) can be used to depict the goals and constraints of the work domain that the observation data produce.

Cognitive task analysis (CTA) (Rasmussen 1986) can reveal and describe the cognitive work that operators perform as they confront the constraints that work domain analysis reveals. A CTA lays out the details of interaction between the operator and the environment in terms of the available goals and the means to achieve them.

Process tracing (Woods 1993) can be used to extract patterns from the phenomena collected through CTA, then distill those patterns into descriptions of functions that occur while providing care. The investigators will then use the common features revealed through process tracing.

14.5.3 DEVELOP REPRESENTATIONS

Prototype development can be used to create representations based on relevant work domain semantics and cognitive task elements that were discovered during data collection and revealed through data analyses. The process can model initial concepts for a series of representations in hard copy form using desktop publishing software. The structure of existing data that clinicians and patients already use will provide structure for the representations. Review hard copy examples with clinicians and patients.

Use of rapid design prototyping by repeating this cycle frequently maximizes the fidelity of the representation design to participant needs. Validated representations can then be translated from rough to refined form through the creation of electronic display versions. Using selected software platform(s), representations can be created that embody previously discovered domain semantics. Computer-generated versions of the representations can be reviewed with participants in order to determine how they are perceived compared with earlier exploratory versions.

14.5.4 EVALUATE REPRESENTATIONS

Outcome-oriented evaluation (Rubin 1994) can be used to assess the concepts as they are developed and refined. This participatory design can incorporate representative users in the design process, conducting informal interviews with clinicians and patients who use the rough and refined prototypes. As a complement to participatory design, clinicians and patients can perform actual work tasks in a realistic setting using the representations in an exploratory usability test.

Consultants can provide their expert evaluation of initial hard copy and software concept designs, addressing both usability and interaction design.

Representations can be evaluated in a laboratory test with a small set of clinicians and patients. Assessment usability evaluation in a laboratory setting can determine the effectiveness of a computer-based interactive version of the representations, and be used to determine how well they fit the participants' perceptions and tasks.

Through this process, phenomena involved in the cognitive work that patients, care providers, and other participants perform can be observed in the real world and directly linked to representations. Phenomena are captured and used as data. Analyses extract patterns and meaning from the data. Findings draw conclusions about the real world that the analyses produce. The resulting representations reflect what actually occurs in the consumer healthcare work domain.

14.6 SOLUTION DIRECTIONS

Insightful study of actual work in the real world produces results that serve the needs of those who perform it. In the case of diabetes care management, Klein and Meininger (2004) provide an excellent example of the insightful research that is necessary to learn about the cognitive work that patients confront when caring for chronic conditions, and what impediments they face. Findings from such studies form the foundation for design direction to develop tools that match the needs that research discovers.

Prior work that was cited earlier in this chapter suggests that essential aspects of self-managed care include:

- Current information related to recent developments about disease(s)
- Ability to translate general information about disease(s) to one's own profile
- Collect, store, monitor, and review data on one's own condition and therapy
- Develop relationships with those who can be of benefit, including clinicians, fellow patients, family members, and friends
- Learn how to perform monitoring and therapy skills

Some current Internet resources may evolve in the direction that would support these and other activities that are part of a robust and resilient eHealth system. For example, Hejlesen and colleagues (Hejlesen, Larsen, and Pedersen 2006; Brøndum, Jensen, and Hejlesen 2006) have developed a web-based approach to assist self-management of chronic conditions such as diabetes and cardiovascular disease. Babylink (Lyon et al. 2004) is a web-based tool developed to support communication between clinicians and parents in neonatal intensive care units (NICU). The system searches electronic records for information, such as diagnoses, treatments, and communication, to create encrypted reports for each infant. Parents who are given site access privileges can review reports in addition to photos, messages from the staff, and general information about issues and care of neonates. Brennan (2007) has championed an eHealth agenda that, among many goals, moves beyond goals of learning toward assisted cognition and patient activation. One means to achieve this is the personal portal that provides patient information of interest to those who are approved to visit a secure "Patient Status" web page. This includes an individual's health history and goals, prescriptions, interactive tools such as reminders, and billing and payment. Such personal health record systems that store patient-related health information make it possible for physicians to eventually consult with individuals over the Internet for a fee (Saporito 2005).

14.7   CONCLUSION

Technology-mediated healthcare will differ from traditional face-to-face relationships and changes to perception, and decision making will affect patients as well as their relationship with acute care clinicians. The human factors of consumer eHealth dictate that information tools serve the needs of those who seek to restore individuals to improved physical and mental health. This requires a well-founded understanding of the cognitive work that is performed by all who participate in the process. The "information ecology" for this cognitive work will be the collective result of many changes in technology, organizations, and knowledge. Having an explicit understanding of how these three relate will be essential to make future technology work in concert, rather than work at cross purposes, with its users.

ACKNOWLEDGMENTS

Dr. Nemeth's research during the development of this chapter was made possible by the generous support of the Department of Anesthesia and Critical Care of The University of Chicago and a grant from the U.S. Food and Drug Administration Center for Devices and Radiologic Health.

REFERENCES

Aarts, J., Doorewaard, H., and Berg, M. 2004. Understanding implementation: The case of a computerized physician order entry system in a large Dutch university medical center. *J Am Med Inform Assoc* 11 (3): 207–16.

Adams, K., and Corrigan, J.M. 2003. *Priority Areas for National Action: Transforming Health Care Quality.* Washington, DC: Institute of Medicine. National Academies of Science. 52–53.

Akerkar, S.M., Kanikar, M., and Bichile, L.S. 2005. Use of the Internet as a resource for information by patients: A clinic-based study in the Indian population. *J Postgrad Med* 51:116–18.

American Medical Association (AMA). 2002. "Health literacy introductory kit." American Medical Association. http://www.ama-assn.org/ama/pub/ printcat/8035.html (accessed December 22, 2002).

Anton, B., and Nelson, R. 2006. Literacy, consumer informatics, and healthcare outcomes: Interrelations and implications. *Stud Health Technol Inform* 122:49–53.

Asch, S., Kerr, E., Keesey, J., Adams, J., Setodji, C., Malik, S., and McGlynn, E. 2006. Who is at greatest risk for receiving poor-quality health care? *N Engl J Med* 354:1147–56.

Association of Telehealth Service Providers (ATSP). 2007. Telemedicine: Telemedicine coming of age. Association of Telehealth Service Providers. http://tie.telemed.org (accessed March 26, 2007).

Bamberg, R., Schulman, K., MacKenzie, M., Moore, J., and Olchesky S. 2005. Effect of adverse storage conditions on performance of glucometer test strips. *Clin Lab Sci* 18 (4): 203–9.

Billings, C.E. 1996. *Aviation Automation: The Search for a Human Centered Approach.* Mahwah, NJ: Lawrence Erlbaum.

Brennan, P. 2007. Patient-centered computing: Are the patients ready? Are we? The University of Wisconsin, Madison. http://healthsystems.engr.wisc.edu/papers_presentations/PatientCenteredComputing.ppt (accessed June 12, 2007).

Brøndum, M., Jensen, M.T., and Hejlesen, O.K. 2006. A web-based decision support system for cardiovascular disease management. In *SHI2006 Proceedings: 4th Scandinavian Conference on Health Informatics,* Aalborg University, Aalborg, Denmark, Virtual Centre for Health Informatics, Aalborg University.

Christofferson, K., and Woods, D.D. 2002. How to make automated systems team players. In *Advances in Human Performance and Cognitive Engineering Research*, ed. E. Salas, 1–12. Vol. 2. JAI Press, Stamford, CT: Elsevier.

Clark, N.M. 2003. Management of chronic disease by patients. *Annu Rev Public Health* 24:289–313.

Cook, R.I., Render, M., and Woods, D.D. 2000. Gaps in the continuity of care and progress on patient safety. *BMJ* 320 (7237): 791–94.

Dean, B., North, S.E., Harrison, L.G., and Martin, F.I. 1981. Evaluation of home glucose measuring devices. *Med J Aust* 2 (4): 197–200.

Donn, J. 2006. Study: Most get mediocre health care. *Miami Herald*. http://www.miami.com/mld/miamiherald/living/health/ 14106584.htm (accessed March 15, 2006).

Doyle, D.J., Ruskin, K.J., and Engel, T.P. 1996. Then Internet and medicine: Past, present, and future. *Yale J Biol Med* 69 (5): 429–37.

Dunn, E.V., Conrath, D.W., Bloor, W.G., and Tranquada B. 1977. An evaluation of four telemedicine systems for primary care. *Health Services Res* 77 (5): 748–54.

Elford, R. 2008. Telehealth and healthcare team communications. In *Improving Healthcare Team Communications: Building on Lessons from Aviation and Aerospace*, ed. C. Nemeth, 221–44. Aldershot: Ashgate.

Finkelstein, S.M., Speedie, S.M., and Potthoff, S. 2006. Home telehealth improves clinical outcomes at lower cost for home healthcare. *Telemed J E Health* 12 (2): 128–36.

Freudenheim, M. 2010, February 9. In Haiti, practicing medicine from afar. *New York Times*. D5.

Garg, A.X., Adhikari, N.K.J., McDonald, H., Rosas-Arellano, M., Deveraux, P.J., Beyenne, J. 2005. Effects of computerized clinical decision support systems on practitioner performance and patient outcomes: A systematic review. *JAMA* 293 (10): 1223–38.

Heinzelmann, P.J., Lugn, N.E., and Kvedar, J.C. 2005. Telemedicine in the future. *J Telemed Telecare* 11 (8): 384–90.

Heiser, J., and Tversky, B. 2006. Arrows in comprehending and producing mechanical diagrams. *Cogn Sci* 30: 581–92.

Hejlesen, O.K., Larsen, L.B., and Pedersen, C.F. 2006. Telemedicine supported patient-centered diabetes care. *SHI2006 Proceedings: 4th Scandinavian Conference on Health Informatics: Aalborg University*. Aalborg, Denmark: Virtual Centre for Health Informatics, Aalborg University.

Hutchins, E. 2002. Cognitive artifacts. MIT COGNET web site: http://cognet.mit.edu/ MITECS/Entry/hutchins (accessed July 7, 2002).

Klein, H., and Meininger, A. 2004. Self-management of medication and diabetes: A cognitive control task. *IEEE Trans Syst Man Cybern A* 34 (6): 718–25.

Klein, G., Woods, D.D., Bradshaw, J.M., Hoffman, R., and Feltovich, P.J. 2004. Ten challenges for making IT a team player in joint human-agent activity. *IEEE Intell Syst* 19 (6): 91–95.

Kohn, K.T., Corrigan, J.M., and Donaldson, M.S. 1999. *To Err Is Human: Building a Safer Health System*. Washington, DC: National Academy Press.

Larkin, J.H. 1989. Display-based problem solving. In *Complex Information Processing: The Impact of Herbert A. Simon*, eds. D. Klahr and K. Kotovsky, 319–41. Hillsdale, NJ: Lawrence Erlbaum.

Levinson, W., Kao, A., Kuby, A., and Thisted R.A. 2005. Not all patients want to participate in decision making. *J Gen Intern Med* 20 (531): 531–35.

Lipschitz, R. and Shaul, O.B. 1997. Schemata and mental models in recognition-primed decision making. In *Naturalistic Decision Making*, eds. C.E. Zsambok and G. Klein, 293–303. Mahwah, NJ: Lawrence Erlbaum.

Lyon, A.J., Freer, Y., Coyle, C., and Stenson, B. 2004. 167 Babylink-A web-based tool to improve communication between clinicians and parents in a neonatal unit. *Pediatr Res*. 56: 3.

Mackay, W.E. 1999. Is paper safer? The role of paper flight strips in air traffic control. *ACM Trans Comput Hum Interact* 6 (4): 311–40.

McGlynn, E.A., Asch, S.M., Adams, J.A., Keesey, J., Hicks, J.H., DeCristofaro, A., and Kerr, E.A. 2003. The quality of healthcare delivered to adults in the United States. *NEJM* 348:2635–45.

MedicineNet. 2007. http://www.medterms.com/script/main/art.asp?articlekey=38929 (accessed May 10, 2007).

Nemeth, C. 2003. The master schedule: How cognitive artifacts affect distributed cognition in healthcare. Dissertation Abstracts International 64/08, 3990 (UMI No. AAT 3101124).

———. 2004. *Human Factors Methods for Design*. New York: Taylor & Francis/CRC Press.

———. 2005. Health care forensics. In *Handbook of Human Factors in Litigation*, eds. I. Noy and W. Karwowski. 37-1–37-18. New York: Taylor & Francis.

———. 2007. Healthcare groups at work: Further lessons from research into large scale coordination. *Cogn Tech Work* 9 (3): 127–30.

Nemeth, C., Nunnally, M., O'Connor, M., Klock, P.A., and Cook, R. 2005. Getting to the point: Developing IT for the sharp end of healthcare. *J Biomed Inform* 38 (1): 18–25.

Nemeth, C., O'Connor, M., Klock, P.A., and Cook, R.I. 2006. Discovering healthcare cognition: The use of cognitive artifacts to reveal cognitive work. *Org Stud* 27 (7): 1011–35.

Noel, H.C., Vogel, D.C., Edos, J.J., Cornwall, D., and Levin, F. 2004. Home telehealth reduces healthcare costs. *Telemed J E Health* 10 (2):170–83.

Norman, D.A. 1991. Cognitive artifacts. In *Designing Interaction: Psychology at the Human-Computer Interface*, eds. J.M. Carroll. 17–38. New York: Cambridge University Press.

Obradovich, J. and Woods, D.D. 1996. Users as designers: How people cope with poor HCI design in computer-based medical devices. *Hum Fact* 38 (4): 574–92.

Oh, H., Rizo, C., Enkin, M., and Jadad, A. 2005. What is eHealth?: A systematic review of published definitions. *World Hosp Health Serv* 41 (1): 32–40.

Perednia, D.A. and Allen, A. 1995. Telemedicine technology and clinical applications. *JAMA* 273 (6): 483–88.

Powell, J.A., Lowe, P., Griffiths, F.E., and Thorogood, M. 2005. A critical analysis of the literature and consumer health information. *J Telemed Telecare* 11 Suppl 1:41–43.

Ramnarayan, P., and Britto, J. 2002. Paediatric clinical decision support systems. *Arch Dis Child.* 87:361–62.

Rasmussen, J. 1986. *Information Processing and Human-Machine Interaction: An Approach to Cognitive Engineering*. Amsterdam: North-Holland.

Rasmussen, J. and Pjetersen, A. 1995. Virtual ecology of work. In *Global Perspectives on the Ecology of Human-Machine Systems*, eds. J. Flasch, P. Hancock, J. Caird, and K. Vicente. 121–156. Hillsdale, NJ: Lawrence Erlbaum.

Rubin, J. 1994. *Handbook of Usability Testing*. New York: John Wiley.

Rumley, A.G. 1997. Improving the quality of near-patient blood glucose measurement. *Ann Clin Biochem* 35 (Pt 1): 281–86.

Saporito, B. 2005, June 27. The e-health revolution. *Time* 165 (26): 55–57.

Schillinger, D., Wilson, C., and Bindman, A. 2003. Reply: Who actually has low health literacy? *Arch Intern Med* 163 (14): 1746.

Schwartz, K.L., Roe, T., Northrup, J., Meza, J., Seifeldin, R., and Neale, A.V. 2006. Family medicine patients' use of the Internet for health information: A MetroNet study. *J Am Board Fam Med* 19 (1): 39–45.

Sharpe, V., and Faden, A. 1998. *Medical Harm*. Cambridge: Cambridge University Press.

Smith, P.K., Fox, A.T., Davies, P., and Hamidi-Manesh, L. 2006. Cyberchondriacs. *Int J Adolesc Med Health* 18 (2): 209–13.

Tate, P.F., Clements, C.A., and Walters, J.E. 1992. Accuracy of home blood glucose monitors. *Diabetes Care* 15 (4): 536–38.

Tomky, D.M., and Clarke, D.H. 1990. A comparison of user accuracy, techniques, and learning time of various systems of self blood glucose monitoring. *Diabetes Educ* 16 (6): 483–86.

Vicente, K. 2000. Work domain analysis and task analysis: A difference that matters. In *Cognitive Task Analysis*, eds. J.M. Schragen, S.F. Chipman, and V.L. Shalin, 101–18. Mahwah, NJ: Lawrence Erlbaum.

Von Korff, M., Gruman, J., Schaefer, S.J., Curry, S.J., and Wagner, E.H. 1997. Collaborative management of chronic illness. *Ann Intern Med* 127 (12): 1097–1102.

Wittson, C.L., and Benschoter, R.A. 1972. Two-way television: Helping the medical center reach out. *Am J Psychiatry* 129 (5): 136–39.

Woods, D.D. 1993. Process tracing methods for the study of cognition outside of the experimental psychology laboratory. In *Decision Making in Action: Models and Methods*, eds. G. Klein, J. Orasanu, and R. Calderwood. 228–251. Norwood, NJ: Ablex.

———. 1998. Designs are hypotheses about how artifacts shape cognition and collaboration. *Ergonomics* 41 (2): 168–73.

———. 2000. Behind human error: Human factors research to improve patient safety. Oral testimony at National Summit on Medical Errors and Patient Safety Research. Quality Interagency Coordination Task Force (QuIC) and Agency for Healthcare Research and Quality (AHRQ).

Woods, D.D., and Hollnagel, E. 2006. *Joint Cognitive Systems*. Boca Raton, FL: Taylor & Francis/CRC Press.

Zhang, J., and Norman, D. 1994. Representations in distributed cognitive tasks. *Cogn Sci* 18:87–122.

15 Consumer Behavior of Employees Using Information and Communication Technology Products in an Organizational Setting

Ken Eason

CONTENTS

15.1 EMPLOYEES AS CONSUMERS

The objective of this chapter is to consider the behavior of employees at work in an organizational setting as consumers of information and communication technology (ICT) products. The term "consumer" is more often applied to members of the public and it may seem unusual to apply it to people at work. Indeed, there are several features of being a consumer of a product that may not seem appropriate to a person at work. First, employees may have little say in the purchase of the products that they use at work. Second, it might be presumed that they have little choice about how they use the product; we might assume that they are told what tasks to perform with the product, trained in its use and thereafter expected to meet performance standards in doing work with the product. On this basis, employees might better be classed as "product users": they may make heavy use of a product and, for that reason, product design is very important to them, but they may be regarded as passive users, i.e., their motivation to use the product, whether they like it, value it, have a good experience of it, etc., may seem less important. In this chapter, I hope to show that, in the realm of ICT products, employees behave as consumers in their use of these products in the sense that they exercise a lot of discretion in what they use and how they use it, and that the patterns of use they display have many implications for the design of the products.

The ICT products in use in the working environment are many and various. The hardware may be telephones of many kinds, often with much other functionality, personal organizers, and computer devices in many forms. There is even greater variety in the software applications that are available on these devices, ranging from generic packages such as work processing to applications that are very specific to the tasks that employees perform. There is, of course, a burgeoning literature on the human factors of these devices, much of it related to the use of them in a working environment, e.g., Preece, Rogers, and Sharp (2002), Helander, Landauer, and Prabhu (1997). However, it is not intended to review this literature in this chapter. The focus here is on the behavioral patterns of employees who use these products and the lessons this behavior teaches for product design.

There is a growing literature, much of it ethnographic in its orientation, that points to the active way in which people at work respond to the technical products they use. Pinch and Bijker (1987), for example, use the term "social construction of technology," to describe the way that employees interpret and use technology in local settings. As a result of research that is specifically about the use of ICT products, Orlikowski (2000) describes how users have "a practice lens" through which they determine what they will use and how they will use it in the work they do, i.e., in their "practice." These studies point to emergent behavior in the use of technology that often comes as a surprise to both the designers of the technology and to the senior management that purchased and implemented the technology in the organization. This chapter will review the evidence about the forms this emergent behavior takes, but first it is useful to establish some concepts about the organizational setting in which this behavior is occurring.

15.2 THE ORGANIZATIONAL SETTING: A SOCIOTECHNICAL SYSTEMS PERSPECTIVE

The most common framework used in the human factors literature to describe the factors influencing the behavior of an employee at work is one that focuses on the triad: task-human-technology, i.e., it recognizes that the human has a task to undertake and that successful task performance depends on having available technology that is fit for this task and is in a form that the human is able to use effectively. A fourth variable is the environment in which the task is undertaken, often presented as the physical environment, which may influence the ease with which the task can be undertaken. In an organizational setting, the task and the way it is to be undertaken may be largely prescribed.

This framework has great value in explaining behavior with technical products, especially in the primacy it gives to the task the employee is undertaking. However, as many authors have pointed out, it does not include many variables that are also known to influence behavior in an organizational setting. The task in hand, for example, may be one that the employee shares with other employees and the collective way in which they work together will influence the use of technical products. The individual operator framework draws attention to the tight interdependency between the human being undertaking a task and the technology they use. Vicente (2004), for example, stresses that this interdependency exists at many levels, i.e., at the overall organizational level, the work system level, the level of a work team and of the individual operator. A theoretical framework that emphasizes this interdependency at all levels in the organizational setting is sociotechnical systems theory and it offers concepts that will be used in this chapter to illuminate the behavior of employees with technical products and to explore design implications.

Sociotechnical systems theory originated at the Tavistock Institute of Human Relations in London in the 1950s as a way of understanding what happened in organizations as work processes became more mechanized. As a result of a series of studies in industries such as coal mining (Trist et al. 1963) and weaving (Rice 1958), a theory was developed that showed the strong interdependency of the social and technical systems, which are the two mechanisms whereby work gets done in an organization. In essence, this theory uses the same concepts as the individual operator framework, but at a more macro level. The individual's task now becomes part of the collective task of the organization and the task one person does may be highly dependent on the tasks performed by others. The social system is conceived as a collection of employees in work roles who are dependent on one another through task relationships and through power and control structures. The technical products in use may be stand-alone tools but, because of the need to co-ordinate and control the work process, are more likely to be products shared and used in common ways across the workforce. It is also recognized in sociotechnical systems theory that environmental forces may influence the ease or difficulty of undertaking organizational work, although, in this case, the factors at work may be market forces, the financial climate, etc., as well as the physical environment.

There are several implications of this theory for the behavior of employees with ICT products that arise when the product is implemented and it becomes part of a complex work, producing system with many interdependencies. First, the dominant way that the product will be judged will be by its ability to support the tasks undertaken in each work role. Second, the judgment will be based on the support it gives to the tasks as employees experience them rather than to models of the tasks that may have informed the design of the products. Third, the product may not be stand-alone, but may be better regarded as a technical system serving a range of users, and the extent to which it is closely coupled to the work process and supports people who have to co-operate in their task behavior may influence the use of the system. Lastly, the employees will be part of a work community and cultural practices may grow up around the product as they share knowledge and experience of it. In sum, we may look for emergent behavior patterns with products not as a consequence of the individual views of employees, but as a result of the product being evaluated for its fit with the collective needs of employees within the reality of the sociotechnical system in which they undertake their work.

15.3   ACTIVE RESPONSES OF EMPLOYEES TO TECHNICAL PRODUCTS

How then do employees respond to technical products that they are asked to use to perform their tasks in organizations? In the act of purchasing and implementing technical products, the management of organizations will have expectations about how they will be used to further the ends of the enterprise. No doubt, there are cases when the use of the products goes strictly according to plan, but they are hard to find in the literature. What is reported in the literature is a mixture of the kinds of behavior patterns listed in Table 15.1. Each of these responses will now be explored in more detail.

15.3.1   NON-USE OF A PRODUCT

When computer systems were first introduced into business, it was not uncommon for 40% of the applications to be adjudged failures. The systems were never used or were used for a short period before removal. The striking fact is that, although systems have become very much more

TABLE 15.1
Patterns of Product Use

Behavior Pattern	Description
Extent of Use	
Use as planned	Use as planned by people in a specific work role in the organization implementing the product
Non-use	The non-use of the product by people in a work role who were intended to use it
Partial use	The use of a more limited set of features or functionality of a product by people in a work role than was intended when it was implemented
Type of Use	
New practices	The use of the product to support the work of people in a work role in ways that were not envisaged when it was implemented
Workarounds	The use of the product in unintended ways because normal use would not permit the completion of the task being undertaken

sophisticated and reliable, the results of surveys suggest that the failure rate remains as high as 20% and that most of the other projects that implement systems also have major problems. The Standish Group regularly conducts surveys of computer applications in business and, for 2004, reported 29% successful projects, 18% failures, and 59% 'challenged' projects (in which there are serious cost and time overruns, less functionality delivered and adopted than planned, etc.) (Standish 2007). There are many specific case studies that illustrate the reasons for high levels of failure. Many case studies are the result of formal investigations of public service systems where a public inquiry is required. However, there is every reason to suppose that similar issues lie behind failures in the private sector. One example of a system failure is the attempt by the London Ambulance Service to implement a computerized despatch system that would direct ambulances to an incident and then direct them to appropriate hospitals (Page, Williams, and Boyd 1993). This system was implemented twice and on both occasions led to chaos, with many casualties waiting hours for an ambulance to arrive. The system was scrapped. Some of the reasons why it failed are technical; it did not, for example, have the capacity to cope when the number of incidents built up. However, most of the problems arose because the system did not adequately map onto the processes by which the ambulances and hospital emergency services worked and the system did not, for example, have the kind of up-to-date knowledge of road conditions that the previous human despatchers of ambulances were able to maintain.

It is instructive to look at projects that were successful; they were usually characterized by being small and being introduced through a user-centered process that placed an emphasis on the real nature of the work being undertaken by its users. By contrast, failed projects were usually characterized as large, ambitious, and with the intention of transforming the work processes of employees in a relatively short period of time. In many cases, the discovery that the work process was more complex and more volatile than the planners foresaw led to the realization that the system was not going to be successful.

15.3.2 PARTIAL USE

Modern ICT products often come with a long "tick list" of functions that they can perform. ICT products continue to converge so that what was, for example, just a mobile phone has now migrated to be a camera, a phone, a means of accessing the internet and sending emails and many more things as well. Many specific software applications also offer many different features. When a system is successfully implemented, what use do employees make of the range of features that are available and is it the range of features that they were intended to use?

The answer in general appears to be that they make use of a small percentage of what they could use. I first became aware of this as a result of a study conducted in a national domestic bank (Eason 1984). The bank made available to staff working in its branches a system that gave them access to specific details of each customer's account; they could, for example, get information about the current balance, the latest debit or credit, the highs and lows of the account, and interest paid. In total, there were 36 "codes" that they could use. An analysis of the log of usage of this system showed that four of these codes accounted for 75% of usage, that the average number of codes used by a member of staff was eight, and nearly half of the codes were never used at all. This came as a great surprise to the designers because each code had been created to serve specific tasks that the branch staff had to undertake. To understand the reasons behind these findings, we conducted a study of 125 staff in 15 branches, asking general questions about the use of the system, but also asking what they would do in a number of task scenarios, selected to test the use of codes widely in use and codes very rarely used. The study revealed different patterns of use in different branches:

- In 13 branches, most of the staff had a working knowledge of about eight codes. They were aware of the others but did not understand them or use them. Some of the codes were fairly general purpose, e.g., the last 16 entries to the account provided a mini-statement and this could serve a lot of purposes. When confronted by task scenarios where the system designers expected use of codes they did not know, the staff either used a general purpose code such as the last 16 entries, or they ordered a printed statement that would not arrive until the next day. Reviewing this behavior, the systems designers recognized that these strategies would be effective in most circumstances, but they often introduced delays in progressing the task and they could be risky, e.g., if the specific entry you were looking for was not in the last 16, you might assume it had not been processed whereas it might have been the seventeenth most recent entry.

- In the remaining two branches, the average use of codes was much higher (18) and in many more of the task scenarios members of staff chose the "correct" code, i.e., the one designed for the task. What was different about the these two branches was that, when asked what they would do if they did not know what code to use, many of the staff replied that they would consult a colleague or, in particular, consult the head of the machine room (effectively the local person in charge of computer systems). Further investigation showed that these two branches had established a "community of practice" (Brown and Duguid 2002) in which knowledgeable members of staff were quickly able to provide help and this had increased the average level of working knowledge about the system in the branch. There was no policy in the bank to promote this practice and it had not spread to other branches.

We have found evidence of partial use of technical systems in many other studies. In a study of advanced telephone systems in five organizations, for example, we found that of the many additional facilities by which users could store, bar, forward calls, etc., only one or two were in common use (Eason and Damodaran 1986). Most staff, for example, were aware of some of the facilities, especially the "conference call" facility, and planned to make use of them but had not got round to it. In a recent study of a system that enabled users to locate information in a large store of literature, we found that "the passive majority" used between 1 and 5 of the 22 facilities that were on offer whereas there was an "active minority" who used between 8 and 18 (Eason, MacIntryre, and Apps 2006).

What emerges from these studies is a user perspective on product functionality that can be likened to an iceberg. At the top, above the surface, is a small amount of functionality that the user understands how to use and has an identified use for in the tasks they undertake. This is the part of the product for which they have a working knowledge: it is what the philosopher Heidegger (1962) calls "ready to hand." Indeed, it can be so "ready to hand" that it gets used in an elastic way to serve tasks for which it was not designed. Below the surface is the rest of the functionality of the system, some of which the user may be aware of but has not yet explored how to use and some of which they have no knowledge at all. It can be argued, of course, that this is a state of affairs that is to be expected. In a multi-functional product, especially one designed for a generic market, there could well be much functionality that has no relevance for the user and it is to be expected that they would gravitate toward the use of the functionality that is of value to them. However, analysis of the findings in all of these studies shows that, in addition to functionality that was of no value to the users concerned, there was also functionality that colleagues with similar tasks made good use of and functionality that they were aware of and wanted to use sometime in the future. In each of these studies, we also found that when we explained what was available, users often went on to try previously unexplored functionality and brought it into their "working knowledge." It seems that, under the daily pressure to get work done, each product becomes a limited working tool and much of its potential to serve the tasks of the user may be wasted. We will explore the product design implications of this conclusion in the final section of the chapter.

Other studies have also revealed partial use of a product but for different reasons that are more directly associated with the organizational setting. In a recent study of the use of electronic patient information systems in hospitals, we have found that many records, designed to contain the administrative, demographic, and clinical information about the patient, are largely devoid of clinical details (Eason 2009). The reasons are that the clinical staff responsible for entering this information is concerned about the implications of placing it on record. This may be because they are concerned about the security of the system and that confidential information about their patients may fall into the wrong hands or that they are anxious that their professional judgments may be available for later scrutiny. Administrative and demographic data are much more factual and "safer." In this case, the use of the product is determined by the user's perception of the indirect organizational consequences of product use.

15.3.3 NEW PRACTICES

In some circumstances, as users start to explore the capability of a new technical product, they find that some part of its functionality has a particular value for them and this becomes their frame of reference for the product with the result that their usage patterns are quite different from those that were planned for the product. A well-known example that is not from an organizational setting is the way users, especially the young, discovered a rather obscure function of the mobile phone; that it could be used to send short text messages. This is a fiddly process and designers did not think it would get much use. However, young people found it was very cheap and they even devised their own form of abbreviated English to render it more usable. In organizational settings, it is often found that a product provides some functionality, apparently peripheral to the main function of the product, that helps users with a particular task and this then becomes a major role for the product. In a study we made of the development of an automated product for switching schedules in the electricity industry (Eason et al. 1995), such a user response changed the entire direction of the development. Switching schedules are the sequence of actions performed on an electricity network to isolate a section so that action can be taken to maintain or repair it. The initial aim of the product development was to automate the development of switching schedules, but in the process of its development, a graphical representation of the network was developed that electricity engineers found very useful for many other tasks. They did not like the idea of automating switching schedules for safety reasons and that line of development was discontinued. However, the concept of a portable, graphical, up-to-date representation of the state of the network got enthusiastic support from the electricity engineers and this became the new line of development. We have noted on a number of occasions that the community of users reviews a new product for its usefulness and often takes its use in an unexpected direction because they have particular task needs that it can serve. In current work on the implementation of electronic patient records in the UK National Health Service, we have been fostering this search by encouraging healthcare service delivery teams to review new products for the specific benefits it might yield for them (Eason 2005). In one instance, for example, a mental health delivery team concluded that a function of the record system that allowed it to alert them to patients they should see in the next few weeks was very useful. This is because mental health patients are notorious for missing appointments. It was very difficult for busy staff to check weeks ahead in order to issue reminders and an automated way in which this could be done had great value to them. Of the many functions an electronic patient record system might perform, this was rated very lowly by its developers, but it became its major function for this group of users.

It is always the hope of product designers that the product will inspire its users, but this evidence suggests that the nature of the inspiration may be organizationally specific and quite difficult to predict. There are again interesting implications for product development, which will be explored later in the chapter.

15.3.4 WORKAROUNDS

In many of the examples given above, the employee could exercise some choice about whether to use a product to do their work (or whether to use a particular function of the product). The product in these instances is only "loosely coupled" to the process of task completion, i.e., there are other ways that the task can be undertaken. There are, however, many circumstances where there is "tight coupling" of the technological product to the process of task completion, e.g., the car driver has to use the steering wheel to drive the car but may or may not use the more loosely coupled Satnav. A customer service representative in a call center has much less choice about the use of the telephone system that allocates customer calls to them than (say) a manager who may choose between using the telephone, email, or a face-to-face meeting to set up a communication. In circumstances where the tool is closely coupled to the work process, the employee, we could infer, will behave like an operator—using the technology as required to do the work—and there will be less evidence of consumer choice behavior. There is a growing body of evidence showing that this is often not the case. What happens when employees are faced with a dilemma—the particular task they are confronted by requires different treatment than that enabled by the product they are supposed to use? In these circumstances, the evidence is that employees show a lot of ingenuity in developing "workarounds," i.e., using the product in an unplanned way in order to accommodate the particular demands of the task as they see it.

There are two reasons why employees might find that there is a need for a workaround. First, it may be that the product designers have made assumptions about the tasks to be undertaken with the product that are not appropriate to the organization in which the employee works. Secondly, it may be that the product has been customized or configured within the organization to implement policies and practices that management wish employees to follow. However, employees may conclude that they cannot or will not abide by these practices. We will examine each of these circumstances in turn.

Many software products are designed as general purpose tools for use in many organizations, e.g., accountancy, order processing, and stock control. Abdelnour Nocera (2007), for example, studied the use of an enterprise resource planning (ERP) system in companies in different countries and gives many examples of different patterns of use emerging from the use of the same technical system. In one country, he found that the rules in the system that debarred staff from offering discounts to customers after a certain stage of the ordering process did not fit local business culture where it was the norm to offer discounts at any stage. The local staff found a novel way of doing calculations with the system that meant they could keep offering discounts as required by their business clients.

In a study I made of meeting management systems, I also found examples of assumptions made by systems designers that did not fit the custom and practice of employing organizations. These products are designed to fix the agenda of meetings, manage and distribute relevant papers and documents, facilitate minute taking, etc. In some instances, voting facilities were also included so that each attendee could vote on resolutions put to the meeting and the system would automatically tally the votes. Employees in different companies tried to use these products to serve widely differing types of meetings. Many, of course, never held votes and the users of the products found ways of switching off or ignoring the vote management function of the system (a form of partial use). However, to be used at all, these products demanded that agendas and papers were fixed well before the meeting and no new material could be added, but this did not fit the practices of all but the most formal of meetings. In many meetings, it was the practice to introduce new agenda items, new attendees, and new papers at the meeting. Employees involved in these types of meetings did not just discontinue the use of the meeting manager products, they resorted to various forms of workaround in which the formal meeting (using the product) was reduced to a minimum and more informal meetings developed alongside it to deal with the "real" business.

There are many examples of workarounds in the use of products in healthcare. Berg (1999) reports a study of medical staff in an intensive care unit making use of specialist products to monitor and control a patient's condition. One product enabled the fluid balance of a patient to be monitored. The product worked on half-hour intervals and if staff entered critical data at 10.29 in order to get a fluid balance at 10.30, the system would report it as a reading at 10.00. As this data could be very misleading, the staff developed a workaround that "fixed the clock" so that they always knew the exact time of the reading. In another example (and a case found elsewhere in studies of medical practice), nursing staff used a medication system that required a doctor's "instruction" to give the medication before they could administer it. In many instances, no doctor was available and previous practice was for the nurses to proceed with medication and for doctors to authorize it after the event. In order to preserve this practice, the medical staff found a way of winding back the clock in the product so that it would permit medication "before" authorization.

Examples also abound in the medical literature where managers have requested that products be configured to implement practices and processes that they wished to make standard in the organization only for employees to develop 'workarounds' of those configured products in order to undertake work in a way that they considered more appropriate. One example is the procedure implemented to control access to electronic medical records. In the UK National Health Service, users are issued with a smart card and a pin number that controls both the patient records that can be accessed and how much of the record can be accessed, i.e., doctors can see medical records, administrators only administrative data, etc. This "role-based access" policy was implemented because of widespread concern about the possibility of misuse of confidential patient data. In practice, the use of smart cards has led to widespread reports of workarounds. Medical teams who need to share and discuss information about patients, finding that each person only has a partial view, have taken to sharing the smart card of the person with the fullest access. In an Accident and Emergency Department, the medical team declared it was under too much pressure for each person to be continually gaining access using their smart cards. So they put the smart card of the shift leader in the system at the beginning of the shift and everybody used it thereafter (Collins 2007).

Another example of a medical workaround is the behavior of doctors in a mental health trust in their use of a patient medical records system (Eason 2009). The system provided highly structured files in which the doctors could record the case notes of a patient in a way that would facilitate statistical analysis across all records. The doctors found that recording patient case notes this way was laborious, tended to fragment the "story" of the patient, and did not allow them to record many of the issues they considered relevant. So they developed a workaround in which they used a free text input area, intended as a place to record the occasional note, as a place to write the full narrative of the patient. They ignored the more structured parts of the record. In this way, they met the needs of themselves, the patient, and their immediate colleagues to share information about the patient, but did not serve the managerial needs of their hospital, which needed to know, for example, the number of specific kinds of medical interventions being made.

An example of a workaround in a different kind of public service is provided by Broadhurst et al. (2009) in a study of social workers working in Children's Services in UK local authorities. These social workers are responsible for gathering evidence about children at risk so that decisions can be made about whether they can remain with their families or be taken into the care of the Local Authority. As a result of public concern following some widely publicized cases where children had died after mistreatment by their parents, the process by which the social workers gathered evidence had been formalized and implemented in a computer system. It was a system that required the social worker to input structured information in considerable volume about each child and to make decisions (e.g., that no further action was required or that a full assessment needed to be made) in a fixed period of time. The public outcry had also led to a rising tide of referrals about possible child abuse and, as a result, the workload of the social workers had risen at the same time as they were being asked to input larger volumes of information into the computer by specific deadlines. The researchers found that the social workers were adopting a range of workarounds in order to meet the performance targets they had been set. One example was that, where there was more than one child in a family and they were required to file a full account of each child, they used a "cut and paste" technique, copying the details of one child into the file of another on the principle that most of the information (about family circumstances for example) was the same. They also discovered that if they left most of the fields in the middle of the record blank, the computer system would still accept the record and file it as closed by the deadline.

These are only a few of the examples of what users of products do when they are confronted by a product that requires a specific kind of use but the current requirements of the task require something different. It can be argued that, in many cases, they should not behave as they do, but in practice it is inevitable that human ingenuity will find ways of coping with situations such as these.

15.4 DISCUSSION AND IMPLICATIONS FOR PRODUCT DESIGN

The evidence from the patterns of ICT product use in organizations suggests that employees are far from passive operators and might be better classed as active consumers. Their behavior ranges from outright rejection of the technology to finding new and valuable ways to make use of it with a range of partial use strategies and use through workarounds in the middle. Viewed as a totality, the evidence shows that employees, faced with the daily necessity of dealing with their tasks in the form that they manifest themselves, develop strategies for using the tools at their disposal in ways that meet their specific, local needs. And the result of this is that they may "frame" the product (Pinch and Bijker 1987) in ways that are quite unexpected to both the designers of the product and to the managers of the organization who purchased the product. In some cases, the responses may be quite individual, each employee finding their own distinct way of using the product. However, most of the studies suggest that local groups of users, those who, in sociotechnical systems theory terms share in the undertaking of the collective task, find a common way of framing the use of the products. Brown and Duguid (2002), from a study of photocopier service engineers, have used the term "community of practice" to describe the way that local groups of employees develop their own, shared, ways of assimilating technical products in useful ways into the practices by which they get their work done.

But what are the design implications of these patterns of behavior? In particular, what do they mean for user-centered design? In each of the cases described, users were producing unplanned patterns of behavior, suggesting that whatever process led to the development of the product, it did not include a detailed understanding of the many different contexts in which the users worked. It is instructive to take this data and see what they mean for "re-engineering" the design process so that it can be sensitive to what become the emergent needs of its employee consumers.

The first implication is to question what we mean by the "design process." If it means the process by which the technical facilities are made "ready to hand" for the consumer, then one conclusion is that this process has several stages where some form of design takes place. As a minimum, when we are referring to a product used by groups of employees in different organizations, we can identify three stages of design. The majority of ICT products are developed by suppliers for a mass market and the first design team is within the supplier organization who creates a generic product for a broad but identifiable business market. The second stage of design is in the organization that purchases the product and modifies it for use in their organizational setting. This may be a minor process of customizing facilities for local use or may be a major piece of work commissioned from the supplier to produce a custom-based version of the product. The more coupled it is intended that the product should be into the work process of the organization, the more likely it is that detailed design work will need to be undertaken at this stage. It is important to note that, if the product is sold to many different organizations, the supplier could ultimately be supporting many different versions of the same product. The final focus of design is when the consumers themselves, often working in concert as a "community of practice," "frame" the product in some way relevant to their particular needs and circumstances. As we have seen from the cases presented above, consumers can be active in selecting facilities to use and customizing the way they use them and are, in a real sociotechnical sense, designing the technology and their local practices, to both take advantage of the technological capabilities for the tasks they are performing and to avoid any unwanted consequences. It is instructive to take a closer look at these three stages of design effort to explore what a user-centered approach that spans all three might entail.

15.4.1 GENERIC PRODUCT DESIGN

The developer of a product for the business market hoping to sell the product to many organizations around the world is targeting users in many different roles within many different sociotechnical systems. The nature of the specific tasks that each user undertakes and the interdependencies within each work system mean that there is only a limited degree to which users can adapt their behavior to suit the assumptions of the technology. The technology has to suit what may be widely different user needs and, if it does not, it may not be used, may be used to a limited degree, or may force employees into elaborate workarounds. What then would it mean to be user-centered in this context? The overarching aim of the generic design team, it would seem, would be to take advantage of the later stages of design, which could shape the product in different ways to suit local organizational needs. The task of the generic product design team is then to produce a product with the flexibility to permit later design. The need to create opportunities for local work teams to develop systems to meet their own needs has led, in the sociotechnical systems theory tradition, to the formulation of the "principle of minimum critical specification" (Cherns 1976, 1987), which states that only those variables that have to be fixed for the achievement of overall objectives and design integrity should be fixed when a system is initially developed. All other issues should be left open so that at later stages of design they can be locally determined to meet the needs of local users. In practice, for user-centered design, this means that the design of generic products should match the generic properties of its user population and the invariant characteristics of the tasks that it is intended to support. It should not, for example, include detailed assumptions about how people will run meetings or offer discounts to their customers. The product should also be designed and sold in a way that supports the later stages of design, for example, it may offer users many options so that they can shape its use to their needs or it may resemble a toolbox, which later "designers" can customize in different ways to meet local needs.

But how are designers to establish what is generic and what is specific to a particular organizational setting? In my experience, design teams often have limited access to the business worlds of the users they aspire to create products for. They may gather data by examining similar work environments, e.g., the accounts department, in their own organization and they may use this as their model in the design process. Some design groups have access to customer organizations that have a special relation with the supplier that permits early testing, and user groups in these organizations may become their model. In either situation, there is a danger that the product is designed on the basis of the specific characteristics of these user settings. At the very least there is a need to examine several very disparate user settings and to pay as much attention to the variability between them as to the regularities. Another very important source of user-centered design information is created when the product is in the marketplace and is being adopted within organizations. It is at this stage that evidence becomes available of the working practices of the user population, what they use and do not use, what workarounds they find necessary, what new and exciting practices are being created on the basis of secondary features of the product and so on. Again, in my experience of working with many design teams, while they often get "bug" reports from customers, it is rare for them to have access to information about emergent patterns of user behavior. Since most products go through subsequent design phases and new versions of the product are released, this kind of user-centered information can be invaluable in deciding the scope and functionality of later releases. The use of ethnographic methods to make explicit the emergent patterns of behavior with products and systems would produce a rich source of information to guide future design.

15.4.2 IN-HOUSE DESIGN

The implementation of a new technical product in an organization is often treated as a purely technical matter: it needs installing, running to ensure it is working effectively, users need training, and so forth. However, as the evidence cited above shows, any technical product placed in a working sociotechnical system is going to have implications for the way users perform their tasks, for the relations they have with one another, and so on. What is needed at this stage is a local sociotechnical design process (Eason 2005) that explores the implications of the product for each of the work roles that will be impacted by it. There is a need to understand the potential benefits that the product brings to each work role, and therefore what functionality will be of most value and the form it needs to take to realize its benefit. It is also important to understand potential barriers to use; for example, the value, to the user, of continuing to use products that are already "ready to hand" or the possible dangers to their autonomy of using products that control what they do or make apparent their work behavior to others in the system. The product of an early evaluation of sociotechnical implications can be a design process that utilizes the flexibility in the product to customize it for the organization so that benefits can be achieved and unwanted consequences avoided. The process can also include a social system design process; creating new working practices and organizational arrangements to foster the effective utilization of the product and the sharing of knowledge.

When a product is to be closely coupled into the work process of the organization, it is often the case that there is an internal design process that customizes the product to meet the needs of the organization for an efficient and effective work process. This is often, however, cast as an engineering task in process improvement or as a business process re-engineering exercise (Hammer and Champy 1993) in which the agenda is to maximize the use of the product's potential in realizing efficiencies in the way the collective task of the organization is undertaken. As Mumford (1994) has pointed out, this approach to technical system implementation marginalizes the human contribution to the new system design; it may be reduced, at the end of design phase, to a recognition that there is a need to train the users in the correct operation of the technology to achieve the process improvements. It is these approaches that often lead to the workforce developing workarounds because they find that the assumptions about the tasks they undertake that are built into the way the technology has been implemented are at variance with the everyday experience they have of the tasks they confront. In order to avoid this kind of outcome, there is a need for any process improvement project to be treated as a user-centered, sociotechnical systems design project, i.e., to engage with the users who "man" the process in an exercise of looking for improvements made possible, perhaps, by new technology. At the heart of such a user-centered approach should be a regard for the reality of the task world so that users receive new technology in a form that does not require workarounds. An advantage that the in-house design team has over its generic design team counterpart is that it can engage with the users who will actually use the product (i.e., user-centered design is undertaken with the real users) and, provided the generic team have supplied the flexibility, it should have the tools to customize the product to meet local user needs.

15.4.3 LOCAL DESIGN

Once the users are making use of the product, the process of making it their own can begin. However sophisticated user-centered design has been up to this point, it is still likely that this process will have a local design element; employees will decide what is most useful, share experiences and learning and perhaps find new ways of exploiting the potential of the product that no one thought of during the earlier design stages. In a fast moving, changing world, it is also the case that, no matter how well the earlier stages addressed the issue of matching real task needs, the nature of the tasks will change and there will be an evolving need to re-match the technology to emerging forms of the task. All these factors point to the fact that there will be emergent forms of the sociotechnical system. Many of the earlier examples of emergent systems can be considered in some senses dysfunctional in that they entail elaborate ways of helping the work system to cope or risky ways of cutting corners, etc. An important ambition is to establish a continuing method of recognizing the need for system adjustments and then using flexibility in the product and the learning capability in the social system to bring the more structural features of the system into line with the emergent need. One of the advantages of employees as consumers is that, unlike their public counterparts, they are already in communities of practice, and, as the banking example cited earlier demonstrated, they can share knowledge and develop new forms of practice. If some part of the technical design resource can be maintained in the organization and can continue to evaluate the effectiveness of the product with the user community, this new learning could then influence further customization of the product. In this way, in an evolving world, we recognize that design never ends and that in small incremental changes, a product can continue to evolve in the way it presents itself to its community of users.

15.5 CONCLUSIONS

Treating employees as consumers of technological products is a useful way of focusing on the active ways in which they use and do not use the resources of these products. It avoids the perhaps more passive connotations of labeling them "users" or "operators" and recognizes that in many respects they re-create the product in local work practices that enable them to undertake tasks in their sociotechnical system. This chapter has reviewed many examples of the different patterns of use that result; from straightforward non-use, to the use of only part of the functionality of the product, to the development of complex workarounds when the product does not do what is needed, to a very positive outcome when the product makes possible some new and valuable task behavior that had not been predicted. At the root of the more negative examples, where the users struggle with the product provided, are assumptions about the user and the task world that design teams have built into the product. The organizational world is too varied and the world is changing too fast for detailed assumptions about how tasks need to be undertaken to be fixed in products. In this chapter, we have outlined a user-centered approach to product and system development that recognizes a number of stages of design that can be used to ensure that the product in the hands of employees is matched (and continues to be matched) to their evolving task needs. It is a design process that highlights the need for generic products to be flexible and for local design processes to be capable of using this flexibility to match local needs as they emerge.

REFERENCES

Abdelnour Nocera, J.L. 2007. *The Social Construction of Usefulness: An Intercultural Study of Producers and Users of a Business Information System*. Saarbruecken: VDM Verlag.

Broadhurst, K., Wastell, D., White, S., Hall, C., Peckover, S., Thompson, K., Pithouse, A., and Davey, D. 2009. Performing 'initial assessment': Identifying the latent conditions for error at the front-door of local authority children's services. *British Journal of Social Work* Jan: 1–19.

Brown J.S., and Duguid, P. 2002. *The Social Life of Information*. Harvard: Harvard University Press.

Cherns, A.B. 1976. The principles of socio-technical design. *Human Relations* 29:783–92.

———. 1987. Principles of socio-technical design re-visited. *Human Relations* 40 (3): 153–62.

Collins, T. 2007. NHS security dilemma as smartcards shared. *Computer Weekly* 30 Jan: 1.

Eason, K.D. 1984. Towards the experimental study of usability. *Behaviour and Information Technology* 3 (2): 133–43.

———. 2005. Exploiting the potential of the NPfIT: A local design approach. *British Journal of Healthcare Computing & Information Management* 22 (7): 14–16.

———. 2009. The national programme for information technology: A socio-technical perspective. In *Integrating Healthcare with Information and Communications Technology*, eds. W. Currie, and D. Finnegan, 183–204. Oxford: Radcliffe.

Eason, K.D., and Damodaran, L. 1986. Usable customer interfaces. In *Local Telecommunications II*, ed. J. Griffiths, 178–89. London: Peter Peregrinus.

Eason, K.D., Harker, S.D.P., Raven, R.F., Brailsford, J.R., and Cross, A.D. 1995. Expert or assistant: Supporting power engineers in the management of electricity distribution. *AI Society* 9:91–104.

Eason, K.D., MacIntryre, R., and Apps, A. 2006. A 'joined up' electronic journal service; user attitudes and behaviour. In *Libraries without Walls 6: Evaluating the Distributed Delivery of Library Services*, eds. P. Brophy, J. Craven, and M. Markland, 63–70. London: Facet.

Hammer, M., and Champy, J. 1993. *Re-engineering the Corporation; A Manifesto for Business Revolution*. London: Nicholas Brealey.

Heidegger, M. 1962. *Being and Time*. Translated by J. Macquarrie and E. Robinson. New York: Harper and Row.

Helander, M.G., Landauer, T.K., and Prabhu, P.V., (eds.) 1997. *Handbook of Human-Computer Interaction.* Amsterdam: Elsevier.

Mumford, E. 1994. New treatments or old remedies: Is business process re-engineering really socio-technical design? *The Journal of Strategic Information Systems* 3 (4): 313–27.

Orlikowski, W. 2000. Using technology and constituting structures: A practice lens for studying technology in organisations. *Organisation Science* 11 (4): 404–28.

Page, D., Williams, P., and Boyd, D. 1993. Report of the Inquiry into the London Ambulance Service. Report commissioned by the South Thames Regional Health Authority, London.

Pinch, T., and Bijker, W. 1987. The social construction of facts and artefacts. In *The Social Construction of Technological Systems*, eds. W. Bijker, T. Hughes, and T. Pinch, 17–50. Cambridge, MA: MIT Press.

Preece, J., Rogers, Y., and Sharp, H., eds. 2002. *Interaction Design: Beyond Human-Computer Interaction.* New York: Wiley.

Rice, A.K. 1958. *Productivity and Social Organization; The Amadabad Experiment.* London: Tavistock.

Standish Group. 2007. The CHAOS Reports. http://www.standishgroup.com/chaosresources/chronicles.php (accessed January 8, 2010).

Trist, E.L., Higgin, G.W., Murray, H., and Pollock, A.B. 1963. *Organizational Choice.* London: Tavistock.

Vicente, K.J. 2004. *The Human Factor; Revolutionizing the Way People Live with Technology.* New York: Routledge.

Section III

Case Studies

16 Constraint-Based Approach to Design

Daniel P. Jenkins, Neville A. Stanton, Paul M. Salmon, and Guy H. Walker

CONTENTS

16.1 INTRODUCTION

The way a product or service is used and integrated into a system is often dictated by the constraints imposed in the design process. These constraints may be a result of technical limitations, cost, legal concerns, or imposed by the manufacturers to maximize future profits. This chapter uses an ecological design technique based on cognitive work analysis (CWA; Rasmussen, Pejtersen, and Goodstein 1994; Vicente 1999; Jenkins et al. 2009) to capture the constraints of the iPod and the domain in which it interacts. This constraint-based approach allows us to consider some of the constraints imposed on the product and the impact of removing them.

16.1.1 iPod

The iPod is a portable digital music player (see Figure 16.1) that allows the user to store a large music collection in a single location using a hard drive to store up to 20,000 songs (depending on model; figures based on the 80Gb version).

The product, measuring $103 \times 61 \times 14$ mm (see Figure 16.2), is not much larger than an average mobile phone, allowing it to fit comfortably into most pockets. Stored music can be played back via the supplied headphones or via a computer, Hi-Fi, or car stereo (with appropriate leads or third party gadgets, see Figure 16.3).

FIGURE 16.1   Fourth generation iPod. (Photographed by the author.)

The first iPods were initially developed solely as digital music players with video added as an additional feature on later models. This chapter documents the analysis of a fourth generation iPod (see Figure 16.1). This version of the iPod was the last to remain solely as a music player before the introduction of the color screen and video playback capabilities.

The success of the iPod is irrefutable; since its introduction in October 2001, the iPod has continued to grow in popularity and is without doubt the world's best-selling digital music player, claiming a market share in excess of 62% (Jobs 2007a; based on data for November 2006). In April 2007, Apple announced the sale of their 100 millionth iPod (Apple 2007a). This success is likely to have been influenced by a number of factors, including Apple's strong brand, the product's ease of use, the product's strong design style, and a string of successful marketing campaigns. Apple's design ethos appears to strike a good balance between a focus on the product's technical capabilities and what Redström (2006) terms "the user experience" delivered through the software and the product's interface design.

From a consumer perspective, the iPod is clearly a great product; not only does it dominate the sales market, the iPod unsurprisingly also receives consistently high reviews in independent opinion polls and in magazines' product comparison articles (T3 2006; The Gadget Show

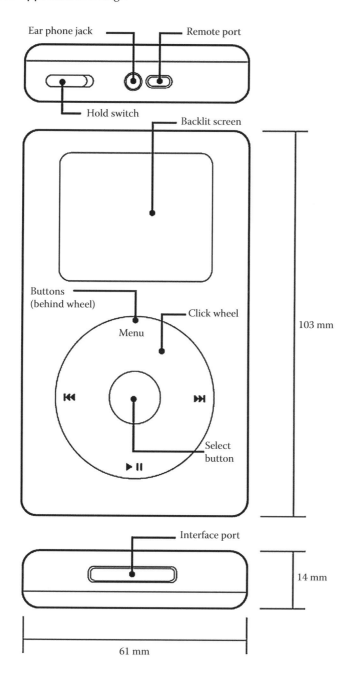

FIGURE 16.2   External attributes of an iPod.

2007). It is also assumed that the Apple Organization has also done exceptionally well from the product. Even without considering the direct revenue resulting from sales of over 100 million iPods, Apple have massively raised their profile and brand awareness through the sale of the product. It is postulated that Apple's decision to make iPods solely compatible with iTunes has also had significant implications for the success of the organization. A huge number of PCs now have Apple software sitting on them for the first time. The free to download software is responsible for over two billion music track downloads (as of January 2007), equating to five million songs sold per day. When considering CD sales alongside downloads, Apple are

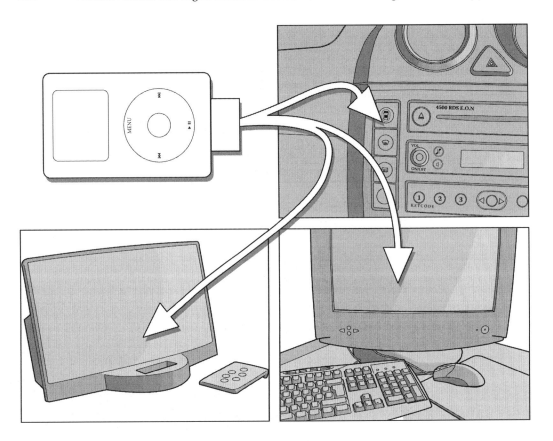

FIGURE 16.3   Common products the iPod is interfaced with.

the fourth largest music reseller in the United States (Jobs 2007a). Initial speculation would suggest that the Apple Organization has done an outstanding job of striking a balance between creating a product that is a great design for the user, as well as a great design for the growth and profitability of the organization.

16.2   THE ANALYSIS

The analysis used to investigate the iPod and the domain that it inhabits was based on a human factors framework called CWA (Vicente 1999). This constraint-based approach was selected due to its formative approach. A formative approach aims to discover, and document, how the system performs when the constraints imposed on it are considered and manipulated, leading to an event and time-independent description of the system. As stated in the introduction, this chapter aims to capture the constraints that control the use of the product. Arguably, other normative or descriptive methods that focus on how the product currently performs are less suitable for this analysis as they may miss some of the constraints or functions that have been deliberately imposed on the system.

Originally developed by Rasmussen, Pejtersen, and Goodstein (1994), CWA was born out of an ecologically based analysis used to consider the suitability of a nuclear power program for the Risø National Laboratory in Denmark. The framework has subsequently been developed and applied for a number of purposes including: system modeling (e.g., Hajdukiewicz 1998); system design (e.g., Bisantz et al. 2003); training needs analysis (e.g., Naikar and Sanderson 1999), training program evaluation and design (e.g., Naikar and Sanderson 1999); interface design and evaluation (Vicente 1999); information requirements specification (e.g., Ahlstrom 2005); tender evaluation (Naikar and Sanderson 2001); team design (Naikar and Saunders 2003); and error management strategy design

(Naikar and Saunders 2003). These applications have previously taken place in a variety of complex safety critical domains including: aviation (e.g., Naikar and Sanderson 2001); process control (e.g., Vicente 1999); nuclear power (e.g., Olsson and Lee 1994); naval (e.g., Bisantz et al. 2003); military command and control (e.g., Jenkins et al. 2009; Salmon et al. 2004); petrochemical (e.g., Jamieson and Vicente 2001); road transport (e.g., Salmon et al. 2007); health care (e.g., Burns, Momtaham, and Enomoto 2006); air traffic control (e.g., Ahlstrom 2005); and manufacturing (e.g., Higgins 1998).

The CWA process starts by assigning a boundary around the system in order to establish the constraints within it. Based on an understanding of the domain constraints, the analyst is then prompted to answer the question of *what* activities are conducted within the domain as well as *how* this activity could be achieved and *who* could perform it. According to Vicente (1999), CWA can be broken down into five defined phases, as shown in Table 16.1. This chapter will focus the analysis on the initial phase of CWA, work domain analysis (WDA); the subsequent two phases, control task analysis (ConTA) and strategies analysis, will also be used. In this particular case, the benefits of extending the analysis into the last two phases, worker competencies analysis and social organization and cooperation analysis, were not perceived to be great enough to warrant the analysis.

16.2.1 Work Domain Analysis

The first phase of CWA, WDA, is used to describe the system domain independently of specific actors or situations. The domain is modeled by considering the reason the system exists, the analysis also captures the physical objects and attributes of the system, modeling what these objects or attributes can afford. The way that the analysis is conducted is dependent on the system. In this case, we have a good understanding of the physical components and we can quite easily extract the functions that these components can afford; it is also possible to assume the overall purpose of the system. The information in this phase is normally displayed in a graphical representation called the abstraction hierarchy. An abstraction hierarchy (see Figure 16.4) can be used to model these assumptions and to create a link between what the system is capable of and what the system's purpose is. At the top of the hierarchy the overall purpose of the domain is recorded, at the base of the hierarchy the physical objects are recorded, between these levels the system is described at a number of levels

TABLE 16.1
Phases of CWA

Phase	Acquisition	Tool
Work domain analysis (WDA)	Document analysis Review by SME	Abstraction decomposition space
Control task analysis (ConTA)	Cognitive walk through Study of work practices	Decision ladder
Strategies analysis	Critical decision method Interaction analysisVerbal protocol analysis	Information flow map
Social organization and cooperation analysis (SOCA)	Communication analysis Interaction analysis	All of the above
Worker competencies analysis (WCA)	Reparatory grid analysis Review of decision ladder	Skills rules knowledge

Source: Vicente, K.J., *Cognitive Work Analysis: Toward Safe, Productive, and Healthy Computer-based Work*, Lawrence Erlbaum Associates, Mahwah, NJ, 1999; Acquisition methods have been added from Lintern et al. Asymmetric adversary analysis for intelligent preparation of the battlespace (A3-IPB). United States Air Force Research Department Report 2004.

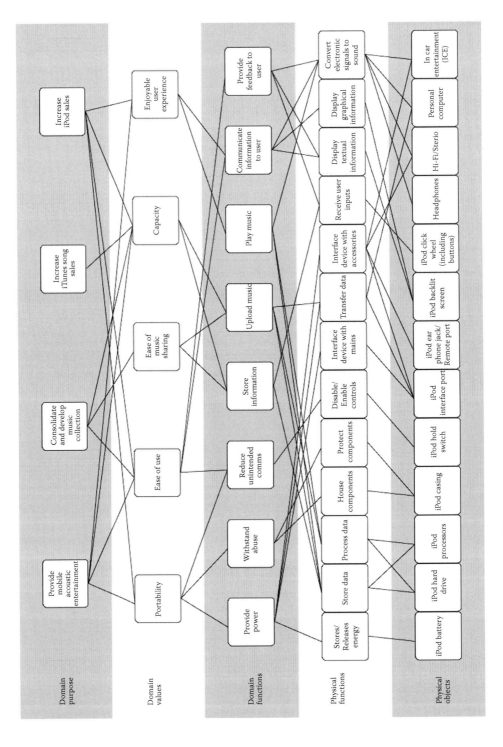

FIGURE 16.4 The abstraction hierarchy for the iPod system.

of abstraction, these levels are linked using structural means-ends links. The number of levels of abstraction is not fixed, however most analyses tend to use five levels.

16.2.1.1 Physical Objects

The physical objects level is found at the base of the hierarchy. At this level, the components that make up the iPod both internal and external are recorded.

- Backlit screen
- Casing
- Click wheel
- Buttons (behind click wheel)
- Select button
- Hold switch
- Remote port
- Interface port
- Ear phone jack
- Headphones
- Hard drive
- Printed circuit board (PCB)
- Battery

These components can be seen graphically in Figures 16.2 and 16.5.

It is also worth considering the other objects that exist in the iPod's domain that make up significant parts of the system. In this case, a personal computer is an integral part of the system; without a

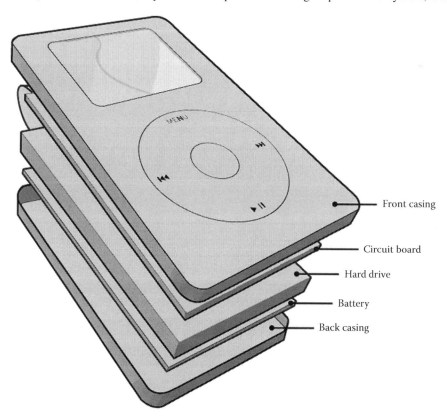

FIGURE 16.5 Exploded view of an iPod showing internal components.

computer, it is impossible to upload (add) music on to the iPod. Also, many users would also consider objects such as car stereos and Hi-Fi systems as integral parts of the system.

16.2.1.2 Physical Functions

The physical functions level resides above the physical objects level. The physical functions can be described as the "things" that each of the components can do. These functions should be independent of the purpose of the system and recorded in very generic terms. An example of a function for the screen would be "to display textual and graphical information" *not* "to show track number." This generic function description is important as it allows the screen to be considered for purposes other than the display of track information.

By adding links between the objects and the functions they can afford, it is apparent that some components can afford many things. It can also be observed that some functions require more than one physical object to perform, or that some functions can be conducted by more than one object, indicating some level of redundancy. As an example, the interface port can be seen to allow a number of interface-related functions to be carried out. Also, the function of converting electrical signals to sound can be conducted by a number of physical objects, including headphones, Hi-Fi, and car stereo.

16.2.1.3 Domain Purpose

The domain purpose level is the highest level within the abstraction hierarchy. The task of determining the domain purpose(s) of the iPod is a fundamental consideration as this will influence the rest of the hierarchy. The purpose set should be independent of any particular activity and it should capture the reason for the design or procurement of the system. By examining the functionality of the first generation iPod, it can be assumed that the product was developed for the sole purpose of consolidating a music collection into a portable device. This purpose focuses quite heavily on a physical user-centered definition of the iPod. There is no consideration of any of the purposes placed by the Apple Organization relating to brand image, revenue, or market share, which are important to almost all business organizations. Without modelling these constraints this analysis would be incomplete. There appear to be constraints placed on the iPod that cannot be explained purely by a user-centered definition of the product. From a technical perspective, there seems to be no real reason why the iPod should work solely with Apple's own software, iTunes. Many of the iPod's competitors do not have this constraint and allow their users to interact their products with a host of software packages. Another key constraint on the iPod is the ability to copy and transfer music from one computer to another. From a technical perspective this should be possible as it is in many other digital storage devices, however, within the iPod this is a very difficult thing to do. According to Jobs (2007b), much of the complexity of this issue lies within the contracts between Apple and the "big four" music companies: Universal, Sony BMG, Warner, and EMI. Since Apple does not own or control any music itself, it must license the rights to distribute music from others. As part of their agreement to allow music to be distributed electronically, these music companies enforce a requirement that the downloaded music be protected, limiting its distribution, in Apple's case to five computers. The agreements in the contracts between Apple and the music companies explains the constraint that music downloaded through iTunes can only be played on an iPod and the constraint that legally protected downloads from other services are not compatible with the iPod; however, it does not fully explain Apple's decision to make the iPod solely compatible with iTunes, as Jobs (2007b) points out an estimated 97% of the music on the iPod is unprotected.

It seems apparent that there are additional factors influencing the decision to add in these constraints to the iPod system, other than a user-centered design. From a user perspective, the purpose of the iPod could be captured as "to provide mobile acoustic entertainment" as well as "to consolidate and develop music collection." From an Apple Organization perspective, the overall purpose could be assumed to be "to increase growth and profitability for the organization," this could be broken down to "increase iPod sales" and "increase iTunes song sales."

16.2.1.4 Domain Values

The domain values level sits directly below the overall domain purpose level. Here, information is captured that can indicate how well the system is achieving its purposes. The criteria here should be defined in terms that can be measured, for example, the capacity in Gigabytes or subjective and performance ratings for "ease of use." The values are linked to their relating domain purposes using means-ends links. Portability is an important function in providing mobile acoustic entertainment; this in turn increases iPod sales. The ease of use of the product influences all the functional purposes with the exception of increasing iTunes song sales. The ease of music sharing is important in the development of music collections. Capacity has an influence on all the domain purposes because a suitable capacity is required for acoustic entertainment, for developing and consolidating a music collection, increasing iTunes sales, and increasing iPod sales. An insufficient capacity would have a detrimental effect on each of these purposes. Finally, the user experience effects the acoustic entertainment as well as iPod sales.

16.2.1.5 Domain Functions

The domain functions level sits in the middle of the hierarchy; this level links the purpose independent, physical functions at the base of the hierarchy to the more abstract ideals of the system at the top of the hierarchy. Using means-ends links, the Abstraction Hierarchy (AH) can be connected up; any cell can be selected as *what* is trying to be achieved, the connected cells above should answer the question of *why* this should be achieved, the connected cells below should answer the question *how* it could be achieved.

Inspection of the hierarchy reveals a partial conflict between the user's domain purpose of developing their music collection and the Apple Organization's perceived requirement to increase music sales through iTunes downloads. Technologically, there are a number of additional ways in which a user could easily increase their music collection. This could be done by making it easier to copy music across a number of computers; another method would include some level of synchronization between iPods either by cable or wireless technology. It is postulated that the reasons for Apple not including this level of functionality are also due to music piracy laws; however, other data storage devices and digital music players do not have these restrictions in place. It is also hypothesized that the organization's profit as a result of iTunes downloads was a significant factor in this decision.

16.2.2 Control Task Analysis

We have seen the benefits of viewing the domain independently of activity in the WDA captured in the abstraction hierarchy (see Figure 16.4). In order to understand the domain further, it is advantageous to look at the known recurring activities that occur within this domain. The second phase of the analysis, ConTA, models these known recurring tasks, focusing on what has to be achieved independent of how the task is conducted or who undertakes it.

Naikar, Hopcroft, and Moylan (2005) describe the contextual activity template for use in this phase of the CWA (see Figure 16.6). This template is one way of representing activity in work systems that are characterized by both work situations and work functions. According to Naikar, Hopcroft, and Moylan (2005), the work situations (situations decomposed by schedules or location) are shown along the horizontal axis and the work functions (activities characterized by its content independent of its temporal or spatial characteristics [Rasmussen Pejtersen, and Goodstein 1994]) are shown along the vertical axis of the contextual activity template. The dashed boxes indicate which situations the work functions *can* occur in, whereas the circles and whiskers indicate where the work functions *typically* occur.

In this case, the situations have been defined by how the iPod system is configured at any one time. The particular situations will have an effect on the functions that can be conducted. The situations under consideration are: the iPod connected to the mains on AC charge; the iPod using headphones while on the move; the iPod synchronized with the computer running iTunes; the iPod synchronized with a Hi-Fi system; and the iPod synchronized with some form of in-car entertainment (ICE). Taking

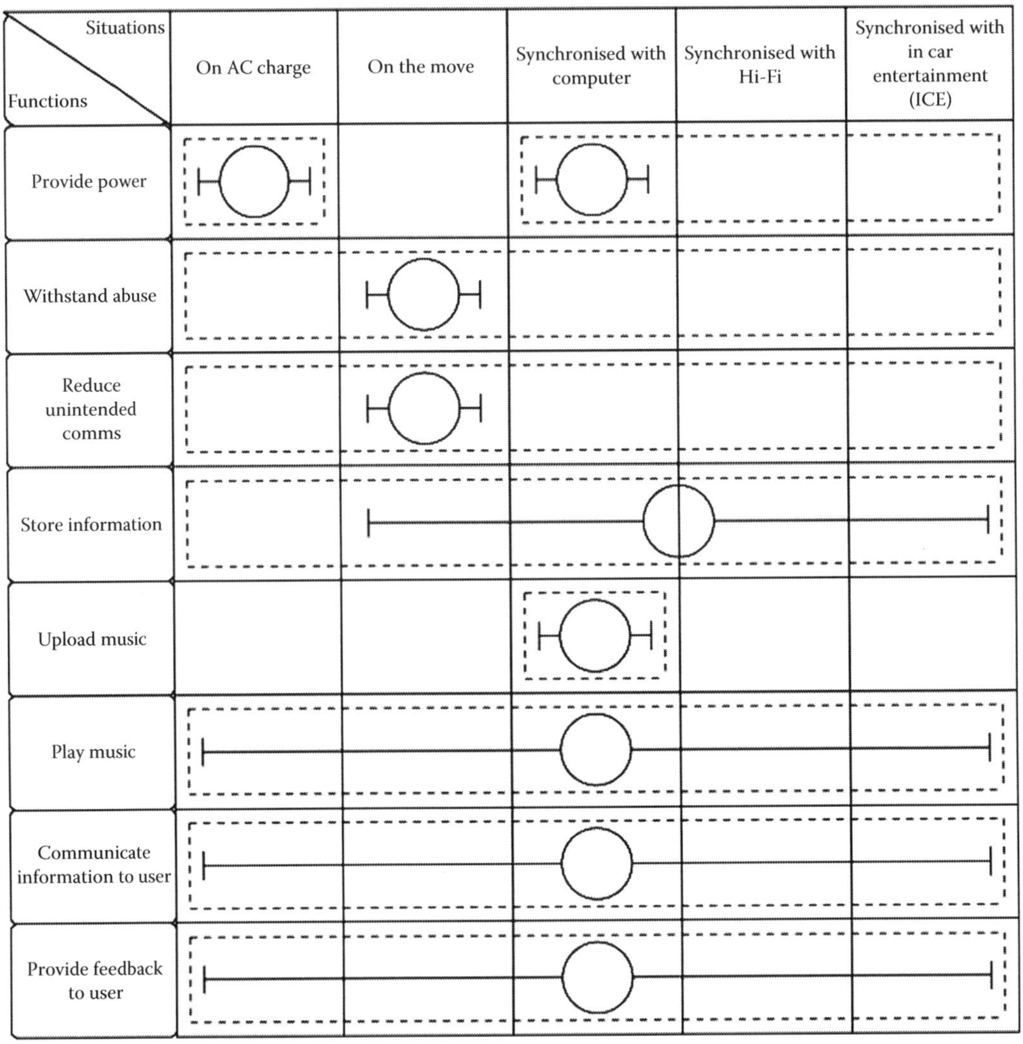

Functions / Situations	On AC charge	On the move	Synchronised with computer	Synchronised with Hi-Fi	Synchronised with in car entertainment (ICE)
Provide power					
Withstand abuse					
Reduce unintended comms					
Store information					
Upload music					
Play music					
Communicate information to user					
Provide feedback to user					

FIGURE 16.6   Contextual activity template for the iPod system.

the top function "provide power" as an example, it can be seen that the iPod *can* receive power to charge the battery in all situations except on the move. The situations "synchronized with Hi-Fi" and "synchronized with ICE" are shown with only a dotted line to indicate that they *can* receive power in these situations; however, unlike while connected to a computer or connected to the mains, they may not *typically* receive power, depending on how the system is configured. Further development of the system into how the battery could be charged while on the move may be of significant benefit to many users of this product who have limited access to the situations where the iPod battery can be charged (i.e., mains, computer, Hi-Fi, or ICE).

From Figure 16.6 it is also apparent that the only situation where music can be uploaded to the iPod is while connected to a computer. We can gain further information about why this is by revisiting the constraints captured in the abstraction hierarchy (Figure 16.4). By examining the links down from the node in the center of the abstraction hierarchy "upload music," it can be observed that the only way of achieving this is to use a combination of a personal computer and an interface port. This current setup limits the iPod's suitability to people who own or have regular access to a PC that they can load music onto. If the iPod was to contain functionality to record directly from a Hi-Fi or a car stereo, then the product suitability could be extended.

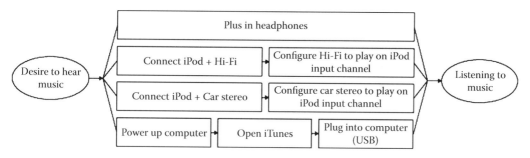

FIGURE 16.7 Strategies analysis for listening to music.

The remaining functions show a high level of flexibility within the system across the defined situations. Many of the functions (play music, communicate information to user, and provide feedback to user) *can* and *typically* occur in all situation. The iPod *can* "withstand abuse" and "reduce unintended comms" at all times, although *typically* this is only required when the user is "on the move."

16.2.3 STRATEGIES ANALYSIS

Strategies analysis is used to look in more detail at known recurring activities. This stage of the analysis considers the tasks analyzed in the contextual task analysis phase and considers the strategies that are likely to be used to complete them. The strategy adopted by an actor at a particular time may vary significantly. Different actors may perform tasks in different ways; also, the same actor may perform the same task in a variety of different ways.

Figure 16.7 shows a strategies analysis for listening to music. Here the tasks are described in more detail than the ConTA to show the steps required to get from the start state "desire to hear music" (shown on the left of the diagram) to the end state "listening to music" (shown on the right of the diagram). The analysis does not aim to prioritize or comment on the best strategy, it merely aims to capture the flexibility within the system.

In Figure 16.8, strategies are considered for the task of charging the iPod battery. Current strategies are shown at the top of the analysis (in white), the shaded boxes at the base show strategies for developing a method of charging the iPod batteries while on the move. These include third-party accessories that could include solar panels, or some kind of motion-to-electricity transformer possibly fitted to footwear.

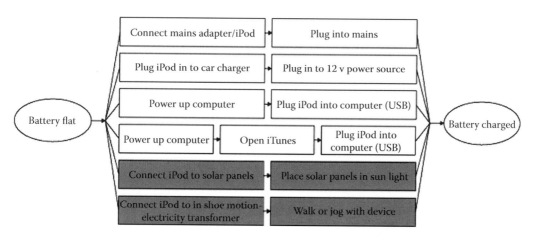

FIGURE 16.8 Strategies analysis for charging the battery.

Figure 16.9 shows the strategies analysis for uploading music. The white boxes show two of the current methods for uploading music, both methods require the user to connect the iPod to a computer with iTunes installed on it. The first strategy shows a system configured to automatically synchronize the iPod with the computer. The second strategy is for a manual update. Two additional strategies are shown as shaded boxes; these strategies were briefly described in the ConTA phase. The strategies cover recording audio files directly from a Hi-Fi or car stereo. This additional functionality would allow users to plug their iPods directly into a Hi-Fi or car stereo output and record tracks from other formats, such as vinyl records, directly onto the iPod. A further strategy is also included at the bottom of Figure 16.8, which allows two users to connect their iPods together. Once connected, the iPod could list non-duplicate songs, the user could then select from this list the songs that they wished to copy across. The connection could be made physically by a cable or wirelessly via Bluetooth or WLAN technology.

Freedom and flexibility within a system allows users to adapt and select a way of achieving an end state that is most appropriate to them in a given situation. A flexible system is also likely to support and promote formative thinking and emergent behavior. In some cases, a tightly constrained system with only one way of achieving an end state may be beneficial. The presence of a choice requires decision making, which can slow down the process of conducting the task. A well-designed interface that clearly presents the most common strategy but still allows expert users to adapt their behavior, is often the best way to tackle this conflict.

16.3 CONCLUSIONS

This chapter has introduced a constraint-based approach for considering the design and flexibility of a product or service. While this chapter has focused on the chosen example of the Apple iPod, it is contended that the same, or similar, approach can be applied to almost any product or service. The exact analysis undertaken will be informed by the characteristics of the domain in question. Larger, more social systems may benefit from the inclusion of the social organization and cooperation phase of the analysis.

This analysis has modeled the iPod system in terms of the constraints governing its use and operation. The analysis has described the relationship between what the physical objects in the system can perform, and what the system is required to perform. The analysis goes on to describe situations in which the iPod is constrained in its activity, such as charging the battery on the move and uploading music while away from a computer. In looking at these tasks in more detail, it has been possible to suggest potential strategies for resolving these.

From interpretation of the captured constraints, it has been possible to postulate the reasons behind their existence with specific reference to the constraints benefits to the user and their benefits to the Apple Organization. As stated in the introduction to this chapter, it is evident from the sales

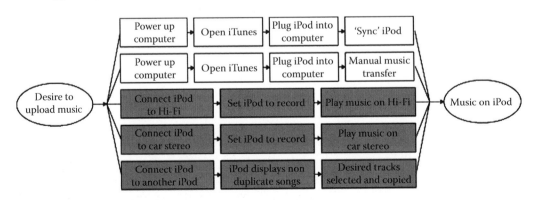

FIGURE 16.9 Strategies analysis for uploading music.

figures of iPods, independent product ratings, and the number of track downloads, that Apple seems to have struck a successful balance between the requirements of the user and the requirements of the organization.

Putting aside the requirements of the Apple Organization, the analysis makes it possible to suggest some changes to the design of the iPod system. These changes would remove some of the constraints placed on the operation of the iPod, providing users with additional flexibility and functionality. It is predicted that some of these changes would have an impact on the hypothesized requirement of the Apple Organization "to increase sales of iTunes downloads." The recommendations are detailed in the following sections.

16.3.1 iTunes Exclusivity

Increasing the support of music packages other than iTunes, allowing users who have a preference for another music software package to still use an iPod including the support of other music formats (the iPod does not support all music formats on the market, i.e., WMA, Ogg, SHN, and FLAC).

16.3.2 Music Sharing

It is also hypothesized that increased functionality to support the sharing of music would allow users to develop their music collections. Of particular benefit would be a system that could determine and display non-duplicate songs possibly through an iPod to iPod music transfer function, such as that built into the "Microsoft Zune" music player.

Landmark cases such as the agreement between Apple and EMI (Apple 2007b) to remove digital rights management (DRM; the protection encoding that stops users placing the music on more than five computers) indicate that the music industry is rethinking the strict restrictions it places on Apple and its competitors. Only time will tell if Apple chooses to open up the iPod to other music management software, exposing if the decision to limit the iPod to iTunes was solely due to copyright agreements with the music companies or an attempt to entice more customers into the iTunes music store.

ACKNOWLEDGMENTS

It should be noted that the analysis in this chapter is in no way supported or endorsed by the Apple Organization. Any design decisions discussed in this chapter have been hypothesized by analyzing the final product, press releases, and public statements made by the company. This analysis was completed using the Human Factors Integration Defence Technology Centre (HFI-DTC) CWA software tool.

REFERENCES

Ahlstrom, U. 2005. Work domain analysis for air traffic controller weather displays. *Journal of Safety Research* 36:159–69.

Apple. 2007a. 100 Million iPods sold. Press release. http://www.apple.com/pr/library/2007/04/09ipod.html (accessed December 12, 2007).

———. 2007b. Apple unveils higher quality DRM-free music on the iTunes Store. Press release. http://www.apple.com/pr/library/2007/04/02itunes.html (accessed December 12, 2007).

Bisantz, A.M., Roth, E., Brickman, B., Gosbee, L.L., Hettinger, L., and McKinney, J. 2003. Integrating cognitive analyses in a large-scale system design process. *International Journal of Human-Computer Studies* 58:177–206.

Burns, C.M., Momtaham, K., and Enomoto, Y. 2006. Supporting the strategies of cardiac nurse coordinators using cognitive work analysis. In *Proceedings of the Human Factors and Ergonomics Society 50th Annual Meeting*, 442–46.

Hajdukiewicz, J.R. 1998. Development of a structured approach for patient monitoring in the operating room. Masters Thesis, University of Toronto.

Higgins, P.G. 1998. Extending cognitive work analysis to manufacturing scheduling. In *Proceedings 1998 Australian Computer Human Interaction Conference*, eds. P. Calder and B. Thomas 236–43. OzCHI'98, November 30–December 4, Adelaide, IEEE.

Jamieson, G.A., and Vicente, K.J. 2001. Ecological interface design for petrochemical applications: Supporting operator adaptation, continuous learning, and distributed, collaborative work. *Computers and Chemical Engineering* 25:1055–74.

Jenkins, D.P., Stanton, N.A., Walker, G.H., and Salmon, P.M. 2009. *Cognitive Work Analysis: Coping with Complexity*. Aldershot, UK: Ashgate.

Jobs, S. 2007a. Macworld San Francisco 2007. Keynote address. http://www.apple.com/quicktime/qtv/mwsf07/ (accessed December 12, 2007).

———. 2007b. Thoughts on music. http://www.apple.com/hotnews/thoughtsonmusic/ (accessed December 12, 2007).

Lintern, G., Cone, S., Schenaker, M., Ehlert, J., and Hughes, T. 2004. Asymmetric adversary analysis for intelligent preparation of the battlespace (A3-IPB). United States Air Force Research Department Report.

Naikar, N., Hopcroft, R., and Moylan, A. 2005. Work domain analysis: Theoretical concepts and methodology. DSTO-TR-1665 System Sciences Laboratory: Edinburgh, Australia.

Naikar, N., Pearce, B., Drumm, D., and Sanderson, P.M. 2003. Technique for designing teams for first-of-a-kind complex systems with cognitive work analysis: Case study. *Human Factors* 45 (2): 202–17.

Naikar, N., and Sanderson, P.M. 1999. Work domain analysis for training-system definition. *International Journal of Aviation Psychology* 9:271–90.

———. 2001. Evaluating design proposals for complex systems with work domain analysis. *Human Factors* 43:529–42.

Naikar, N., and Saunders, A. 2003. Crossing the boundaries of safe operation: A technical training approach to error management. *Cognition Technology and Work* 5:171–80.

Olsson, G., and Lee, P.L. 1994. Effective interfaces for process operators. *The Journal of Process Control* 4:99–107.

Rasmussen, J., Pejtersen, A., and Goodstein, L. 1994. *Cognitive Systems Engineering*. New York: John Wiley.

Redström, J. 2006. Towards user design? On the shift from object to user as the subject of design. *Design Studies* 27 (2): 123–39.

Salmon, P.M., Stanton, N.A., Regan, M., Lenne, M., and Young, K. 2007. Work domain analysis and road transport: Implications for vehicle design. *International Journal of Vehicle Design* 45 (3): 426–48.

Salmon, P., Stanton, N., Walker, G., and Green, D. 2004. Future battlefield visualisation: Investigating data representation in a novel C4i system. In *Weapons, Webs and Warfighters* eds. V. Puri, D. Filippidis, P. Retter, and J. Kelly Proceedings of the Land Warfare Conference, DSTO, Melbourne, 2004.

T3. 2006. Review of Apple iPod 80GB. http://www.t3.co.uk/reviews/entertainment/mp3_player/apple_ipod_80gb (accessed December 12, 2007).

The Gadget Show. 2007. Best Buys.... MP3 players – 30/3/07. Online article. http://gadgetshow.five.tv/jsp/5gsmain.jsp?lnk=501&featureid=352&description=MP3%20players (accessed December 12, 2007).

Vicente, K.J. 1999. *Cognitive Work Analysis: Toward Safe, Productive, and Healthy Computer-based Work*. Mahwah, NJ: Lawrence Erlbaum Associates.

17 Development of a Media Player: Ambition and Constraints in the Consumer Electronics Industry

John Jansen and Bruce Thomas

CONTENTS

17.1   INTRODUCTION

At the Nielsen/Norman User Experience event in 2004, Donald Norman spoke about adding pleasure and fun to interface design. After his presentation, someone in the audience asked why nobody had managed to develop a digital audio player for a Microsoft platform to rival Apple's success with the iPod. In his reply, Norman indicated that, in his opinion, only a small number of companies were in a position to pull off such a feat; one of them was Philips. This chapter provides a glimpse of how Philips answered this challenge.

In the 10 years since Jordan, Thomas, and Taylor (1998) described human factors work in Philips, both the company and the human factors involvement have changed. The consumer electronics business of today is dominated by extremely short development cycles and outsourcing of much of the development work. Philips (2004) has stated that outsourcing is a key part of the company's supply management strategy. It is argued that outsourcing allows Philips to "add value through reduced operating and capital costs" and, at the same time, "it increases flexibility and decreases time-to-market and time-to-volume." By this means, Philips "can leverage its core competencies to best effect" and provide access to the technologies and skills of its suppliers. In such a development environment, it is difficult to provide extensive user input and multiple iterative cycles of design, and a thorough user-centered product development as reported, for example, by Thomas and van Leeuwen (1999), is no longer feasible.

Some of the main changes in human factors work at Philips since the Jordan, Thomas, and Taylor (1998) paper include:

- Organization changes. The part of Philips known as "Philips Corporate Design" in 1998 is now called "Philips Design." This reflects a change in the organizational role; Philips Design is now an independent business unit within Philips, responsible for its own budget and activities. Nevertheless, human factors remains an integral part of the design process, and this work continues to examine intangible aspects of use as well as direct usability. There is also a continuous drive to update methods to ensure user involvement in design (see e.g., Kersten, Thomas, and Docampo Rama 2007).
- Role of human factors. Because of the changes in the product creation process as described above, activities such as the analysis of user needs become much more of a background research activity, which is carried out so that there is a strong body of knowledge directly available as needed for individual product development projects (see e.g., Rameckers and Un 2005; Bueno and Rameckers 2007). Interaction design focuses on implementation using standard solutions intended to create a strong and consistent image for the company. These standard solutions are developed independently of individual product developments. Evaluation continues to be based on ISO 9241 (see e.g., Thomas and van de Ven, Chapter 9 this volume).
- Relationships. Despite the changes, the relationships between disciplines remain crucial to effective human factors involvement. The faster paced and more focused activities in current developments mean that such relationships are even more crucial to success.

A further fundamental change in recent years has been the drive throughout the company to fulfill the brand promise of "Sense and Simplicity." This places an additional emphasis on providing real and validated user benefits based on a process of insight generation and validation.

The case study described here illustrates a way of working in the consumer electronics industry. It highlights some of the problems user interface designers and human factors specialists currently face in large consumer electronics companies. Due to time constraints, the integration of effort was critical to the outcome, which has not changed since the development of mobile phones described by van Leeuwen and Thomas (1999). Thus, this chapter focuses on the overall user interface development with implicit (sometimes explicit) human factors involvement.

17.2   DESIGN PROCESS

17.2.1   AMBITION

The project described in this case study was set up to achieve some very ambitious goals in a very short time of six months.

The main focus areas for value proposition differentiation were

- "The pickup test"—to create a distinctive design and finishing to "wow" the user.
- A great user experience—to create a "best in class" product with a unique input device and an intuitive user interface.
- Service differentiation—to set up a service package to support the product, to be accessed through a network-based "media manager," which would assist and encourage the user to discover and share new music.

Further, for the Philips Personal Infotainment business this project represented a "make or break" situation. In order to maintain Philips presence as a key player in this particular market, it was essential to develop a product that could stand alongside some very successful competitor products, particularly the Apple iPod.

These ambitions meant that "the pressure was on!"

17.2.2   ANALYSIS

17.2.2.1   Understanding Heritage

In a large organization like Philips, a project of this nature does not start in a vacuum. There was a long history of the development of consumer entertainment products leading up to the current development. Philips' heritage already provided a rich source of internal knowledge as well as considerable assets in terms of intellectual property and brand identity.

At the time of development, Philips already had several digital audio players on the market. These products already had a common interaction design in the form of list-based hierarchical browsing, and featured the Philips-owned method of scanning long content lists, "SuperScroll™."

Finding a specific item in a long content list can be a particular problem for users of a digital audio player, which could contain several thousand songs. In order to assist the user to quickly get to a particular point in the list, the SuperScroll™ navigation enables the user to scroll up or down with increasing and decreasing speeds. Speed is controlled by the pressure applied to the scroll keys, so that pressing lightly enables the user to see items as they scroll by relatively slowly, while pressing more accelerates the scrolling rate. At the highest scrolling speed, the user is no longer able to read the individual entries in the list, so the current first letter is superimposed on the screen. Reducing the pressure again returns to the slower scrolling speed.

Navigation control enhancement such as SuperScroll™ is essential to provide user controlled effective highlight navigation in large, ordered data spaces. SuperScroll™ has the significant potential to be one of the recognizable signature elements shared among Philips user interfaces. Already it is consistently used in all Philips digital audio products and marketing communication:

> Double action search keys allow easy speed control when scrolling through large libraries. You can choose to move through your playlist one song at a time or just press and hold down a little harder to SuperScroll™ swiftly to your desired track.

The benefit of being able to navigate swiftly, precisely, and effortlessly through menus and content libraries is seen as a key user interface quality in operating devices that contain large amounts of content.

The new development had to take into account the existing navigation structure, but should include a new "inspirational" user interface direction. However, before the project started—besides some inspirational work—no new and proven user interface concepts had been developed that were considered suitable. Therefore, new concepts needed to be developed, taking into account the state-of-the-art technology available. An early direction adopted was to use light to follow finger movement. Further, it was decided to review and improve the SuperScroll™ mechanism.

17.2.2.2   Competitor Analysis

A competitor analysis was carried out in order to learn what other manufacturers were doing in this market. This provided the team with information regarding good common solutions, problems to avoid, and, most particularly, where it was necessary to differentiate the new product.

It became very clear from the competitor analysis that apart from Apple, none of the competitors offered a particularly good solution with respect to usability. Apple's success in designing an appealing user interface can be attributed to

- Focusing on the primary functionality. The user is not confronted with a large number of features, which means that the navigation structure can be kept very simple.
- The control wheel, which offers something more engaging than "point and click" for navigation and speed control. It provides a direct relationship between highlight location and finger displacement on the input device (wheel).

Identifying Apple as the key competitor enabled the team to focus on some key objectives, particularly to keep the user interface simple and to find a method for navigation and speed control that was different from, but at least as engaging as the control wheel.

17.2.2.3   Usability and Capability

There is a strong tendency in the consumer electronics industry to add ever more features to improve the competitive strength of a product, particularly for products that already have a well-established market. At the same time, it is well known that consumers complain about the complexity that often arises as a result of this "featuritis." It would seem, therefore, to be counter-productive to continue to add features that consumers apparently don't need and that add complexity to the product. However, it also seems that consumers, particularly if they are uncertain about precisely which product features they require, will often buy the product with the greatest number of features (see Rust, Thompson, and Hamilton 2006).

"Feature fatigue" occurs because both manufacturers and consumers believe that more features make a product more capable and, therefore, more valuable in use. However, most consumers don't use anywhere near the full functionality and then have a problem with reduced usability. This creates a business threat, since poor usability endangers business through loss of repeat sales and negative word-of-mouth propaganda.

Nevertheless, manufacturers continue to add features to consumer products for a number of reasons:

- Adding software features (seemingly) costs next to nothing
- Multi-functional products cater for a larger market audience
- Feature-rich products attract a wide market; feature-few products attract a narrow market
- Developers love power users
- Power users are seen as cool, the confused masses are seen as losers
- Marketing models predict that features steer product appeal
- Each positively valued attribute is assumed to increase the net utility to the consumer
- Each positively valued feature will increase a product's market share relative to products without that feature

This cycle continues because consumers knowingly fall for features appeal. They behave like children in a candy store—the belly ache comes after eating the sweets. Consumers realize that products with more features are harder to use, but before purchasing, they value capability more than usability. Only after purchase, once a product is used, usability suddenly matters very much.

Thus, the dilemma for the manufacturer lies in the danger of giving people what they want. The results of getting this wrong are

- Products being returned: many products are returned not because of faults, but because of show-stopping usability issues ("fault not found")
- Dissatisfied customers go elsewhere
- Disgruntled voices have global reach: in the internet era, a single customer's complaints can reach everyone, not just the shop assistant (consumers find each other in the "new market of conversations")

The problem of large numbers of features is particularly challenging in the mobile phone industry, where the device is small and the number of keys and screen space available to present the features to the user is extremely limited. Nokia have developed the "usability knee" (Lindholm, Keinonen, and Kiljander 2003), which shows that the perceived usability and required learning curve for a product are directly related to the number of functions and task complexity. The usability knee visualizes the trade-off between rich features and the usability of the device, the knee or breakpoint of a user interface (UI) paradigm ("ease-of-use") concept that will be reached at a level of functionality or function complexity, e.g., features that allow users to handle multiple phone calls, phonebook management, time management, rich media and messaging, and office applications.

A graph can be drawn of the number of features and the level of usability achieved for a particular user interface. The size of the screen and the number of buttons determine the number of features that can be supported in a usable manner. The usability knee is the point at which the functions provided place too much demand on the interface available.

Philips' products must perform in the competitive environment, so some features needed to be included. The challenge for the development team, therefore, was to make a judicious selection of features that would enhance the offering, but not compromise the overall simplicity of the navigation structure.

17.2.3 CONCEPT

17.2.3.1 Features and Functions

The challenge was to gain the highest possible perceived usability, i.e., it makes a "promise" to be simple to use (this promise is made through the affordances and marketing communication) while supporting a high quality of experience for the targeted functionality and function complexity (deliver on promise and exceed on expectations). To define the targeted functionality for the product proposition, now less technology driven, a process of user insight development and validation is used. The usability knee has been used as a tool to visualize and discuss with product management the trade-off of their request to make it "as simple as the iPod," while specifying a very different number of functions and level of functional complexity. The tool made it possible to emphasize the need for a reduction of functions and to sharpen the value proposition and positioning with respect to competitor products.

The basis of the increased function complexity, compared to the reference product, was the unique functionality in terms of services and contextual options intended to encourage the user to explore, and to discover and share music using advanced features and services.

The functions directly on the product should assist the user to find something appropriate to listen to, e.g., through association. One such feature was "Like Music," which enables the user to ask for more of the same kind of music currently playing. Another feature would be to browse colorful images and select music via visual cues such as album art.

In order to enable the user to discover new music, the product should facilitate recommendation and sharing. A music-sharing feature, available via an internet media manager, would enable users to send each other music recommendations. Users could also receive suggestions based on their listening behavior.

In addition, a number of contextual options would provide more possibilities to go further than the primary functionality. This would not interfere with the basic simplicity of operating the product to play music, but would open up new possibilities at a later time. Such options included rating the music so that particular songs are more or less likely to be played when in a random playback mode, creating and organizing playlists "on the go," flagging new songs for purchase, obtaining information on the track, album, or artist currently playing, etc.

17.2.3.2 Appropriate User Interface

Although the development team was striving to create an innovative user interface, it should nevertheless be intuitive. In order to achieve this, four principles were adopted:

- Familiarity. The existing Philips navigation structure should be retained so that the required learning curve would be minimal.
- Appropriate interface. This would require an understanding of the task complexity trade-off, as well as understanding the context. The chosen UI paradigm would need to simply support the core functionality, while the task complexity should be monitored and kept to a minimum.
- Personalization. Users should be able to express themselves through the use of the product.
- Rich experience. The product should fulfill and exceed expectations and enable "pleasure of use."

Above all, the product should score highly on perceived usability. Design articulation would therefore set the expectations. Actual "ease of use" would be achieved through defining an appropriate user interface paradigm. It was fundamental to the concept that initial experiences should be of successful operations with a minimal learning curve for basic usage.

Finally, the design should express the identity of a Philips product. It should have a Philips "signature," which would be achieved through the distinctive design articulation and Philips owned characteristic user interface principles.

17.2.3.3 Engagement

Gaver et al. (2005) argue that in non-work settings, rich, ambiguous, and open-ended design is more important than usefulness and usability per se, since they offer users the opportunity "to appropriate, explore and interpret." The user should engage with the product; that is, the product should stimulate the user's interest, so that a relationship is built up between the product and the user. To achieve this, it was felt that the functionality for the new product should be so hidden that users only become aware of it when it is relevant making them curious about the possibilities available.

Concepts were developed to engage the user. These included hiding the display. When switched off, the display does not show through the black surface of the product. When switched on, the background color of the display matches the color of the product, so that there is an invisible transition between the physical surface and the display. Another idea was to make the whole product front surface a touchable area that comes to life when the user touches it. The buttons are touchable areas rather than physical keys, and their positions are marked by LEDs that only light up when the product is in use.

A further idea was to stretch the touch area on the product surface beyond that of a simple touch pad through direct feedback on the touch area to the fingertip and a free gesture controller, with which the hand can freely interact. This would provide a more expressive user interface than simple button pushing.

In order to add further emotional qualities, the product should express itself through sound, light, and color, making use of real-world physics behavior designed to enhance the pleasure of the experience, thereby creating a magical quality to the interaction.

Ultimately, the user interface should flirt with the user, only showing what might be possible when it is possible. It should tease the user, inviting the user to touch and explore, only revealing its capabilities when approached a piece at a time rather than everything at once.

17.2.3.4 Gesture

An important element in achieving a "great user experience" was the idea of touch and gesture as the input principle. For this to be effective there should be no compromise in interpreting touch gestures, the input device should be able to interpret the "intentions" of the user by distinguishing the different energies of the user's strokes and taps. Further, there should be immediate high-quality multi-modal feedback so that the user can recognize its behavior and learn intuitively the control language.

How far the use of gesture might go was shown in the film "Minority Report." While it was too late to capitalize on the hype that the film generated, it showed a direction that was clearly different and more fun than just "point and click." Gesture makes use of the expressivity of motion, enabling the user to interpret action, providing a new dimension in control.

Using touch and gestures as the input principle has a number of advantages. They are new and so provide an interesting way to interact with the product. They can also be powerful, especially for character input, continuous movements, and absolute positioning.

However, gestures might require the user to invest time in learning how to use them. There might be limited feedback, no visual cues before the user starts to make the gesture, complete freedom of movement would be difficult to explain, and there was little experience with the use of gesture as an input mechanism. Further, the technology was not mature, and since this is critical to the use of gestures, there would be some risk of creating a poor impression when it did not work reliably, so not delivering on the promise and not exceeding expectations.

It was concluded that although gesture would be desirable, it was too soon to develop a product with fully free gesture control. Instead, the new product would be a conscious interim step between conventional input through button pressing and full gesture control, with the intention to plan further enhanced propositions for future generations. Gesture control would be used primarily for vertical scroll behavior, where conventional solutions are either indirect or require repetitive control actions.

17.2.3.5 Flywheel Concept

The gesture control designed was based on the concept of a flywheel. The idea was that navigating through a long list would be something like spinning a wheel. The user would provide impetus to the "wheel" by stroking the front of the product. The wheel would then spin according to the amount of force applied (the intention of the user) on the stroke (the energy given to the wheel).

The flywheel is a clear step toward gesture-based interfaces. The Apple wheel and Philips' own SuperScroll™ do offer more than point and click for speed control, but still require the user to maintain contact with the control device. With the Apple wheel the user has to keep rotating. With SuperScroll™ the user has to hold down a button. With the flywheel, the user makes a gesture and releases the control. Even though the gesture sometimes needs to be repeated, it does not require constant contact to work.

The flywheel control brings the control much closer to "real-world" control. While the metaphor we have used is based on rotating a wheel, it can also be compared to thumbing through the pages in a book (an application in the real world that is very similar to looking through an "album" of pictures or songs). Rapid movements bring you close to where you want to go, followed by slower more deliberate movements to zero in on what you are looking for.

The flywheel concept is different, therefore attention getting. This can be a show-off factor, "Look what this can do." This is strongly reinforced with the behavior of the LEDs underneath the touch input device reacting on the input of the user and giving emphasized feedback on the flywheel behavior.

The flywheel is not a barrier to traditional, familiar interaction. Familiar control actions (e.g., simple press to move highlight down one action or long press to scroll highlight down) can be used without additional user actions or new learning. Therefore, the addition of a flywheel control really is a new dimension.

17.2.4 Usability Testing

17.2.4.1 Set Up

The short timescale and very rapid translation of design into implementation meant that there was little opportunity to test the overall user interface. However, because this project introduced some fundamentally new methods of interaction, it was necessary to obtain input from potential users regarding their usability and acceptability. In order to obtain the greatest benefit from user involvement, it was important to identify the key questions early in the development process and to get a very quick answer.

The project constraints did not allow for extensive set up or formal recruiting of representative users. The methodology used was therefore somewhat quick and dirty. Nevertheless, quick and dirty methods provide a means to try out new ideas economically, and they make it possible to obtain some direct input from users rather than relying on the opinion of a single expert (see Thomas 1996).

The test was carried out in Philips Design's usability laboratory in Eindhoven, the Netherlands. Eighteen owners and users of digital audio players (10 female, 8 male) took part in the test. They were recruited from the families and friends of Philips Design's internal staff. The age range was from 15 to 30 years. Despite the informality of the recruiting process, a reasonable match to the target population was achieved.

The test procedure was composed of five parts:

- The participants were first shown three variations of the flywheel input mechanism and asked to perform a number of tasks navigating a list of 3000 song titles shown on a computer screen.
- The participants ranked the operability of each flywheel implementation and gave feedback on their usability.
- The participants were presented with a screen-based simulation of the user interface and asked to perform a series of 22 tasks.
- The participants filled in a questionnaire concerning the ease and pleasure of use.
- A closing interview was held with each participant to explore their experiences during the test.

The flywheel software algorithms were continuously updated during the course of the test as the strengths and weaknesses of the implementation were identified. By the end of the test, it was felt that the responsiveness of the interface had been greatly improved.

17.2.4.2 Results

The test results showed that the concept did indeed achieve a perception of simplicity. The features provided were considered to be necessary and not excessive, so that they did not get in the way of primary functions while still giving "feature freaks" what they wanted.

A particularly interesting finding was that participants in the test considered the up-down movement for scrolling to be more intuitive than rotating a wheel (referring to the Apple wheel) for navigation in a vertically oriented list.

The behavior of the flywheel was generally well received, although it was noted that some kind of instruction would be needed to explain how to optimally make the variety of swipes and taps necessary to operate it. The testing also showed that tactile feedback on touch gestures was essential, and it was particularly necessary to feel and see the ends of the vertical touch-strip. The reason was that the tap and hold functions had to be applied right at the ends of the touch-strip. The test was conducted with a completely flat top surface and it was found that users had difficulty in positioning their fingers on the end of the strip.

17.2.4.3 Redesign

Following the test, a number of modifications were made to improve the details of the navigation structure. These were comparatively minor, since no fundamental problems with navigation had been detected.

Design modifications were made to the front panel to simplify the handling of the flywheel, especially to improve the feel at the ends of the touch-strip. These included dimples to identify the end points where the user could tap for particular functions. Nevertheless, the fundamental problem of needing a physical stop at the end of the stroke remained, since the chosen design direction still called for a completely flat surface.

The algorithm specification for the swipe and hold were adjusted, both during and after the test, which eliminated some awkward and undesirable characteristics of the flywheel behavior.

17.2.5 REALIZATION

17.2.5.1 Final Product

The final product is illustrated in Figure 17.1. Despite having a strong product concept, the final product did not fully live up to the ambitions of the project team. The ambition of service differentiation through a "Music Messenger" to discover and share new music was not implemented and a

FIGURE 17.1   The Philips GoGear HDD 6330.

number of essential simplicity targets were not solved. The ambition of providing a great user experience with a unique input device was not fully realized; despite the fact that an intuitive (familiar, pleasurable) user interface had been tested and described in detail, the final implementation could not match the specification.

However, the product did meet the ambition of the "pickup test." The design and finish were distinctive and consumer tests showed that users did pick this product among several competitors lying on a table. The product design and the user interface worked harmoniously in identity and control, but tactile and behavioral details to enhance the use of the flywheel experience were not satisfactorily solved.

There were two main reasons for the somewhat disappointing end result:

- Limitations of the technical product platform and missing ecosystem synergy (i.e., PC manager and connected services) meant that some key elements of the concept were missing, such as waking up the device when the user's finger hovered over the control surface without actually touching it, as well as some key features such as "Like Music" that—without the ecosystem in place—could not be implemented with a satisfying level of experience.
- Limitations in the time for implementation meant that some functions had to be left out and serious compromises had to be made to the flywheel concept, so that a key signature element for the product was compromised. In general, much of the richness in the detail providing the targeted user experience was not included.

Since the release of the GoGear HDD 6330, Philips has produced a number of successor products using the same (and improved) user interface, e.g., GoGear SA9100, GoGear SA9200, GoGear SA9325, and GoGear SA9345.

17.2.5.2 Evaluation of the Process

Key members of the development team worked very closely together in the Philips Hong Kong offices, even though several members of the team were based elsewhere (e.g., the Netherlands and India). This "one room" approach significantly shortened communication lines as well as creating a strong team spirit.

The high time pressure worked well in the initiation and concept phase of the project. It facilitated decision making, the creation of flow and momentum, and it contributed to a strong team commitment. Later in the project, it all became less realistic. The specification became very difficult to write; multiple rewrites became necessary as details had to be changed. In the implementation phase there was too little time to get details right. Later decisions were made with the priority of just getting it on the market in time rather than fulfilling the initial ambitions.

Weaknesses in the key partnerships, required for developing the product, also created problems. It proved to be particularly demanding to get the external manufacturer to understand the emotional aspects of the design. In the end, the manufacturer failed to deliver the intended user experience and richness of the user interface due to priority setting on getting the basic technical components right. The final solution would have been much better if the development team had known from the start the restrictions and constraints of the product platform. Negotiations also failed with the network service provider, which meant that many unsatisfactory work-arounds had to be implemented on the product itself to compensate for limitations in using a standard PC music manager solution.

The product proposition was highly dependent on a strong interrelationship between user experience design and hardware and software platform requirements. Before the project started, there was little to no integrated user experience design or product roadmap development, thus platform requirements were missing. The initial high ambitions should have come with a clear picture of interesting "next steps" as well as a good insight into their feasibility and verification of concepts.

Although the project had very strong ambitions in creating the right user experience, these were not conclusively and formally expressed in terms of quality targets. Besides time pressure, a problem here was a lack of metrics regarding user experience qualities.

17.2.5.3 Evaluation of Proposition

In the end, the final product was optimized to perform well as a stand-alone device that had the capability to work within existing third-party system constraints. It therefore addressed only the parts of the system related to the stand-alone product itself, rather than the full user experience concerning the seamless enjoyment of digital music "anywhere and anytime," neglecting a need for "ecosystem" thinking, addressing all touch points where the user comes into contact with the consumer entertainment system.

A seamless and synergetic functional integration between the PC Digital Media Manager, connected services, and the portable digital audio player has become the key to success. It is here where simplicity starts. Even more when music services such as enhanced recommendations features (e.g., making use of listening behavior and sharing) become part of the product proposition, defining the ecosystem becomes crucial to get right.

How important the strategic alliance with the PC music manager is has become painstakingly clear, seeing the amount of unnecessary functions and awkward interactions found on the digital audio player just because essential media management cannot be done during synchronization. For example, the function "delete playlist" is needed on the portable player because the standard media manager will not delete old on-the-go playlists at synchronization.

17.2.5.4 Next Steps

Although compromises were made in the project reported here, the ambition remains. In the drive for relevant innovation, Philips CE management recognizes the increased importance of radically improving on user experience and there is now a new and clear directive. This directive addresses a fundamental change in approach by first specifying the user experience on clear and validated user insights and then defining the technical platform requirements to drive and steer the technical platform development.

User experience and design—truly delivering on our brand promise of Sense and Simplicity—will become the Philips brand differentiator. In an interview with the *Guardian* newspaper (Schofield 2007), Rudy Provoost, chief executive of Philips Consumer Electronics, said,

> You should be able to see it's a Philips design even if the brand name is not on it. It has to permeate through user interfaces, the packaging, the online experience. We have to create the Philips experience at every touch point with the consumer,... Consumer electronics has been too much electronics and not enough consumer. I'm in the consumer experience business. It's all about offering consumers great experiences, rather than the 'tech and spec' game. Philips is now a very different company in that sense.

17.3 DISCUSSION

17.3.1 USER INVOLVEMENT

User involvement in the design and development process is a cornerstone of human factors. Current product development cycles, however, are becoming so short that the opportunity to spend time obtaining consumer feedback is extremely limited. There is a danger, therefore, that the involvement of potential users in the development process will be skipped altogether, or that the input from users will come too late to have any substantial effect on improving the product.

One solution to this problem is to conduct a series of "quick and dirty" trials. Indeed, this is the general approach recommended by Nielsen (2000). However, in this project, even this approach would have meant continually trying to catch up with the development team, who were already implementing the design as it was formulated.

In this project, new ideas for interaction were introduced. It was critical to gain an understanding of the usability of these new interactions, but much less important to understand the existing navigation structures, which were re-used from previous product developments. Therefore, the usability tests focused on these new interactions. In a project of this kind, the human factors specialist needs to identify at a very early stage what is really new in the project and focus on this from the beginning.

At the same time, there is a need to have a sound human factors background knowledge concerning standard solutions, which can be introduced immediately to steer development. Thus, user involvement can still be carried out as a general activity in developing and refining these standard design elements.

17.3.2　Concept Generation

A good background knowledge of human factors not only contributes to the evaluation of solutions, but can also be a strong influence in continually assisting the development team to make choices regarding user interface directions.

In this case study, information was provided to the project team concerning user interfaces and consumer behavior through:

- Benchmarking similar consumer products and solutions
- Theoretical knowledge on human capabilities
- Experiments with the technology
- Mapping of solutions and defining the area of interest e.g., the usability knee

Further, the concept was based on an existing internal harmonization standard. The fundamental structure and appearance of the user interface was defined using this standard. Concept generation then explored how to first stretch the style so that it was still recognizable, and then how to fully adapt the style to accommodate the new interactions proposed. This adaptation of the style influenced the direction not only for interaction, but also for visual appearance, e.g., the use of icons within the overall graphical direction.

Thus, this case study shows that the development process is facilitated by generating knowledge both within the project and drawing on knowledge gathered before the project started. Critical here was that the development, exploration, and knowledge generation was an ongoing process, which had to be carried out quickly and efficiently as information, ideas, and solutions were required.

17.3.3　Iteration

Compressed development timescales also mean less opportunity for an analysis and synthesis approach, as described in Jordan, Thomas, and Taylor (1998). By the time a project starts, the analysis must already be in place. This means that more background knowledge and background research must be carried out and accumulated before starting a concrete development project. Therefore, there is a requirement in a consumer product development team to carry out continuous background research to support any specific product development. At Philips Design, there is continuous human research activity intended to support all of the company's businesses (see, e.g., Rameckers and Un 2005; Bueno and Rameckers 2007).

The case study reported here started out with the ambition to be revolutionary, but quickly had to bow to the constraints of existing company standards and the evolutionary development of original design manufacturer (ODM) technology. This meant that some of the project's objectives were not fully reached. Truly revolutionary projects require a revolutionary scope and the resources to carry out major explorations, as well as the freedom to deviate from company policies, neither of which was truly possible in this case.

Since the project team was ultimately constrained to using existing user interface concepts and standards, they could rely on tests of those concepts that had been carried out before this specific product development, so that the development team could focus on implementation rather than iteration. In a rapid development environment, internal company standards are essential to ensure a good user experience.

17.3.4 RELATIONSHIPS

One of the major obstacles in this project was working with external organizations, both in terms of partners in networking as well as with external manufacturers. In these cases, there was often a feeling that the project team was not in control, e.g., having to work to their standards. This did not benefit the proposition, which was particularly constrained by the demands of the media manager, which required too many work-arounds.

Internally, however, in this case study, as in many other projects, there was no fixed boundary between disciplines. This truly facilitated the development, but required very close cooperation with the people in each of the disciplines. This is an experience that has been shown time and again (see e.g., Thomas and van Leeuwen 1999; Jordan, Thomas, and McClelland 1996). With the more exacting development conditions described here, this cooperation between disciplines was more critical than ever.

What worked particularly well in this project was the implementation of short communication lines. A "one room" approach was adopted where the key members of the team, including software engineers, hardware engineers, product managers, marketers, designers, etc., all worked together at one location, which truly facilitated the close cooperation described above. This meant that the members of the team were in constant discussion and were therefore on the same learning curve throughout the development.

ACKNOWLEDGMENTS

The authors would like to thank all our colleagues in Philips who worked with us on this project. Without their collaboration, it would not have been possible to present the case study reported here. In particular, we would like to acknowledge the cooperation of Aninda Dasgupta, Arun Sundaram, Bart Mantels, Fu Ho Lee, Jasper Vervoort, Karthik Chandrasekar, Keith KF Leung, Len Lim, Michael Jerome, Low Cheaw Hwei, Vinod Rs, and Wietske Rodenhuis.

REFERENCES

Bueno, M., and Rameckers, L. 2007. Research for Innovation: Fitting the design process at Phillips Design. In *Market Research Best Practice: 30 Visions for the Future*. eds. P. Mouncey and F. Wimmer, 623–44. Chichester: Wiley.

Gaver, W., Boucher, A., Pennington, S., and Walker, B. 2005. Evaluating technologies for Ludic engagement. Paper presented in the workshop "Innovative Approaches to Evaluating Affective Interfaces" CHI 2005, Portland, OR.

ISO-DIS 9241-11. Ergonomic requirements for office work with visual display terminals (VDTs): Part 11: Guidance on usability.

Jordan, P.W., Thomas, B., and McClelland, I.L. 1996. Issues for usability evaluation in industry: seminar discussions. In *Usability Evaluation in Industry*, eds. P.W. Jordan, B. Thomas, B.A. Weerdmeester, and I.L. McClelland, 237–43. London: Taylor & Francis.

Jordan, P.W., Thomas, B., and Taylor, B. 1998. Enhancing the quality of use: Human factors at Philips. In *Human Factors in Consumer Products*, eds. N. Stanton, 147–58. London: Taylor & Francis.

Kersten, B.T.A., Thomas, B., and Docampo Rama, M. 2007. Multi stakeholder co-assessment: A process to guide the selection of personal healthcare innovations. Paper presented at the International Conference on Inclusive Design, Royal College of Art, London, April 1–4, 2007.

Lindholm, C., Keinonen, T., and Kiljander, H. 2003. *Mobile Usability: How Nokia Changed the Face of the Mobile Phone*. New York: McGraw-Hill Professional.

Nielsen, J. 2000. Why you only need to test with 5 users. http://www.useit.com/alertbox/20000319.html (accessed February 2008).

Norman, D.A. 2004. Expectation design: The next frontier. Paper presented at the Nielsen/Norman User Experience Event, Amsterdam, November 1–5, 2004.

Philips. 2004. Supply markets. www.philips.com/About/businessesandsuppliers/Suppliers/Section-14746/Index.html

Rameckers, L., and Un, S. 2005. People insights at the fuzzy front of innovation. Paper presented at the ESOMAR Qualitative Conference, Barcelona, November 3–15, 2005.

Rust, T.R., Thompson, D.V., and Hamilton, R.W. 2006. Defeating feature fatigue. *Harvard Business Review* 84 (2): 98–107.

Schofield, J. 2007. 'Philips is now a very different company'. *Guardian*, Thursday May 24.

Thomas, B. 1996. 'Quick and dirty' usability tests. In *Usability Evaluation in Industry*, eds. P.W. Jordan, B. Thomas, B.A. Weerdmeester, and I.L. McClelland, 107–14. London: Taylor & Francis.

Thomas, B., and van Leeuwen, M. 1999. The user interface design of the Fizz and Spark GSM telephones. In *Human Factors in Product Design*, eds. W.S. Green and P.W. Jordan, 103–12. London: Taylor & Francis.

18 Interaction Design for Mobile Phones

Wen-Chia Wang, Mark S. Young, and Steve Love

CONTENTS

18.1 INTRODUCTION

Nowadays, mobile devices are ubiquitous in our lives. Products such as mobile (cell) phones, personal digital assistants (PDAs), music players, handheld games consoles, book readers, and even video players are popular commodities. With such products, the term "mobile" should be descriptive of the users, not the device or application (Ballard 2007). This chapter focuses on how users interact with arguably the most common mobile device—the mobile phone.

From the earliest mass-market models in the mid-1990s, which only delivered voice and text communications, mobile phones have evolved almost beyond recognition. The introduction of the iPhone to the mobile phone market in 2007 set a precedent for the new "smartphone" generation, with its innovative mode of operation and the concept of "applications" radically changing the relationship between the handset and the user. At the time of writing in 2010, the iPhone 4 has just been released, representing the latest generation of smartphone, which not only has the basic attributes necessary for communication, but also offers schedule management, 3G, and even 4G technologies for emailing and surfing the internet, video calling, movement sensors, GPS navigation, and a wealth of gadgets and applications for interacting with the device. Physically they are different, too. Although smaller than their ancestors, most are notably larger than their immediate predecessors—mainly due to the constraints of screen size for web browsing. Furthermore, data entry has moved on from a number keypad, with the market largely polarizing into models with touch screens and those with QWERTY keyboards. With users increasingly turning to their smartphones rather than their PCs to access the internet (Hinman, Spasojevic, and Isomursu 2008), the requirements for hardware, software, and services pose a challenge to interaction design for mobile phones (Hinman, Spasojevic, and Isomursu 2008; Isomursu and Ervasti 2009; Kaasinen et al. 2009).

In the following discussion, we place the user at the center of the design process to understand the impact of individual differences, methodologies for understanding users, and evaluation of the design.

18.2   DESIGN PROCESS

While the specific terminology of the design process may vary among authors, in general there are three to four stages in common across all descriptions, such as: (a) analysis, design, and realization (Rakers 2001), (b) understanding users, developing prototype designs, and evaluation (Jones and Marsden 2006), or (c) identifying user needs and requirements, meeting those requirements through alternative designs, presenting the designs by interactive versions, and evaluation (Sharp, Rogers, and Preece 2007). Although the design process may seem a straight line from idea to tangible product, progress along it is not necessarily linear (Löwgren and Stolterman 2004), and stages might repeat according to the designer's working habits, business objectives, or consumers' requirements until the goal is achieved (see Figure 18.1). The process involves engineers, industrial designers, and user representatives' working together, and throughout the process ideas may appear, fail, or be replaced by better concepts (Jones and Marsden 2006).

The importance of understanding users in the interface design process cannot be understated. The design director of Nokia, Antti Kujala, points out that the first step in the Nokia design process is for anthropologists and psychologists to observe people's behavior and communication patterns (BusinessWeek 2007). The purpose of the interface between the user and the device is not only to integrate part of the elements of the system and its users, but also for those elements to impart information to each other (Barfield 1993). Thus emerges the related fields of interaction design and user-centered design (UCD), over and above the term interface design.

18.2.1   INTERACTION DESIGN

A broad definition of "interaction design" is designing spaces for humans to communicate and interact with computing (Winograd 1997) and the understanding of why and how humans interact with interactive products in everyday life (Thackara 2001; Sharp, Rogers, and Preece 2007). It covers the design of every digital object (i.e., objects enabled by electronic technology); digital services (those enabled or enhanced by electronics, such as the internet), and digital experiences (the design contexts) (Moggridge 2005). As can be seen in Figure 18.2, interaction design is based on human factors and tends toward human qualities and digital design. It provides a clear distinction of interaction design with other dimensions.

Interaction design should support "an interaction of a person with the artefact," or "an interaction among people that is mediated by the artefact" (Erickson 2005, 302). It is about how to create user experiences for improving people's work, communication, and interaction with devices (Sharp, Rogers, and Preece 2007).

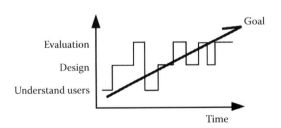

FIGURE 18.1   Interface design process.

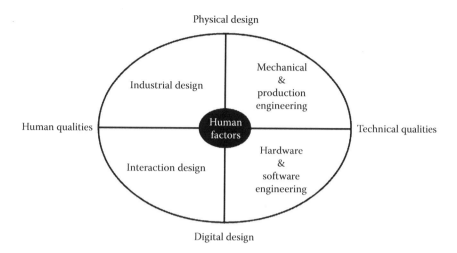

FIGURE 18.2 Interaction design and its disciplinary neighbors. (From Moggridge, B., *Theories and Practice in Interaction Design*, 271–86. Lawrence Erlbaum Associates, Mahwah; London, 2005.)

18.2.2 User-Centered Design

User-centered design, a term coined by Norman and Draper in 1986 (Norman and Draper 1986), is one of the most important perspectives of interaction design. From this perspective, design should make things easy to determine and visible, easy to evaluate the current state (of a system), and easy to operate by natural mapping between the user's intentions/actions and the consequent results. In the meantime, designers should make sure that the user understands what to do and let the user know what is going on (Norman 1998).

ISO 13407 (1999) provides guidance on the UCD approach. Figure 18.3 presents UCD as a loop activity. It integrates human factors, ergonomics knowledge, and techniques for improving efficient design and enhancing use, while avoiding negative effects on human health, safety, and performance. In order to apply and interpret the UCD process appropriately, though, it is necessary to have some understanding of the human user.

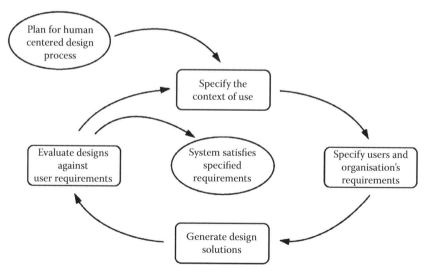

FIGURE 18.3 ISO 13407 human-centered design processes for interactive systems. (From ISO/IEC 13407, Human-centered design process for interactive system teams, Switzerland: International Organization for Standardization, 1999.)

18.3   HOW DO HUMANS INTERACT WITH A DEVICE?

In interacting with the world around us, what human beings are seeking is "an experience of being alive" (Campbell 1988, 1). According to Anderson (2000, 44), Norman describes the importance of experiencing a product as "the entire experience, from when I first hear about the product to purchasing it, to opening the box, to getting it running, to getting service, to maintaining it, to upgrading it." Moreover, product design includes diverse aspects like industrial design, graphic design, instructional design, usability, and behavioral design. Generally, user experience attempts to address users' pragmatic and hedonic level; it is subjective, highly situated, and dynamic in nature (Väänänen-Vainio-Mattila, Roto, and Hassenzahl 2008). While academic research emphasizes user experience theories, models, and frameworks, industry focuses on practical user experience work in product development, functionality, usability, and novelty.

Understanding user experience differs from usability evaluations, which typically calculate task completion times or test users' memory for icons (Kuniavsky 2003). Although user experience can be evaluated by general methods, such as observation, questionnaire, focus groups, and so forth, UCD should be kept in mind for understanding users' needs and values (Väänänen-Vainio-Mattila, Roto, and Hassenzahl 2008). Experience prototyping is one of the approaches for understanding user experience and applying the results to the design process (Buchenau and Suri 2000). Prototyping criticizes the design by understanding existing experiences, exploring design ideas, and communicating design concepts. One study adopted field trials to capture users' experience for understanding how people access the internet from their mobile phone (Isomursu and Ervasti 2009).

The user experience evaluation process at Nokia begins by examining users' needs and behaviors by new concepts, which are innovated based on users' reflections from market insights and technological opportunities. Problems are then identified, as well as any gaps, and then new ideas are generated. It is a cycle of feedback and development. Nokia requires that evaluation results of user experience come out as if on an assembly line, and expect that the results can be applied for different types of products and prototypes, finding out the pros and cons for each product. Their aim is to use the results to achieve excellent user experience rather than technical quality alone (Roto, Ketola, and Huotari 2008). With the smartphone revolution in mind, Kaasinen et al. (2009) revealed that understanding users and their usage of the mobile internet, improving services and how to discover those services, improving the software and hardware of the device, and improving infrastructures are crucial aspects for optimizing mobile internet users' experience.

When usability and the user experience are optimized, reacting to a device such as a mobile phone only seems to take a second without thinking. However, there is much more going on behind the scenes in terms of human information processing. According to Card, Moran, and Newell (1983), the process of human interaction with a device includes three subsystems: the perceptual system, the motor system, and the cognitive system. Each system has its own memories and processors. As can be seen in Figure 18.4, the perceptual processors of sight and hearing receive signals

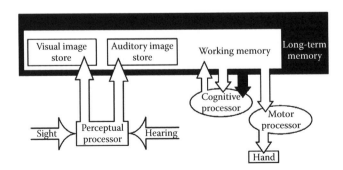

FIGURE 18.4   The model human information processor. (From Card, S., Moran, T.P., and Newell, A., *The Psychology of Human–Computer Interaction*, Lawrence Erlbaum Associates, Hillsdale, 1983.)

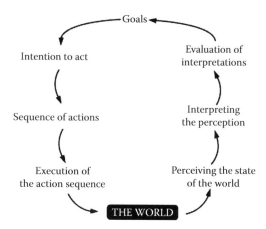

FIGURE 18.5   Norman's seven-stage action. (From Norman, D.A., *The Design of Everyday Things*, MIT Press, London, 1998.)

and then send these signals on to the brain, which stores the visual and auditory image individually. The cognitive processor retrieves information from long-term memory, and in turn, working memory communicates with the cognitive processor to make decisions and effect actions via the motor processor.

Norman (1998) suggests seven stages in the human's action cycle of completing a task, from intention to action, execution, and interpreting the result (Figure 18.5). Figure 18.6 applies Norman's action cycle to a typical task on a mobile phone. For instance, the goal is to change the ring tone on the phone. The intention triggers the actions to access the menu, review the ring tones and select one ring tone from the list. The user should get feedback from the system, such as a confirmation. According to their understanding of the feedback, the user should know whether the ring tone has been changed or not. However, not all mobile phone interfaces and systems are the same, meaning feedback may be inconsistent between handsets. When these various formats of interface come up against the wide variety of human individual differences, the objective of making a mobile phone usable becomes increasingly difficult.

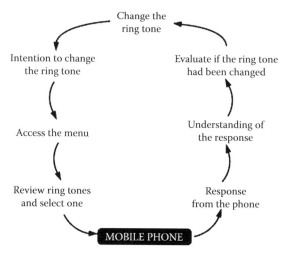

FIGURE 18.6   Applying a task to Norman's seven-stage action.

18.4 UNDERSTANDING USERS

From a psychological point of view, it is more important to consider users individually rather than as a social agent when psychology meets computer users (Moran 1981). Benyon, Crerar, and Wilkinson (2001) refer to human individual differences as physiological, psychological, and sociological aspects. Needless to say, these categories are not mutually exclusive, but in the following sections we draw out some of the key individual differences under each heading.

18.4.1 PHYSIOLOGICAL DIFFERENCES

Gender is one of the fixed physiological characteristics that can influence performance. For instance, a study on childrens' use of mobile phones (Ziefle and Bay 2008) showed that boys took advantage of a specific and successful navigation strategy of operating the mobile phone menu. Furthermore, boys were more willing to explore and interact with the phone, and they worried less than girls about adopting a trial and error approach.

Age also affects performance, and in general, older people perform poorly on a wide range of cognitive tasks (e.g., letter comparison, pattern comparison, paired associates, free recall, series completion, the Wisconsin Card Sorting Test [WCST] number categories, object assembly and block design; Salthouse, Fristoe, and Rhee 1996). Decreasing spatial abilities and navigation behavior might affect older people in operating a hierarchical menu (such as typically used on a mobile phone) (Ziefle and Bay 2004, 2008). Bay (2003) suggests that providing the spatial information of the menu in the manual might improve the performance of users who are under 50 years old, but not for those who are more than 50 years old.

More widely, studies show that most older users' dissatisfactions with mobile phones are exemplified by the small size of the handset, including the screen, buttons, characters, text sizes, the complexity of the menu, and the superfluous functions on the mobile phone (Kurniawan 2008). According to a contextual research study that aimed to understand mobile phone users who were over 60 years old, their particular requirements for a phone can be categorized as physical factors (e.g., vision, hearing, dexterity of fingers), cognitive factors (e.g., clues for operation), and cultural and psychological factors (e.g., creating usage culture for emotional needs) (Kim et al. 2007).

Although research has identified factors that might affect older people's operation of mobile phones, it seems that the market has not yet launched a mobile phone specific for that age group. Design improvements for older users should include bigger and fewer keys, basic functions for calling, harmonious color design, simple and square shape of the physical phone, and an emergency key (Wang 2008).

18.4.2 PSYCHOLOGICAL DIFFERENCES

According to Norman (1983), important elements for interaction behavior include: what is the target system; what is the conceptual model of that target system; what is the user's mental model of the target system; and what is the scientist's conceptualization of the mental model. Based on Norman's observations, mental models are incomplete, unstable, unscientific, parsimonious, and have no clear boundaries for different devices and operations. The limitations of mental models mean that a visualized structure of a device will be helpful for presenting "meaningful links and logical interrelationships" and integrating all information for controlling a device (Zwick 2005, 76).

Ziefle and Bay (2004) used a card-sorting approach to discover that older adults' mental models of a mobile phone are not necessarily hierarchical, but instead may be more linear and less interconnected. The reasons behind such cognitive structuring might be reduced memory capacity and spatial abilities, and less experience of operating high tech products. A further study used the handset of a PDA and a mobile phone to explore the relationship between mental models and navigation performance. The results showed that a correct mental representation of the system structure is helpful for users' performance of the device. Furthermore, the correct mental model can even reduce the impact of age (Ziefle, Arning, and Bay 2006). But the development of such mental models can depend on another psychological individual difference—cognitive style.

Green (1985, 3) quotes the definition of cognitive style from Messick (1969) as "stable and relatively enduring consistencies in the manner or form of cognition." According to Allinson and Hayes (1996, 119), Messick (1976, 5) defined it as the "consistent individual differences in preferred ways of organizing and processing information and experience." Cognitive style is a unique mode for each person to perceive, to think, to remember, and to solve problems (Green 1985). Despite researchers using different terms to define cognitive style, they all point out that an individual's cognitive style is the sole approach and process of organizing and portraying information (Riding and Rayner 1998; Benyon, Crerar, and Wilkinson 2001).

Table 18.1 shows the various types of cognitive styles. Although cognitive style is one of the human differences that has been applied in many diverse studies, including mobile phone interface design (Kim and Lee 2007), information systems (Mullany 2001), and web directories (Magoulas, Chen, and Dimakopoulos 2004), only a few types of cognitive style are well understood (such as the holistic vs. serialistic and visualize vs. verbalize dimensions). Whereas the understanding of user behavior is crucial to interaction designers (Cooper 1999), further research in this area should set out to establish an alternative to user segments by understanding the connection between cognitive styles, preferred interface patterns, and operation behavior.

Studies of cognitive abilities with mobile phones also involve elements of physiological and sociological differences. For instance, Western and Eastern users tend to have different cognitive styles and prefer different formats of interface on a mobile phone (Kim et al. 2007). According to Masuda and Nisbett (2001), Japanese users pay more attention to the relationship between an object and the circumstance than Westerners. Their study revealed that Japanese users tend to explain events by the interaction between the object and the situation, while Westerners tend to explain events by referencing the character of the object. Therefore, Nisbett and colleagues (Nisbett et al. 2001;

TABLE 18.1
List of Cognitive Styles

Researchers and Published Year	Cognitive Styles
Kretschmer (1925)	Dissociate/integration attention style
Goldstein and Scherer (1941)	Abstractness/concreteness of information
Jung (1923)	Extraversion/introversion
Klein and Schlesinger (1951)	Tolerance for instability/tolerance for unrealistic experience
Bieri (1955); Messick (1976)	Conceptual articulation: complexity/simplicity
Pettigrew (1958)	Broad/narrow thinkers
Gardner et al. (1959)	Equivalence range/breadth of conceptualization/category
	Constricted/flexible control
	Range of scanning
Rokeach (1960)	Open-mindedness/flexible control
Harvey, Hunt, and Schroder (1961)	Conceptual complexity: abstract/concrete
Kagan, Moss, and Sigel (1963)	Conceptual differentiation styles (description, relational, and inferential)
Messick and Fritzky (1963)	Field articulation: element/form articulation
Hudson (1966)	Convergent/divergent thinkers
Rotter (1966)	Locus of control: external-internal
Kagan (1966)	Reflectivity/impulsivity style
Pask (1969)	Holistic/serialistic
Parlett (1970)	Syllabus-bound/syllabus-free style
Paivio (1971); Richardson (1977)	Visualizer (imager)/verbalizer dimension
Pask (1972); Pask and Scott (1972)	Holist-serialist
Goldstein and Blackman (1978)	Cognitive complexity/cognitive simplicity

Norenzayan and Nisbett 2000) carried out a series of studies for exploring the relationship between culture and cognitive styles. They labeled East Asian users as holistic in thought while Westerners tend to be analytic in thought. Similarly, Unsworth (2005) examined Chinese and Western adults based on a categorization task between Chinese and American children (Chiu 1972). Both studies presented similar outcomes in that Chinese users prefer to categorize objects by the similarity and the relationship between objects, while Westerners prefer to refer objects by similarity.

Such studies provided a foundation for the research of Kim, Lee, and You (2007), who aimed to discover the correlation between cognitive styles of categorization and mobile phone menu structure. They showed that Korean users prefer a thematically grouped menu whereas Dutch users prefer a functionally grouped menu. One of the studies in this series (Kim and Lee 2005) also noted that Korean and American users have a significant difference when it comes to icon recognition on mobile phone interfaces. All of this suggests that cultural impact should be a key consideration for interface design on the mobile phone.

18.4.3 Sociological Differences

Accessibility and usability problems can occur due to different languages (e.g., left-to-right versus right-to-left input and reading, characters, numbers, and format of date and time) and culture (Benyon, Crerar, and Wilkinson 2001). The user's understanding of such representations means that the cultural context is key to interaction design (Bourges-Waldegg and Scrivener 1998). Hofstede's (1997) cultural dimensions, applied in various fields, have five facets: (a) power-distance, (b) individualism versus collectivism, (c) uncertainty avoidance, (d) masculinity vs. femininity, and (e) time orientation (orientation to the past, present, and future).

A comparison of mobile phone users in the UK and Sudan applies Hofstede's (1997) cultural dimensions to find that people in the UK are more comfortable than Sudanese when using a mobile phone in public (e.g., transportation, walking on the street). Meanwhile, Sudanese are more willing to switch off the phone for some specific occasions like worship, lectures, and meetings (Khattab and Love 2008).

The impact of cultural dimensions also affects the way people learn how to use a mobile phone (Honold 1999). This study suggests that Chinese users might need to communicate with sales persons, watch an animation, video and online help with step-by-step instructions, an emphasis on pictorial information about operation procedures, and larger fonts of the important information to achieve efficient learning. However, German users are satisfied with traditional user manuals that stress pictorial information for a clear overview of the functionalities of the phone, and give a detailed index in the manual. The cultural dimension of collectivism vs. individualism was also used to interpret variables of users' behavior patterns on mobile phones between Korean and U.S. users by observation, surveys, and focused group interviews (Kim et al. 2003). Their results imply that different products might be associated with other cultural dimensions.

Cultural difference not only affects the way that people interact with a mobile phone, it also brings an alternative method for communication with each other. Because of the economic situation in South Africa, people use beeping (paging) instead of calling (Donner 2007) to save the cost of communication via mobile phone. Such beeping can be categorized into three types: call back beeps (e.g., customers calling business contacts); pre-negotiated instrumental beeps (e.g., students reminding each other to attend classes and examinations); and relational beeps (e.g., greetings). People even design different types of beeping for different events. Mobile phone technology in this area stretches beyond the purpose of communication; it also improves people's business and the methods they use to trade.

In recent years, global industries have established research laboratories in targeted countries for obtaining information from local residents (e.g., Nokia). Previous work that we have mentioned in this chapter has used questionnaires, interviews, focus groups, and usability tests for exploring the impact of cultural difference on the usage of mobile phones. Both qualitative and quantitative methods provide detailed information for understanding users' interaction behavior with a mobile phone in a UCD process.

18.5 METHODOLOGIES FOR UNDERSTANDING USERS' INTERACTION WITH A MOBILE PHONE

A variety of methodologies can be used to gain an insight into people's attitude and behavior when they interact with mobile phone services and applications. These can range from laboratory-based studies where all the variables are controlled as much as possible (e.g., gaining an insight into specific navigation issues with a mobile phone interface), to assessing the impact of a new mobile social network service in a field study where the data are collected in situ and the researcher has no control over the way the application is used. Other important factors to consider when devising a methodology include understanding the organization if the service is being evaluated in a commercial or industrial context, or whether or not the service would benefit from a heuristic analysis as part of the usability evaluation. In addition, identifying specific behaviors as testable usability goals are also beneficial for a deeper understanding of users (Sharp, Rogers, and Preece 2007).

In relation to this, Love (2005) suggests that task completion times and the number of errors participants make are useful approaches for data collection when the user operates a task. Sharp, Rogers, and Preece (2007) and Love (2005) both mention the benefit of using video or audio recording to observe users' operation processes with mobile devices. This may be more problematic (for both ethical and practical reasons) when the video or audio recording occurs in public places and might take a long time in these contexts to collect the data needed, as well as the time needed subsequently for the data analysis.

One method that has been used to try and overcome the issues identified above is the diary method. One approach to the diary method involves broad general questions to help participants to report their experience and behavior to a specific event (e.g., connection problems when using an application in specific public places). In terms of duration, some researchers suggest that several days is long enough to get information from participants (Jones and Marsden 2006; Love 2005). However, a major drawback of this approach is ensuring that participants remember to continue to make diary entries for the duration of the study.

Log files are another method used to help understand users' using behavior via system records and has been applied to website design in recent years. For example, these data are helpful for personalizing electronic newsletters (Carvalho, Jorge, and Soares 2006), web services (Koch, Ardö, and Golub 2004), and so forth. In terms of calculating the frequency of accessing the menu, key pressing, changing the setting of the volume, using time, storage, and the context, Jeon et al. (2008) suggest that this approach provides very accurate information to categorize mobile phone users into different usage categories (e.g., communicative use, entertainment use, and restricted use).

It is important to notice that the triangulation method is necessary for obtaining both objective and subjective information from mobile phone participants in order to gain a better understanding of their attitude and behavior (Love 2005). Table 18.2 lists the pros and cons of using several well-known usability evaluation approaches for understanding user behavior and attitude.

This suggests, more than ever before in the area of human–computer interaction, that the advent of ubiquitous computing has led to a whole new range of methodologies that can be applied to usability evaluation.

18.6 EVALUATION

ISO 9241-11 (1998, 2) defines usability as the "extent to which a product can be used by specified users to achieve specified goals with effectiveness, efficiency and satisfaction in a specified context of use." It can take place at any time during the design process and is helpful for discovering problems earlier (Thomas 2001). It is important to understand why, what, where, and when to evaluate an interactive product (Sharp, Rogers, and Preece 2007).

When selecting an evaluation technique to assess the interface design of a mobile phone, who will examine the interface, where will this evaluation take place, and how the results will be

TABLE 18.2
Pros and Cons of General Evaluation Approaches

Methods	Results	Pros	Cons
Interviews	Mostly qualitative	Less expensive than ethnographic observation; getting people's experience by their words	Time consuming, interviewer's experience might influence the result
Questionnaire	Quantitative and qualitative	Cheap, easy for analyzing data rapidly, good for specific questions	Not flexible as interviews, questionnaire design is crucial
Observation	Mostly qualitative	Direct to get the information about people's natural behavior	Context might influence the participant's performance
Audio and video data recording	Mostly qualitative	Repeat reviewing of the data	Time consuming, might be limited in public places
Diary study	Mostly qualitative	Fill the gap when the observer is not around the participant	Might need more time to analyze a large amount of data from participants
Focus group	Mostly qualitative	Allow to raise diverse issues	Time consuming for analyzing the recorded conversation
Field research	Mostly qualitative	Understand people's natural behavior in their daily life in the context	Time consuming and high cost for collecting information; sample size might small
Performance measures	Quantitative and qualitative	Provide both quantitative and qualitative data for analyzing users' interaction behavior with the system	Principles of calculating times, errors, interrupts, and silences are essential
Think aloud protocol	Mostly qualitative	Understanding the participant's thoughts during operating a device	It might be difficult for participants to talk out loud and operate the device at the same time; time consuming for transcribing the audio recording
Log file analysis	Quantitative	Collecting data without interrupting the system performance	It might incur ethical considerations due to observing participants without their perception
Ethnography	Qualitative	Understand details of people's life	Might take weeks even years to collect data

analyzed are essential considerations that must be taken into account (Jones and Marsden 2006). For instance, sometimes usability experts will be needed to participate in the evaluation instead of end users, and at other times the evaluation of the mobile system has to take place in the context of where it will be used by individuals (e.g., busy cafe), instead of the controlled environment of a usability laboratory.

As can be seen from Table 18.2, general approaches of usability evaluation, including questionnaires, interviews, observation, and testing a design by the prototype method, are also important (Sharp, Rogers, and Preece 2007; Jones and Marsden 2006). In addition, applying multiple evaluation plans and strategies are also essential (Grinnell and Unrau 2005; Love 2005; Hammersley 1997).

Before testing target users, the Keystroke-Level Model (KLM) (Card, Moran, and Newell 1983) is a way to apply the user's behavior to a practical design. It is helpful to predict the task completion time and to compare different designs for the same goal. This approach has been applied to interaction with both computers and handsets. Due to the different operation modes of the PC and the mobile handset (e.g., user's attention might be split between the phone display and real-world distractions such as walking down a busy street), Holleis, Hußmann, and Schmidt (2007) suggest extending the limitation of prediction time of the KLM. Therefore, their study inserts additional operators to the KLM, such as the time for making actions, finger movement, gestures, initial action, attention shift from the keypad, display, or hotkeys to each other, and the time for the user's distraction by the context. The outcome suggests that the mobile phone KLM is especially useful for predicting different user groups' operation time for completing a task. It also suggests that the prediction time has to be adjusted along with different and new interactions in the future.

Recently, repertory grids (Kelly 1955) have been applied to product evaluation (Baber 1996), to capture the user's inner understanding of the world. The approach applies personal constructs of cognitive models with different degrees of experience and expertise for understanding people's perception of products. The first stage of the repertory grid is to define elements about the product. Based on these constructs, participants are required to evaluate the comparative elements by using a Likert scale. This approach can reveal which elements are important for users and how the user measures the product (Stantion and Young 1999). Brinkman and Love (2006) applied the repertory grid approach for developing an instrument to investigate the connection between user's attitude, social norms, and the preference for an interface design. The results from this study indicated that people evaluate a public or private application with different criteria and different importance based on social pressures (e.g., what would their friends think of their choice of application).

The concept of placing "the user" at the center of design has been gaining importance in recent years. Based on our literature review, the evaluation of a mobile phone should not only emphasize the operational behavior with the hardware and software, but also the user's personal view of the product. It seems that applying psychological approaches might be the solution for understanding the user and their point of view for developing a usable design.

18.7   CONCLUSION: THE FUTURE OF MOBILE PHONE INTERACTION

The mobile phone has become an essential part of people's lives. The market is constantly and rapidly changing because of improvements in technology, from the traditional 12 number keypad to the touch screen. The trend of 3G and 4G technology is potentially influencing people's needs and behaviors in interacting with a mobile phone. Its reach extends to various services and applications, from personal assistant to entertainment, healthcare, mobile learning, and the internet (Love 2005).

Nokia presents a concept about "relationships with objects" for future generation products and services (Nokia 2009). Via sensors to collect and transfer data to the device that we rely on, the eco-system will integrate all information for establishing our personal digital profile. This idea will push the mobile phone from a passive position to an active role. In recent years, designers are not only creators of products, but also enablers of experiences (Press and Cooper 2003). How to improve on or create a new mobile communication system by understanding people's natural abilities, their differences, operation behavior, and experience is a field ripe for further exploration. It is essential to consider how to integrate mobile technology into our life without the process of "taking over" (Nokia 2009). After all, it is the user who is supposed to be at the center of the interaction design process.

REFERENCES

Allinson, C.W., and Hayes, J. 1996. The cognitive style index: A measure of intuition-analysis for organizational research. *Journal of Management Studies* 33:119–35.

Anderson, R. 2000. Organizational limits to HCI: Conversations with Don Norman and Janice Rohn. *Interactions* 7 (3): 36–60.

Baber, C. 1996. Repertory grid theory and its application to product evaluation. In *Usability Evaluation in Industry*, eds. P.W. Jordan, B. Tomas, B.A. Weerdmeester, and I.L. McClelland, 157–65. London: Taylor & Francis.

Ballard, B. 2007. *Designing the Mobile User Experience.* Chichester: John Wiley.

Bay, S. 2003. Cellular phone manuals: Users' benefit from spatial maps. In *CHI '03 Extended Abstracts of the 2003 Conference on Human Factors in Computing System*, eds. G. Cockton and P. Korhonen, 662–63. Ft. Lauderdale, Florida, April 5–10, 2003. New York: ACM Press.

Barfield, L. 1993. *The User Interface: Concepts & Design.* Wokingham: Addison-Wesley.

Benyon, D., Crerar, A., and Wilkinson, S. 2001. Individual differences and inclusive design. In *User Interfaces for All: Concepts, Methods and Tools*, ed. C. Stephanidis, 21–46. London; Mahwah: Lawrence Erlbaum Associates.

Bieri, J. 1955. Cognitive complexity-simplicity and predictive behaviour. *Journal of Abnormal and Social Psychology* 51:263–68.

Bourges-Waldegg, P., and Scrivener, S.A.R. 1998. Meaning: The central issue in cross-cultural HCI design. *Interacting with Computers* 9 (3): 287–309.

Brinkman, W.P., and Love, S. 2006. Developing an instrument to assess the impact of attitude and social norms on user selection of an interface design: A repertory grid approach. In *Proceedings of the 13th Eurpoean Conference on Cognitive Ergonomics: Trust and Control in Complex*, eds. A. Rizzo, G. Grote, and W. Wong, 129–36, Zurich, Switzerland, September 20–22, 2006. New York: ACM Press.

Buchenau, M., and Suri, J.F. 2000. Experience prototyping. In *Proceedings of the Conference on Designing Interactive Systems: Processes, Practices, Methods, and Techniques*, eds. D. Boyarski and W. A. Kellogg, 424–33. Brooklyn, New York, USA, August 17–19, 2000. New York: ACM Press.

BusinessWeek. 2007. Nokia's Global Design Sense. [Online] (Updated 10 August 2007). http://www.business-week.com/innovate/content/aug2007/id20070810_686743.htm (accessed January 10, 2010).

Campbell, J., and Moyers, B. 1988. *The Power of Myth.* New York: Doubleday.

Card, S., Moran, T.P., and Newell, A. 1983. *The Psychology of Human-Computer Interaction.* Hillsdale, NJ: Lawrence Erlbaum Associates.

Carvalho, C., Jorge, A.M., and Soares, C. 2006. Personalization of e-newsletters based on web log analysis and clustering. In *WI 2006. Proceedings of the 2006 IEEE/WIC/ACM International Conference on Web Intelligence*, eds. T. Nishida, Z. Shi, U. Visser, X. Wu, J. Liu, B. Wah, W. Cheung, and Y.M. Cheung, 724–27. Hong Kong, China, December 18–22, 2006. Washington, DC: IEEE Computer Society.

Chiu, L. 1972. A cross-cultural comparison of cognitive styles in Chinese and American children. *International Journal of Psychology* 7:235–42.

Cooper, A. 1999. *The Inmates are Running the Asylum: Why High-Tech Products Drive Us Crazy and How to Restore the Sanity.* Indianapolis, IN: Sams; Hemel Hempstead: Prentice Hall.

Donner, J. 2007. The rules of beeping: Exchanging messages via intentional "missed calls" on mobile phones. *Journal of Computer-Mediated Communication* 13 (1): 1–22.

Erickson, T. 2005. Five lenses: Toward a toolkit for interaction design. In *Theories and Practice in Interaction Design*, eds. S. Bagnara, and G.C. Smith, Ch. 21, 301–10. Mahwah; London: Lawrence Erlbaum Associates.

Gardner, R.W., Holzman, P.S., Klein, G.S., Linton, H.B., and Spence, D.P. 1959. Cognitive control: A study of individual consistencies in cognitive behaviour. *Psychological Issues* 1 (4): 1–186.

Goldstein, K., and Scherer, M. 1941. Abstract and concrete behaviour: An experimental study with special tests. *Psychological Monographs* 53 (1): 1–9.

Goldstein, K.L., and Blackman, S. 1978. *Cognitive Style: Five Approaches and Relevant Research.* New York: John Wiley.

Green, K.E. 1985. Cognitive style: A review of the literature. *Technical Report 1*, 1–38. Chicago: Johnson O'Connor Research Foundation, Human Engineering Lab.

Grinnell, R.M., Jr., and Unrau, Y.A. (eds.) 2005. *Social Work Research and Evaluation: Quantitative and Qualitative Approaches.* New York: Oxford University Press.

Hammersley, M. 1997. *Reading Ethnographic Research: A Critical Guide.* London: Longman.

Harvey, O.J., Hunt, D.E., and Schroder, H.M. 1961. *Conceptual Systems and Personality Organization.* New York: John Wiley.

Hinman, R., Spasojevic, M., and Isomursu, P. 2008. They call it surfing for a reason: Identifying mobile internet needs through PC internet deprivation. In *CHI '08 Extended Abstracts on Human Factors in Computing Systems*, eds. M. Czerwinski, A. Lund and D. Tan, 2195–2208. Florence, Italy, April 5–10, 2008. New York: ACM Press.

Hofstede, G. 1997. *Cultures and Organizations: Software of the Mind.* New York, London: McGraw-Hill.

Holleis, P., Hussmann, H., and Schmidt, A. 2007. Keystroke-Level Model for advanced mobile phone interaction. In *Proceedings of the SIGCHI Conference on Human Factors in Computing Systems*, eds. M.B. Rosson and D. Gilmore, 1505–14. San Jose, CA, April 28–May 3, 2007. New York: ACM Press.

Honold, P. 1999. Learning how to use a cellular phone: Comparison between German and Chinese users. *Technical Communication* 46 (2): 196–205.

Hudson, L. 1966. *Contrary Imaginations.* London: Methuen.

ISO 9241-11. 1998. Ergonomic Requirements for Office Work with Visual Display Terminals (VDTs) – Part 11: Guidance on Usability. Brussels: CEN.

ISO/IEC 13407. 1999. Human-centered Design Process for Interactive System Teams. Switzerland: International Organization for Standardization.

Isomursu, M., and Ervasti, M. 2009. Touch-based access to mobile internet: User experience findings. *International Journal of Mobile Human Computer Interaction* 1 (4): 58–79.

Jeon, M.H., Na, D.Y., Ahn, J.H., and Hong, J.Y. 2008. User segmentation & UI optimization through mobile phone log analysis. In *Proceedings of the 10th Conference on Human-Computer Interaction with Mobile Devices and Services, Mobile HCI 2008*, eds. G. Henri ter Hofte, I. Mulder and B.E.R. de Ruyter, 495–96. Amsterdam, the Netherlands, September 2–5, 2008. New York: ACM Press.

Jones, M., and Marsden, G. 2006. *Mobile Interaction Design*. Chichester: John Wiley.

Jung, C. 1923. *Psychological Types*. Translated by H. Godwin Baynes, 1949. London: Routledge & Kegan Paul.

Kaasinen, E., Roto, V., Roloff, K., Vaananen, K., Vainio, T., and Maehr, W. 2009. User experience of mobile internet: Analysis and recommendations. *International Journal of Mobile Human Computer Interaction* 1 (4): 4–23.

Kagan, J. 1966. Reflection-impulsivity: The generality and dynamics of conceptual tempo. *Journal of Abnormal Psychology* 71 (1): 17–24.

Kagan, J., Moss, H.A., and Sigel, I. 1963. Psychological significance of style of conceptualization. In Basic cognitive processes in children. *Monographs of the Society for Research in Child Development* vol. 28, eds. J.C. Wright and J. Kagan, 73–112. Lafayette, IN: Child Development Publications.

Kelly, G.A. 1955. *The Psychology of Personal Constructs*. London: Routledge in association with the Centre for Personal Construct Psychology, 1991.

Khattab, I., and Love, S. 2008. Mobile phone use across cultures: A comparison between the United Kingdom and Sudan. *International Journal of Technology and Human Interaction* 4 (2): 36–50.

Kim, J.H. and Lee, K.P. 2005. Cultural difference and mobile phone interface design: Icon recognition according to level of abstraction. In *Proceedings of 7th International Conference on Human Computer Interaction with Mobile Devices and Services*, eds. M. Tscheliqi, R. Bernhaupt and K. Mihalic, 307–10. Salzburg, Austria September 19–22, 2005. New York: ACM Press.

Kim, H., Heo, J., Shim, J., Kim, M., Park, S., and Park, S. 2007. Contextual research on elderly users' needs for developing universal design mobile phone. In *Universal Access in HCI, Part I, HCII 2007, LNCS*, ed. C. Stephanidis, 950–59. Berlin, Heidelberg: Springer-Verlag.

Kim, J.H., and Lee, K.P. 2007. Culturally adapted mobile phone interface design: Correlation between categorization style and menu structure. In *Mobile HCI '07: Proceedings of the 9th International Conference on Human Computer Interaction with Mobile Devices and Services*, eds. A.D. Cheok and L. Chittaro, 379–82. Singapore, September 11–14, 2007. New York: ACM Press.

Kim, J.H., Lee, K.P., and You, I.K. 2007. Correlation between cognitive style and structure and flow in mobile phone interface: Comparing performance and preference of Korean and Dutch users. In *HCI International 2007, 12th International Conference, with 8 Further Associated Conferences*, eds. Stephanidis, C., Jacko, J., Smith, M.J., Aykin, N., Duffy, V.G., Harris, D., Shumaker, R., Schuler, D., Schmorrow, D.D., and Dainoff, M.J., Vol. 4559, 531–40. Beijing, China, July 22–27, 2007. Berlin, Heidelberg: Springer-Verlag.

Kim, S.W., Kim, M.J., Choo, H.J., Kim, S.H., and Kang, H.J. 2003. Cultural issues in handheld usability-are cultural models effective for interpreting unique use patterns of Korean mobile phone? In *Proceeding of UPA (Usability Professionals' Association)-Advanced Usability Professionals Conference*. Scottsdale, AZ, June 20–27, 2003.

Klein, G.S., and Schlesinger, H.J. 1951. Perceptual attitudes toward instability: I. Prediction of apparent movement experiences from Rorschach responses. *Journal of Personality* 19:289–302.

Koch, T., Ardö, A., and Golub, K. 2004. Browsing and searching behaviour in the Renardus web service a study based on log analysis. In *International Conference on Digital Libraries: Proceedings of the 4th ACM/IEEE-CS Joint Conference on Digital Libraries*, eds. H. Chen, H. Wactlar, C.C. Chen, E.P. Lim and M. Christel, 378–378. Tucson, AZ, June 7–11, 2004. New York: ACM Press.

Kretschmer, E. 1925. *Physique and Character*. London: Kegan Paul.

Kuniavsky, M. 2003. *Observing the User Experience: A Practitioner's Guide to User Research*. San Francisco; London: Morgan Kaufmann.

Kurniawan, S. 2008. Older people and mobile phones: A multi-method investigation. *International Journal of Human-Computer Studies* 66 (12): 889–901.

Love, S. 2005. *Understanding Mobile Human-Computer Interaction*. Oxford: Butterworth.

Löwgren, J., and Stolterman, E. 2004. *Thoughtful Interaction Design: A Design Perspective on Information Technology*. London: MIT Press.

Magoulas, G.D., Chen, S.Y., and Dimakopoulos, D.A. 2004. A personalised interface for web directories based on cognitive styles. In *User-Centered InteractionParadigms for Universal Access in the Information Society: 8th ERCIM Workshop on User Interfaces for All*, Vienna, Austria, June 28–29, 2004. Revised selected papers: *Lecture Notes in Computer Science*, eds. C. Stary and C. Stephanidis, 3196, 159–66. New York, Inc. Secaucus, NJ: Springer-Verlag.

Masuda, T. and Nisbett, R.E. 2001. Attending holistically versus analytically: Comparing the context sensitivity of Japanese and Americans. *Journal of Personality and Social Psychology* 81 (5): 922–34.

Messick, S. 1969. Measures of cognitive styles and personality and their potential for educational practice. In *Developments in Educational Testing, Vol 1*, ed. K. Ingendamp, 329–41. London: University of London Press.

———. 1976. *Individuality in Learning*. San Francisco: Jossey-Bass.

Messick, S., and Fritzky, F.J. 1963. Dimension of analytic attitude in cognition and personality. *Journal of Personality* 31:346–70.

Moggridge, B. 2005. Interaction design: Six true stories. In *Theories and Practice in Interaction Design*, eds. S. Bagnara, and G.C. Smith, Ch. 19, 271–86. Mahwah; London: Lawrence Erlbaum Associates.

Moran, T.P. 1981. An applied psychology of the user. *Computing Surveys* 13 (1): 1–11.

Mullany, M.J. 2001. Using cognitive style measurements to forecast user resistance. In *NACCQ (National Advisory Committee on Computing Qualifications) Proceedings of the 14th Annual Conference*, 95–100. Napier, New Zealand, July 3–5, 2001. Dunedin: Wickliffe Press.

Nisbett, R.E., Peng, K., Choi, I., and Norenzayan, A. 2001. Culture and systems of thought: Holistic vs. analytic cognition. *Psychological Review* 108:291–310.

Nokia Conversations. 2009. Future technologies, ideas & opinions: The opportunities for the future. [Online] (Posted November 11, 2009). http://conversations.nokia.com/2009/11/11/the-opportunities-for-the-future/ (accessed January 5, 2010).

Norenzayan, A., and Nisbett, R.E. 2000. Culture and causal cognition. *Current Directions in Psychological Science* 9:132–35.

Norman, D.A. 1983. Some observations of mental models. In *Mental Models*, eds. D. Gentner and A.L. Stevens. Ch. 1, 7–14. Hillsdale; London: Lawrence Erlbaum Associates.

———. 1998. *The Design of Everyday Things*. London: MIT Press.

Norman, D.A., and Draper, S.W. eds. 1986. *User Centered System Design: New Perspectives on Human-Computer Interaction*. Hillsdale, NJ: Lawrence Erlbaum Associates.

Paivio, A. 1971. *Imagery and Verbal Processes*. New York: Holt, Rinehart and Winston.

Parlett, M.R. 1970. Strategy competence and conservations as determinants of learning. In *The Ecology of Human Intelligence*, ed. L. Hudson, Ch. 18. New York: Penguin Books.

Pask, G. 1969. Strategy competence and conservations as determinants of learning. *Programmed Learning* October: 250–61.

———. 1972. A fresh look at cognition and the individual. *International Journal of Man-Machine Studies* 4: 211–16.

Pask, G., and Scott, B.C.E. 1972. Learning strategies and individual competence. *International Journal of Man-Machine Studies* 4:217–53.

Pettigrew, T.F. 1958. The measurement of category width as cognitive variable. *Journal of Personality* 26: 532–44.

Press. M., and Cooper R. 2003. *The Design Experience: The Role of Design and Designers in the Twenty-First Century*. Aldershot: Ashgate.

Rakers, G. 2001. Interaction design process. In *User Interface Design for Electronic Appliances*, eds. K. Baumann and B. Thomas, Ch. 3. London: Taylor & Francis.

Richardson, A. 1977. Verbalizer-visualizer, a cognitive style dimension. *Journal of Mental Imagery* 1:109–26.

Riding, R., and Rayner, S. 1998. *Cognitive Styles and Learning Strategies: Understanding Style Differences in Learning and Behaviour*. London: D. Fulton.

Rokeach, M. 1960. *The Open and Closed Mind: Investigations into the Nature of Belief Systems and Personality Systems*. New York: Basic Books.

Rotter, J.B. 1966. Generalized expectancies for internal versus external control of reinforcement. *Psychological Monograph* 80, 1–28.

Roto, V., Ketola, P., and Huotari, S. 2008. User experience evaluation in Nokia. Presented in *Now Let's Do It in Practice: User Experience Evaluation Methods for Product Development, Workshop in CHI'08*. Florence, Italy April 6, 2008. http://www.cs.tut.fi/ihte/CHI08_workshop/papers/Roto_etal_UXEM_CHI08_06April08.pdf (accessed December 15, 2009).

Salthouse, T.A., Fristoe, N., and Rhee, S.H. 1996. How localized are age-related effects on neuropsychological measures? *Neuropsychology* 10:272–85.

Sharp, H., Rogers, Y., and Preece, J. 2007. *Interaction Design: Beyond Human-Computer Interaction*. 2nd ed. Chichester: John Wiley.

Stantion, N.A., and Young, M.S. 1999. *A Guide to Methodology in Ergonomics*. London: Taylor & Francis.

Thackara, J. 2001. The design challenge of pervasive computing. *Interactions* 8 (3): 47–52.

Thomas, B. 2001. Usability evaluation. In *User Interface Design for Electronic Appliances*, eds. K. Baumann and B. Thomas, Ch. 16. London: Taylor & Francis.

Unsworth, S.J. 2005. Cultural influences on categorization processes. *Journal of Cross-Cultural Psychology* 36 (6): 662–88.

Väänänen-Vainio-Mattila, K., Roto, V., and Hassenzahl, M. 2008. Towards practical user experience evaluation methods. In GCOST (European Cooperation in Science and Technology) *Proceeding of the International Workshop on Meaningful Measures: Valid Useful User Experience Measurement* (VUUM), eds. E.L-C. Law, N. Bevan, G. Christou, M. Springtt, and M. Lárusdóttir, 19–22. Reykjavik, Iceland, June 15, 2008. Toulouse, France: Institute of Research in Informatics of Toulouse (IRIT).

Wang, Q. 2008. The effects of interface design about mobile phones on older adults' usage. In *Wireless Communications, Networking and Mobile Computing, 2008. WiCOM '08. 4th International Conference*, eds. L. Cuthbert and Z.T. Wang, 1–4. Dalian, China, October 12–14, 2008. Piscataway, NJ: IEEE Press.

Winograd, T. 1997. From computing machinery to interaction design. In *Beyond Calculation: The Next Fifty Years of Computing*, eds. P. Denning and R. Metcalfe, 149–62. New York: Springer-Verlag.

Ziefle, M., and Bay, S. 2004. Mental models of a cellular phone menu: Comparing older and younger novice users. In *Mobile Human-Computer Interaction – Mobile HCI 2004*, eds. S. Brewster and M. Dunlop, 25–37. Berlin: Springer.

———. 2008. Transgenerational designs in mobile technology. In *Handbook of Research on User Interface Design and Evaluation for Mobile Technology* [e-book], ed. J. Lumsden, Ch. 8. Hershey, PA: Information Science Reference. http://reference.igi-online.com/custom/title.asp?ID=170 (accessed December 5, 2009).

Ziefle, M., Arning, K., and Bay, S. 2006. Cross platform consistency and cognitive compatibility: The importance of users' mental model for the interaction with mobile devices. In *CHI 2006 Workshop: The Many Faces of Consistency*, eds. K. Richter, J. Nichols, K. Gajos, and A. Seffah, Vol.198, 75–81. Montreal, Canada, April 22-23, 2006. http://ftp.informatik.rwth-aachen.de/Publications/CEUR-WS/Vol-198/ (accessed October 23, 2009).

Zwick, C. 2005. *Designing for Small Screens: Mobile Phones, Smart Phones, PDAs, Pocket PCs, Navigation Systems, MP3 Players, Game Consoles*. Lausanne; London: AVA.

19 Human Factors in Protective Headgear Design

Roger Ball

CONTENTS

19.1 INTRODUCTION

Throughout human history, people have worn head coverings as protection against weather and physical danger. Some of the oldest forms of headgear are armored helmets, intended to protect soldiers from rocks, arrows, and spears. The modern military continues to be a driving force in the development of protective headgear. The technologically sophisticated ballistic helmets of today respond to ever-greater challenges of armed conflict, applying advances in composite materials, optics, and electronics.

The use of protective helmets for non-military applications is a relatively recent innovation. Around the turn of the twentieth century, football players adopted a form of leather "head harness" to protect against abrasions and "cauliflower ear." Hard hats for work applications appeared as early as 1919, manufactured out of canvas and glue, to give miners protection from falling rocks and, more importantly, as a place to mount a light for hands-free pick axing. Protective helmets for motorcyclists were created soon after the first motorcycle appeared in the 1900s, although they were initially not much more than a leather hat intended to keep the head warm and the hair tidy.

Today, a wide variety of protective helmets is available for many different sports and working conditions. The demand for specialized headgear continues to grow, driven by athletes, scientists, and the military, seeking to extend the limits of human performance.

19.2 DESIGNING PROTECTIVE HEADGEAR

The design of protective helmets requires an understanding of head anatomy, anthropometry, biomechanics, fit, helmet design, materials, manufacturing, and testing standards. In order to create well fitting, stylish, and protective helmets, many variables must be understood and addressed. In order to design a piece of headgear, a series of questions need to be answered; What type of helmet is being designed? What is the anatomy of the human head? What are the types of impacts and mechanisms of injury? What are the performance standards that it must satisfy? What is the intended user population? How do you create a comfortable, well-fitting helmet?

19.2.1 TYPES OF HELMETS

There are generally four main categories of modern helmet, as defined by their different characteristics in use: the hard hat, the crash helmet, the sports helmet, and the environmental helmet (Figures 19.1–19.3).

FIGURE 19.1   Construction of a hard hat. This helmet protects primarily from falling objects and minor impacts.

FIGURE 19.2 Sports helmet. This whitewater helmet is designed to protect against the hazards of a fast flowing river, rocks, and debris, at the same time allowing water to escape quickly.

19.2.1.1 Hard Hats

Hard hats are used by construction workers, meat packers, and steel workers. A "blue collar" symbol of unionization and industrial work, the hard hat communicates toughness and durability as the badge of physical labor. Produced in vast numbers, hard hats are the least expensive type of helmet available. Some models use as few as three or four molded plastic components to create a complete helmet. In all models, a hard, dome-shaped outer shell is connected to an adjustable hatband by

FIGURE 19.3 A sports helmet. This snowboard helmets needs to perform well at low temperatures in a hostile mountain environment.

means of a suspension rigging system. Invented in the 1950s, the suspension concept was a striking innovation at the time, combining ventilation with impact protection. Suspension systems today are no longer state of the art, outside of hard hats, and are used mainly as a component in more advanced multi-material approaches to protection.

The suspension system was surpassed in terms of impact protection by the introduction of expanded foams in the 1970s. Expanded foams made possible the creation of "crash" helmets: one-time-use helmets that absorb force, thereby protecting the user's head. Bell Helmets pioneered the use of lightweight industrial packaging materials in their Bell Biker model bicycle helmet. It provided the breakthrough that inspired the entire modern bicycle helmet industry.

19.2.1.2   Crash Helmets

Crash helmets provide effective head protection in activities that are prone to high-impact "crashes," including auto racing, motorcycle riding, equestrian riding, skiing, and bicycle riding. A modern crash helmet is constructed of a thin, hard, outer shell covering a thicker inner liner of dense expanded foam, normally expanded polystyrene (EPS). At the moment of impact, the rigid outer shell structure spreads the narrow force of impact, avoiding a concentration of "point loaded" impact forces onto a small area of the head. Spread across a wider area, the force of impact proceeds to crush the foam cell structure of the rigid inner lining, permanently compressing it. This crushing action absorbs and dissipates the energy of the blow, lessening the strength of the forces that penetrate to reach the head of the wearer. Because the compression of the foam is permanent, any helmet that has been involved in a crash must be discarded, as it will no longer provide protection in the future.

The lightweight design of a modern bicycle helmet is an impressive feat of engineering. A 7-oz piece of styrofoam can save a 100-kg rider travelling at 20 km/hr from serious head injury or even death.

19.2.1.3   Sports Helmets

Sports helmets differ from crash helmets in that they are intended for repeated abuse under lighter impact forces. In sports such as ice hockey, American football, lacrosse, baseball, cricket, and skateboarding, a single session of play can involve multiple impacts and blows to the head. Sports helmets also employ a rigid outer shell lined with a softer inner foam, but the inner foam of a sports helmet does not permanently compress under force. Sports helmets use foams that can withstand repeated impacts. They compress to absorb the force of an impact and then re-expand to return to their original shape, thickness, and effectiveness. Repeat impact foams are characterized by a distinctive "squishy" texture quite unlike the rigid styrofoam of a crash helmet. While they can handle more than one impact, they absorb less energy than crash helmets.

Although resilient, repeat impact foams do not offer an indefinite lifespan. However, in sports like ice hockey, some players become attached to their helmets, and have been known to use the same one continuously for several years, even as long as 20 years. This is not a good idea, as all helmets are made of plastics, and plastics by their nature are impermanent. Plastics are sensitive to the chemical compounds found in paint, adhesive sticker glue, sweat, and household cleaners; as well as to the effects of exposure to UV rays and high temperature. Deterioration of the chemical bonds in a plastic helmet can severely reduce its mechanical strength in both the shell and liners. The visual appearance of the helmet may not change, but it will no longer perform properly. Helmet manufacturers do not closely define the life cycle of their products, though some suggest that the helmet be replaced after 5 years of continuous use. Little is known about the effect of aging on the protective abilities of a helmet years after it was manufactured.

19.2.1.4   Environmental Helmets

The fourth major type of helmet is the survival or environmental helmet. These helmets form part of a larger system that protects a user in the kind of hostile or oxygen-depleted environments found in high altitude flight, space exploration, or deep sea diving. Because they need exact fit, survival helmets must often be custom designed for each individual user, and can be prohibitively expensive. The cost of a complete Apollo space suit in the 1960s was reported at about US$2,000,000 (ILC Dover Inc. 1994). These types of helmets often lead the way in pioneering the use of advanced materials and manufacturing innovations, such as today's composite materials and mass customization manufacturing methods. After such innovations have been developed for extreme conditions, they often make their way into mainstream consumer helmets.

19.2.2 Anatomy of the Human Head

The human head is comprised of a wide variety of physical elements, with the central bony skull the main structure (Gray 1977). The skull surrounds and protects the brain, which can be described as a mass of nerve fibers supported by blood vessels and glands. At the front of the head, the critical features of the eyes, nose, mouth, teeth, tongue, and throat are grouped together in the sensitive structures of the face.

The very top of the head consists of the skull and brain, covered by a thin layer of skin and only a very few superficial muscles. This is the area of the body that conforms most closely to our underlying skeletal structure. No mass of muscle tissue lies between the skin and bone to create shape diversity as in other areas of the body. It is impossible to change the shape of the head through muscle exercise. The head shape of body builders does not change with the rest of their physique.

The head may be hard, but it is also sensitive. At the front of the head, the face contains a disproportionate number of significant sensory organs. Vision, hearing, and taste are all located in or near the face. These delicate organs are exceptionally vulnerable, so that all head protection must pay diligent attention to their protection. The inelasticity and fragility of the head, in combination with the critical neural functions and senses housed there, helps to explain the importance of protective helmets for human safety, and the need to understand the physical structures of the head.

19.2.2.1 Skull Anatomy

The human skull is composed of two sections: the rounded cranium, made up of eight bones; and the forward face, containing 22 bones (Gray 1977). The bones of the face include the maxillary bones of the jaw, of which the mandible or inferior maxillary is the part that moves.

19.2.2.2 Skull Growth

In newborn babies, the bones of the top and the sides of the skull are not yet joined together, but are separated by membranes. The flexible skull of the baby permits it to compress in order to pass through the constricted human birth canal at birth (Roberts 1999). This flexibility is unique among primates, and is linked to our relatively short gestation period. When compared to other primates, humans are technically embryos at birth, and remain embryo-like for the following 9 months, as the brain continues to grow at rapid fetal rates, eventually becoming twice as large as it was at birth. The smaller brains of other primates are fully formed at birth or soon after (Gould 1977). It is not surprising that head protection for infants has always been an important concern for parents.

Only 63% of its mature size at birth, the skull reaches 88% after 2 years. By the age of ten, the skull will have grown to 95% of its final size, and by age eighteen the skull will reach more than 98% of its final size (Nellhaus 1968). As a baby matures, the skull bones begin to join together. Generally, before the age of two, the last of the open fontanelles or "soft spots" will have closed. The development of the rigid interpenetrating "sutures" connecting the plates of the skull takes far longer. In old age, the sutures become entirely rigid.

19.2.2.3 Brain Anatomy

Protection of the brain is the principal goal of the skull, and of the modern safety helmet. Contemporary medical research continues to reveal how central the brain is to human well-being in every way. Even mild brain injuries can have devastating impacts on cognitive and mental functions.

Detailed understanding of the anatomy of the brain is outside the scope of this chapter; however, some understanding of its physical qualities will help to appreciate the role of the helmet in protecting the brain against physical injury. A colorful description of the shape and form of the brain is given by helmet authority, James Newman (2007, 256), who wrote:

> Think of your head as an odd-shaped container covered with a thick rubbery skin and filled with warm Jell-O. The Jell-O is not entirely consistent in density and strength but is more or less so. The inside of the container, your skull, is not smooth everywhere and in some places is quite jagged. Your brain, the warm Jell-O, is contained within a membrane, called the dura, much the same as the membrane an egg is within, inside its shell. In order for it to be injured some component of it must be damaged.

19.2.2.4   Upper Spine and Neck

At the base of the head, the skull is attached to the spine at the first cervical vertebrae of the spinal column, C1. The spine supports the physical structure of the body, and protects the nerves of the spinal cord. At the point where it joins to the head, the soft tissues seamlessly mesh blood vessels, nerves, endocrine system, muscles, cartilage, and skin. Disproportionately heavy for its small muscular support, the head and neck area is extremely vulnerable to physical forces. Accidents to the neck can have a drastic effect on the body overall, and such accidents cannot be prevented by helmets. Helmets protect only the head.

19.2.3   HEAD INJURY

Researching the biomechanics of injury can be done through medical journals dedicated to studying the effects of injury. Databases of emergency room admissions provide a detailed overview of the reality of how people are injured as they list the activity and injury type. Private firms who specialize in injury study can be contracted to prepare in-depth research on injury mechanism.

19.2.3.1   Causes

The head may be injured by impacts from moving projectiles, or by crash situations created when the body impacts a stationary object. These injuries vary according to different activities, so that all specific helmet designs must seek to address different risks. A construction helmet must protect its wearer from falling objects on a construction site. A sports helmet protects its wearer from skin lacerations, as well as from potential skull fractures caused by the impact of moving objects, such as baseballs, cricket bats, or ice hockey sticks. Skull fractures can also occur when a head in motion impacts an unyielding surface, such as when a cyclist launches over the handlebars onto the hard pavement.

19.2.3.2   Effects

The effects of head injury can be severe. As James Newman (2007, 57) describes it:

> A skull fracture can occur when the skull bones bends more than it is capable of doing without breaking. Injury to the brain can occur if any part of it is distorted, stretched, or compressed, or if it is torn away from the interior of the skull.
>
> An impact to the head can cause the skull to deform and, even if it does not fracture, the underlying brain tissue can be injured as it distorts under the influence of the deforming skull. Even if the skull does not bend significantly, but the head as a whole is caused to move violently, distortion of the brain within the skull will occur. This typically leads to generalized diffuse injury such as concussion, and in extreme case, coma.

19.2.3.3   Injury Prevention

No helmet can prevent all injuries. Indeed, there is no protective device at all that can preserve a motorcyclist from the risk of serious injury or death in a high-speed collision. However, under less extreme conditions, helmets have been proven to be extremely effective in preventing much serious injury involving both the body and cognitive brain functions. They do so by distributing impact forces across a wider area of the head, and by absorbing energy that would otherwise reach the head directly.

By reducing the amount of relative movement between the parts of the head, the probability and/or severity of injury can be reduced. This is accomplished with a helmet reducing the forces on the head that would otherwise be produced on impact, by distributing the blow over a greater area than would otherwise be the case.

A helmet should never be seen as a substitute for caution, however. Wearing a protective helmet does not make one invincible, or reduce the dangers of risk taking. Any activity that requires wearing protective equipment is inherently dangerous. An attitude of personal responsibility toward danger is still the most essential tool of injury prevention.

19.2.4 TESTING STANDARDS

The first step in designing a piece of headgear is to understand the nature of the forces that act on it and the injury mechanisms of its particular application. In construction helmets, impacts are typically on top from falling objects, where as a football helmet can expect impacts at any location, often with multiple impacts at multiple locations at the same time. Motorcycle helmets will experience crashes at high speeds whereas baseball helmets will experience impacts while the player is stationary or running at relatively slow speeds. This seems a daunting task, however there are guidelines that can help to categorize these criteria in the form of testing standards.

Different sports or professions all demand different safety standards, because they each present users with very different risks. A cyclist who hits an obstacle will typically shoot forward over the handlebars, while snowboarders tend to fall backward, striking the back of the head in the so-called "heelside slam." Firefighting and kayaking helmets must be water resistant, which is not necessary for many other activities. Helmets designed for different activities target their protection at different types of accidents.

19.2.4.1 International Standards

Standards vary between Europe and North America because they are issued by different governing bodies, such as the the American Society for Testing and Materials (ASTM), Comité Européen de Normalisation (CEN), or the British Standards Institute (BSI). The differences between their requirements tend not to be large, and are mainly in terms of how the tests are physically handled. European test guidelines usually specify that the test headform and helmet must free-fall inside a guided wire cage, allowing the headform and helmet to drop under the force of gravity (European Committee for Standardization 2007). North American test facilities prefer to use a monorail drop system, in which the headform and helmet are attached to an armature as they drop the length of a fixed rail (ASTM 2007a). In both cases, the distance traveled in the test depends on the type of headgear being tested. A motorcycle helmet must drop from a height of approximately 2 m, while a bicycle helmet is dropped from 1 m (Snell Memorial Foundation 2005; ASTM 2007b). These varying heights reflect the different forces likely to be encountered during the two activities (Figures 19.4 and 19.5).

All helmet safety standards typically include a number of different elements. These will include a definition of the type of activity that the helmet is intended to address, a specification of any physical requirements for the helmet structure, and a description of a testing method, including the performance results required.

Currently, standards exist for more than a hundred different categories of helmet around the world. Well-known helmet standards include the ASTM F1447-06 Standard Specification for Helmets Used in Recreational Biking or Roller Skating and the EN1077 Standards for Alpine Skiers. More obscure would be something like the ASTM F2400-06 Standard Specification for Helmets Used in Pole Vaulting. Still under development are helmet standards that will set safety parameters for rodeo horse riding, and the use of electric assistive devices like the Segway.

FIGURE 19.4 North American helmet impact rig. The helmet with the headform attached travels down a guided monorail system.

19.2.4.2 Coverage Requirements

Physical requirements for the helmet itself may include recommendations for specific materials and retention systems, and the provision of mandatory labeling and marking. The standard will also specify "helmet coverage," or which areas of the head must the helmet cover for it to function correctly. A bicycle helmet has "high coverage" because it only needs to cover the high front and top of

FIGURE 19.5 European helmet test rig. The helmet with the headform attached falls freely inside the steel cage.

the head against the one-time effect of flying forward over the handlebars or falling to the side. An ice hockey helmet has "low coverage," extending downward to the neck and forward to cover the ears, to protect its wearer from the multiple impacts of pucks, sticks, fists, ice, boards, and Plexiglas walls, coming from all directions. For measurement purposes, coverage areas are specified in terms of planes and angles relative to the standard head reference plane known as the Frankfort plane (sometimes referred to as the "basic plane") (ASTM 2007c). Measuring the coverage requirement for a helmet involves placing the helmet on a headform marked with reference points and planes, and visually judging whether the required underlying planes are covered.

19.2.4.3 Helmet Conditioning

The central part of a helmet standard is a detailed description of the required performance testing. Testing may begin with a required "helmet conditioning" stage to ensure that the helmet is being tested under conditions that approximate its environment in use. For example, snowboard helmets must perform at the low temperature of $-20°C$, but will not be used at hot temperatures like $30°C$, thus a standard may require the helmet to be chilled before testing (European Committee for Standardization 2007). Conversely, bicycle helmets must perform at $30°C$, but are not normally used at $-20°C$, so a bicycle helmet might be heated before testing (ASTM 2007b). Construction workers are subject to extreme environmental conditions, and so the CAN/CSA Z94.1 Standard Specifications for Industrial Protective Headwear requires a construction helmet to be submersed in water for two hours before testing, to verify that all the materials function when wet. A bicycle or snowboard helmet will not have to undergo water conditioning prior to testing.

The actual tests themselves will focus on a helmet's ability to resist the kind of impact forces normally experienced during use. Some specialized helmets must also pass other tests. Construction helmets are required to demonstrate resistance against flammability and electric current (Canadian Standards Association 2009a).

19.2.4.4 Hazard Anvils

Impact testing begins with a drop test, in which the helmet, attached to a test headform, is dropped onto a standard flat-surfaced metal anvil. Some helmets are also tested against additional specific hazard anvils, which simulate objects or environmental structures specific to a particular activity. For example, in a cycling accident, the rider might fall into the sharp edge of a street curb. For this reason, bicycle helmets are tested against a radiused 45-degree steel impact anvil to simulate the impact of a curbstone against the head (ASTM 2007a).

19.2.4.5 Penetration Test

Some helmets are also given penetration tests (Canadian Standards Association 2009b). Penetration tests subject the outer surface of a helmet to a variety of dynamic challenges. These simulate the impact of objects that might come into contact with the helmet, such as a hockey puck, ski edge, or horse's hoof. In the Canadian CSA Z262.1-M90 standard, a machined steel bar is used to simulate the end of a hockey stick. The bar must not be able to penetrate any opening in the helmet to reach the head. A motorcycle helmet must resist the impact of a machined steel dart dropped from a height of 1–2 m, depending on the standard. Such penetration requirements effectively limit the size of ventilation holes that can be designed into a helmet. The commercial success of a helmet depends on the creative balance of user needs for comfort with the restrictions imposed by safety requirements.

19.2.4.6 Standards Organizations

Researching the individual standards requirements is best done through contacting the testing organizations that develop and publish the performance standards. Standards committees are responsible for each individual standard category and they are composed of a team of experts from industry, medicine, and the testing laboratories. It is recommended to contact the organizations not only to purchase a copy of the relevant standard, but also to understand what future developments are planned. Standards are always being refined and improved and it is important to know where they are going. Becoming a member of organizations such as the ASTM, the CE, or the CSA, allows access to a highly developed network and a database of specialized knowledge. These organizations hold meetings bi-annually and membership is open to anyone.

Once an understanding of the biomechanics of head injury and a thorough review of the testing standard has been completed, it is time to move on to the actual design of the product. The next consideration is how to create a piece of headgear that "fits" its intended users. But what exactly do we mean by fit?

19.2.5   Fit Factors

In its broadest sense, a head-centric product like a helmet can be understood to "fit" when a user who is trying one on feels that it is the correct size and shape, physically comfortable, emotionally satisfying, and stylistically appropriate. None of these issues is straightforward. The criteria for physical comfort, for example, vary widely for different products. Products that touch the body may be "loose fit" like clothing, "firm fit" as in helmets, or "precision fit" as in military aviation helmets, where even a few millimeters of misalignment can throw off delicate helmet-mounted displays. A single product may require a different fit in different places, as when a pair of sunglasses must rest lightly on the bridge of the nose, but grasp the temples of the head tightly in order to stay put. All these types of fit require a thorough understanding of the size and shape of the underlying body. This is particularly important when inelastic parts of the body, like the head, are in contact with inelastic products like safety helmets. Even small variations of size and shape are crucial under those conditions.

19.2.5.1   Emotional Factors

In terms of the end user, acceptance of helmets essentially relies on what marketing and sales people call the "mirror test," where a customer visits a helmet store and uses a mirror as the primary tool to make a purchase decision. In this complex process, the users first try on the helmet to assess the interior shape. Is it the correct size and shape? Then they evaluate the comfort aspects. Are the interior pads soft enough? Do the materials irritate the skin? Does it feel hot? Finally, they judge the emotional aspects of the color, materials, and the form of the product as it appears on themselves, and assess it in terms of personal vanity, and for the social messages communicated to its intended audience of other people. Does it look good on me? Will my friends laugh at me? The product will also be mentally compared to competitive products for factors like "value" and "style."

Professional approaches to the problem of fit come from essentially two different directions: academia and industry. Recent academic research into the area of comfort and design and the emotional value of products has begun to establish, for the first time, a theoretical underpinning and technical vocabulary for the multiple issues involved (www.designandemotion.org).

19.2.5.2   Technical Fit

Currently, research has not achieved deep penetration into what might be called the area of "technical fit," in which industry, governments, and testing organizations have established quantifiable guidelines for the size, shape, and performance of manufactured head-mounted products. In most cases, technical fit guidelines have emerged slowly, in a process of historical development made in response to needs for user safety. However, technical fit remains the "front line" of commercial application, directly affecting the lives of millions of people around the world.

19.2.5.3   Head Shape

The most important variable to creating good technical fit is an understanding of the shape and dimension of the head, and the size range of intended user group. Manufacturers develop their

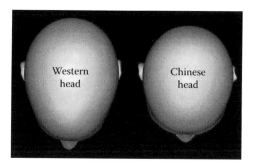

FIGURE 19.6   Head shape variation. The median shape of the Western (European) head on the left and the Chinese head on the right, as seen from above.

own proprietary versions of a shape that makes a comfortable fitting helmet. These shapes are not necessarily accurate or successful. Poor fit remains the problem that plagues most headgear.

The size and shape of the head you are intending to fit with a headgear is of critical importance. Head shape varies according to the intended population. For example, Chinese heads are rounder than European heads (Ball 2009; Ball et al. 2010). A difference of a few millimeters in width can make the difference between a well fitting and an uncomfortable fitting helmet. Age is another important variable. In the case of infant bicycle helmets, the intended user may be as young as 2–3 years old, which means a special smaller size range will need to be created. Although there is little scientific research about the differences between male and female head shapes, evidence shows that women's heads are smaller in circumference (Farkas 1994); some manufacturers offer helmets that are designed specifically for women (Figure 19.6).

Understanding head anthropometrics is critical to successful design, but where do we find the information? Head shape data can be collected by traditional anthropometric measurements or more advanced 3D scanning, but these surveys are time consuming and require significant sample sizes

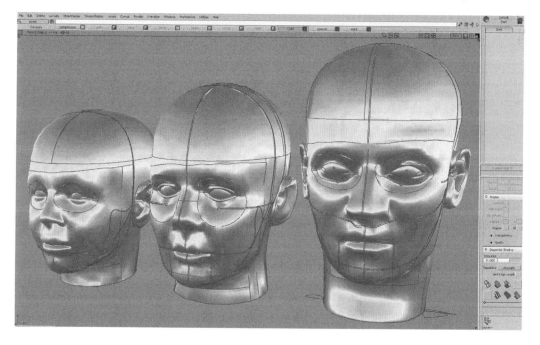

FIGURE 19.7   Reference headform. This set of facially featured reference headforms is used in the evaluation of optical products. Three sizes are used to represent children, youth, and adults.

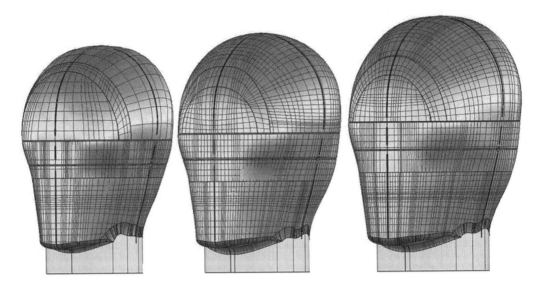

FIGURE 19.8 Reference headform. This set of reference headforms represents the head shape geometry above the Frankfort plane. These headforms are used to establish helmet coverage.

and training in proper anthropometric data collection. Tools, called headforms, are commercially available, which represent the median shape of a human head.

19.2.5.4 Headform Types

There are many different types of headforms commercially available for use in the testing of protective equipment. Some headforms represent just the upper part of the head and others are facially featured. The headforms come in a variety of materials and hardnesses. Different standards specify different headforms and some standards use two different types of headforms within the same test protocol. There are testing headforms that are used to physically test a helmet and reference headforms that are used to visually check a helmet for coverage (Figures 19.7 and 19.8).

19.2.5.5 Reference Headforms

Reference headforms represent the median shape of a specific population and are suitable for use in the design of the interior shape of a helmet. The EN960 European headforms are the most widely accepted reference headforms used in most helmet standards. They come in a set of 13 sizes as measured by circumference, starting from 500 cm circumference. They are commercially available and are suitable for casting and measuring as a base to develop an interior shape that is technically accurate (CADEX 2009).

The first set of 10 Chinese reference headforms developed from the SizeChina 3D anthropometric survey (following the EN960 protocol for reference headforms) has recently been developed and are commercially available (Certiform 2008). This new set of headforms is intended for use in developing "Chinafit" head products (Figure 19.9).

19.3 CONCLUSIONS

The design of headgear requires knowledge of a variety of factors. A thorough review of the testing standards is critical to understanding the types of impacts and environmental hazards posed by each activity. An understanding of head anatomy and the mechanisms of injury are essential before beginning work on a new helmet design. A clearly defined user profile is needed to gather

FIGURE 19.9 SizeChina headforms. This set of ten reference headforms represents the full adult Chinese population. These headforms are suitable for use as design tools for products seeking to create "Chinafit."

the anthropometric data that will match the intended user groups head shape. The compilation and application of this information prior to the design phase will shorten development time, improve accuracy of internal fit, and reduce errors and revisions in the final product.

REFERENCES

ASTM 2007a. Standard test methods for equipment and procedures in evaluating the performance characteristics of protective headgear. In *Annual Book of ASTM Standards*, Vol. 15.07. USA: ASTM International.
———— 2007b. Standard specification for helmets used in recreational bicycling or roller skating. In *Annual Book of ASTM Standards*, Vol. 15.07. USA: ASTM International.
———— 2007c. Standard specification for headforms. In *Annual Book of ASTM Standards*, Vol. 15.07. USA: ASTM International.
Ball, R. 2009. 3-D design tools from the SizeChina project. *Ergonomics in Design: The Quarterly of Human Factors Applications* 17 (3): 8–13(6).
Ball, R., Shu, C., Xi, P., Rioux, M., Luximon, Y., and Molenbroek, J. 2010. A comparison between Chinese and Caucasian head shapes, *Applied Ergonomics* 41 (6): 832–39 (Available online 12 March 2010).
CADEX. 2009. Catalogue – EN960 headforms. http://www.cadexinc.com/ (accessed June 30, 2009).
Canadian Standards Association. 2009a. *Standard Specifications for Industrial Protective Headwear-Performance, Selection, Care and Use*, 5th ed., Vol. CAN/CSA Z94.1-05. Canada: Canadian Standards Association.
————. 2009b. *Standard Specifications for Ice Hockey Helmets*, 5th ed., Vol. CAN/CSA 262.1-M90. Canada: Canadian Standards Association.
Certiform. 2008. The chinese head and face 3D digital database. www.certiform.org (accessed October 5, 2008).
European Committee for Standardization. 2007. *Standard Specification for Alpine Skiers and Snowboarders*, Vol. EN 1077. 19. Brussels: European Committee for Standardization.
Farkas, L.G. 1994. *Anthropometry of the Head and Face*, 2nd ed. New York: Raven Press.
Gould, S.J. 1977. *Ever since Darwin: Reflections in Natural History.* Harmondsworth: Penguin.
Gray, H.F.R.S. 1977. *Anatomy, Descriptive and Surgical.* New York: Gramercy Books.

ILC Dover, Inc. 1994. Space suit evolution: From custom tailored to off-the- rack, NASA. http://history.nasa. gov/spacesuits.pdf (accessed June 30, 2009).

Nellhaus, G. 1968. Head circumference from birth to eighteen years. *Pediatrics* 41 (1): 106.

Newman, J.A. 2007. *Modern Sports Helmets: Their History, Science, and Art.* Atglen, PA: Schiffer.

Roberts, J.M. 1999. Diagram C. In *The Illustrated History of the World,* Vol. 1: Prehistory and the First Civilizations, 155. New York: Oxford University Press.

Snell Memorial Foundation. 2005. *Standard for Protective Headgear for use with Motorcycles and Other Motorized Vehicles.* North Highlands, CA: Snell Memorial Foundation.

20 Creating a User-Centered Supermarket Checkout

Matthew Lyons and David Hitchcock

CONTENTS

20.1 ONE MAN'S VISION

In March 2000, Mike Wemms, Tesco Operations Director, was in his last year before retirement. Mike had enjoyed a long and successful career at Tesco, having been with the company "man and boy," and he had seen many changes in how Tesco sold its produce. He had been there in the days of "pile it high and sell it cheap" and had seen the introduction of new products such as washing machines and 28" flat screen televisions coming through the checkouts.

However, as a retailer, and acutely aware of the needs of his staff, he knew that the checkouts Tesco had been using in its stores for the past 20 years, were past their "sell by date." The staff were complaining about them, and he knew that if they were complaining, they wouldn't be giving as good a service to customers as they would if they weren't complaining! (It's retail not rocket science.)

Mike wanted to be able to offer a better service to customers at the checkout. He had a vision to develop a new checkout that would be better for customers, simpler for staff, and cheaper for Tesco. He wanted to do this for his colleagues, and friends in the stores, before he retired.

Developing a large and complicated piece of equipment like a checkout is a complex task, especially starting again from first principles as happened with this project. It is much easier with high-level sponsorship for a project of this kind to succeed. Mike was involved throughout the

FIGURE 20.1   Existing Tesco checkout, the "Mk5 standard belted checkout."

development of the checkout, and was able to unlock many doors and issues that could otherwise have "shipwrecked" the project.

Tesco has a well-established, in-house design management team, which leads the development of all retail fixture and fittings that are installed in its stores. The design team sits in the Property Services division of the business, based in Welwyn Garden City, and it was this team that were called on to lead the development of the checkout (Figure 20.1).

20.2   CREATING A SPECIFICATION

The brief given to the design team was very simple; to create a checkout that was right for the Tesco of the twenty-first century. One where the customer assistant (CA; cashier) using the checkout had everything they needed to hand in order to serve the customer. One where the bigger shopping loads of the twenty-first century could be fully accommodated on the belt, and one where the customer could receive exceptional service.

Key to the service was the need to have the CA standing when serving the customer. The standing CA (checkout operator) is freer to move around, and is at a more personal level to the customer. The CA would be able to pack bags more easily if they were standing. Furthermore, the manual handling risks should be reduced by a standing operator as twisting and overreaching can be avoided through movement of the feet rather than contortion of the trunk.

To develop a complex piece of equipment like a checkout, a useful approach is to use a detailed design specification that captures all the design requirements of all the "users" of the checkout. In this case, users were considered to be anybody who would have any interaction with the checkout during its whole lifetime in the store.

Based on the product design specification outlined in *Total Design* (Pugh 1991), 28 headings for the design specification were identified. The design manager adapted the design specification and developed a first draft that was to be used to promote debate with the project team and sponsors. It is often easier to give somebody an opportunity to say why something is wrong, than to ask them what is right! The design specification headings were

1. Aesthetics/appearance and finishes
2. Cleaning
3. Competition
4. Customer-focused requirements

   5. Documentation
   6. Electrical
   7. Equipment purchasing
   8. Ergonomics
   9. Health and safety
  10. Installation
  11. IT services
  12. Key features of design
  13. Layout planning
  14. Legislative standards and specifications
  15. Life in service
  16. Maintenance
  17. Manufacturing
  18. Materials
  19. Mechanical
  20. Packaging
  21. Performance
  22. Political and social implications
  23. Security
  24. Shipping
  25. Size
  26. Staff-focused requirements
  27. Timescales
  28. Trading law

This design specification was a dynamic, constantly changing document, being updated as new ideas and design requirements became apparent. However, it served as a benchmark to ensure that the project stayed on track, and nothing was forgotten.

Each of the headings was further broken down into individual specification requirements, under the following headings.

What the specification requirement is
A simple single sentence that described a single requirement for the design.

Ideas we have now
It is usual for people to have ideas when thinking about what they want from a design, this gave them the opportunity to express these ideas and to capture them, not lose them.

Comments
A field for any other information to be captured, such as observations from competitors or things that had been trialed before.

Priority
The priority of each specification requirement was identified so that trade-offs could be made if conflicting specification requirements were identified.

20.3   INFORMATION GATHERING

Once the draft specification was in place, information gathering began with the intention of adding as much flesh to the bones of the specification as possible. The aim was to identify every aspect of the design that needed to be accommodated, so that the checkouts could be designed ideally to meet them all. Information gathering took place through both formal and informal processes.

From a formal perspective, the project team of over 20 people were pulled together to review the specification and add their opinions and options to their areas of expertise, and to comment on other areas. This provided useful insights into the needs of the users of the checkouts. For instance, the cleaners did not like polished surfaces, as these were difficult to clean. So, we were able to capture these requirements in the specification before beginning the detailed design process.

More informally, the design manager leading the project was trained in how to use the existing checkout design in store. This enabled a wealth of information to be captured and added to the specification, i.e., identifying that there was no effective storage area for plastic bags within the checkout. It also enabled the design manager to begin developing a rapport with a team of CAs, who would then provide support, encouragement, and advice throughout the project.

Another important source of information for the design specification was the observation of competitor checkouts. The UK is unusual in that many retailers have their own custom design of checkout, which is used by them exclusively. Other, non-UK-based retailers buy their checkouts from standard ranges provided by checkout manufacturers.

The result of this custom approach to designing UK checkouts is that there are many differences between them, addressing different design challenges. Most interesting were the various approaches to bag packing; some had introduced frames or mechanisms to make it easier for the customer to bag pack, with differing levels of success (Figure 20.2).

Tesco's checkouts are made by a number of different checkout manufacturers, so it has good, open relationships with many of the key players in this area. This enabled access to review competitors' checkouts during the manufacturing process, and discuss the processes used by the competition, and the ideas behind some of the peculiarities observed.

The end result was a thoroughly comprehensive specification document, with in excess of 100 individual specification requirements. The process identified aspects of the design that would not have been obvious if this approach had not been taken.

For example: The material used for the signing area of the checkout was too hard, thereby preventing a ballpoint pen from operating properly when signing the credit card receipt. Therefore, a specification requirement was to provide a suitable surface for signing on.

FIGURE 20.2   One retailer's approach to assisted customer bag packing.

To control undesirable postures and movement, particularly during goods handling, it is best to operate a checkout while standing. However, the existing checkout design did not allow for stowage of the checkout seat if the CA wanted to operate the checkout in the desired standing position. The specification requirement was for the chair to be easily stowed so that the CA could operate the checkout while standing.

When maintaining checkouts, the access panels to the old checkout design were at the side of the checkout. This meant that when maintenance was required, the checkout aisle was closed down, closing two checkouts rather than just the one being maintained. The specification requirement was captured that the checkouts should be maintainable without closing down the checkout aisles.

20.4   APPOINTING A DESIGN AGENCY

While forming the design specification, the search for the right design support began. The in-house Tesco team of design managers are experts in the process of design and effective retail design; however, they are not able to get too involved in the detailed design development of equipment or environmental designs.

Tesco's experience of using product designers was limited. The main approach to design was to work with a number of specialist retail design consultancies that tended to be more focused on interior and graphic design. This project needed a different approach, so the decision was made to use a product design consultancy.

The Chartered Society of Designers was approached and a shortlist of companies provided. Each company was considered, visited, and the project was shared with them. The company appointed to the project was selected because its approach appeared to best suit the design approach that Tesco used and it was committed to incorporating ergonomics principles throughout the design project. Ergonomics was seen as core to the project, not least in light of concerns about musculoskeletal disorders among checkout workers (Mackay et al. 1998).

20.5   BRIEFING THE DESIGNERS

Key to the success of any design project is the initial kick-off meeting; this meeting sets the context for the whole project. Tesco has a very clear way of working, which it has embedded into its culture, and it has invested heavily in training its people to work that way.

Therefore, at the initial meeting, the design manager and head of design, Jeremy Lindley, was able to set the scene for how the project was to proceed. For example, the "Tesco Way" was explained; how Tesco works, how it structures its meetings, what it expects from its suppliers and its design consultants. All these things were clearly explained and the expectation was that they had to be adhered to.

Also key at this stage was the mantra "Better, Simpler and Cheaper," which was being used to drive all design development work used by Tesco. All designs had to be able to demonstrate being "Better for Customers, Simpler for Staff and Cheaper for Tesco."

Finally, the key document shared at this meeting was the design specification. The design specification was reviewed in detail along with images of the existing checkout, so that the designers could begin to understand the checkout and what it should do.

The design manager was conscious that there was a need to get the designers up the learning curve as quickly as possible. So, almost immediately, time was arranged for the designers to spend time working on checkouts in store, and also visiting the checkout manufacturers.

20.6   ERGONOMICS OBSERVATION

In addition to gaining experience of using the checkouts and the insight that this gave, ergonomics observation was also initiated at the earliest stages. This enabled the design team to gather information on how the existing checkouts were being used in store over a long period of time. The information gathered was then used to add additional detail to the design specification.

This observation identified a number of issues with the checkout that had not been picked up before. It was clear that large bottled carbonates were a big problem for the checkouts, as they were seen rolling around on the checkouts, and were difficult to control. There was no area for folding clothes—customers and staff struggled to find a space on the checkout where they could effectively fold clothing before purchase.

The existing design clearly required the CA to unacceptably twist excessively when operating the checkout, to use the weighing scales, and to access the till. It was also clear by observing the CAs using the checkout that the storage within the checkout for carrier bags or till rolls was inadequate; CAs were seen struggling to find the correct bag that they required to serve the customer.

This observational work and additional insights from the designers, largely finalized the design specification. The specification was then presented to the full project team for buy in and approval to proceed with the design work.

20.7   INITIAL DESIGN WORK

With the design specification in place, the designers were left to generate as many design ideas as they could. Their approach was very much to generate a quantity of ideas based on some or all of the elements of the specification. For example, they may focused on bag packing as an area, and develop as many design options as they could; the focus may have been modularity, and as such they would develop as many ideas around a modular checkout as they could (Figure 20.3).

The designers developed a large volume of ideas, and also identified what they believed were the key design challenges for the project. These design challenges became the key focus for their attention when developing the equipment. These design challenges were as follows.

Customer assistant centricity (or CA centric design)
The CA centric approach focused on the needs of the CA when using the checkout. A CA centric approach ensured that the CA's needs were always at the forefront of the designers work. The physical ergonomics of the checkout, the interaction of the CA with the customer, the access to carrier bags, how the checkout assistant interacted with the checkout controls. All could be summed up by the term CA centric.

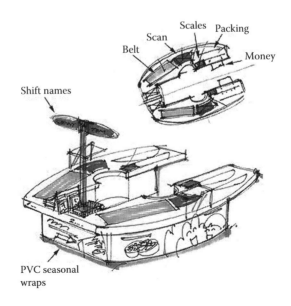

FIGURE 20.3   An example of an initial design concept drawing.

Modular construction
Modularity became clear as one of the key design challenges for the checkout. If it was to be manufactured in bulk, transported, and efficiently and easily maintained, then a modular approach was key.

Work flow
From both customer and staff perspectives, it appeared that the process that goods followed through the checkout was all wrong. Much of the twisting suffered by the CA, and the difficulties experienced by the customer when paying for goods, could be attributed to a poorly understood assessment of the work flow of the checkout.

Checkouts in pairs
Another key observation was that the checkouts were developed as individual checkouts, and placed in pairs in a store. This was driven by the need to occasionally have odd numbers of checkouts in store. So, for the occasion that a single checkout needed to be installed in store, all the checkouts were developed as singles. It was clear that if the checkouts were developed as pairs, for the occasions when a single was required, the checkout could be developed independently and a significant overall saving could be achieved. This meant, for instance, that cash conveyancing could be simplified, as only one vacuum tube would be required where previously there would have been two.

The designers were able to distil the design specification into these five key design challenges, which made their job much easier at this early stage of development.

20.8 CONCEPT DEVELOPMENT

The various design concepts were collected into five discreet design concepts, which were given the go-ahead to be developed further. Each concept captured elements of the design specification with different weightings, and the designs could be graded from the "radical" to the "conservative" in their design.

The designs were shared with the project team, and with the checkout assistants, with whom a relationship had been established at the start of the project. The designs were assessed against the original design specification, and it became clear that there were two favorite designs (Figure 20.4).

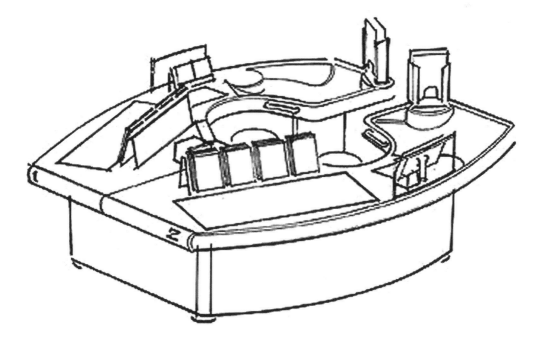

FIGURE 20.4 The "traditional" checkout design concept.

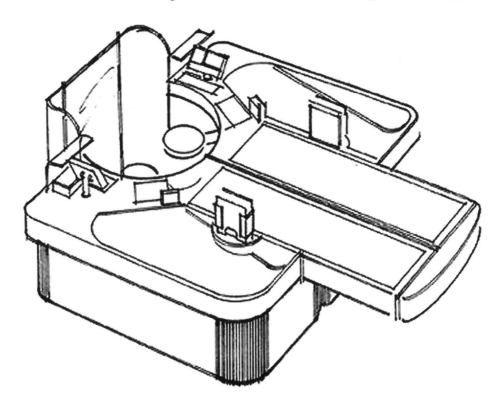

FIGURE 20.5   The "radical" checkout design concept.

The first, and the favorite of the CAs, was a "traditional" design with a front and rear conveyor belt. This design addressed the key issues of ergonomic improvement and stowage in the checkout for all checkout consumables.

The second design was a "radical" design, which only had one belt, relocating the packing bay to the mid-point of the checkout and the payment area of the checkout to the rear of the checkout. This appeared to make sense from the perspective of workflow. The customer unpacked, the CA scanned the product, the customer packed, pushed their trolley to the rear of the checkout (thereby letting the next customer begin to unpack), paid, and left. The radical design was the favorite of the rest of the project team because it offered potentially very attractive benefits to the business (Figure 20.5).

If the "radical" approach could be made to work, it delivered a far superior ergonomic operation for the CA. It saved a lot of money on the cost of the checkout, as a large component cost of the checkout is in the conveyor belt used. But the exciting potential came from the fact that the checkout was much shorter than the old checkout design, and yet still had a longer input belt.

If the checkout was shorter, it could free up space in store that could be used to sell from; potentially, an additional 30 modules of retail space could be available, generating in the order of £20,000 of additional sales per week.

The decision was made to work up the two favorite designs into full-size MDF prototypes, which would enable presentations to take place, comparing the two designs, and for ergonomic assessments of the two designs to take place.

20.9   FIRST MOCK-UPS

The MDF versions of the two checkout designs were developed quickly and installed in the Tesco prototype center in Welwyn Garden City for review, along with an example of the existing checkout design for direct comparison.

FIGURE 20.6   MDF mock-up of the "radical" checkout design.

Each prototype was reviewed by the project team, and the "benefits and concerns" of each discussed. The checkout assistants from the local Tesco store in Welwyn Garden City were also given the opportunity to review the checkout models, and pass comments. All these comments were captured and acted on where appropriate (Figure 20.6).

These early mock-ups enabled the design of the checkouts to evolve, and rough positioning of equipment and "design assumptions" could be checked. In addition, the size of the input belts could be compared with typical shopping baskets, ensuring that the belts could accommodate the sizes of shop that were required. Early ergonomic testing could also take place at this time, by checking the operation of the scanning process. It was clear at this early stage that the "radical" design had the potential to provide a far superior scanning operation.

The mock-ups were mainly used to illustrate the design solutions identified through the design process, and how these were to be addressed. With these models, the designers were able to demonstrate the principles of CA centric, the way goods would flow through the checkout, how the footprint compared to an existing checkout, and how easy it was for the CA to support the customer in packing their bags and providing a good service.

The key meeting at this phase was the time spent with Mike Wemms, who had the "casting vote" on which of the two designs to choose. Forever the retailer, the lure of the radical design was obvious, and his choice was to "give it a go." The possible success of this checkout was too great an opportunity to miss.

20.10   FIRST STAGE ERGONOMICS TESTING

Fitting trials were necessary to accurately place the equipment within the checkout (Figure 20.7). A fitting trial is an experimental study in which a sample of participants use an adjustable mock-up of a workstation (in this case, the checkout) in order to make judgments as to whether a particular dimension is "too big," "too small," or "just right" (Pheasant 1986).

Four items of equipment were identified as being particularly important to healthy posture and movement: the cash till, printer, scales, and "SNIKEY" (keypad/display panel/card swipe). Each of

FIGURE 20.7 Ergonomic data collection from customer assistants.

these was modeled to actual size. Through the process of fitting trials with 30 representative operators, commonality of positioning was identified.

Compromises to the optimum placement of some items were necessary. For example, the scales could not be positioned precisely as identified because of fixing limitations within the checkout surface. The fitting trials data were used to determine the consequences of altering positions. This enabled decisions to be made that minimized the number of potentially excluded users. Even after careful ergonomic appraisal through user trials, in certain circumstances it still may be a matter of reasonable choice where something is positioned (Brown, Cross, and Walker 1983).

20.11   DESIGN ENGINEERING

With ergonomic data to support them, the designers passed the development of the checkout over to their engineering team. Working with detailed reference to the design specification, and basing the design on the proposal produced by the creative design team, the engineers had the unenviable and complex task of managing the process of development and compromise, which is the task of the development engineer.

Over the course of several weeks, and with weekly updates with the Tesco design team, the design evolved, as the challenges of attending to all the elements of the design specification were felt. Without the design specification, the project could easily have drifted, and some of the key objectives could have been missed; the design specification provided a constant and unambiguous reference for the design team.

Some of the many features of the design at this stage included:

- A single belt, which significantly reduced the overall cost of the checkout, and was made as a module, so it could be swapped out of the checkout quickly, if the checkout belt broke down.
- The checkout was designed to be "flat packed" so it could be assembled at the store where it was to be used, allowing many more checkouts to be transported, thereby greatly reducing delivery costs.
- The checkout furniture was designed to be assembled in a similar fashion to office furniture, having supporting legs onto which are hung identical "modesty" panels. This made the checkout modular lighter and quicker to assemble at the store. Previously, all checkouts had a metal sheet or plywood skirt supporting the weight of the checkout top.
- The checkout was designed to be a pair, rather than an individual checkout, which meant that unnecessary duplication of components could be avoided, such as "air tube" cash conveyancing equipment or conveyor belt controls, further reducing the cost and complexity of the design.

Once the technical design was completed by the engineers, the design was passed over to the manufacturer involved in the project team to date, who was able to take the designs and refine them a little to ensure that they were manufacturable in "reality" and able to be produced by their manufacturing capabilities. The manufacturer was able to pass an eye over the designs to ensure that, from their experience, the checkouts were going to be fit for purpose and operate effectively in store. The manufacturer was, throughout this process, open to the ideas that were coming from the designers; not threatened by the proposals produced by the designers, but always willing to "try." It is essential to find a manufacturer with this approach when trying to reinvent such a well-established design concept.

The manufacturers then made two fully operational prototypes of the checkout. Each checkout was a different "hand," so that the operation of the checkout could be tested from a right- and left-handed CA's perspective, and from a customer perspective (Figure 20.8).

To review the checkout prototypes, a number of different workshops were organized. These included different styles of workshop for the project team, the CAs from the local store and, for the first time, with actual customers.

The workshop with the entire project team involved over 20 people, so careful management of proceedings was essential, making the required style quite formal. People are always willing to be very critical if given the opportunity, collectively jumping onto any concerns and criticisms. To prevent this occurring in this situation, the project team were asked to focus on their area of specialism, as captured in the design specification, and to use "Post-It" notes to capture their concerns, sticking them onto the checkout at the location of concern. Once this was done, the team reviewed the notes collectively, ensuring the concerns were understood. These notes were then used to modify the design where necessary before the next stage of checkout manufacture.

A similar process of "Post-It" notes evaluation was used for the CAs, along with a representative from the union (USDAW). The CAs were given time alone with the checkout to try it out together. This allowed the CAs to assess the checkout without feeling intimidated; a much more informal style of workshop.

For the customers, a research agency was enrolled to ask the customers consistent questions related to their experience of using the checkout and how they felt about the new checkout design. The CAs who had also been involved provided the "customer assistance" to the customers.

The trials provided useful detail on how the checkout should be developed, and how the CAs and customers responded to the checkout. The feedback from the customers and CAs did raise some concerns over how customers would react to the radical new "flow" of goods through the checkout, but not to a level that meant Tesco did not want to pursue this design.

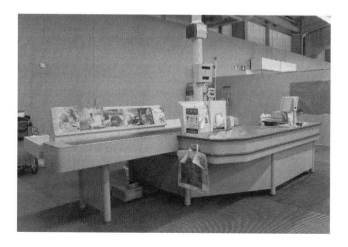

FIGURE 20.8   The first working prototype of the "radical" checkout.

The benefits identified for the checkout at this stage included:

- A folding table was included for folding clothes on.
- A lean seat was provided so that the standing CA could rest while remaining "standing."
- Belt override allowed large items on the belt to be forced passed the input belt "magic eyes," which stop the belt when the goods reach the CA. This made it easier for large and heavy items to be handled by the CA.
- Easier to exit the checkout to offer customers help, and easier to bag pack.
- Facilitated team working, by using a "back to back" design of checkout pairs.
- Improved ergonomics and better working postures.
- Integrated bag dispensers made access to bags easier for customers and CA.
- Quicker throughput of goods when scanned.
- Simple cable trunking was included in the checkout, to reduce exposure to messy wires under the checkout.
- Simpler to manufacture.
- Smaller checkout and cheaper to manufacture.
- Storage within the checkout for all checkout consumables.

20.12 WORKING PROTOTYPES

Following these trials, permission was granted by Tesco to trial the checkouts in an operating store. The checkout manufacturer refined the designs, working with the Tesco design manager and the designers, to produce two pairs of checkouts to install in the trial store. Before the two checkout pairs were installed in the store, however, they were set up again in the Tesco mock-up center for further evaluation (Figure 20.9).

Once again, the project team, CAs, and customers were shown the checkout and feedback captured. Minor changes were suggested and these were actioned before the checkouts were installed in the trial store. The mock-up also provided an opportunity to develop training material for the new checkouts, and to train the CAs from the trial store (Figure 20.10).

In December 2000, the "radical" checkout design was installed in a Tesco store. This date was chosen to give the checkout a test in the "worst case" scenario; checkouts are worked hardest over the Christmas period.

FIGURE 20.9 The second working prototype of the "radical" checkout design.

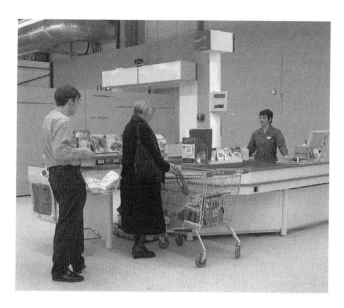

FIGURE 20.10   The "radical" checkout design undergoing customer trials.

20.13   STORE "TRIALS"

Almost immediately, it became apparent that there was a problem with the radical design (Figure 20.11). Despite the desire of the project team to trial the radical design because of the many benefits it could bring to Tesco, the customers were unable to see how the checkout benefited them!

The fundamental problem was the change in flow of the goods. Despite the logic of relocating the packing bay of the checkout, customers were unable to understand this fundamental change.

FIGURE 20.11   The "radical" checkout design installed for store trials.

FIGURE 20.12   The "radical" checkout design installed in use during store trials.

Customer after customer walked to the rear of the checkout, expecting that their shopping would arrive there, despite the fact that, if they had looked at the checkout, there was no visible means of getting it there! It became apparent that the habits formed in the customer over years of using checkouts in supermarkets, were very hard to overcome.

The customers were very forthright in expressing their feelings about the checkout. The CAs explained how to use the checkouts, but it was necessary to introduce a "complaint card" for the customers to feedback directly to the design team, as the CAs were finding customer concerns too time consuming to deal with.

When some of these complaints were followed up, and the reasoning behind the checkout explained, the customers immediately came round to the idea and were "happy" with the new design. However, it very quickly became clear that the radical design was too radical; a step too far for customers to deal with. After 6 weeks in store, the checkouts were removed (Figure 20.12).

While the checkouts were in store, the project team reviewed them and checked the operation of the design against the design specification. This enabled all aspects of the design to be reviewed, overcoming the "background noise" of the problems with the customer flow. The design manager collated all this feedback and produced a review document, reviewing how successfully each element of the design specification had been achieved in the design.

Although the trial was cut short, it did reveal a lot of useful information that could then be used to improve the next development of the checkout. Four main concerns were apparent:

- Apart from not understanding the checkout flow, customers were concerned that the products they were buying went out of reach when passing behind the packing bay. Although they hadn't actually paid for the products, they felt that they "owned" them, and were concerned about the apparent lack of control they felt when this happened.
- The panels on the side of the checkout, fitted like office furniture, did not reach the floor. This meant that items within the checkout, such as litter or footstools were sticking out under the checkout skirting. In addition, the footstool provided for CAs choosing to sit, was causing an unacceptable tripping hazard.

- Staff members using the checkout were complaining about working on the checkout, claiming they were suffering from back pain, etc. The physical ergonomics of the checkout had been checked so it was unlikely that any biomechanical risk was being realized. Rather, it was thought that the psychosocial factors of change and dealing with complaining customers were having a negative impact on the CAs, and they simply didn't want the hassle.
- The folding table was too time consuming to pull out and use, so it was unused throughout the trial.

But some aspects of the checkout were successful. The checkouts felt very spacious due to the orientation of the input belt, the extra storage, longer belts, improved leaflet displays, and modular manufacturing process were all considered to be successful.

20.14   BACK TO THE DRAWING BOARD

Following the limited success of the radical design of checkout, the project team met up again to review the design specification and decide on a new course of action. The design specification was revised following the learnings from the trial, and the earlier "traditional" design dusted off and assessed against this revised specification.

Progress on the traditional design of checkout was rapid. Much of what had been tested and learned in the development of the radical design was still valid for the revised design. New fitting trials were able to be conducted without delay, and the manufacturing knowledge acquired to design a modular checkout was still to hand. These common aspects of the design allowed the new "simple" checkout to be developed at pace.

20.15   MOCK-UPS

Due to the confidence that the project team had in the traditional design, and the successful elements of the radical design, the designers very quickly developed the design for the traditional checkout, maintaining the factors for good ergonomics (usability, safety, and performance) defined and refined in the earlier stages. This was quickly manufactured as a full-size timber mock-up, and taken to the various approval groups at Tesco to be approved (Figure 20.13).

The same CAs were used to review the checkout design, and generally the comments were much more positive. Thankfully, the concerns over customer flow were now removed.

FIGURE 20.13   MDF mock-up of the "simple" checkout design.

FIGURE 20.14   Design team considering the challenges of the "simple" checkout design.

It was easier to assess this design as the design manager was able to directly compare the unsuccessful aspects of the radical design with the new design, and so feel confident that the issues were resolved (Figure 20.14).

Within 3 months, another fully working prototype checkout was manufactured and installed in the Tesco prototype center. Once again, the checkout was tested with CAs and customers.

The results of the trials with customers and CAs were very positive. The customers were not concerned in any way about the operation of the new checkout, in some cases totally unaware of the changes in the checkout. But the CAs were very positive, able to understand and buy in to the benefits that had been designed into the checkout (Figure 20.15).

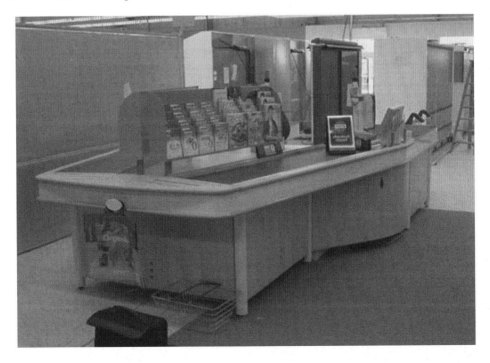

FIGURE 20.15   First working prototype of the "simple" checkout design.

20.16 A NEW BROOM?

It was at this point that Mike Wemms retired, and a new director, David Potts, took on the mantle of operations director and sponsor for the project. David was happy with the progress made, but had a valuable insight regarding training in stores. David's view was that training was not adequately delivered in stores, and as such any new equipment that was introduced should be "intuitively" designed. So, no training material was to be supplied with the new checkout design. This was a difficult design challenge, and not all elements of the checkout could be given absolute affordance. The design team addressed any shortfalls by using simple labels on the checkout.

So, if anything lacked immediate clarity about what it was and how it was to be used, it was labeled to identify what it was for. In this way, no training material would be required.

At this time, Tesco was undergoing a rebranding exercise in store, emphasizing the red, white, and blue of the Tesco logo. One aim of this exercise was to use these colors more in the store, in signage, fixtures, etc., so the checkout was styled by the design agency leading this work to be more in line with the new brand identity (Figure 20.16).

20.17 A NEW STORE TRIAL

A new trial store was identified and the decision was made to install a full bank of "simple" checkouts in the store. This was a new store, and to install the checkouts was a considerable risk. To ensure that this risk was minimized, a set of replacement checkouts was also manufactured and held in storage, just in case they needed to be replaced in a hurry.

As the store was new and not yet trading, the store team was able to test the new checkouts thoroughly before operating with customers. This meant that they were able to get to grips with them easily, and the design manager was able to be involved in the process of training participants for the trials, ensuring that the CAs and those who trained the CAs understood the features and benefits of the new checkout (Figure 20.17).

FIGURE 20.16 Manufacturing of the restyled (branded) prototype of the "simple" checkout.

FIGURE 20.17   "Simple" checkout installed for store trials.

The new store opened, and it was immediately clear that the checkouts were working well. Customers were able to navigate the checkouts effectively, and the CAs were able to use the checkout without difficulty.

Productivity research was undertaken to assess the speed of throughput at the checkout, and it was found that customers were being processed more quickly through the checkout. On average, it was taking customers 3 seconds less to get through the checkout, which equated to a potential saving of £1.7 million in checkout costs per annum across the business. These time savings were attributable directly to the benefit of having all the checkout consumables immediately available to hand.

Again, the project team were invited to visit the trial in store and review the checkout operation against their original design specification. A report was produced, again based on the design specification, and apart from a few minor modifications, the design was "signed off" by the project team.

FIGURE 20.18   A bank of "simple" checkouts in a "typical" Tesco store.

20.18 CONCLUSION: THE SUCCESSFUL DESIGN

It took 2 years to develop the Tesco "simple" checkout. The successful design met the vast majority of the requirements as specified in the design specification. It was quicker to use, cheaper, ergonomically superior, and contained more storage than the previous design. Customers were generally ambivalent to it, but appreciated the better service that the CAs were able to give them (often a sign of good ergonomics). The checkout enabled improved customer service, rather than preventing it, achieving its fundamental objective.

The key drivers of this success were Tesco's willingness to take a risk and trial something different; having a senior supporter for the project who was willing to be involved; a relentless attention to a design specification that gave due consideration to ergonomics and prevented the design from going off track; and a varied and committed project team who were enthusiastic about the project from beginning to end (Figure 20.18).

Finally, the benefit of 3 seconds meant that a business case could be pulled together to put these checkouts into stores. Just 3 seconds... but as the Tesco advertising campaign states "Every Little Helps"!

ACKNOWLEDGMENTS

The authors would like to express their thanks to all those involved in the project, including Tesco (design management team and project participation), Renfrew Associates (designers), David Hitchcock (ergonomist), Clares Retail Systems (manufacturers), and USDAW (union representation).

REFERENCES

Brown, S., Cross, N., and Walker, D. 1983. *Ergonomics in Design*. Design Processes and Products. The Open University. 118–30.

Mackay, C., Burton, K., Boocock, M., Tillotson, M., and Dickinson, C. 1998. *Musculoskeletal Disorders in Supermarket Cashiers*. Norwich: Her Majesty's Stationery Office.

Pheasant, S.T. 1986. *Bodyspace: Anthropometry, Ergonomics and Design*. London: Taylor & Francis.

Pugh, S. 1991. *Total Design – Integrated Methods for Successful Product Engineering*. Wokingham: Addison Wesley.

21 Design and Development of an Interactive Kiosk for Customers of Jobcentre Plus

David Hitchcock, James O'Malley, and Peter Rogerson

CONTENTS

21.1 THE COMMISSION

In 2002, Jobcentre Plus commissioned a customer internet access project to develop and trial a new technology solution aimed at enhancing the electronic delivery of services to working-age members of the public (Figure 21.1).

FIGURE 21.1   Close-up of customer internet access kiosk.

Building on the success of its award-winning touch-screen job search kiosks, the business wanted to explore the benefits of providing free access to a range of internet websites for working-age customers in the newly modernized Jobcentre Plus office environment. In this way, service users would be able to perform self-help tasks, such as searching for job vacancies from recruitment agencies, and looking up information about careers, training, transport, housing, and benefits. In addition, it was envisaged that the solution would also support the completion of web-based job applications and online claims forms.

The business benefits identified by the project were

- More efficient and effective support for Jobcentre Plus customers in assisting them to find information to help them move from benefit and into work.
- Directing customers who are able to help themselves to use this service would allow staff more time to concentrate on the harder-to-help customer groups.
- Increasing the number of job entries that could be put on the system would provide more opportunities for customers to find work.
- Enhanced customer perception of Jobcentre Plus and its services—the system would help to meet the increased expectations of its customers, including employers.
- Helping to meet e-government targets by providing wider access to Jobcentre Plus web-based information and services, and potentially allowing customers access to a wide range of e-government initiatives in other departments and local authorities (e.g., access to local housing benefit information and advice).

Of particular interest was the extent to which the service could accommodate the needs of disabled and disadvantaged customers. It is this requirement that provides the main focus of this case study.

21.2   BACKGROUND

As part of its response to the UK's twenty-first century modernizing government agenda, the then Employment Service invested heavily in computer technology to transform the way it delivered job

FIGURE 21.2   Vacancy card displays in an employment service job center.

broking, one of its core business activities, to service users. In 2001, the organization took what was widely regarded as the bold step of replacing job vacancy cards in its public offices with touch-screen job search kiosks, known as "jobpoints."

This network of single function kiosks enabled customers across the Employment Service estate, and some external partner locations, to access comprehensive details of all vacancies on its national jobs database. Prior to this change, job seekers were only able to see summary information on jobs in their local area, subject to limits on available display space (Figures 21.2 and 21.3).

Despite some obvious concern about customer acceptance of the new technology, the kiosks were a great success, as judged by customer perceptions and increased job search activity. For example, a post-implementation evaluation revealed (Figures 21.4 and 21.5):

- 81% of job seekers were satisfied with jobpoints
- 92% found jobpoints easy to use
- 48% of job seekers reported an increase in satisfaction with services

FIGURE 21.3   Jobpoints in-situ.

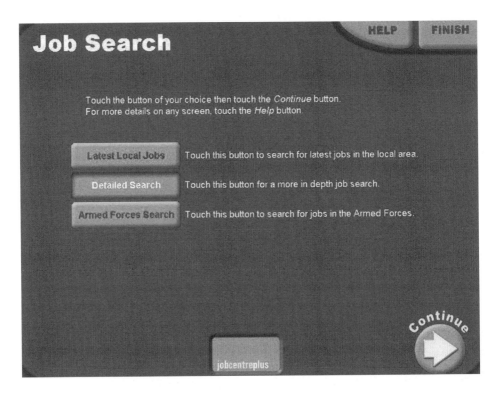

FIGURE 21.4   Jobpoint menu screen.

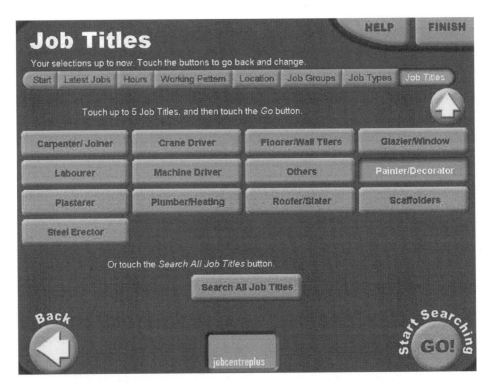

FIGURE 21.5   Jobpoint job title selection screen.

- Between 31% and 48% of job seekers reported an increase in
  - Number of vacancies looked at
  - Types of jobs looked at
  - Geographical areas searched
  - Number of job applications made
  - Knowledge of the types of jobs on offer
  - Confidence in finding work

21.3   INTRODUCTION OF CUSTOMER INTERNET ACCESS

With the formation of Jobcentre Plus (a merger of the Employment Service and Benefits Agency), the business decided that the new service should also offer free internet access for customer use. Initially, the newly refurbished public offices were equipped with one or more dedicated customer PCs, each providing access to a small number of relevant government websites.

Early evaluation findings were very disappointing and the service was quickly withdrawn. Specific problems included:

- Equipment misuse (e.g., theft of computer mouse, by-passing web filtering software)
- Performance issues (e.g., poor reliability, operating speeds)
- Limited functionality (e.g., small number of approved websites; absence of search engines and email facilities)
- Limited public awareness

Despite this early setback, senior business managers remained committed to the idea of providing a customer internet access service. From a strategic perspective, it became clear that Jobcentre Plus should do this via an interactive kiosk solution, and so a dedicated project team was assembled to manage all aspects of its design, development, deployment, and evaluation. The project team comprised employees of Jobcentre Plus and their IT partner, EDS.

21.4   WHAT IS AN INTERACTIVE KIOSK?

Kiosks have come a long way since the term was first used to describe exotic garden pavilions and temple structures found in thirteenth-century Middle Eastern and Indian societies. Over time, kiosks have become typically associated with any small, indoor or outdoor booth or stand where people can obtain relatively low cost, everyday products and services. Newspapers, lotto tickets, and sweets are fairly typical purchases, along with payments at car park booths and requests for information at on-street information points. Common to each of these examples is an element of human interaction: between purchaser and seller; requester and giver.

The type of products and services deliverable through kiosks has changed significantly with the growth of computer technologies; kiosks have become electronic self-service units, where transactions no longer depend on a human–human interaction. At their simplest, putting coins in a machine slot returns a stamped ticket or a fizzy drink. At the other end of the scale, a person can now manipulate, print, and pay for digital photographs in retail outlets without any intervention from a member of staff.

The growth of interactive kiosks is influenced by the needs of organizations and businesses to reduce operating costs, increase customer reach, and improve customer service. One way of achieving easier, speedier, and cost-effective access to services is through self-service kiosks. The extent to which self-service kiosks are accepted and exploited by consumers is no doubt influenced by the quality of the kiosk experiences.

21.5   KIOSK BASICS

Kiosks come in many different shapes and sizes, offer different features to their users, and can be located indoors and outdoors. Good kiosk design takes into account four key contextual factors:

- Nature of the user task(s)
- Characteristics of the user population
- Physical and social environment considerations
- Technology platforms

Whatever its purpose, the kiosk comprises both hardware and software components. Hardware covers not only the materials used to house the computer and associated internal workings, but also the physical input and output devices that mediate the interaction. Software relates to both the "hidden" computer operating system and the "visible" graphical interface that provides the application instructions, controls, and feedback.

Common input devices include:

Keyboards: The keyboard offers perhaps the most accommodating means of interaction with the kiosk, allowing free text to be entered and navigation to be controlled by cursor keys.

Non-keyboard input devices (NKIDs): NKIDs provide an easy and efficient means of moving the cursor around the screen. Trackballs and touch screen are the most typical devices, although other types exist, such as the mouse, touch pad, and joystick.

Touch screen: The touch screen offers a particular method of interaction to the keyboard and other NKIDs. The user simply has to touch the item of interest on the screen to activate it. Although not as flexible—realistically it can only offer limited interaction—the touch screen can be especially helpful for those unfamiliar with, unsure about, or unskilled using a keyboard.

Common output devices include:

Printers: The printer reproduces what the user sees on the screen and delivers it through a conveniently situated and safe printer slot. The quality of the output can vary greatly depending on the printer settings, paper type, etc.

Speakers: Aural outputs serve a number of important functions. Often complementing visual and tactile outputs, they can alert users to an event (e.g., return of a debit card), enhance the interaction (e.g., when playing games), and increase accessibility for users with a range of sensory, cognitive, and communication impairments.

Screen displays: Kiosk monitors display visual input and output signals that govern much of the interaction between the kiosk and users. Graphical interfaces can be "read" by assistive technologies, such as screen readers and magnification software.

The main features of the kiosk service piloted in a total of 40 Jobcentre Plus offices between 2005 and 2007 are described below. The "Jobcentre Plus Internet Service" was the official name for the customer internet access project in the public offices.

21.6   HARDWARE

The kiosk featured a 17" touch screen and keyboard-trackball input devices, fixed in a predominantly metal casing that housed the computer processor, printer, and printer consumables. The maintenance port for accessing the internal workings was at the front, underneath the keyboard shelf (Figures 21.6 through 21.8).

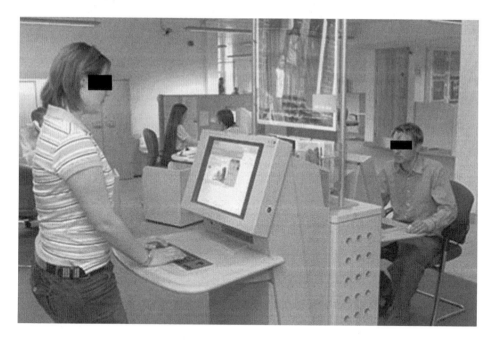

FIGURE 21.6   Standing and seated versions of the kiosk.

21.7   SOFTWARE

The software interface was menu driven and had a customized web browser. The principle function was to enable customers to identify and select external websites to meet a variety of work-focused and welfare-related information needs. Secondary functionality related to user help, location changes, and news feeds.

Users were able to search for information under six main categories (Figure 21.9):

- Directories and listings
- Education and training
- Government and local councils

FIGURE 21.7   Touch-activated monitor.

FIGURE 21.8 Integrated keyboard and trackball.

- Social and financial support
- Transport
- Work

Selections could be narrowed down at a second level before the web links were selected at the third and final level. On making their final selection, the user effectively left the Jobcentre Plus system until they either started a new search or ended their session (Figures 21.10 and 21.11).

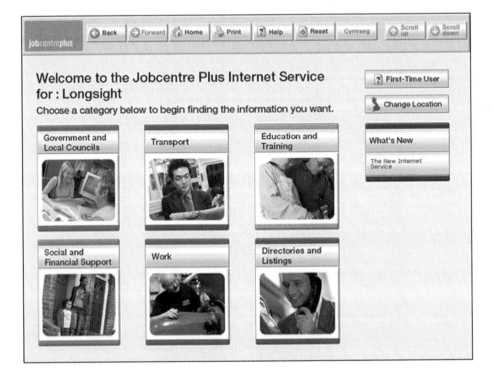

FIGURE 21.9 The Jobcentre Plus internet service home page.

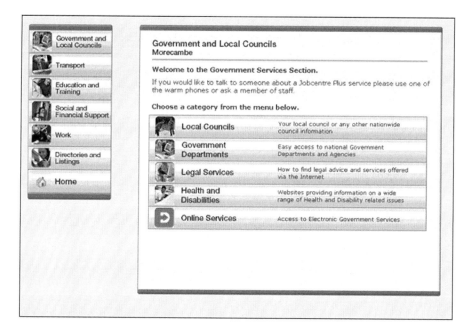

FIGURE 21.10   A second level menu for government and local councils.

21.8   REQUIREMENTS

The ergonomics aspects of kiosk design are known to influence take-up and usability (Maguire 1999), so these were initially reviewed as part of the selection process when comparing four different designs of kiosk. This was achieved using the "FFP score" technique of determining fitness for purpose (FFP); an iterative appraisal that considers stakeholder requirements and the physical ergonomics aspects.

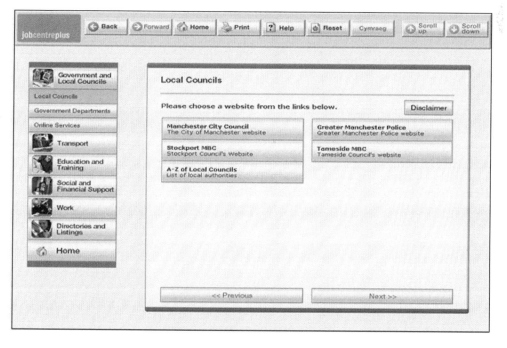

FIGURE 21.11   Web links at the third menu level.

For the purpose of this case study, a practical definition would imply that a design is fit for purpose if the kiosk:

1. Performs in the manner intended by the designer
2. Performs in the manner expected by the user
3. Is accessible and usable for all its target users
4. Is so designed that its size and shape facilitates good fit and reasonable comfort for the user
5. Presents no undue risk to the user
6. Presents no undue risk to other people who interact with it, including members of staff, members of the public, children, installers, and maintainers
7. Allows for reasonably foreseeable misuse (e.g., it is not intended to be used as a "climbing frame" but could be by an unattended child, thus it should not topple over under such circumstances)
8. Continues to operate satisfactorily for items 1–7 throughout its intended life cycle

21.9   FITNESS FOR PURPOSE SCORE: MEASURING FITNESS FOR PURPOSE

FFP can be assessed in a descriptive, piecemeal way, but it can be difficult to determine the impact of improvement measures. Inclusive design—and the requirements of FFP—demands consideration of a host of factors. It is not uncommon in such multi-factorial situations that an improvement in one aspect can have a detrimental effect in another. For example, providing headphones on a lead to support screen reader output for the visually impaired may increase the risk of strangulation of children or other vulnerable members of the public. In an attempt to resolve this difficulty and to manage and monitor the FFP criteria through all stages of the design cycle, the authors used the FFP score technique, first developed for the selection of call center furniture by Hitchcock and O'Connell (1999).

The FFP score is based on input from all those involved in a project, from the originator to the end user. It draws on information and data that is available through the literature and project-specific trials to assess issues, such as dimensions and ranges of adjustment, usability of controls and displays, maintenance provision, and suitability for use.

The FFP score has five stages; the first three establish the criteria against which the equipment will be assessed, the fourth is the appraisal itself, and the fifth stage is intended to enable monitoring of the design as it evolves and is implemented and used in-service.

Stage 1: Creating a stakeholders' wish list

It is generally accepted that if a kiosk design is to succeed, the stakeholders should be involved throughout the design or selection cycle. The first stage of the FFP score process is, therefore, to ask representatives of all those involved in the project (designers, IT, accessibility experts, customers, human factors specialists, etc.) to provide a "wish list" of all the features they would like to see in the kiosk. Each representative is also asked to use the following scale to rate how important they consider each criterion:

5: Essential (non-negotiable, must be included)
4: Very important
3: Important
2: Slightly important
1: Not important (but would like to be included anyway)

The mode (most frequently occurring) rating is identified for each criterion and a single prioritized wish list representing the consensus view is presented back to the stakeholder representatives for

TABLE 21.1
FFP Score Stage 1 Output—Stakeholders' Wish List

Design Criteria	Rating	Importance
Ease of maintenance	5	Essential
Safe	5	
View the full screen	5	
Wheelchair access	5	
Be obvious as to what it is	4	Very important
Comfortable use of input devices	4	
Ease of access to printout	4	
Fit in with Jobcentre Plus requirements	4	
Intuitive to use	4	
Provide feedback of actions	4	
Quality of output	4	
Robust	4	
Use comfortably sitting or standing	4	
Facilitate privacy in use	3	Important
Future proofing	3	
Heat control	3	
Left-handed or right-handed operation	3	
Noise control	3	
Multiple user viewing	3	
Positionability	3	
Provide space for reading material	3	
Provide space to write	3	
Screen size	3	
Sound output	3	
Telephone facility	2	Slightly important
Touch-screen facility	2	
Personal storage	1	Not important
Voice input	1	
Webcam facility	1	

comment and review. It is most convenient and beneficial to conduct this stage in a group setting as it allows for interactive discussion, which is particularly helpful when reviewing the consensus wish list to confirm priorities in the case of "surprises" and to decide whether additional items should be added to the list. Table 21.1 shows the prioritized consensus stakeholder wish list produced as part of the development of the kiosk.

Stage 2: Developing a human factors specification

By its very nature, the consensus wish list is somewhat basic and in need of development if it is to be of value to the project in considering and monitoring individual design factors. Therefore, the second stage of the FFP score is for the human factors personnel with their specialist knowledge of usability and accessibility to categorize and develop the list into more specific criteria. As an example, drawing from the prioritized wish list shown in Table 21.1, "comfortable use of input devices" can be broken down to include criteria such as

- Keyboard should have a QWERTY layout for familiarity
- Keyboard should be designed according to ergonomics principles
- NKIDs (e.g., trackball) should be designed according to ergonomics principles

- Touch screen should enable ease and accuracy of use
- There should be a speech activation facility

Progressive redescription can continue until each design aspect is clearly specified to the level required. For example, the ergonomics principles for each aspect of keyboard design could be listed separately (key spacing, key depression force, material finish, etc.). It is, however, usually sufficient to only break down the wish list to the first level. Furthermore, a very large and unwieldy set of criteria is likely to result from further redescription.

Redescription of the stakeholders' wish list resulted in the 46 factors shown in Table 21.2. To decide on the relative importance of each of these specific criteria, the human factors expert applies the same rating scale as before. For example, as the QWERTY keyboard was considered to be an essential part of the requirement for "comfortable use of input device," it received a rating of 5.

The product of the stakeholder wish list rating and the human factors breakdown list rating provides an overall weighting of importance on a scale of 1–25; 1 (1 × 1) being not important, 25 (5 × 5) being essential.

The conclusion of this second stage is for the weighted breakdown list to be verified by the stakeholder representatives. When the list is finally agreed, it forms the human factors specification (HFS) for the kiosk. From this point, all design and selection decisions should derive from or be compared to this HFS.

21.10 APPRAISALS

The four kiosk designs under consideration were appraised for each of the applicable 46 factors and were rated according to the following scale:

7: Very good—equivalent to "ergonomics best practice"
6: Good
5: Slightly good
4: Minimum criteria satisfied
3: Slightly poor
2: Poor
1: Very poor—equivalent to "fails all ergonomics criteria"

The outcomes of the appraisals are shown in Table 21.3. The advantage of using the FFP score was that it not only allowed small differences to be identified between design features, but it also pinpointed particular shortfalls and their relative importance.

At the conclusion of the appraisals, the project team decided that designs 2(a) and 2(b) should proceed to trials. The designs of these kiosks were fundamentally the same, but version (a) was intended for use by a standing operator and version (b) for a seated operator.

A number of changes to these kiosks were discussed for potential improvement. These discussions were developed further to consider the findings of fitting trials to validate some of the physical design features and the findings of a consumer risk assessment. Furthermore, the project team ultimately decided to remove some of the initially proposed design factors; considering them to be beyond the remit and resources of the project (e.g., speech activation facility, telephone facility).

21.11 WHEELCHAIR AND SITTING EVALUATION

In order to validate the dimensions of the designs, reference to anthropometry data were supplemented by fitting trials, which were conducted according to the recognized principles described by Pheasant (1986). The findings of this work indicated a number of potential reach and clearance

TABLE 21.2
Example FFP Score Stage 2 Output—Human Factors Specification (HFS)

Category	Weighting	Description
Input devices	25	Keyboard has QWERTY layout
	25	Design of keyboard
	25	Design of input device
	25	Provision of screen reader
	25	Touch-screen facility and quality
	25	Speech activation facility
	25	Smart card reader facility
Safety	25	Free from sharp edges
	25	Free from protrusions
	25	Free from exposed cables
Screen	25	Size of the screen
	10	Viewing arc for two users
Distinction	16	Aesthetic appeal
	16	Signage facility on kiosk
	8	Differentiation from jobpoints
Intuition	20	Affordance of input devices
	20	Affordance of output devices
Feedback	20	Tactile feedback of input
	16	Provision of Braille labels and locators
	16	Aural feedback
	12	Feedback through button lights
Maintenance	12	Position of maintenance ports
	12	Access clearance of maintenance ports
	12	Security of maintenance ports
	12	Cleaning considerations
Posture	20	Height of the screen
	20	Viewing distance of the screen
	20	Angle of the screen
	20	Angle of the keyboard
	20	Height of input devices
	20	Reach to input devices
	20	Position and retention of printout
	20	Forward wheelchair access
	20	Height of leg clearance
	20	Depth of knee clearance
	20	Depth of feet clearance
Documents	12	Reading/writing space and location
	12	Facility on kiosk for writing tools
	6	Waste receptacles
Sound output	15	Quality of sound output
	15	Headphone facility and quality
	9	User volume control
Telephone	10	Telephone facility
Personal storage	3	Position of storage space
	3	Access to storage space
	3	Security of storage space

TABLE 21.3
Summary of the FFP Score Appraisal

Kiosk	Overall FFP Score (%)	Factors Below Minimum Requirement (%)	Major Factors Below Minimum Requirement (%)
Design 1	56	52	24
Design 2(a) (to be used standing)	59	46	21
Design 2(b) (to be used sitting)	66	39	17
Design 3	61	48	18

problems for wheelchair users, so it was decided to conduct specific trials to assess these issues for the "seated" version of the kiosk.

Earlier work revealed a number of pertinent issues requiring further investigation:

- The thigh clearance beneath the kiosk would be sufficient for normal seated use, but would probably be unsuitable for users in forward-facing wheelchairs.
- Space for comfortable positioning of feet would be sufficient for most people (although larger people might find it tight), but would probably not be adequate for those in forward-facing wheelchairs.
- Reach distance to activate the touch screen appeared to be excessive and would more than likely be problematic, if not impossible, for forward-facing wheelchair users.
- Like the touch screen, the keyboard distance seemed somewhat excessive and might benefit wheelchair users from also being slightly raised.

Because the requirements of the investigation were specifically in relation to reach and clearance issues, the decision was taken to conduct a series of user trials ensuring that the range of user sizes was represented. Project constraints made it unrealistic to recruit actual wheelchair users representing such a range of sizes. However, as it was possible to easily simulate typical wheelchair use, a group of subjects of suitable sizes was recruited to take part in the trials. These subjects received clear instructions on how to maneuver, position, and use the wheelchair, and as a secondary test, a powered scooter. To provide further information about the suitability of the "seated" kiosk, each subject also conducted the trial using a representative office seat and from a standing position.

The trial method was straightforward. Each subject rated out of 10 the convenience of use to activate nine different aspects of the kiosk, selected to represent "worst case" and "common use"—seven points on the touch screen and two on the keyboard (Figure 21.12).

This activity was conducted for each of the six different positions (in different orders for each subject):

- Wheelchair face-on
- Wheelchair side-on
- Scooter face-on
- Scooter side-on
- Office seat face-on
- Standing face-on

In the case of the wheelchair and scooter, the subjects performed the trial with both "freedom of movement" (i.e., could twist, bend, and reach) and with "restricted movement."

Both the touch screen and the keyboard appeared to be in suitable positions at the "seated" kiosk for those using the seat or opting to stand. The same was true for wheelchair and scooter users, but

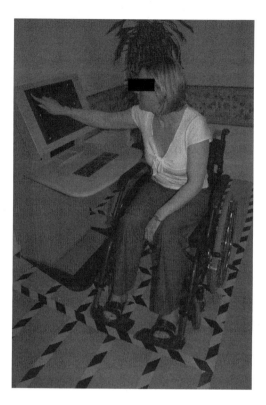

FIGURE 21.12   Wheelchair accessibility.

only if they approached the kiosk sideways-on (a common and acceptable practice). Limited horizontal clearance beneath the kiosk meant that wheelchair users could not get close enough to use the touch screen when sitting face-on. It may be obvious that scooter design exacerbates the reach difficulty, but Figure 21.13 illustrates the extent of this. To further emphasize the problem, the subject pictured has a greater than average female extended arm length (59th percentile).

21.12   CONSUMER RISK ASSESSMENT

There are typically two main user groups of interactive kiosks:

1. Members of the public, including people with disabilities and children. Although children are not the primary users of the kiosk, their interaction with the equipment must be considered when appraising any risks that may be presented. Children are relatively poor at recognizing hazards and assessing risks. If risks are reduced or removed for children, then there should be a parallel reduction in risk for the adult users.
2. Staff who may assist members of the public using the kiosk, and/or clean and maintain the equipment.

The kiosk should not expose either of these groups to risks to their health, safety, and well-being. Suitable and sufficient risk assessments need to be conducted to identify and address any hazards.

Since the introduction of the Health and Safety at Work etc. Act 1974, employers have been required to manage "so far as reasonably practicable" risks arising from their undertakings. The concept of identifying hazards and implementing control measures to manage risks, rather than implementing prescribed legislative criteria, has been a specific provision of health and safety legislation since the early nineties.

FIGURE 21.13 Emphasized scooter accessibility difficulties.

Jobcentre Plus has its own risk assessors to address the health and safety of staff. However, the assessment of risks to consumers was considered by the authors, including:

- Mechanical hazards including entrapment, entanglement, choking/suffocation, edges, projections, and corners, moving parts, hazardous heights, and stability and footholds.
- Chemical hazards resulting from contact with, ingestion or inhalation of materials and substances that may have a harmful effect.
- Fire, thermal, and electrical hazards including flammability and hot surfaces.

Among a small number of minor risks, the principal risk identified (albeit of a "low risk" rating) was that of the printout port, which allowed potential, deliberate or accidental access to hazardous electrical and mechanical hazards inside the kiosk. Simply reducing the front outlet size of the paper slot would impede paper flow when in use, therefore, the designers opted to install an angled (restriction) plate to reduce overall aperture size and restrict movement up behind the paper slot toward the paper cutter mechanism. This plate reduced the front aperture height from ~45 to ~35 mm. As the plate bends behind the front aperture, the accessible gap was reduced further to ~27 mm. According to the anthropometry source Childata (Norris 1995), reducing the aperture height to ~35 mm is not especially effective in and of itself—the 5th percentile (small) wrist depth of 17½–19 year olds is given as only 34 mm. However, the reduced point of 27 mm created by the restriction plate is more effective in that (Figure 21.14):

a. Access requires a fair degree of hand/arm articulation.
b. The size should not now allow access to 5th percentiles over the age of 6½ years.
c. The plate edge is rounded and of sufficient gauge as to not present a sharp edge.

21.13 SOFTWARE DEVELOPMENT

The project team assembled a multi-disciplinary user interface design group to work in parallel with the physical ergonomics design group. Membership included a range of human factors, IT,

FIGURE 21.14   Paper cutter risk—before and after control measure.

and business domain experts, including graphic designers, information architects; field managers; marketing officers, diversity and equality advisers; accessibility specialists; and human–computer interaction (HCI) practitioners.

In keeping with the project team's human factors strategy, all software design activity was required to adhere closely to the established principles of user-centered design (ISO 13407) and inclusive/accessible design guidelines (e.g., NDA AccessIT).

The group were tasked with producing software to deliver the following functional system requirements:

- Access to web links relevant to the needs of customers in defined local office catchment areas
- An animated attractor sequence
- Online help
- News feeds

Non-functional requirements for the system indicated that it should be

- Easy-to-use
- Accessible
- Usable
- Secure

The process began with the graphic designers being asked to produce a sample of screen designs to be used to stimulate discussion about the desired "look and feel" (appearance and navigation) for the user interface.

While none of the initial designs captured the imagination of the group, they certainly helped crystallize the group's desire for a graphical interface that would encourage touch-screen interaction, but would still support keyboard-trackball operations for those preferring to work in that way. Not only would the majority of target users be more familiar and comfortable with the touch screens, it was also felt to be the most efficient way of navigation through a short-sequence menu hierarchy.

Drawing inspiration from educational and games websites with their creative use of graphics, color, button shapes and sizes, etc., the second iteration screen designs were very much closer to what the team envisaged for the interface.

As can be seen in Figures 21.15 through 21.18, the shift from tab metaphor to big, visible touchable buttons simplified the menu-driven navigation. The graphical images added non-verbal meaning to the information category labels, reduced the amount of unnecessary (and often unwanted) onscreen text, and increased the overall visual appeal of the page.

Once the group was satisfied with the basic "look and feel" of the proposed design, it was subjected to the first of a number of validation exercises involving prospective system users. Typically,

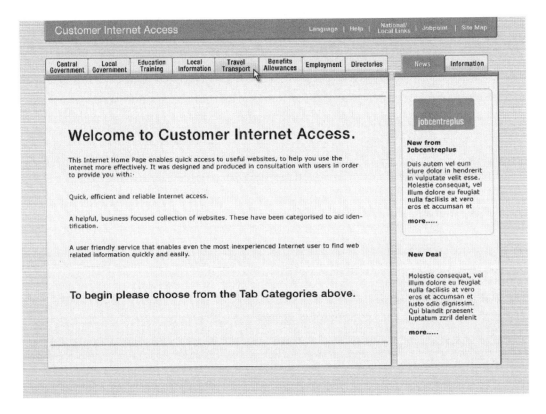

FIGURE 21.15 Example "home page" for one of the earliest designs.

customers attending Jobcentre Plus offices were approached to see if they would be willing to take part in a 20–30 minute informal feedback session, to be held in a quieter part of the office or in a private room.

Volunteers performed scenario-based tasks using example screen shots—in either paper or electronic formats—and were then asked for their opinions and suggestions. Observational and self-report data collected during these sessions were used to refine the system navigation, button labeling, system feedback messages, aesthetic qualities, and so on.

The recruitment of volunteers was informed by knowledge of Jobcentre Plus' major customer demographics. It wasn't always that easy to achieve a representative user sample and so, at times, feedback was often provided by whoever was kind enough to give their time.

Jobcentre Plus advisers had been using browser technology to access a web links database from their PCs for a couple of years before the decision was taken to extend the facility to its customers. As one of the business benefits of the customer internet service was to save adviser time by encouraging greater customer self-help, it made sense to see how this proven information repository could be adapted for customers to use themselves.

Information categories used by staff were organized under the following headings:

- Directories
- Employment
- Media/publicity
- Public sector
- Search engines
- Support services
- Trade unions

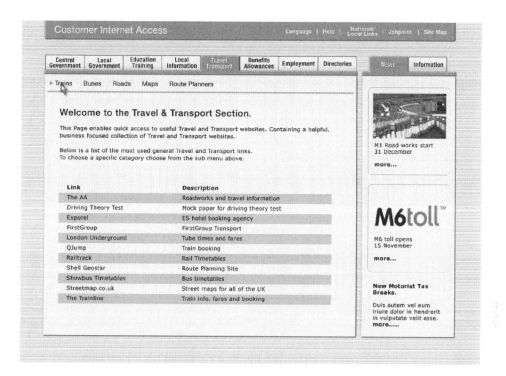

FIGURE 21.16   Example "lower level" page for the same design.

- Training/education
- Transport/travel

An analysis of the web links showed some were relevant nationwide, whereas others would be of interest in specific locations only. Therefore, the interface design had to be able to present national and local links relevant to the location of each participating office.

A revised set of information categories were agreed (originally eight, but later reduced to six) and, in keeping with the commitment to a user-centered design process, the nominated project co-ordinator in each of the kiosk locations was asked to run focus groups in their office to enable customers to

- Validate the main and sub-category information structures
- Generate lists of additional web links to include in the database
- Suggest information needs that the project could translate into suitable websites

Information content was managed through a separate piece of software held by the project team. The content management system served two main functions: firstly, it allowed information categories to be modified and new web links added on request; and secondly, it enabled the vetting of websites to check they were likely to be relevant to customers' information needs, business requirements and would not cause undue upset, embarrassment, or adverse publicity.

A commercial web classification and blocking tool was used for vetting purposes. Some categories were easy to identify and block, such as adult material, games, and gambling. Others were easy to allow (e.g., education, government, travel, and job search). In a few cases, a whole website category could be blocked when the reason for wanting to access a particular site within that category might be entirely reasonable. For example, the shopping category was blocked but estate agent sites (a sub-section) could be appropriate for someone planning to relocate areas to take up a new job. As a result, the task of approving and blocking websites could become quite time consuming.

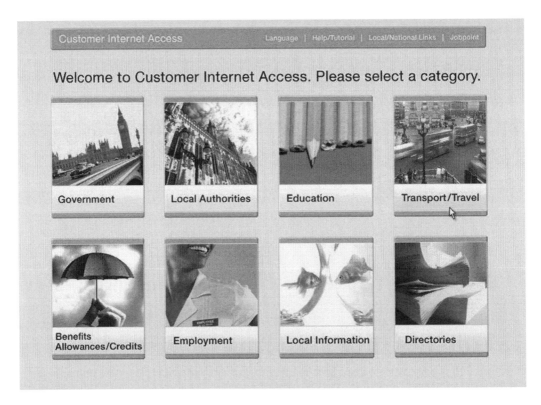

FIGURE 21.17   Second iteration of the "home page" design.

By the end of the pilot exercise, a total of 1200 links could be accessed from the network of 40 participating offices. This included "duplicate" national links that featured across many or all kiosk locations.

Three versions of user help were produced for the kiosks, although only two were actually implemented during the trial period.

The basic design comprised a "new user" help button on the welcome page and a help button accessible from the browser bar on all other screen pages. Selecting help took the user to a menu of help information covering topics such as the purpose of the kiosk, descriptions of the key functions, how to navigate, and use of the controls. Feedback during the early phase of the pilot indicated that not many users were accessing the help information and among those that did, most were put off by the largely text-based content. User reaction was more positive when this was replaced with more graphic-rich content (Figures 21.19 and 21.20).

Building further on this feedback, some exploratory work was done to create a multi-media tutorial based around a human-form avatar completing a typical information search task. Visual, aural, and text output explained the purpose of the kiosk and showed how each of the input and output devices worked. The design could not be added during the live trial because of technical compatibility issues, but it would be seriously considered for any future development to address the needs of people who are unwilling or unable to engage with more static text and images (Figure 21.21).

The idea for the news feeds was to be able to identify messages and stories that could be brought to customers' attention in a prominent way. Some might be of national relevance (e.g., the introduction of a new welfare benefit), while others could be of more local interest (e.g., the opening of a new store or a job placement success story).

As can be seen from Figures 21.9 and 21.17, the original eight information categories were reduced to six in an attempt to accommodate the news feed facility. Unfortunately, late in the design

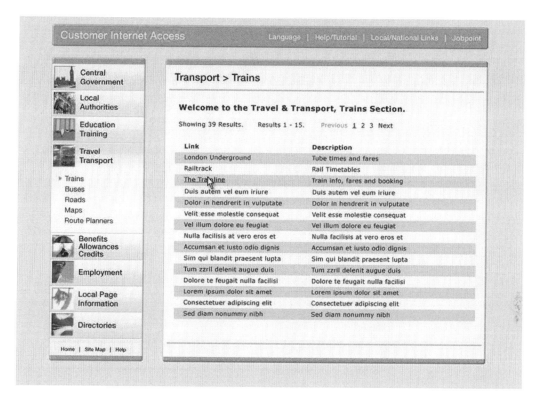

FIGURE 21.18   Second iteration "lower level" page design.

phase it was discovered that the news feed facility could not be activated for technical reasons, but it was too late to remove it from the interface.

Given the relatively simple functional requirements and an experienced design team, most decisions were made quite easily and confidently. Involving users provided important validation of the emergent design and highlighted modifications to be made to improve ease of use. Almost inevitably, technical and business constraints influenced some aspects of the final design, and the evaluation served to consider the impact of such compromises. Examples included:

- Having been advised that it wouldn't be feasible from a technical perspective to include a conventional vertical scroll bar in a touch-screen activated internet page display, the "agreed" solution was to provide scroll-up and scroll-down buttons in the browser bar and monitor user reactions to this arrangement during the trial.
- The marketing and communication teams responsible for corporate branding informed the group it needed to replace the graphics selected to represent the six information categories with ones considered more representative of a people-focused business.
- Requests from users to increase the utility of the kiosk service by including the facility to enter URLs, access search engines, and send emails were rejected on the grounds that it could compromise the security of the system. Of all the constraints imposed on the system design, this probably had the most significant effect on the customer experience.

21.14   ACCESSIBILITY TRIALS

Accessibility refers to the relative ease or difficulty with which a person can make use of buildings, transport, consumer products, and, in this case, a kiosk-delivered government service. A number

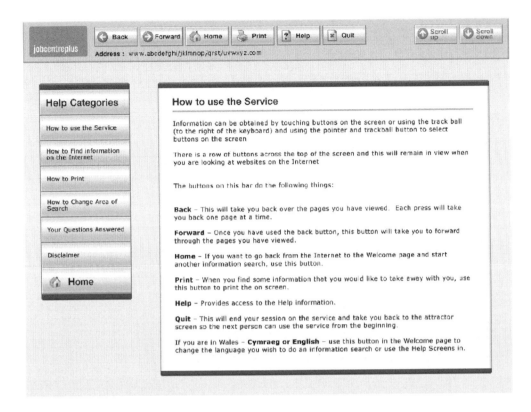

FIGURE 21.19   An original help page.

of design decisions were taken by the internet kiosk project team in recognition of the different capabilities of a large, diverse user population, many of whom experience a disability or social disadvantage that makes them a priority customer group for the organization.

With a view to ensuring that Jobcentre Plus met its obligations under the Disability Discrimination Act (1995), trials were arranged to investigate the extent to which users with physical, sensory, cognitive, and intellectual impairments could interact successfully with onscreen information and be able to reach and activate input and output devices. One of the key outputs of the study was the highlighting of any shortfalls requiring attention through design changes, adaptation, or human support.

Implicit within the trials' design, and in accordance with ISO 9241-11, was the requirement that an accessible design should also be usable by all in the target group(s)—supporting the effective and efficient completion of tasks, with the required degree of satisfaction to encourage future use.

The accessibility trials focused on four major disability categories: partial sightedness; dyslexia; impaired motor control; and general learning disability. The needs of a fifth group, wheelchair users, had already been addressed as part of the ergonomic hardware design.

The rationale for choosing these four types was that they represent disability groups likely to visit Jobcentre Plus offices in significant numbers who might reasonably expect to be able to use the kiosks independently. Moreover, some characteristics associated with these impairments can also be relevant to other disabilities.

An impairment specifically excluded from the study was total blindness. This was for two main reasons: firstly, the condition affects a comparatively small customer group whose specific needs have led to adviser interviews and telephony services becoming recognized as their primary service channels; and, secondly, while the internet is a technology well suited to and exploited by a great

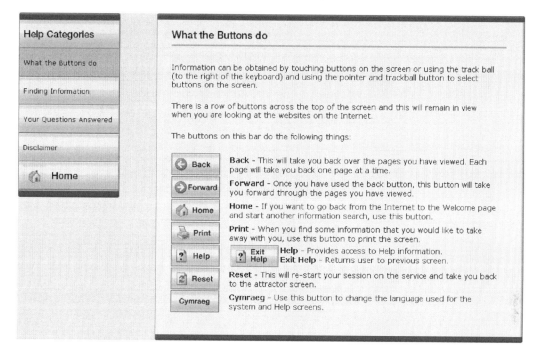

FIGURE 21.20   The re-worked help page.

many blind people, it was hard to identify one suite of assistive technology products (screen readers, magnifiers, etc.) that would be familiar and acceptable to anything more than a minority of this group.

Opportunities to assess a fully working prototype were limited during the development phase, so it was agreed that the trials would take place in the live environment. This had the advantage of providing the most realistic testing without any risk that the findings would be too late to be incorporated in a nationally implemented kiosk solution.

With the help of Jobcentre Plus' disability employment advisers (DEAs) and several local disability organizations, 23 participants were recruited for the trials. Our aim for four to six subjects per group (a number beyond which it was thought there would likely be diminishing returns in terms of providing new information) was achieved for all but the dyslexic group. A small payment was made to compensate participants for their time and travel expenses.

The relatively informal 45-minute user sessions involved participants carrying out self-directed and prompted information searches, including at least one job search. Observational and user feedback data were collected to shed light on issues such as ease of use, confusion and error, and perceived satisfaction.

Some findings were very much in line with the expectations of the evaluators, yet there were sufficient new insights provided to make it a very worthwhile and necessary exercise.

In terms of general findings, it is worth noting that people with the same primary disability may present different degrees of symptoms. Contrary to our original thinking, this will have a bearing on how many people are recruited to future trials. Also, many user difficulties highlighted usability issues that are independent of specific disabilities or disadvantage.

None of the participants had used the internet access kiosk prior to the trials, even though many had used the touch-screen jobpoints in the office. Irrespective of any personal difficulties with use of the kiosks, almost all had enjoyed taking part and thought the kiosks were a good idea.

Key findings for each of the four disability groups are summarized below.

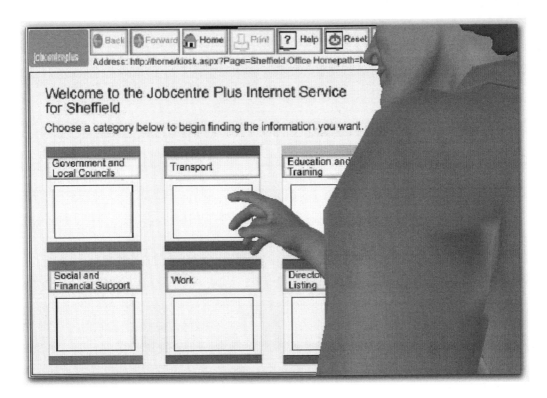

FIGURE 21.21 Human avatar demonstrating an information search.

21.14.1 SIGHT IMPAIRED

The sight-impaired group, including conditions such as nystagmus, macular degeneration, and retina pigmentosis, found the kiosk least accessible. As a minimum, they would require modifications such as

- A facility to customize software settings (color schemes, screen illumination, font sizes, timeouts, cursor size/speeds, etc.)
- A modified keyboard (key size, lettering, etc.)
- More flexible keyboard controls
- Reduced glare and reflection (matt surfaces, kiosk located away from windows, etc.)
- Improved printout quality

Some believed screen readers and magnifiers would make it easier for them to access the kiosk, although they also recognized the challenge of identifying a universal solution from among the many products on the market.

21.14.2 DYSLEXIA

Dyslexic users, though fewer in number, appeared to experience many of the customization issues mentioned by the sight-impaired users. In addition, this group felt that the organization of screen display elements was a particularly important issue in aiding information processing, as was the facility for spell checkers to minimize the effect of spelling difficulties when entering text into onscreen fields.

21.14.3 MOTOR CONTROL

Users in this group reported wide-ranging conditions, including arthritis, dyspraxia, and multiple sclerosis. The severity of the condition can vary greatly, and participants reported that there was less likelihood of them wanting to make use of the kiosks on one of their "bad days." Most used standard PC peripherals at home and they had few difficulties with the kiosk input devices. Where difficulty was noted, it tended to be that the keyboard keys felt a little on the small size.

21.14.4 LEARNING DIFFICULTIES

Users typically reported special education backgrounds and many, though by no means all, had some degree of literacy and numeracy deficits. The menu navigation presented few difficulties and, to a large extent, the errors they made were no different to those made by users in the other groups. That said, their general approach to tasks gave the impression that they would benefit from additional human support to interact successfully with some external websites. A paradox of simplifying and modifying aspects of the user interface (browser bar, URL field, etc.) for this group was that the interface looked and behaved differently from the one they tended to use at home or elsewhere, e.g., one user kept trying to enter a web address in the read-only URL field.

21.15 USER TRIALS

User trials, conducted in Jobcentre Plus offices, focused on posture and consumer safety during a series of early site visits, where 55 users were observed interacting with the kiosks and consulted about their opinions.

In much the same way that interface design must be evaluated to determine the software usability, so must the hardware be evaluated to assess its usability. It is unlikely that user exposure at a kiosk will ever be sufficient to cause musculoskeletal disorders, such as backache, "ULD" (upper limb disorders), and the like. However, compromised posture may lead to discomfort, dissatisfaction, inaccessibility, restricted use by those with minor impairments, and ultimately under-use (or even non-use).

From the observations, there appeared to be little difference between the use of the "standing" and "sitting" kiosks. In the user consultation exercise, customers were asked which kiosks they had used, and the responses were fairly even; 34% said "standing," 39% "seated," and 26% "both." Most users spent 15 minutes or less at the kiosk.

The postures adopted and the movements made while using the kiosks were observed by members of the human factors team. The benchmark for a comfortable, healthy posture was assumed to be the "neutral" one, in which least loading is placed on the musculoskeletal system. The observations, therefore, focused on key areas of conflict with the neutral posture—head position, elbow height, and wrist height. Back and lower limb positions were also included for the "seated" kiosk. The results of these observations are shown in Table 21.4.

TABLE 21.4
Poor Upper Limb Postures and Movements During Kiosk Use

	"Standing" Kiosk (%)	"Seated" Kiosk (%)
Head too high	25 ($n=9$)	0 ($n=0$)
Head too low	0 ($n=0$)	0 ($n=0$)
Elbow too high for touch screen	0 ($n=0$)	0 ($n=0$)
Elbow too low for touch screen	0 ($n=0$)	16 ($n=3$)
Wrist too high for keyboard	53 ($n=19$)	0 ($n=0$)
Wrist too low for keyboard	0 ($n=0$)	0 ($n=0$)

Other noteworthy observations regarding posture were

- Many of the users of the "standing" kiosk used the keyboard shelf to lean against—this was an appropriate height for nine out of ten users.
- Eight out of ten users were able to adopt healthy lower limb postures when using the "seated" kiosk.

Six percent of the users reported actual discomfort when using the kiosks, although almost 30% said that they would like greater comfort. Interestingly, one in three of these users suggested that comfort issues would affect their use of the kiosks, particularly for long periods. Monitoring of the hazards identified in the consumer risk assessment is a necessary part of the safety management process.

The original consumer risk assessment revealed a number of risks to be monitored. As already stated, the main risk was identified as being an excessively wide paper slot that could enable access to internal components. This was suitably corrected in advance of the trial and assessed again; proving to be satisfactory.

The other risks identified related to: sharp edges, protrusions, exposed cables, entrapment hazards, entanglement hazards, choking hazards, hazardous heights, stability, and chemical hazards. These were considered to present a very low risk to users, but were nevertheless monitored throughout the user trials by the project team and Jobcentre Plus staff at the offices taking part. No accidents were reported.

21.16 EVALUATION STRATEGY

Evaluation formed a significant part of the Jobcentre Plus Internet Service pilot. In addition to the accessibility and user trial evaluations already reported, the evaluation of the kiosk service focused on five broad project themes and used a variety of data collection methods.

Data were collected throughout the life of the pilot, with the experiences of customers and staff collected mainly during a 12-week period, a few weeks after "go-live" in the 11 offices that made up the first phase of the pilot.

Initially, a period of time was allocated at the end of phase one for the changes suggested by the evaluation to be made to the kiosk hardware and software prior to the start of the second phase of the pilot, when another 29 offices were to be added. In reality, only very minor changes could be accommodated in the time available, which meant that the evaluation of the phase two offices concentrated more on live running issues, such as system performance, monitoring usage levels, etc.

The evaluation topics and data collection tools are described below.

21.16.1 CUSTOMER PERCEPTIONS AND BEHAVIOR

- Customer comment slips placed by the kiosks and at reception captured ad-hoc user feedback, including suggested web links.
- Floorwalkers and reception staff kept weekly logs to monitor customer usage, requests for assistance, and other noteworthy behaviors.
- Observational and interview data were collected by a team of independent, external evaluators over a 2-day period at each of the participating offices. Semi-structured interview schedules covered topics such as awareness, usability, accessibility, utility, internet behaviors, and demographic information.

21.16.2 Business Preparation and Implementation

- Opinions about the works order, communications, guidance, and marketing products were collected from the relevant business teams as part of the project team's ongoing stakeholder engagement plan. Review workshops were organized for representatives of the participating offices every 3 or 4 months.

21.16.3 Staff Perceptions and Behavior

- All office teams completed a weekly or monthly team log to capture any perceived impacts on their job roles, customer relationships, and performance outcomes.
- The external evaluators carried out one-to-one interviews with staff in job roles that brought them into closer daily contact with the kiosks and kiosk users—floorwalkers, onsite IT support, and office managers. Topics covered included kiosk maintenance, queue management, customer support, quality of help desk services, and siting of kiosks.

21.16.4 Technical Issues

- Technical issues were logged and managed on an ongoing basis in line with standard project management arrangements. Issues were typically identified from an analysis of help desk calls (software issues, equipment failure, etc.), system performance logs (speed, availability, etc.), and web-blocking statistics.

21.16.5 Wider Contextual Issues

- The degree of aesthetic and environmental fit with Jobcentre Plus office branding was assessed during site visits made by the department's design consultants.
- The potential impact of other community-based internet provisions on the demand for a Jobcentre Plus kiosk service was explored in staff and customer interviews and supplemented with research carried out by the evaluation team during site visits.

21.17 RE-LAUNCH OF THE SERVICE

In an attempt to increase lower than anticipated kiosk activity levels, the service was re-launched mid-way through the pilot. A workshop was arranged for all participating offices, at which a re-launch strategy was agreed. The exact form of the re-launch was left to each participating office, drawing on a range of options generated at the workshop.

Exit surveys were the agreed method for obtaining comparison measures of awareness and usage before and after the re-launch. A representative sample of Jobcentre Plus customers ($n \sim 380$) in six offices were approached and invited to answer a short set of mainly closed questions as they were leaving the office. The six offices were selected to reflect different office characteristics, such as urban/rural location, size, and customer demographics. The surveys took place in September 2005 and in February 2006.

21.18 CONCLUSIONS AND THE FUTURE

The Jobcentre Plus Internet Service pilot ended in March 2007. In addition to managing the design, implementation, and a 2-year live-running of a credible kiosk system, the project team delivered a comprehensive evaluation that will inform the development of future kiosk solutions as part of the organization's developing e-channels strategy. In this regard, it is probably fair to say that the project was a great success. It is acknowledged, however, that the customer internet service had relatively low uptake among the target user population, and there were few perceived benefits for employees in the participating offices. Therefore, in terms of informing Jobcentre Plus' future thinking and plans, the following evaluation messages were particularly pertinent.

- **Increasing the utility of the kiosk service should help address the low take-up among the target user group.** The kiosk service must be something people want or have to use. Providing access to information from the internet was not enough for most customers, even if it was free. It is likely that improved usage figures would have been achieved if the kiosk software had offered internet functionality to enable customers to access search engines, send emails, and complete online job applications and claims forms. Then again, the increasing number of people with internet access at home suggests that the kiosk service will need to host services that can't really be delivered in any other way, or be justifiable in terms of smaller numbers of people expected to use it.

- **Staff engagement requires a good fit with business processes and recognition of the personal/business benefits it will bring.** Despite the best efforts of the champions in some office locations, on the whole neither of these requirements was clearly established during the pilot. As a result, it was hard to find evidence of real ownership and promotion within the participating offices.

- **Judgments about the usability, accessibility, and business value of the kiosk service would benefit from establishing clear, measurable acceptance criteria during the design phase.** With regard to accessibility and usability, the general conclusion was that most users found the service easy to access, learn how to use, and were able to accomplish basic information-seeking tasks effectively and efficiently. That said, no more than 6 in 10 rated the kiosk as good or very good (exit survey, February 2006) and a sizeable minority of customers experienced access and/or usability issues that deterred them from either using it at all, or having a second go.

  Similarly, the findings in the same survey that 4% of customers had used the kiosks more than once, and a further 17% only once, appears very disappointing, but could be viewed more positively depending on what is known about the user and non-user profiles, the use made of the information obtained, the availability of this information elsewhere, and the time and monetary costs involved for the business. One in six customers using the service seems quite a low figure, but if the users were mainly people without access to the internet elsewhere, and if the information they found helped them get a job, a college place, or the correct welfare benefits, then it would be more likely to be considered a worthwhile business investment.

  Defining success in terms of agreed and measurable acceptance criteria would help the business to be clear about what it needs and provide a useful reference point for designers during the design phase. It would also help ensure greater objectivity when making sense of the evaluation data.

- **Incorporating accessibility tools into the software design would help overcome many of the access issues experienced by users with sight impairments, specific learning and communication disabilities.** It would also lessen the complications of having to meet the different needs of a diverse user population in a single interface design. In hindsight, far less attention was paid to ensuring software accessibility than was given to physical accessibility, and this is reflected in the type of accessibility issues reported by users. The UK Disability Discrimination Act requires all new service provision to complete a Disability Equality Impact Assessment. The need for customization tools and other assistive technologies, such as screen readers and voice input/output, should be identified as part of this process.

- **Good quality user help should be provided to assist service users who lack experience and confidence using internet and kiosk technologies.** As the intention was to promote self-service among Jobcentre Plus customers, the kiosks needed to provide this support largely through online help and leaflets. Investing in multi-media help resources and testing the usability of the help documentation more fully in the design phase should pay dividends.
- **The need for human support will be influenced by the nature of the kiosk service.** Notwithstanding the provision of good quality online help, there will always be some people who will require direct assistance to get the most from this type of technology. Greater user support was one of the two most frequent suggestions for improving the service (better advertising/promotion was the other), although some of this could be addressed by improved physical and software design. The need for user support may be more evident where service users have no option other than to use the kiosks.
- **Depending on the nature of the tasks, kiosks may need to support standing and seated postures.** An adjustable kiosk design may be a more flexible and economical option in the future. It would benefit people with a range of physical needs that may require them to sit, stand, or alternate between the two postures. It would also suit the wider user community who might be happy to stand for relatively brief tasks, like a job search, but may prefer to be seated if completing a lengthy online form.
- **The locating of kiosks is an important consideration**. Users report a need for privacy when working on tasks of a personal nature, and so locating kiosks in quieter parts of the office where there is less passing traffic appears to be the most obvious solution in such circumstances.

Despite the transformation of the fabric of Jobcentre Plus offices, a number of customers still feel uncomfortable in this environment and prefer to spend as little time on the premises as possible. As this is bound to have an adverse effect on the appeal of discretionary kiosk services, it would be worth exploring the scope for installing them in more acceptable public spaces. Internet usage in public locations is on the increase, so there may be some merit in providing easy access to e-government services in this type of environment.

REFERENCES

British Standards Institution. 1998. BS EN ISO 9241-11:1998 Ergonomic Requirements for Office Work with Visual Display Terminals (VDTs) – Part 11 Guidance on Usability. Milton Keynes: BSI.

———.1999. BS EN ISA 13407:1999. Human-centred Design Processes for Interactive Systems. Milton Keynes: BSI.

Dix, A., Finlay, J., Abowd, G.D., and Beale, R. 2004. *Human-Computer Interaction,* 3rd ed. Harlow: Pearson Education.

Hitchcock, D.R. 1999. *The Importance of the Ergonomic Environment*. Call Centre Solutions Conference Proceedings, 212–38. Advanstar Communications, 26-28 April 1999, RAI International Exhibition & Congress Centre, Amsterdam, The Netherlands, Chester, UK.

Lynch, P.J., and Horton, S. 2002. *Web Style Guide: Basic Design Principles for Creating Web Sites*. New Haven, CT: Yale University Press.

Maguire, M. 1999. A review of user-interface design guidelines for public information kiosk systems. *International Journal of Human-Computer Studies* 50 (3): 263–86.

National Disability Authority. n.d. Public access terminals. Dublin: National Disability Authority. http://accessit.nda.ie/it-accessibility-guidelines/public-access-terminals (accessed August 2, 2007).

Neilson, J. 2000. *Designing Web Usability: The Practice of Simplicity*. Indianapolis, IN: New Riders.

Norris, B. 1995. *Childata: The Handbook of Child Measurements and Capabilities: Data for Design Safety*. London: Department of Trade and Industry.

Pearrow, M. 2000. *Web Site Usability*. Rockland, MA: Charles River Media.

Pheasant, S.T. 1986. *Bodyspace: Anthropometry, Ergonomics and Design.* London: Taylor & Francis.

Preece, J., Rogers, Y., and Sharp, H. 2002. *Interaction Design: Beyond Human–Computer Interaction.* New York: John Wiley.

Spool, J.M., Scanlon, T., Schroeder, W., Snyder, C., and DeAngelo, T. 1999. *Web Site Usability: A Designer's Guide.* San Francisco: Morgan Kauffman.

Disability Discrimination Act. 2005. (c.13). London: HMSO.

22 Participatory Method for Evaluating Office Chairs: A Case Study

Lia Buarque de Macedo Guimarães

CONTENTS

22.1 INTRODUCTION

The basic functions of a product were identified as long ago as in antiquity, when the roman architect and engineer Vitruvio (1955) proposed that a product be perceived, at least, according to three functions: the practical (*Utilitas*), the aesthetic (*Venustas*), and the technical (*Firmitas*). Subsequently, Löbach (1981) proposed the three functions generally accepted in design: the practical, the aesthetic, and the symbolic, which was later redefined by Bürdek (1994) into two: the practical and that of language (which incorporates the aesthetic and the symbolic). More recently, Guimarães (2006) suggested the addition of another function: the ecological.

Office chairs are abundant in terms of quality and price, but it is not always the case that the highest price ensures the highest quality and vice versa. Studies focusing on the comfort of seats are frequently based on the practical function, which encompasses technical issues (including aspects such as strength and durability) and issues of use, especially related to ergonomics (to which elements such as practicality, stability, and comfort converge). There has been great progress in this area, especially with the development of technologies that allow greater precision in making seats at a more affordable cost (Brienza et al. 1996; Gyi, Porter, and Robertson 1998).

For Iida (1990), the best criterion from the ergonomic point of view for the evaluation of workstations and, therefore, of work seats, is the biomechanical one, which evaluates the posture and physical effort required of workers. For this, there is a need to determine the main points of concentration of tensions that tend to cause discomfort, aches and pains and/or pathologies associated with the work. But the biomechanical proposal for comfort transcends the question of distributing pressure and also encompasses the perception of stability (Hänel, Dartman, and Shishoo 1997) and temperature. Besides these objective questions, it is important to remember that comfort can also have subjective factors embedded in it that vary from individual to individual, and, moreover, depend on the context of use. Grandjean (1973) evaluated comfort, both in a subjective way (through questionnaires) and physiologically (by recording body movements and the pressures generated in both the seat and the chair back). It was concluded that direct questioning is more reliable than physiological assessments, corroborating the studies by Rieck (1969, *apud* Grandjean, 1973) on car seats, and those of Schakel, Chidsey, and Shipley (1969a, *apud* Grandjean 1973) on seats for general use, which did not find any relationship between perceived comfort and movement of the body while the seat was being used. Grieco et al. (1997) proposed an approach to make an ergonomic evaluation of work chairs, in line with criteria that were both objective and subjective: safety, adaptability, comfort, and practicality. Corlett (1995a) describes a method proposed by Schakel, Chidsey, and Shipley (1969b) that involved: (a) ranking the chairs for preference while sitting on them, but not being allowed to see or touch them; (b) long-term sitting while working, with regular completion of a general comfort rating and body part discomfort ranking; and (c) completion of a chair feature check list at the end of the session. Similar to the other studies, this proposed procedure does not stress important subjective criteria, such as the symbolic and aesthetic, since people could not see or touch the chairs.

According to Vergara and Page (2001), studies focusing on the comfort of seats are frequently conducted without considering the interaction between subjective and objective factors. However, comfort, practicality, and durability are not always objective parameters, and this makes selection difficult. An interesting case that occurred in a call center company studied by the Center for Design, Ergonomics, and Safety of the Laboratory for Products and Processes Optimization of the Graduate Program in Industrial Engineering, of the Federal University of Rio Grande do Sul (NDES/LOPP/PPGEP/UFRGS) was the repugnance for new office chairs bought from the same manufacturer, and of the same model as the old one, generated in workers who preferred the old ones. The difference was that the old chairs were upholstered with natural fabric (wool) and the new ones with synthetic fabric. Without realizing the difference between the fabrics, the users said they preferred the old ones because they were easier to adjust, more practical, in short, better. But it was exactly the same chair, with the same mechanism, but the perception of thermal discomfort invalidated the other qualities.

Within the scope of the projects that the NDES runs in partnership with several companies, one of the specific needs has been support for the purchase of work seats. In the petroleum distributor, the refinery, and the agency of the judiciary authority, there was a need to purchase a large number of office chairs and they requested a method for identifying what chairs to acquire. However, the specification of appropriate furniture is complicated by the diversity of needs as a result of the characteristics of the work being performed and the preferences of the user. Given the frequent business need to be able to identify the parameters that will define the requirements expected of office furniture, especially seats, and that such parameters are not yet to be found in the literature, the NDES team put forward the suggestion of drawing up an in-depth study, involving users, to formulate a method that would indicate a seat type or model that takes into account not only anthropometric and biomechanical aspects (which are covered by the rules and recommendations of the literature), but also symbolic, aesthetic, and comfort needs, as perceived by users. The idea is to set the ergonomic parameters while taking into account the criteria laid down in the literature and map users' perception regarding the comfort and discomfort/pain they experienced during the working day, so that seats that would match their needs might be specified. To do so, work was undertaken:

- To evaluate the spontaneous preference for/rejection of seats and the perception of the criteria for quality in work seats.
- To conduct experiments to evaluate the preference/rejection tested based on comparing the models of office chairs when used in the workplace.

This chapter describes how the research study was developed, introduces the participatory method proposed for evaluating seats, gives the main results, and draws conclusions.

22.2 OFFICE CHAIRS UNDER STUDY

Twenty-three different office chairs were evaluated throughout a working day by thirty-four volunteers in the scope of the study of Silva (2003) in three different companies: (1) twelve office chairs were evaluated in the petroleum distributor; (2) another twelve (including six from the petroleum distributor) were tested in the oil refinery (in four departments: HR/training, three chairs; logistics, six chairs; civil engineering and process engineering, twelve chairs), both located in the city of Canoas, Rio Grande do Sul; and (3) a public agency of the judiciary authority in the city of Porto Alegre, RS (five chairs).

The twelve chairs used in the study in the petroleum distributor were divided into two groups of six chairs, heterogeneously within groups and homogeneously outside them. In other words, chair "X1" was similar to chair "Y1" and different from the chairs of their group ("X2," "X3," "X4," "X5," and "X6").

Five chairs were evaluated in the agency of the judicial authority, among which was the current work chair ("A"), this model being the same for the five women volunteers. The fifth chair was included in the trials at the request of an employee of the judicial office. It should be noted that chairs "B" and "D" were the same as "Y3" and "X5" from the case study in the distributor. Three chairs were evaluated in the HR/training department of the refinery, chairs "Y2" and "Y3" being the same as "Y2" and "Y3" of the petroleum distributor.

Given the type of activity of the operators in the logistics department of the refinery, three high-backed chairs and three low-backed chairs were trialed, all of them with arm supports, castors, and a mechanism for adjusting the height of the seat. They differed as to whether they did or did not have mechanisms for adjusting the height of the arm, backrest and tilt of the backrest. Six chairs were used in the experiment in the logistics department of the refinery, with chairs "A1," "B2," and "B3" being the same as "X4," "Y4," and "Y3" in the case study in the petroleum distributor and chair "A2" was the same as "C" in the study in the agency of the judiciary authority.

22.3 PARTICIPATORY METHOD USED IN ALL STUDIES AND PROPOSED FOR EVALUATING SEATS

The proposed method (which was called "Dance of the Chairs") fits into the participatory focus of macroergonomic design (Fogliatto and Guimarães 1999; Guimarães and Fogliatto 2000), which incorporates qualitative and quantitative statistical techniques in order to identify the ergonomic demand for products, which enables assessment of the employees' preference for the seats before and after the trials, assessment of satisfaction with the evaluation criteria, and assessment of the sensation of comfort and discomfort/pain caused by the chairs. The seats were assessed in the three companies in line with the following steps:

1. Direct observations and indirect observations (interviews, film-tapes, etc.).
2. Applying the questionnaire on the demand for seats.
3. Spontaneous preference: open interview to identify the spontaneous preference for/rejection of the seats before the experiment.
4. User's perception as to the importance of the attributes (safety, adaptability, comfort, practicality, and aesthetics) in a work seat.
5. Feedback of the questionnaires and discussion of results.
6. Introduction of the experiment and pre-test of questionnaire on discomfort/pain.
7. Questionnaire to evaluate the sensation of discomfort/pain at the end of a week, before the experiment.
8. Preference tested: Experimental use of the seats to compare the models in a real work situation. Basically, the steps of the experiment are
   8.1. A questionnaire to evaluate the difference in the feeling of discomfort/pain between the start and end of the working day
   8.2. A questionnaire to evaluate satisfaction with the seat being trialed
   8.3. A questionnaire to evaluate the degree of importance of the six criteria for evaluating seats
   8.4. Evaluating the preference for seats based on the degree of satisfaction, taking account of the weights of importance of the criteria
   8.5. Evaluating the preference for/rejection of the seats after the experiment by applying the indirect pair-wise comparison test

22.3.1 METHOD OF DEALING WITH DATA

The preference for different seats was established by evaluating the interaction of the four attributes identified (comfort, practicality, safety, and adaptability) that, according to the literature reviewed, shape the idea of comfortable seats, as do aesthetic considerations, which were also taken into account. These were discussed with the research study participants and, based on their understanding of each criterion, the seats were then subjectively evaluated according to their perception of adaptability, general quality, and comfort. Despite following the nomenclature of the literature, the criteria for the qualitative judgment of comfort were in accordance with the user's understanding, and mapping the perception of comfort was undertaken based on the user's assessment of discomfort/pain during the working day, with office chairs. The following sections detail the method.

22.3.1.1 Questionnaires Referring to Sensation of Discomfort/Pain, Demand for the Chair, and Degree of Importance of Evaluation Criteria for Chairs

The questionnaires concerning the sensation of discomfort/pain, seating wants, and the degree of importance of the evaluation criteria of chairs have one feature in common: each volunteer answered only once, i.e., there was no repetition. The statistical analysis used, therefore, was the non-parametric Multiple Comparison test of Averages of Kruskal–Wallis (Daniel 1978). When there was a significant difference between the averages, the Kruksal–Wallis test of Multiple Comparison of Averages was used to verify which items differed significantly from each other. The level of significance set was 5%.

22.3.1.2 Questionnaire to Evaluate the Difference in Sensation of Discomfort/Pain between the Start and End of the Working Day

In order to process the data, a calculation was made of the difference between the results that each volunteer gave to the sensation of discomfort/pain between the start and end of the working day in the places that present discomfort/pain among the twenty-nine points of the regions of the body map, adapted from Corlett (1995b). After obtaining the differences of the sensations of "discomfort/pain" in the parts of the human body that resulted in pain/discomfort for each of the seats trialed, an analysis of variance (ANOVA) was conducted with the following controllable factors in order to determine their effects on the sensation of discomfort/pain: "seat," "volunteer," "day," and "site of the sensation feeling of discomfort/pain," as well as the interactions between the factors of "seat" versus "location of the sensation of discomfort/pain" and "volunteer" versus "site of the sensation of discomfort/pain." The variable response of this analysis was the difference between the feeling of discomfort/pain at the start and end of the working day.

For the factors of "seat" and/or "site of the sensation of discomfort/pain," whenever there were significant differences, to the significance level of 5%, the Duncan test of Multiple Comparison of Averages was conducted. This test was chosen because, as per Montgomery (1991), the value used for the comparison considers the distance between the averages, so the error rate (the value used to check significant differences) changes. It is a test, in principle, which requires balancing and it is approximate, because the averages ranked are not independent ones (because they are the same chairs and the same people).

22.3.1.3 Questionnaire for Evaluating Satisfaction with Seats Being Trialed

With the results of satisfaction/dissatisfaction with the seats trialed in relation to the six evaluation criteria, ANOVA was used with the following controllable factors: "chair," "volunteer," and "day." An analysis was undertaken for each of the following response variables: "comfort," "security," "adaptability," "practicality," "suitability for the job," and "appearance or aesthetics." When there was a significant difference for one of these response variables with regard to the chairs trialed, the multiple comparison of averages test based on Duncan's test at the level of significance of 5% was conducted.

22.3.1.4 Weighting of the Criteria of Satisfaction

The results of the questionnaires on satisfaction/dissatisfaction with the chairs trialed with regard to the six evaluation criteria were weighted, based on the averages of importance attributed to each of the criteria. This was done so that the importance that the volunteers attributed to the criteria might be considered and not only to their satisfaction with them. After obtaining the weighted results of satisfaction/dissatisfaction with the seats trialed with regard to the six evaluation criteria, ANOVA was carried out on the variable factors of "seat," "volunteer," and "day," so as to obtain a general result of satisfaction/dissatisfaction with the chairs trialed, measured against the weighted criteria.

It is emphasized that, in this chapter, regarding the statistics, only the most important results are presented, namely, the relationship between the seat evaluated, satisfaction, and the perception of discomfort pain without detailing the results for the variables of "volunteer" and "day of week," which can be found in Silva (2003).

22.3.2   DIRECT AND INDIRECT OBSERVATIONS TO EVALUATE THE WORK DONE IN THE COMPANIES

During the initial phase of the survey in the companies, direct, systematic, and asystematic observations were made, as were indirect ones, the objective of which was for the investigator to reach an understanding on the needs and requirements of the production and human systems in operation. The researchers' direct observations enabled basic knowledge on the work performed in the departments studied to be gathered, which was complemented by open interviews conducted with the volunteers. The indirect observations consisted of filming for the record and for later (mainly cinesiologic) analysis of postures assumed. The factors considered were the environment, the workstation, and the organization of work.

22.3.3   APPLYING THE QUESTIONNAIRE ON WHAT IS WANTED FROM A WORK SEAT

In the initial phase of the evaluation method for seats, even during the initial phase of the general survey or ergonomic assessment in the company, a specific questionnaire was applied to identify the users' wants with respect to the objective attributes of their work seat. This questionnaire contains items that, when evaluated by the user, are very important in establishing ergonomic design parameters. These items are: seat cushioning; swivel seat; mechanisms for adjusting seat height and tilt, backrest height and tilt; castors, arm support; and footrest.

The questionnaire is structured to identify the degree of importance attributed, by each user, to the items presented. The questions are presented so that users can mark their level of importance on a continuous 15 cm scale, as proposed by Stone et al. (1974), with two anchors at the ends: *of little importance* (or 0) and *very important* (or 15). To minimize the effect of concentrating responses near the anchors, no marks whatsoever were made on the line. The analysis of the results is direct, based on the arithmetic mean of the individual values.

In the petroleum distributor and process engineering department, the Kruskal–Wallis test showed no significant difference between the importance of the items wanted ($\chi^2=8.500$; $p=0.386$ and $\chi^2=11.053$; $p=0.272$, respectively), all of them being averages of importance above the average on the scale from 0 to 15. For HR/training staff, descriptive statistics led to the impression that "adjusting the tilt of the seat" was considered the least important item, while the item "castors" was considered the most important. However, based on the Kruskal–Wallis test, there was no significant difference between the importance of these wanted items ($\chi^2=9.993$; $p=0.266$). This may arise from the small number of women volunteers who participated in the experiment and justifies the use of a non-parametric statistical technique.

In the agency of the judiciary authority, the wants regarding chair items showed a significant difference between each other ($\chi^2=17.278$; $p=0.045$), there having been evidence in the multiple comparison for averages that the items considered most important in the chair were "a mechanism for adjusting seat height," "castors," "a mechanism for adjusting the height of the back," "swivel seat," "cushioning," and "arm support." The item considered least important was the "foot support."

In the logistics department of the refinery, the Kruskal–Wallis test also showed that there was a significant difference ($\chi^2=20.231$; $p=0.017$) between the importance attributed to the items, the most important being "a mechanism for adjusting seat height," "swivel seat," "cushioning," and "castors," while the items "foot support," "mechanism for adjusting height of the backrest," and "adjusting tilt of the backrest" were considered the least important. In the civil engineering department of the refinery, there were also significant differences between the averages of importance of the items ($\chi^2=18.741$, $p=0.027$), whereas by the test of multiple comparison for averages, the items "castors" and "cushioning" were considered very important and differed significantly from the item "mechanism for adjusting the tilt of the seat," which was not considered important.

In general, it can be considered that, despite the people and the jobs they do differing, adjustments and mobility are important for most employees, who tend not to value the footrest.

22.3.4 OPEN INTERVIEW TO IDENTIFY THE SPONTANEOUS PREFERENCE FOR/ REJECTION OF THE SEATS (BEFORE THE EXPERIMENT)

The first specific step of the comparative study of seats was to conduct an open interview to identify the employees' immediate preference for/rejection of the seating that they were shown. It was an assessment of spontaneous preference/rejection.

Prior to the open interviews, a brief explanation was given to the employees about the objectives and data collection techniques, emphasizing how important it was for them to express their preferences and what the concepts were for developing the research study. At this point the various types of seats were introduced, and they were left on exhibit in a designated place while the interviews were being held, so that they could be observed and even handled. In order to minimize the effect of pre-established views, the seats were arranged in a row and "mixed," that is, by alternating the various models that were to be analyzed. They were laid out as if they were in a "showroom" of seats and the users had come to select among them for purchase. The chairs were made available by the manufacturers and came in different colors and textures, which was good for the research study, because the differences helped with regard to the impact of their aesthetic-symbolic functions.

After being shown the seats, each employee, individually, from each company, was invited to a room adjoining the room in which the chairs were being exhibited, where they answered the following questions:

1. What seat do you prefer? Why?
2. What seat do you least prefer? Why?
3. What do you look for in a work seat?

Each interview took on average 10 minutes, and the responses as to the preference for seats are recorded manually in a field diary.

Figures 22.1 through 22.7 present the results of the spontaneous preference/rejection of the office chairs. Basically, the items that the chairs chosen had in common were: castors and a mechanism for adjusting seat height and support for the arms with a mechanism to adjust height. Items that the chairs rejected had in common were: a high, rounded backrest, with no mechanism for adjusting tilt and a rounded seat.

Volunteer	1	2	3	4	5	6
Weight (Kg)	78	51	69	80	98	75
Height (m)	1.60	1.50	1.64	1.80	1.67	1.67
Spontaneous preference		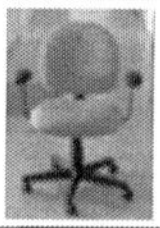	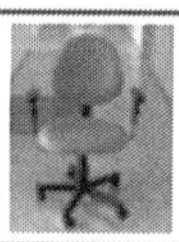	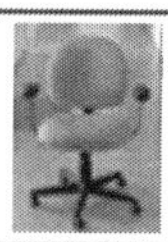		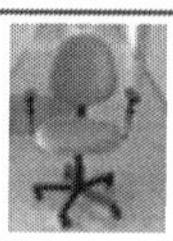
	X1	X2	X3	X2	X1	X3
Spontaneous rejection		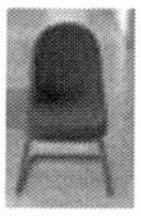			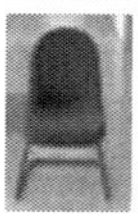	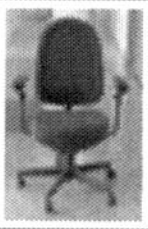
	X6	X6	X6	X4	X6	X4

FIGURE 22.1 Spontaneous preference/rejection of the six chairs of group X displayed in the petroleum distributor.

Volunteer	1	2	3	4	5	6
Weight (Kg)	58.6	63.5	79	96	73	82
Height (m)	1.70	1.68	1.67	1.80	1.68	1.68
Spontaneous preference		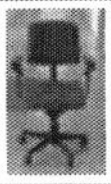		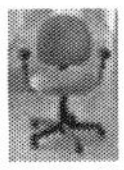		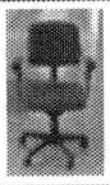
	Y1	Y4	Y1	Y3	Y1	Y4
Spontaneous rejection				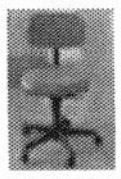	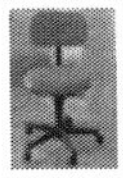	
	Y6	Y6	Y4	Y5	Y5	Y6

FIGURE 22.2   Spontaneous preference/rejection of the six chairs of group Y displayed in the petroleum distibutor.

22.3.5   OPEN INTERVIEW TO OBTAIN USER'S PERCEPTION AS TO THE CRITERIA FOR ASSESSING WORK SEATS

This interview had two objectives: (i) to identify the importance given by staff to the criteria that would be used to evaluate the work seat trials; and (ii) to make the employees verbalize the concepts that each of them had regarding the criteria for evaluating seats suggested by the literature, namely, comfort, safety, practicality, and adaptability (Grieco et al. 1997), to which was added the criterion of "aesthetics."

4. What is a comfortable work seat?
5. What is a practical work seat?
6. What is a safe work seat?
7. What is an adaptable work seat?
8. What are the aesthetics of a work seat?

Volunteer	1	2	3	4	5
Weight (Kg)	55	66	70	80	52
Height (m)	1.60	1.59	1.66	1.70	1.58
Spontaneous preference					
	B	B	B	B	E
Spontaneous rejection				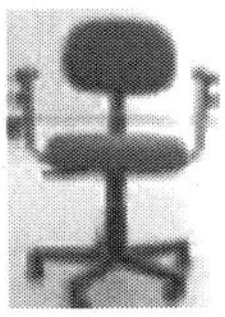	
	E	E	E	A (in use)	C

FIGURE 22.3   Spontaneous preference/rejection by the women volunteers from the agency of the judiciary authority of the five chairs displayed.

Volunteer	1	2	3
Weight (Kg)	50	63	93
Height (m)	1.53	1.73	1.62
Spontaneous preference	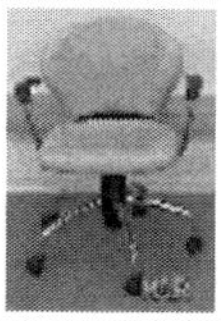		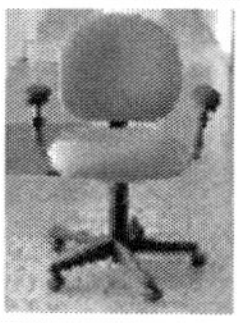
	Z1	Y3	Y3
Spontaneous rejection	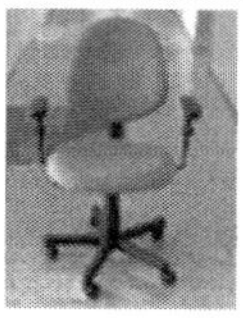	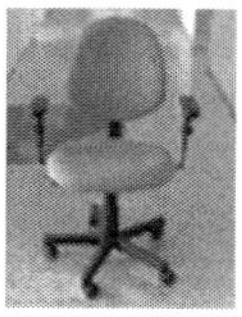	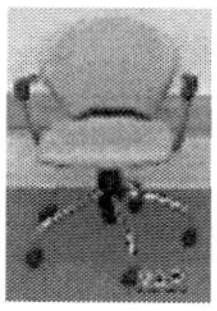
	Y2	Y2	Z1

FIGURE 22.4 Spontaneous preference/rejection of the three chairs displayed in the HR/training department of the refinery.

9. What is an uncomfortable work chair?
10. Do you prefer small or large chairs?

The objective of Questions 4 to 8 was to obtain a definition, from the users, as to the criteria that would be used, *a posteriori*, to evaluate each of the seats. Thus, the evaluation was anchored on the user's definition rather than on the definition pre-determined by experts.

Question 9 was included because, according to Helander and Zhang (1997), Stracker (1999), and Zhang, Helander, and Drury (1996), the concepts of comfort and discomfort are not opposites of one and the same continuum. What was not asked was: "what is discomfort in an office chair?" soon after the question "what is a comfortable working chair?" so that the interviewee was not induced to respond that an uncomfortable chair is one that is not comfortable. Through Question 10 of the interview, the intention was to check if the "small" volunteers, i.e., those of lesser height and lower weight, preferred smaller chairs, while the "large" volunteers, those of greater height and weight, preferred larger chairs, as suggested by Helander et al. (1987).

Volunteer	1	2	3	4	5	6
Weight (Kg)	101	75	83	84	66	85
Height (m)	1.75	1.69	1.81	1.80	1.67	1.73
Spontaneous preference	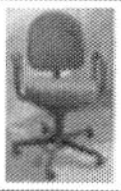				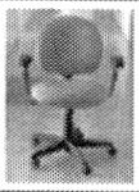	
	B1	A1	B1	A3	B3	A3
Spontaneous rejection			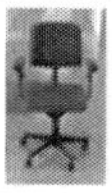			
	B2	B2	B2	-	B2	A2

FIGURE 22.5 Spontaneous preference/rejection of the six chairs displayed in the logistics department of the refinery.

Volunteer	1	2	3	4
Weight (Kg)	72	85	77	85
Height (m)	1.58	1.72	1.78	1.82
Spontaneous preference			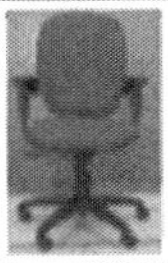	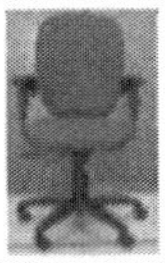
	D	B	C	C
Spontaneous rejection			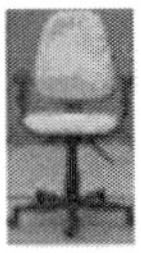	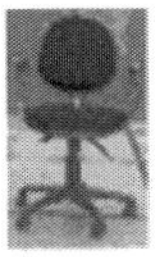
	F	A (in use)	L	J

FIGURE 22.6 Spontaneous preference/rejection of the 12 chairs displayed, by the volunteers from the civil engineering department of the refinery.

It should be noted that, generally, colloquial language was opted for, despite its inaccuracy (e.g., "do you prefer less"), in order to facilitate communication with the employees. The answers were recorded in the field diary or taped and later transcribed as narrated with the possibility of intervention by the researcher so as to facilitate understanding of the situation and/or add aspects observed that were relevant to the study. The data were then tabulated for analysis and to allow comparison, *a posteriori*, with the preferences in relation to the seats trialed.

22.3.6   Feedback to Users on the Definitions of the Criteria

The definition of each criterion, as given by the employees, was then discussed with the participating groups before starting the next stage of trialing the seats. The purpose of the discussion was to ensure that the subjects had a full understanding of each criterion to be evaluated. It should be emphasized

Volunteer	5	6	7	8
Weight (Kg)	65	56	79	76
Height (m)	1.73	1.65	1.77	1.83
Spontaneous preference	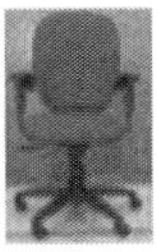	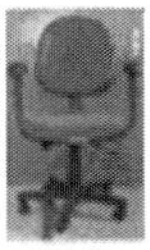	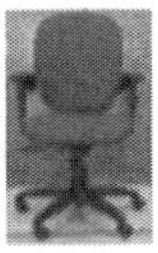	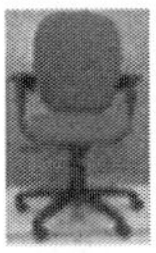
	C	I	C	C
Spontaneous rejection		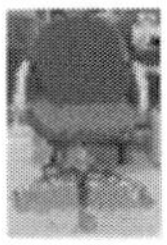	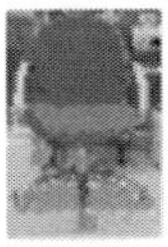	
	E	G	G	E

FIGURE 22.7 Spontaneous preference/rejection of the 12 chairs displayed, by the volunteers from the process engineering department of the refinery.

that great care was taken to ensure that the attributes were understood by the users and not defined in advance by the researchers. It was not necessarily the case that the users' understanding was identical to that of the specialists. The comparative theoretical framework for the study was based on what had been set out by Grieco et al. (1997), Helander and Zhang (1997), and Bomfim (1998).

Based on the answers to Questions 4 to 8 formulated in the interview, it was possible to define the criteria of comfort, safety, practicality, adaptability, and aesthetics, as per the employees' accumulated perception, which could be compared with the concepts developed by the specialists. Figure 22.8 allows a comparison to be made between the perceptions of the office workers who trialed the office chairs and the concepts developed by the specialists as to the criteria of comfort, practicality, safety, aesthetics, and discomfort.

As shown in Figure 22.8, the cumulative perception of the workers generally converged with that of the specialists'. Moreover, by analyzing the perceptions of the individual workers on the comfort, practicality, safety, and aesthetics of the work seat as described above, it was certain that there was no consensus among the concepts attributed. The specialists related discomfort to "fatigue and pain," while the office staff related it to a "chair that is too hard or too soft, that has no adjustment mechanisms and that is not adequate for posture."

It was established that some employees found it difficult to verbalize their perception as to the issues addressed, perhaps because of the subjectivity of the answers or because they were unfamiliar with what the meaning of an issue was, and some appeared to be nervous. Based on previous experience in open interviews that had objective variables only, it was concluded that the verbalization of subjective factors led to difficulties. The concepts of practicality (being practical), adaptability (being adaptable), and aesthetics were not clear to the employees. In some cases, practicality and adaptability were used as synonyms and in others they were not even defined. The question on the aesthetics of the work seat was the one that caused most confusion: most did not know how to conceptualize it (they were afraid to express erroneous conceptions and there was omission). Nevertheless, it is clear that the employees, although not uniformly, had clearly defined what they wanted from a work seat. They were looking for a comfortable seat that combined availability of support for the back (a backrest) with an appropriate position for their legs (which avoids their remaining bent) and with there being a mechanism to adjust the height.

Based on the results, it can be seen that when the employees made a decision as to their preference for/rejection of the seats and on the attributes sought in a work seat, this was based on their previous experience of chairs. It was also noted that the justifications/arguments presented by the employees for selecting or rejecting a particular type of seat (Answers 1 and 2) ratified the attributes that, in their perception, were important for a work seat (size, appearance of softness, support for their back and feet, the tilt of the seat). It can be concluded that for this particular group of people, the subjective aspects that mattered when selecting a work seat (in this case, comfort, ease of use, and relaxation) could be served by a single, objective variable: support.

Another aspect observed was that not all employees, by their choice, trialed (sat on and/or handled) the seats, i.e., they answered the interview questions based only on a visual assessment.

22.3.7 PRESENTATION OF THE EXPERIMENT AND PRE-TESTING THE QUESTIONNAIRE ON EVALUATING DISCOMFORT/PAIN

This step was set aside for the presentation of the experiment to be carried out, and thus for explaining the objectives and the adopted method to the employees. A pre-test of the questionnaire for evaluating discomfort/pain was conducted (adapted from Corlett 1995b), similar to the one that would be answered, later, in the course of the experiment at the start and end of the working day with a view to checking on how staff completed it, thus making it possible to make any necessary corrections. The questionnaire adapted from Corlett (1995b) (the head was added and the discrete scale replaced with a continuous scale) consists of a map of twenty-nine body regions, distributed over six main areas: head, trunk, upper limbs (left and right), and lower limbs (left and right). The

Criteria	Specialists' perception	Accumulated perceptions of the workers who tested the office chairs
Comfort	The chair and its components must have upholstery, contouring and regulating mechanisms that meet the needs and physiological characteristics of different "shapes and curves of the body" (Grieco et al. 1996). Comfort is the feeling of well being (feeling relaxed) and is a sensation linked to aesthetics. Comfort is not the absence of discomfort. (Helander and Zhang, 1997)	That which allows proper posture and does not cause pain. Should provide a feeling of rest. Should fit into the table, have a good back rest, support for the arms, have adjustable seat height and back rest, castors and provide well-being independent of time spent in it, be soft. Should be adaptable, be of adequate height for the job and allow tasks to be carried out.
Practicability	The chair and its components should be easy to regulate/adjust; the covering material should be hygienic (Grieco et al. 1996).	Should be light, easy to move, to allow rapid moving around. Should be resistent, have castors and the adjustable devices needed, which should be easy to action, with no need to use a lot of levers. The material should be easy to clean. Should be light and compact and be suitable for both reading and for tasks at the computer. The arm support should not get in the way.
Safety	The chair cannot be a source or cause of accidents (Grieco et al. 1996).	The chair must be resilient for the job, should be strong, resistent, rigid and firm and not have assembly problems. It should not break, crumble nor expose any part that may hurt the user or cause accidents. Base of the chair should give good support, may not have problems with components, specially the castors and the tilt of the seat and backrest.
Adaptability	The chair and its components must have the correct dimensions and adjustment controls so as to meet the anthropometric needs of a wide range of users (normally at least 90% of potential users) (Grieco et al. 1996).	This should make it possible to have multiple adjustments, to adapt to the height of the user; and to be adaptable to various people, and to the job.
Aesthetics	From the Greek "*aisthesis*", means perception, sensation. The most common meaning for the concept of aesthetics is that it is concerned with what is beautiful, pleasant, harmonious, ugly, etc. (Bomfim, 1998).	This is related to its appearance, finishing and upholstery. It refers to the environment in which the chair is inserted, to modern design and pleasant colours. That which suits the workplace (especially with regard to the colour of the upholstery). Should have a good stuffing and a neutral color and design and beautiful finishing and appropriate to the environment.
Discomfort	Is related to fatigue and pain (Helander et al. 1987).	The uncomfortable chair has no adjustment controls, no arm support, is hard or too soft (gives way) and heavy, it is not possible to find a comfortable position and is not reclinable, and is not firm. Is not suitable for posture and causes pain.

FIGURE 22.8 Accumulated perceptions of the workers who tested the office chairs, and the specialists' opinion on comfort, practicality, safety, aesthetics, and discomfort.

questionnaire adopts the same format as the questionnaire on what is wanted from a seat (Section 22.3.3). The respondents should mark on the adjacent scale each one of the parts of the body, and the intensity of pain/discomfort perceived for each region body. Because of the size of the paper used (A4 landscape, with the model human figure divided into the twenty-nine body parts in the center and scales on the left and on the right), the scale for this questionnaire ranged from no discomfort/pain (or 0) to very uncomfortable/painful (or 9) instead of from 0 to 15.

When applying the pre-test, the volunteers were advised to mark the intensity of discomfort/pain on the continuous scales, only on the locations shown on the body map, which generate discomfort/pain and to note the occurrence of discomfort/pain in the last week of work and not their opinion on the perception of discomfort that day. Thus, instead of an instantaneous evaluation, a general assessment of the working week was obtained prior to the experiment with the traditional seat. It should be emphasized that the main objective of the pre-test was to familiarize the user with the tool for analyzing discomfort/pain to ensure that users understood how the questionnaires should be filled in. However, the pre-test also allowed mapping of the normal work situation, thus identifying complaints of discomfort/pain, which were then matched against the postural analysis conducted in the step of direct/indirect observations. The results were tabulated on an electronic spreadsheet, and the averages obtained were plotted on the graph.

In the petroleum distributor, there was a significant difference ($\chi^2 = 12.971$, $p=0.024$) as there was in the process engineering department of the refinery ($\chi^2 = 17.342$, $p=0.008$), the highest intensity of discomfort/pain being in the regions related to the trunk, namely, the cervical region, lower back, upper back, and neck, even though at an intensity below the average, and even lower in the arms, head, legs and feet. These results were expected during the course of observations in the offices, since the same posture and flexion of the neck was required to be maintained during periods of computer use. In the case of the logistics department of the refinery, the Kruskal–Wallis test did not show a significant difference ($\chi^2 = 13.728$; $p=0.056$) between the averages of the feelings noted of discomfort/pain of parts of the body, but it was emphasized that they were a little higher (around the average) than in the other departments investigated, which had also been expected as a result of the activity, and this occurred not only in the office job of the transfer hall, but also in the tank area (including the effort to open the valves).

22.3.8   Preference Tested: Trialing the Use of the Seats

The study consisted primarily of evaluating the seats in a given work environment. For the experiment, plans were made of the number of seats, volunteers, and days, in order to make a significant result possible of how many chairs trialed satisfied the participants as to the evaluation criteria and did not cause discomfort or pain.

In this stage, an explanation of the experiment was given to the participants, detailing the objectives and the method adopted. Trialing the use of the seats was conducted by applying a set of questionnaires to assess: (i) if there was discomfort/pain (see Section 22.3.7); and at the end of the experiment (ii) the degree of satisfaction with each seat trialed, considering the criteria previously presented and discussed (see Section 22.3.6); (iii) the importance attributed to each one of the criteria (see Section 22.3.3); (iv) the weighted degrees of satisfaction; and (v) preference/rejection based on indirect pair-wise comparison.

With the objective of facilitating the conduct of the experiment for the volunteers, individual notebooks were compiled, containing: (i) guidelines on the experiment; (ii) planning of the experiment; (iii) a list with the perception of the employees on the evaluation criteria; (iv) a questionnaire on evaluating discomfort/pain (two per day: one for the start of the shift; the other for the end of the shift); (v) a questionnaire to assess the employees' satisfaction with respect to the seat used, in accordance with pre-defined criteria (one per day); and (vi) a questionnaire to evaluate the degree of importance of each evaluation criterion, to be answered on the last day.

On the first day, notebooks on the experiment were distributed with the questionnaires. At that time, the employees were advised about the need to make adjustments to each seat so as to allow a suitable posture to be adopted. The employees were given orientation, especially on the use of the platforms for the footrest (equal for all seats, dimensioned as a compromise solution). Thus, for each chair tested, each volunteer received two questionnaires in order to assess:

- Satisfaction with each chair trialed, considering the criteria previously presented and discussed
- Any discomfort/pain at the beginning and end of the shift

During the experiment, the questionnaires are collected, daily, at the end of the shift and the measurements taken of the height to which the volunteer adjusted his/her chair. In order to facilitate the conduct of the experiment for the volunteers, one participant was selected as project coordinator in order to collect, as from the second day of the experiment, the questionnaires of the start of the shift referring to the sensation of discomfort/pain.

Individual data from this experiment were kept confidential, with no identification. All that was necessary was that each volunteer was identified by a number during the trialing period, for the purposes of pair-wise tabulation of the data.

22.3.8.1   Questionnaire to Assess Discomfort/Pain between the End and Start of the Working Day

For identifying the occurrence of complaints of discomfort/pain with the seat being trialed at the start and end of the working day, the employees filled out a questionnaire to assess discomfort/pain in order to measure the differences between the perception of discomfort/pain at the start and end of the working day. The questionnaire used was the same as that used in the pre-test.

The results of each questionnaire (from the beginning and from the end of the shift) were tabulated on an electronic spreadsheet, with the final value of each individual being taken as the difference between the degree of sensations at the end and those at the beginning of the shift, for each body region marked. Thus, the final variable of discomfort/pain was the mean of the differences (final discomfort/pain less the initial one) of all the points scored in the questionnaire. Thus, in the experiments, the level of discomfort was obtained per body region, per volunteer, seat, and day, which enabled the existence of any correlation to be identified between these differences and the seats used.

22.3.8.2   Questionnaire to Evaluate Satisfaction with the Seat Under Test

The questionnaire was structured to measure the degree of satisfaction of each user in relation to the five evaluation criteria discussed: comfort, safety, adaptability, practicality, and aesthetics. In this phase of the study, it became crystal clear that a further criterion should be added, bearing in mind that it was fundamental in the design of workstations: the sixth criterion of suitability for the work. The questions were presented in such a way that the respondents could mark their degree of satisfaction with regard to each criterion by means of a 15 cm continuous evaluation scale, with two anchors at the ends (not satisfied/very satisfied) on which the subject should mark his/her satisfaction with the criterion. Thus, after testing the seat, the volunteers were invited to complete a questionnaire that started with the following question: *At the end of the day, after having trialed your seat, place a mark on the line at the point that best represents your opinion.* Thereafter, the six evaluation criteria were presented: comfort, safety, adaptability, practicality, suitability for the work, and appearance, each with a continuous scale immediately below, as in the following example:

After having used your trial chair, place a mark on the line that best represents your opinion:

*1. Thinking that a seat should be **comfortable,** are you* [dissatisfied/very satisfied]

Not satisfied   *Very satisfied*

The results of the questionnaires were tabulated using an electronic spreadsheet, resulting in the mean degrees of satisfaction relative to each evaluation criterion for each seat.

22.3.8.3 Questionnaire to Assess the Degree of Importance of the Evaluation Criteria

This questionnaire was structured to measure the importance of each criterion of evaluation, in the opinion of the users of the seats. It kept the same format as the other questionnaires, inviting the volunteers to mark the degree of importance on a 15 cm continuous line, ranging from the far left *not at all important* (0), to the far right *very important* (15), as in the example:

Mark on the scale what degree of importance you assign to the following questions relating to a work chair:

*1. The seat should be **comfortable***

Not important   *Very important*

The averages obtained were normalized, i.e., the value corresponding to each criterion was divided by the sum of the averages of all criteria, yielding a percentage value. This value was converted into a weight and the sum of the weights was equal to 1. The results as to the importance of the weights assigned to the criteria for evaluating office chairs are shown in Table 22.1.

22.3.8.4 Assessment of the Preference/Rejection of the Chairs after the Experiment: The Indirect Pair-wise Comparison Test

The indirect pair-wise comparison test aims to establish an ordering of the preferences of users, by comparing the alternatives available in relation to a pattern. In this study, the test was applied as a qualitative reinforcement to the assessment made during the experiment. Because it had been applied to the same employees who participated in the experiment, this test incorporated the perception arising from that experience.

To perform the test, each employee made a comparison, in relation to a seat defined as standard, with all others. The results were tabulated and submitted to statistical validation, thereby obtaining a second order of preference. The first ordering was made at the beginning of the study, in the stage of evaluating the spontaneous preference/rejection.

TABLE 22.1
Mean Degrees of Importance Attributed to Each Criterion in the Petroleum Distributor's, in the Three Departments of the Refinery, and the Agency of the Judiciary Authority

Criteria	Comfort		Safety		Adaptability		Practicality		Suitability		Aesthetics	
Distributor	12.48	0.19	12.44	0.19	11.00	0.17	11.13	0.17	12.40	0.19	5.78	0.09
Refinery HR/ training	13.9	0.20	13.5	0.19	11.2	0.16	11.1	0.16	14.1	0.20	7.2	0.10
Refinery logistics	12.98	0.20	12.22	0.19	12.1	0.19	11.26	0.17	10.66	0.16	5.68	0.09
Refinery, civil eng,	12.95	0.18	11.59	0.16	12.49	0.17	10.50	0.15	13.29	0.18	11.44	0.16
Refinery, process eng.	14.02	0.19	13.49	0.18	14.35	0.19	13.61	0.18	13.03	0.17	7.75	0.10
Judiciary power	13.58	0.18	11.92	0.16	13.42	0.18	12.66	0.17	14.24	0.19	8.14	0.11

Note: The mean degrees of importance attributed to each evaluation criterion on a scale of 15 cm and the weight of the criterion in the column immediately adjacent.

It should be mentioned that it was not possible to incorporate the assessment of the seat currently in use in all stages of evaluation. However, in order to be able to relate the seats trialed with the current one, the volunteers who took part in the experiment were asked to place the seats trialed and their current one in order of preference so as to obtain a parameter.

The volunteers were instructed how to mark, on a continuous scale, the order of preference of the seats trialed. They wrote down the descriptor of the chair and marked an X on the scale as per the relationship of this chair with their current one. The 15 cm scale has three anchors 0 (*much worse*), 7.5 (*their current work chair*), and 15 (*much better*), as in the example:

Mark on the scale your order of preference of the chairs tested in relation to your current work chair:

Much worse   *Seat in use*   *Much better*

22.4   PLANNING THE EXPERIMENTS

The experiments were planned with the aim of enabling all the seats to be evaluated by the same group of employees for a minimum period of time. Due to the need to offer a response in a short period of time, it could be assumed that the volunteers would remain seated for at least three hours, since, as per Helander and Zhang (1987), a chair can be experienced, without errors being made, in this period of time. However, to facilitate the conduct of the experiments, the option was taken, in all studies, that the users would test the chairs over a working day (i.e., all the seats were tested for one day by each person). Taking into consideration the objective of evaluating the chairs in a minimum period of time, and eliminating the individual effects and the effects of the days of the week, the methodology of design and analysis of experiments (Montgomery 1991) was used. An attempt was made to organize the factors of "volunteer," "day," and "chair" based on a Latin square design, e.g., five volunteers, carry out a trial for five days, on five chairs, one per day, resulting in a $5 \times 5 \times 5$ matrix. This methodology allows the statistical analysis to consider the effects of factors that are not being evaluated, but that, in some way, affect the experiment conferring greater reliability to the results. On the last afternoon of the experiment, the research team returned to the study sites to collect the notebooks of the experiment. At that time, the experiment was taped on film so as to record the experiment and the postures assumed with the chairs being trialed.

22.5   RESULTS

22.5.1   Result of the Questionnaire to Evaluate Discomfort/Pain with the Office Chairs (Difference between the Final and Start of the Working Day)

In all the studies with the office chairs, it was noted that the intensity of discomfort occasioned by use of the chairs was small, the maximum discomfort being in the order of 1.5, on a scale from 0 (none) to 9 (a lot).

22.5.2   Result of the Evaluation of Satisfaction in Relation to the Chairs Trialed

The results of the questionnaires were tabulated on an electronic spreadsheet, thus obtaining the mean degrees of satisfaction relative to each criterion of evaluation for each chair. Tables 22.2 through 22.8 and Figures 22.9 through 22.15 show the results in the offices studied.

After the test of use, the results showed what the chairs chosen had in common: castors, a mechanism for adjusting the height of the seat, and an arm support. However, when identifying wants, these items were not always considered more important than others. This demonstrates that the volunteers were unaware of the consequences of having a seat without castors, without a mechanism for adjusting seat height, and without an arm support.

The experiment produced evidence that except for the two fixed chairs, all other chairs met the needs of the employees, especially with regard to comfort, which was the item cited as to what they

TABLE 22.2
Mean Degrees of Satisfaction Relative to Each Criterion of Evaluation and Averages with the Weighted Criteria for Each Chair X

Criteria	Mean						Weighted Mean					
	X1	X2	X3	X4	X5	X6	X1	X2	X3	X4	X5	X6
Comfort	7.42	10.73	7.23	6.43	8.52	7	1.42	2.05	1.38	1.23	1.63	1.34
Safety	9.14	10.53	8.87	7.3	10.14	6.9	1.74	2.01	1.69	1.39	1.93	1.32
Adaptability	8.22	9.38	8.57	6.93	8.88	5.45	1.39	1.58	1.44	1.17	1.5	0.92
Practicality	7.8	9.15	8.33	7.63	9.18	7.65	1.33	1.56	1.42	1.3	1.45	1.31
Adequacy	8	9.92	7.23	7.37	7.18	4.25	1.52	1.89	1.37	1.4	1.36	0.81
Aesthetics	9.54	9.98	9.8	7.97	9.96	10.2	0.85	0.88	0.87	0.71	0.88	0.9
Sum	50.12	59.69	50.03	43.63	53.86	41.4	8.25	9.97	8.17	7.2	8.75	6.65

TABLE 22.3
Mean Degrees of Satisfaction Relative to Each Criterion of Evaluation and Averages with the Weighted Criteria for Each Chair Y

Criteria	Mean					Weighted Mean				
	Y1	Y2	Y3	Y4	Y5	Y1	Y2	Y3	Y4	Y5
Comfort	10.85	8.3	11.32	6.34	9.64	2.08	1.59	2.17	1.21	1.84
Safety	11.18	11.28	12.06	8.02	0.72	2.13	2.15	3.3	1.53	1.85
Adaptability	9	7.03	10.68	8.9	8.34	1.52	1.19	1.8	1.5	1.41
Practicality	9.85	8.55	11.66	8	7.68	1.68	1.46	1.99	1.37	1.31
Suitability	6.85	5.38	10.18	6.88	8.92	1.3	1.02	1.93	1.31	1.7
Aesthetics	11.03	10.83	11.04	9.36	8.88	0.98	0.96	0.08	0.83	0.79
Sum	58.76	51.37	66.94	47.5	44.18	9.69	8.37	11.27	7.75	8.9

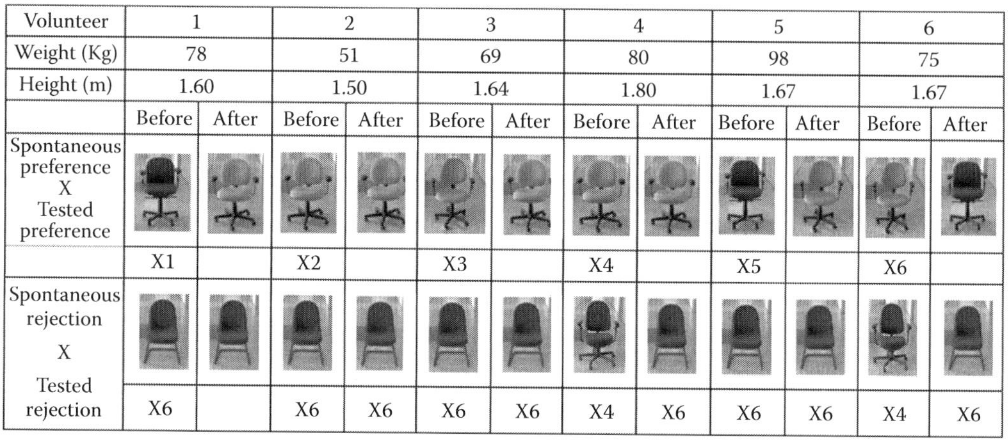

Volunteer	1		2		3		4		5		6	
Weight (Kg)	78		51		69		80		98		75	
Height (m)	1.60		1.50		1.64		1.80		1.67		1.67	
	Before	After	Before	After	Before	After	Before	After	Before	After	Before	After
Spontaneous preference X Tested preference												
	X1		X2		X3		X4		X5		X6	
Spontaneous rejection X Tested rejection												
	X6		X6	X6	X6	X6	X4	X6	X6	X6	X4	X6

FIGURE 22.9 Spontaneous and trialed preference/rejection for the six chairs of group X presented in the petroleum distributor.

Volunteer	1		2		3		4		5		6	
Weight (Kg)	58.6		63.5		79		96		73		82	
Height (m)	1.70		1.68		1.67		1.80		1.68		1.68	
	Before	After	Before	After	Before	After	Before	After	Before	After	Before	After
Spontaneous preference X Tested preference	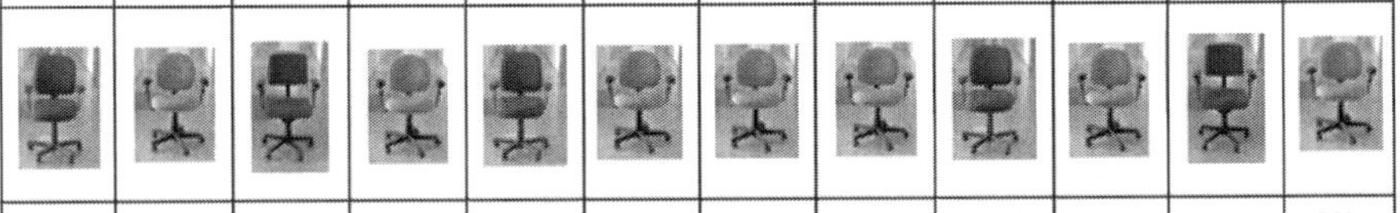											
	Y1	Y3	Y4	Y3	Y1	Y3	Y3	Y3	Y1	Y3	Y4	Y3
Spontaneous rejection X Tested rejection	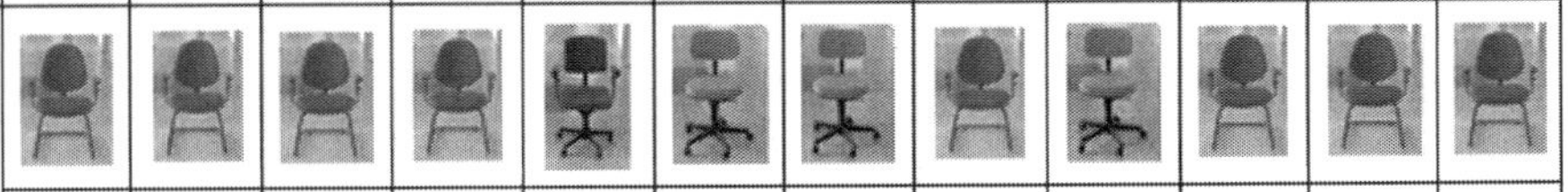											
	Y6	Y6	Y6	Y6	Y4	Y5	Y5	Y6	Y5	Y6	Y6	Y5

FIGURE 22.10 Spontaneous and trialed preference/rejection for the six chairs of group Y presented in the petroleum distributor.

looked for in a work chair. Employees also showed that they were satisfied with the items: practicality, safety, adaptability, and the aesthetics of these chairs.

The correlation was analyzed between the variables of comfort, safety, adaptability, practicality, suitability, appearance, and discomfort/pain adapted from Corlett (1995b). It was verified that the strongest positive correlation occurred between the variables of adaptability and practicality ($\rho=0.897$, $p<0.001$). This is justifiable since the two concepts are similar, as per Table 22.1. There was a negative correlation only between discomfort/pain and the five other variables. The only non-significant correlation was between discomfort/pain and comfort that led to Question 9 of the interview, which was included to evaluate if comfort and discomfort were or were not opposites of one and the same continuum.

As to Question 8, in all studies, with the obvious exception of the study in the HR department of the refinery (with three people), correlations could not be demonstrated between the physical stature of the volunteers and their favorite chairs. The shortest employees did not always reply that they preferred smaller chairs and the largest employees did not always reply that they preferred larger chairs, which diverges from what Hellander et al. (1987) suggested. For these authors, "small" people, those of least height and least weight, preferred smaller chairs because the seat was very long and the support for the spinal column was high, while "big" people, those who were tallest and heaviest, rejected small chairs because the seat was very short and the support for the spinal column was small. It might be noted, however, that there was a correlation between their favorite chairs and the activities carried out.

TABLE 22.4
Mean Degrees of Satisfaction Relative to Each Criterion of Evaluation and Averages with the Weighted Criteria for Each Chair Trialed in the Agency of the Judiciary Authority

	Mean					Weighted Mean				
Criteria	A	B	C	D	E	A	B	C	D	E
Comfort	8.78	11.66	5.82	8.46	8.4	1.61	2.14	1.07	1.55	1.54
Safety	10.83	10.58	9.88	11.76	8.98	1.74	1.71	1.59	1.9	1.45
Adaptability	5.95	8.62	5.4	8.54	6.98	1.08	1.56	0.94	1.55	1.27
Practicality	6.87	7.42	5.14	9.36	5.5	1.18	1.27	0.88	1.6	0.94
Suitability	9.04	8.4	4.22	9.56	4.08	1.74	1.62	0.81	1.84	0.79
Aesthetics	5.65	8.2	13.08	9.78	14.08	0.62	0.9	1.44	1.08	1.55
Sum	47.12	54.88	43.54	57.46	48.02	7.97	9.2	6.73	9.52	7.54

Volunteer	1		2		3		4		5	
Weight (Kg)	55		66		70		80		52	
Height (m)	1.60		1.59		1.66		1.70		1.58	
	Before	After	Before	After	Before	After	Before	After	Before	After
Spontaneous preference X Tested preference	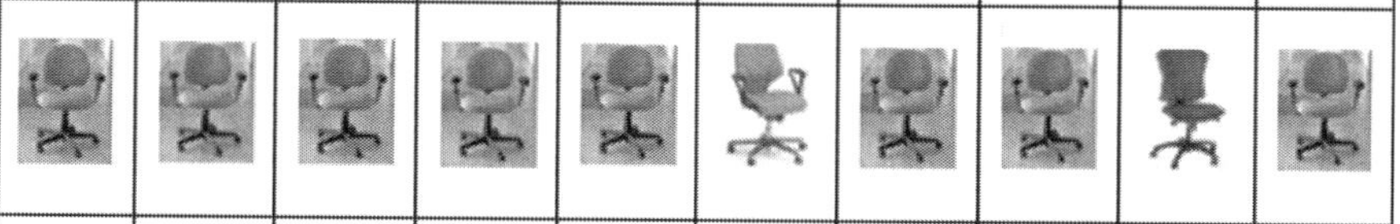									
	B	B	B	B	B	C	B	B	E	B
Spontaneous rejection X Tested rejection	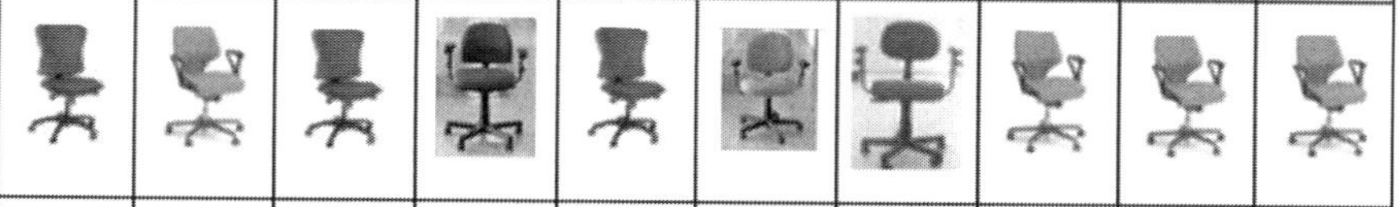									
	E	C	E	D	E	B	A	C	C	C

FIGURE 22.11 Preference/rejection of the women volunteers of the agency of the judiciary authority for the five chairs.

22.5.3 RESULT OF THE PREFERENCE FOR OFFICE CHAIRS: INDIRECT PAIR-WISE COMPARISON TEST

After the trials, the questionnaire was applied on the order of preference of the chairs when the volunteers marked, on a 15 cm scale, the order of the seats trialed in relation to their current work chair, which was situated in the center of the scale.

Figures 22.16 through 22.22 show the order of preference of the chairs after the experiment in the petroleum distributor agency of the judiciary authority, HR/training, logistics, civil engineering, and process departments of the refinery.

In the petroleum distributor, the order by indirect pair-wise comparison did not confirm the order of preference obtained by the previous analysis of comparison of averages, which pointed to the preference for chairs Y1, Y3, and Y4 in this order. However, it did confirm that chair Y6 was very bad, despite it being better than the current one.

In the agency of the judiciary authority, the statistical analysis did not point to differences between the preferences for chairs, but the indirect pair-wise comparison indicated a ranking of

TABLE 22.5
Mean Degrees of Satisfaction Relative to Each Criterion of Evaluation and Averages with the Weighted Criteria for Each Chair Trialed in the HR/Training Department of the Refinery

	Mean			Weighted Mean		
Criteria	Z1	Y2	Y3	Z1	Y2	Y3
Comfort	15	12.1	14.6	2.9	2.6	2.7
Safety	15	13	12.5	2.9	2.5	2.4
Adaptability	12.5	13.1	8	2	2.1	1.3
Practicality	9.8	13.4	10.2	1.5	2.1	1.6
Suitability	15	13.5	13.7	3	2.7	2.7
Aesthetics	3.1	7.9	10.5	0.3	0.8	1.1
Sum	70.4	73	69.5	12.6	12.8	11.8

Volunteer	1		2		3	
Weight (Kg)	50		53		93	
Height (m)	1.53		1.73		1.62	
	Before	After	Before	After	Before	After
Spontaneous Preference X Tested Preference				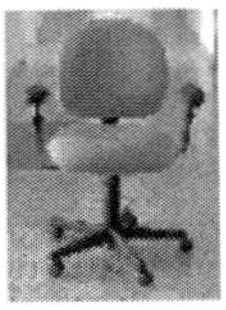	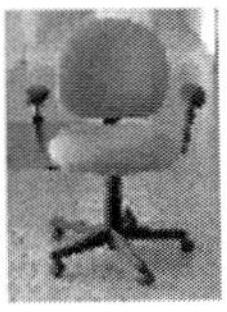	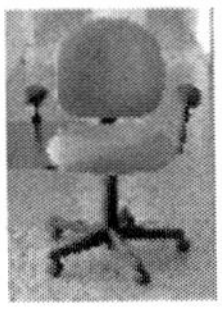
	Z1		Y3		Y3	
Spontaneous Rejection X Tested Rejection	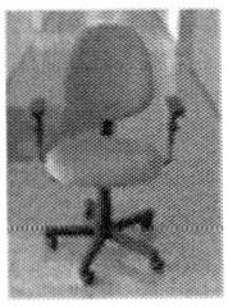	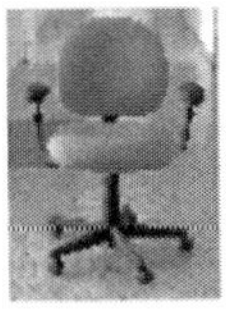	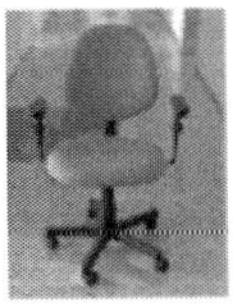	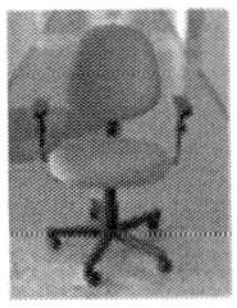	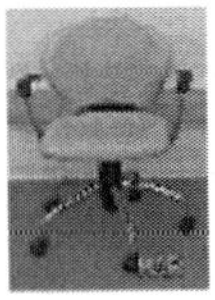	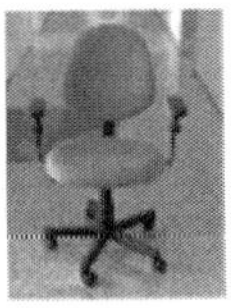
	Y3*		Y2		Y2*	

*Different opinions from the one before the tests

FIGURE 22.12 Results of the spontaneous and trialed preference/rejection in the HR/training department of the refinery.

preference for chair B followed by chair D. The current chair A was at the average of satisfaction and chairs C and E below the average.

In the HR/training of the refinery, the three chairs were all considered good by the three volunteers, and the preference was kept the same as before the test: "Y3" being the most accepted and "Y2" being the third choice. However, direct observation showed that the legs of chair "Z1" projected far from the seat border, which might impose some difficulties in accessing the feet support and, also, generate accidents. Therefore, the NDES team proposed the acquisition of either chair "Y3" or "Y2."

In the case of the logistics department of the refinery, the results confirmed the multiple comparison of averages, which had already shown that the averages of satisfaction with chairs "A3," "B3," and "A2" were the highest and differed significantly from the averages attributed to chairs "A1," "B1," and "B2," which received the lowest averages. It should be pointed out that the current

TABLE 22.6
Mean Degrees of Satisfaction Relative to Each Criterion of Evaluation and Averages with the Weighted Criteria for Each Chair Trialed in the Logistics Department of the Refinery

	Mean						Weighted Mean					
Criteria	A1	A2	A3	B1	B2	B3	A1	A2	A3	B1	B2	B3
Comfort	7.89	10.68	12.27	8.12	3.38	11.66	1.58	2.14	2.45	1.62	0.68	2.33
Safety	8.72	10.72	12.48	8.66	6.24	11.68	1.63	2	2.33	1.61	1.16	2.18
Adaptability	8.51	10.97	12.38	8.06	5.16	8.32	1.4	1.8	2.03	1.32	0.85	1.37
Practicality	7.98	9.69	12.29	9.24	3.92	9.23	1.38	1.68	2.13	1.6	0.68	1.6
Suitability	8.27	10.69	12.41	9.02	3.84	7.88	1.56	2.01	2.34	1.7	0.72	1.48
Aesthetics	6.96	12.19	13.14	7.92	2.23	9	0.61	1.07	1.15	0.69	0.2	0.79
Sum	48.33	64.94	74.97	51.02	24.77	57.77	8.16	10.7	12.43	8.54	4.29	9.75

Volunteer	1		2		3		4		5		6	
Weight (Kg)	101		75		83		84		66		85	
Height (m)	1.75		1.69		1.81		1.80		1.67		1.73	
	Before	After	Before	After	Before	After	Before	After	Before	After	Before	After
Spontaneous Preference X Tested Preference												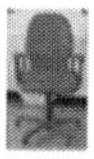
	B1	A3	A1	–	B1		A3	A3	B3	A3	A3	A3
Spontaneous Rejection X Tested Rejection	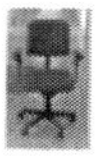		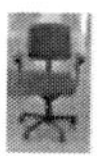		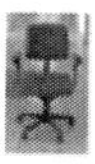	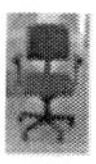		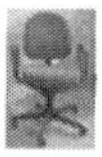			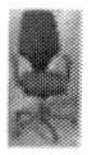	
	B2	B2	B2	–	B2	B2	–	B1	B2	B2	A2	A1

FIGURE 22.13 Preference/rejection of the six chairs trialed before and after the tests in the logistics department of the refinery.

work chair was judged, by the volunteers, to be better than chairs "B1" and "B2," and worse than the others.

After the trials in the civil engineering department, indirect pair-wise comparison also confirmed the multiple comparison of means, which had shown that chair "C" obtained the highest average of preferences and chairs "K" and "E" were the most rejected. In the indirect pair-wise comparison, these employees' current work chairs were judged better than chairs "K" and "E."

TABLE 22.7
Mean Degrees of Satisfaction Regarding Each Criterion of Evaluation for Each Chair (Above) and Averages with the Weighted Criteria (Below) for Each Chair Trialed in the Civil Engineering Department of the Refinery

					Mean							
Criteria	A	B	C	D	E	F	G	H	I	J	K	L
Comfort	11.13	10.1	12.59	11.18	6.51	9.38	7.5	7.27	10.53	11	9.41	9.12
Safety	11.29	11.69	12.84	12.42	8.62	8.28	10.86	7.65	11.06	10.82	6.33	11.25
Adaptability	10.83	10.35	12.57	11.76	2.72	11.48	4.22	7.56	10.26	12.06	6.05	10.9
Practicality	10.7	10.7	13	11.7	5	11.8	5.1	4.3	12.3	10.6	7.3	9.9
Suitability	11.18	9.1	13.05	11.06	3.71	9.28	5.18	5.78	10.69	9.67	5.13	9.39
Aesthetics	10.22	8.21	13.48	9.97	7.74	9.8	10.5	9.12	10.29	10.79	7.06	12.78
Sum	65.35	60.15	77.53	68.09	34.3	60.02	43.36	41.68	65.13	64.94	41.28	63.34
					Weighted Mean							
Criteria	A	B	C	D	E	F	G	H	I	J	K	L
Comfort	2	1.81	2.26	2	1.17	1.68	1.34	1.3	1.89	1.97	1.69	1.63
Safety	1.81	1.89	2.06	1.99	1.38	1.33	1.74	1.23	1.77	1.74	1.02	1.8
Adaptability	1.87	1.79	2.17	2.03	0.47	1.98	0.73	1.31	1.77	2.08	1.05	1.88
Practicality	1.55	1.55	1.88	1.7	0.73	1.72	0.74	0.63	1.79	1.54	1.06	1.43
Suitability	2.06	1.67	2.4	2.03	0.68	1.71	0.95	1.06	1.97	1.78	0.94	1.73
Aesthetics	1.62	1.3	2.13	1.58	1.22	1.55	1.66	1.44	1.63	1.71	1.12	2.02
Sum	10.91	10.01	12.9	11.33	5.65	9.97	7.16	6.97	10.82	10.82	6.88	10.49

TABLE 22.8
Mean Degrees of Satisfaction Regarding Each Criterion of Evaluation for Each Chair (Above) and Averages with the Weighted Criteria (Below) for Each Chair Trialed in the Process Engineering Department of the Refinery

						Mean						
Criteria	A	B	C	D	E	F	G	H	I	J	K	L
Comfort	7.21	6.44	12.25	4.77	2.13	10.02	6.91	8.86	12.95	7.4	6.76	3.45
Safety	5.95	6.96	12.39	4.13	5.59	10.06	7.51	4.88	12.86	11.24	7.66	3.54
Adaptability	3.83	6.28	11.15	7.09	1.18	8.5	3.85	8.16	11.82	6.42	7.91	6.56
Practicality	5.81	6.68	11.48	4.54	1.18	11.46	5.62	3.08	13.43	10.29	6.17	6.88
Suitability	5.22	6	11.96	4.8	0.94	10.75	4.63	7.78	10.57	6.4	6.75	3.78
Aesthetics	9.54	3.55	12.19	7.06	4	11.53	10.03	11.35	10.98	7.12	7.54	7.61
Sum	37.56	35.91	71.42	32.39	15.02	62.32	38.55	44.11	72.61	48.87	42.79	31.82
						Weighted Mean						
Criteria	A	B	C	D	E	F	G	H	I	J	K	L
Comfort	1.33	1.18	2.25	0.88	0.39	1.84	1.27	1.63	2.38	1.36	1.24	0.63
Safety	1.05	1.23	2.19	0.73	0.99	1.78	1.33	0.86	2.27	1.99	1.35	0.63
Adaptability	0.72	1.18	2.1	1.33	0.22	1.6	0.72	1.53	2.22	1.21	1.49	1.23
Practicality	1.04	1.19	1.05	0.81	0.21	2.05	1	0.55	2.4	1.84	1.1	1.23
Suitability	0.89	1.03	2.04	0.82	0.16	1.84	0.79	1.33	1.81	1.09	1.15	0.65
Aesthetics	0.98	0.36	1.24	0.72	0.41	1.17	1.02	1.15	1.12	0.72	0.77	0.77
Sum	6.01	6.17	8.83	5.29	2.38	10.28	6.13	7.05	12.2	8.21	7.1	5.14

In the process engineering department of the refinery, the current work chair was judged better than the chairs "B," "L," "D," and "G," chair "E" being the worst. The better chairs were "I" and "C."

22.6   DISCUSSION OF THE RESULTS

Comparing the results of this study raises important questions for the evaluation and selection of products. It can be inferred that while the evaluation of satisfaction was conducted using an

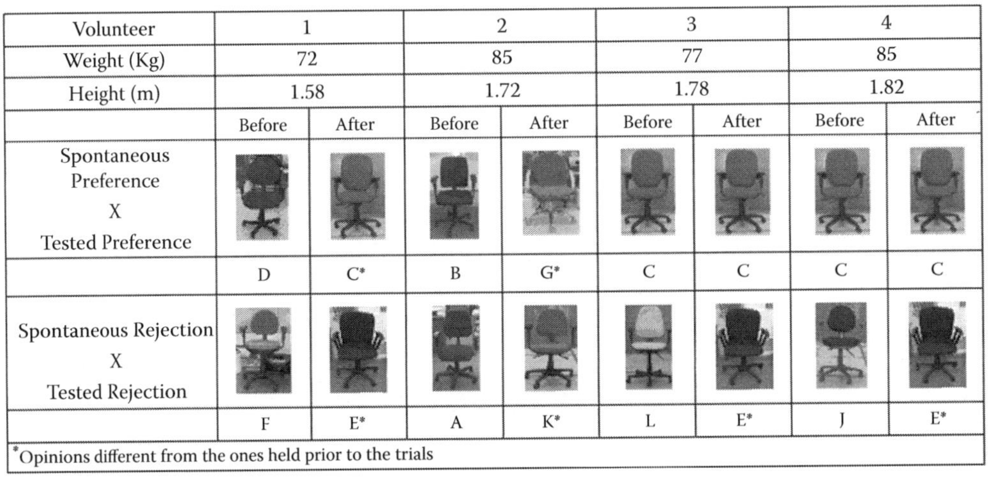

Volunteer	1		2		3		4	
Weight (Kg)	72		85		77		85	
Height (m)	1.58		1.72		1.78		1.82	
	Before	After	Before	After	Before	After	Before	After
Spontaneous Preference X Tested Preference								
	D	C*	B	G*	C	C	C	C
Spontaneous Rejection X Tested Rejection								
	F	E*	A	K*	L	E*	J	E*
*Opinions different from the ones held prior to the trials								

FIGURE 22.14   Preference/rejection of the volunteers from the civil engineering department of the refinery for/of the 12 chairs evaluated before and after the trials.

Volunteer	5		6		7		8	
Weight (Kg)	65		56		79		76	
Height (m)	1.73		1.65		1.77		1.83	
	Before	After	Before	After	Before	After	Before	After
Spontaneous Preference X Tested Preference								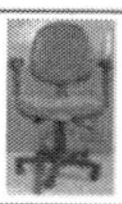
	C	I*	I	I	C	I*	C	I*
Spontaneous Rejection X Tested Rejection	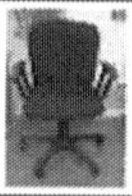	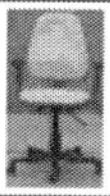						
	E	L*	G	G	G	E*	E	E
*Opinions different from the ones held prior to the trials								

FIGURE 22.15 Preference/rejection of the volunteers from the process engineering department of the refinery for/of the 12 chairs evaluated before and after the trials.

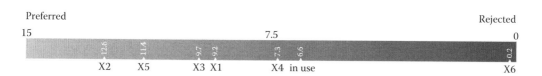

FIGURE 22.16 Order of preference of the office chairs X trialed in the petroleum distributor.

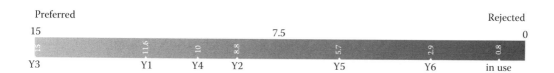

FIGURE 22.17 Order of preference of the office chairs Y trialed in the petroleum distributor.

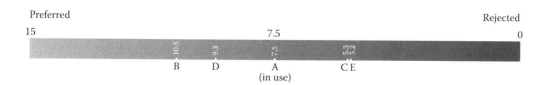

FIGURE 22.18 Order of preference of the office chairs trialed in the agency of the judiciary authority.

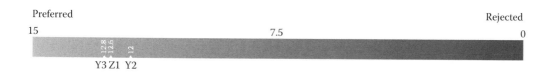

FIGURE 22.19 Order of preference of the office chairs trialed in the HR/training department of the refinery.

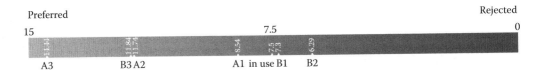

FIGURE 22.20   Order of preference of the office chairs trialed in the logistics department of the refinery.

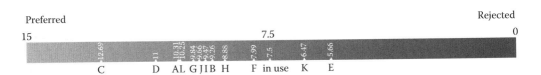

FIGURE 22.21   Order of preference of the office chairs trialed in the civil engineering department of the refinery.

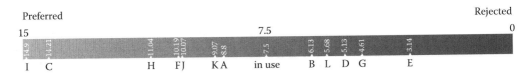

FIGURE 22.22   Order of preference of the office chairs trialed in the process engineering department of the refinery.

objective focus, with criteria previously defined by the volunteers themselves, its results were probably influenced by aspects of a subjective order. The results indicate that the first impression, revealed in interviews, influenced the steps that followed. Although the volunteers attached greatest importance to the criterion of comfort and the least to the criterion of aesthetics, which is directly linked to appearance, comfort tended to be evaluated as a result of what comfort looked like to the volunteers (see answers to Questions 1 and 2).

The two fixed chairs trialed were not accepted by the staff and did not meet the item on adjusting the height, which was essential for ergonomic appropriateness. They can be considered as not having been approved. On the other hand, all the mobile chairs can be considered as having been approved: in terms of comfort, all of them satisfied the needs of the companies' employees; had the mechanisms for adjusting height that enabled them to adapt to diverse users; and they all met Brazilian Ergonomics Norm NR 17 (MTE 2007) and the Norms of ABNT (1997, 1998) for office furniture (NBR 13962 and NBR 14110) and were from manufacturers who already had models tested and approved in mechanical tests in specialized laboratories. It should be clear, however, that the use of a footrest is essential by all staff (in the test, the one that was used had a 15 degree tilt with the ground surface). On the other hand, not all employees wanted an arm support. Thus, it can be recommended that the purchase of chairs with an armrest should be undertaken at the convenience of each employee, but this is not a priority item, not just by individual preference, but sometimes because there is little space available at the workstation and because it holds up the work being done when there is a lot of moving of arms. It was also noted, in several companies and not necessarily in this particular study, that especially shorter women do not like armrests because they sit with their legs on the seat, yoga-style, as a way to ease tension in the hips. This issue is important when selecting chairs because chairs without armrests are (around 20%) cheaper than those with armrests and very often there is unnecessary expenditure, and what is worse, on an item that is not wanted by the user.

22.7 CONCLUSIONS

This chapter has presented a study of a participatory method used in three different companies to find out how people perceived the comfort of seats at work while using them in a real work situation. An analysis was made of the preferences for office chairs in order to understand the real needs of the users, taking into consideration the criteria of safety, adaptability, comfort, and practicality suggested in the literature (Grieco et al. 1997) to which the criteria of aesthetics, suitability for the job, and discomfort were added. The concern with criteria for the purchase of seats is well founded, given the growing demand from companies for recommendations. In addition, it implies multifactor evaluations, bearing in mind the complexity of the relationship between the user and their pieces of equipment and work environments, which translates into comfort perceived and sensed. The results show that the criteria were not very clear and that these were confused in the process of evaluating comfort. For instance, aesthetics was placed alongside "softness" and this seemed to interfere with the subject's first impression of the comfortability of a seat. Aesthetics (i.e., bright colors, "fluffiness," and the size of the seat) were the features that proved to attract the greatest number of first preference votes for the choice of a seat (spontaneous preference) before the seat was trialed over a period of time (tested preference). The main items in an office chair are a mechanism for regulating the height of the seat and a support for the feet, besides mobility (castors and a swivel seat) to allow the maximum adjustments to meet anthropometric differences and to suit the activities in various workstations.

Office chairs, provided they meet the norms, that are adjustable and are not fixed, generally meet the needs of users who also give value to aesthetics (i.e., colored and rounded) or the appearance of comfort (they appear to be fluffy, larger, etc). Thus, it can be concluded that the chairs that most satisfy are those that enable mobility (i.e., have castors and a swivel seat), have an adjustment mechanism for height of seat, and a support for the back (especially, a lumbar support). Another issue that became clear in this research study was that the use of a footrest is indispensible (though not given value by most of the volunteers), whereas the armrest is not always necessary or even desired by users. As it is an expensive item, it is for the company to acquire chair models that suit each employee, which are in line with appropriate aesthetics and economic affordability.

With the exception of three users from the HR/training department, the study found no correlation between the size of the volunteer and the seat size, and therefore did not confirm the proposal of Helander et al. (1987) that people of small stature do not like large chairs, while large people prefer larger chairs. Based on the results of the study undertaken, and with the chairs trialed, it can be said that the preference for a particular chair depends more on the type of work than on the size of the user.

The analysis of the results obtained with the office chairs raises issues of importance for the evaluation and selection of products. It can be inferred that, although the evaluation of satisfaction was done with an objective focus, using criteria previously defined by the volunteers themselves, the results were probably influenced by aspects of a subjective order. The results indicate that the first impression, revealed in the interviews, influenced the steps that followed. Although the volunteers placed greater importance on the criterion of comfort than on that of aesthetics (or appearance), comfort tends to be evaluated in terms of aesthetics and appearance of comfort. For example, in a spontaneous evaluation, the preference always fell back on the more colorful chairs (the yellow and red ones) to the detriment of the black ones, especially if they seemed larger and more "fluffy."

Regarding the method proposed, it can be said that it is sensitive to the different levels of satisfaction, comfort, and discomfort/pain of the users with the seats trialed, and is easy to apply, the volunteers enjoyed participating and so it was possible to obtain additional information, including users' spontaneous statements on the chairs they were trialing and the importance they attach to the evaluation criteria adopted. These observations and the quantitative results of the experiments can be used to incorporate the user's view into design concepts for seats.

In short, the participatory method proposed enables the evaluation, qualitatively, of

(i) The part(s) of the body in which volunteers feel most discomfort/pain during work
(ii) Which items of demand are considered important in the choice of the chair
(iii) The order of spontaneous preference before the trials and the preference after the trials
(iv) The chairs that caused most and least discomfort/pain for the volunteers during the trials
(v) The satisfaction of the volunteers with the chairs trialed as to the criteria of comfort, safety, adaptability, practicality, suitability for work, and appearance
(vi) The importance attributed to these criteria by the volunteers
(vii) The best and worst chairs as a result of the volunteers' satisfaction with the weighted criteria

The method also enables the identification, qualitatively, of

(i) The volunteers' perception as to the criteria of evaluation
(ii) The spontaneous preference before the trials and the preference after the trials

ACKNOWLEDGMENTS

This chapter was in great part based on the masters dissertation of Eloisa Monteiro Silva who received a grant from Capes and had the collaboration of the statistics teaching staff of PPGEP/ UFRGS (Flávio Sanson Fogliatto, PhD, Dr. Carla ten Caten, and Marcia Echeveste, MSc) as well as Patricia Klaiser Biasoli, a statistician then following a master's degree in production engineering, and João Pedro Aguiar and Cinara Campagna who updated all the graphics. The research team from NDES/LOPP/PPGEP/UFRGS who participated in this survey (Lia Buarque de Macedo Guimarães and Eloisa Monteira Silva) wish to thank the companies Steelcase, Giroflex, Alberflex, and Tradesign who supplied the office chairs and Erghos Sistemas para Escritórios Ltda., who supplied the foot supports. We also thank the companies that allowed this study to be conducted and we especially thank all the volunteers without whom the research study would not have happened.

REFERENCES

Associação Brasileira de Normas Técnicas. 1997. *NBR 13962: Móveis para escritório – cadeiras – características físicas e dimensionais*. Rio de Janeiro: ABNT.

———. 1998. *NBR 14110: Móveis para escritório – cadeiras – ensaios de estabilidade, resistência e durabilidade*. Rio de Janeiro: ABNT.

Bomfim, G.A. 1998. *Idéias e formas na história do design: uma investigação estética*. João Pessoa: UFPB.

Brienza, D.M., Chung, K.C., Brubaker, C.E., Wang, J., and Karg, P.E. 1996. A system for the analysis of seat support surfaces shape control and simultaneous measurement of applied pressures. *IEEE Trans Rehab Eng* 4 (2): 103–13.

Bürdek, B.E. 1994. *Diseño: historia, teoría y práctica del diseño industrial*. Barcelona: Editorial Gustavo Gili.

Corlett, E.N. 1995a. The evaluation of industrial seating. In *Evaluation of Human Work: A Practical Ergonomics Methodology*, ed. J.R Wilson and E.N. Corlett, 621–36. London: Taylor & Francis.

———. 1995b. The evaluation of posture and its effects. In *Evaluation of Human Work: A Practical Ergonomics Methodology*, ed. J.R Wilson and E.N. Corlett, 663–713. London: Taylor & Francis.

Daniel, W.W. 1978. *Applied Nonparametric Statistics*. Boston: Houghton Mifflin.

Fogliatto, F.S., and Guimarães, L.B. de M. 1999. Design macroergonômico: uma proposta metodológica para projeto de produto. *Produto & Produção* 3 (3): 1–15.

Grandjean, E. 1973. *Ergonomics of the Home*. New York: Halstead Press Division.

Grieco, A., Occhipinti, E., Colombini, D., and G. Molteni. 1997. Criteria for ergonomic evaluation of work chair. Fifth Proceedings of Work with Display Unities International Scientific Conference, Waseda University, Tokyo.

Guimarães, L.B. de M. 2006. Funções de um produto: tendências formais e ética de produção e consumo. In *Ergonomia de Produto,* Vol. 2, 5th ed. org. L.B. de M. Guimarães, 3-1–3-29. Porto Alegre: FEENG.

Guimarães, L.B. de M., and Fogliatto, F.S. 2000. Macroergonomic design: A new methodology for ergonomic product design. Proceedings of the 14th Triennial Meeting of the International Ergonomics Association and 44th Annual Meeting of the Human Factors and Ergonomics Society, July 29 – August 4, San Diego, California. Santa Monica, CA: Human Factors and Ergonomics Society. (CD-ROM)

Gyi, D.E., Porter, J.M., and Robertson, N.K.B. 1998. Seat pressure measurement technologies: Considerations for their evaluation. *Appl Ergon* 27 (2): 85–91.

Hänel, S-E, Dartman, T., and Shishoo, R. 1997. Measuring methods for comfort rating of seats and beds. *Int J Ind Ergon* 20:163–72.

Helander, M.G., Czaja, S.J., Drury, C.G., Cary, J.M., and Burri, G. 1987. An ergonomic evaluation of office chairs. *Office: Technology and People* 3:247–62.

Helander, M.G., and Zhang, L. 1997. Field studies of comfort and discomfort in sitting. *Ergonomics* 40 (9): 895–915.

Iida, I. 1990. *Ergonomia: projeto e produção.* São Paulo: Edgard Blücher.

Löbach, B. 1981. *Diseño Industrial: bases para la configuración de los productos industriales.* Barcelona: Gustavo Gili.

Ministério do Trabalho e Emprego – MTE. 2007. Norma Regulamentadora n° 17 (NR 17) – Ergonomia. http://www.mte.gov.br (accessed March 15, 2009).

Montgomery, D.C. 1991. *Diseño y Análisis de Experimentos.* México: Grupo Editorial Iberoamérica.

Rieck, A. 1996. Über die Messung des Sitzkomfortes von Autositzen. In *Sitting Posture,* ed. E. Grandjean, 92–97. London: Taylor & Francis.

Schakel, B., Chidsey, K.D., and Shipley, P. 1969a. The assessment of chair comfort. In *Sitting Posture,* ed. E. Grandjean, 92–97. London: Taylor & Francis.

———. 1996b. The assessment of chair comfort. *Ergonomics,* 12:269–306.

Silva, E.M. 2003. Avaliação da preferência de cadeiras para diferentes tipos de trabalhos de escritório. Masters Diss. Universidade Federal do Rio Grande do Sul.

Stone, H., Sidel, J., Oliver, S., Woolsey, A., and Singleton, R.C. 1974. Sensory evaluation by quantitative descriptive analysis. *Food Technol* 28 (1): 24–34.

Stracker, L.M. 2000. Body discomfort assessment tools. In *The Occupational Ergonomics Handbook,* eds. W. Karwowski and W.S. Marras, 1239–52. London: CRC.

Vergara, M., and Page, A. 2001. Relationship between comfort and back posture and mobility in sitting posture. *Appl Ergon* 33 (1): 1–8.

Vitruvio, M.L. 1955. *Los Diez Libros de Arquitectura.* Barcelona: Editorial Ibéria.

Zhang, L., Helander, M.G., and Drury, C.G. 1996. Identifying factors of comfort and discomfort in sitting. *Human Factors* 38 (3): 377–89.

23 Procedural Pictorial Sequences for Health Product Design: A Study of Male and Female Condoms Conducted with Adults with a Low Level of Literacy in Brazil

Carla Galvao Spinillo, Tiago Costa Maia, and Evelyn Rodrigues Azevedo

CONTENTS

23.1 INTRODUCTION

Visual instructions are widely used in Brazil in printed health material to communicate ways of preventing and treating diseases. Instructions that are mostly represented by pictures are referred to as procedural pictorial sequences (PPSs; Spinillo and Dyson 2001). In these sequences, pictures are the main mode of representing information. Considering that there are about 14.6 million illiterate people in Brazil (IBGE 2010), PPSs seem to be of prime importance in getting the message across in educational material, particularly in the health field.

The contamination of the human immunodeficiency virus–acquired immune deficiency syndrome (HIV-AIDS) among the population is one of the main concerns of the Brazilian Ministry of Health. According to governmental sources of the Brazilian Ministry of Health (2010), approximately 593,000 people in Brazil are infected with HIV-AIDS—208,000 women and 385,000 men. There are several reasons for these high figures, such as the poor quality of the Brazilian healthcare system and individuals' attitude toward HIV-AIDS due to prejudice and sexual taboo. Nevertheless, the role played by information in this scenario seems to be of relevance, and the Brazilian government has invested a considerable amount of funds in campaigns and in the production of educational material on the prevention of HIV-AIDS. Although contamination may occur in various ways, such as blood transfusion, the majority of the Brazilian population considers sexual intercourse as its main cause. Taking this into account, the Brazilian government spends an average of U$2.5 million per year on campaigns for the prevention of HIV-AIDS and distributes about 20 million condoms to the population (www.saude.gov.br). However, little has been investigated on the communicational efficacy of such campaigns, and on whether the population knows how to use condoms. There seems to be an inversely proportional relationship between HIV contamination and years of formal education, although in recent years this is gradually changing (Fonseca et al. 2000; Brito, Castilho, and Szwarcwald 2001; Brazilian Ministry of Health 2010). Brazilians with a low level of literacy are more exposed to HIV-AIDS and less acquainted with the methods of prevention, such as the male and female condoms, than literate Brazilians.

Taking the above facts into account, coupled with the high number of illiterate people in Brazil, this chapter is concerned with the effectiveness of PPSs on how to use male and female condoms by adults with a low level of literacy. Before presenting the experimental study on this topic, it is necessary to consider the development of condoms as products and briefly discuss the design of PPSs on how to use condoms. We will also look at some previous studies in this field.

23.2   CONDOMS: A SUCCESSFUL PRODUCT DESIGN?

Condoms, whether male or female, are a method of contraception and prevention of AIDS/STIs (sexually transmitted infections) and are widely employed in many countries. While female condoms are a twentieth century device/product, male condoms may have been used for at least 400 years. Although the literature on the origin of the male condom cannot say exactly when it was invented, there are records indicating the use of condoms before the fifteenth century in Asia, and later in Europe. The need for devices to prevent pregnancy and, mainly, sexually transmitted diseases (STD) among members of the upper class and the military was the trigger for the invention of condoms. The spread of syphilis in the Middle Age and Renaissance Europe, and sexual diseases among soldiers during the wars in the past centuries led to the development of male condoms as a product able to reduce the number of infected people and casualties.

Despite religious and political reactions condemning the use of condoms, mostly from the Catholic Church and conservative parties, male condoms gradually became a popular product in national and international markets in the nineteenth century (Youssef 1993). To improve its production, several materials and techniques have been employed throughout history. The first condoms were made of oiled silk paper, or lamb intestines in the case of China. In Japan, they were made of tortoise shell or animal horn. In Europe, they were made of lambskin/intestines or fabric saturated with a chemical solution and held together with a ribbon. These were described in Falloppio's treatize on health in fifteenth-century Italy. It was only in the late nineteenth century that rubber was employed to manufacture condoms. Initially, molds wrapped with strips of raw rubber were dipped in a chemical solution to treat the raw material. Then, in the early twentieth century, liquid rubber—obtained by using gasoline or benzene—was used, instead of soaking glass molds. This was known as the cement dipping technique, which evolved to produce latex condoms. Such condoms employed water instead of inflammable chemicals.

With the automation of condom production in the 1930s, the industry grew significantly and in the second half of the twentieth century, male condoms became one of the most popular methods of birth control in the United States and Europe, as well as a device to prevent infection by HIV-AIDS. Nowadays, condoms are still made of latex, although they can also be made of polyurethane and polyisoprene (synthetic latex) among other materials. The quality of condoms as an industrial product has become a governmental issue worldwide. This concerns not only its efficiency in the prevention of STIs/AIDS and as a method of birth control, but also its effects on human health, such as the allergic reactions that condoms may provoke. An example of this is the treatment of latex to lower the level of proteins responsible for allergic reactions to a 10% rate by using Vytex in latex condoms, which was approved by the U.S. Food and Drug Administration in 2009 (Vystar 2009).

With the same purpose as male condoms and made of a similar material, female condoms were made available in the late twentieth century. They have stretchy rings at each end and look like a small transparent pouch, and are an alternative device for contraception and prevention of STIs/AIDS for women. This is particularly relevant when considering the cultural and gender aspects of using condoms. In cultures in which men have the dominant role in society, women may not feel confident and comfortable enough to suggest the use of male condoms to their partners (Gomez and Marin 1996). In some cases, women may get negative and even violent reactions from their partners (Acosta and Barker 2003; Geluda et al. 2006). Hence, female condoms have allowed and even empowered women to have their say in the use of this method of contraception and prevention of STIs/AIDS.

Regarding the effectiveness of female condoms in preventing STIs/AIDS and pregnancy, there seems to be no conclusive research results, although it is considered a reliable device. In comparison to male condoms, they are costly, and the procedures for using a female condom differ from the male condom, therefore they have to be learned. These aspects may have affected the popularity of female condoms among consumers. Nevertheless, due to its importance as a method of contraception and sexual diseases prevention, international institutions promote the use of female condoms worldwide by distributing them among women. For instance, in 2005, the United National Population Fund (UNFPA) launched the Global Female Condom Initiative and by 2008, around 35 million female condoms had been distributed in 93 countries. Although it is a significant figure, it is still lower than the number of male condoms distributed yearly worldwide—about 10 billion (CBS 2010).

Although very effective methods of birth control, male and female condoms have an impact on the environment if not disposed of properly. The materials they are made of and the plastic and foil wrappers of their packages are mostly non-biodegradable. Thus, they damage the environment when inappropriately disposed of. They can be mistaken for food by animals when in contact with litter in public spaces, such as parks. If they end up in the ocean, they may be eaten by fish or cover coral reefs, affecting the sea flora and fauna. Thus, condoms should not only be disposed of in litter containers, but they should also be produced as an environmentally friendly product. Thus, manufacturers are expected to further develop both male and female condoms, accounting for their life cycle, from conception to disposal, and their quality as products designed for users, taking into consideration users' needs and background/lifestyle.

From the design point of view, male and female condoms are simple products in terms of shape complexity and their purpose, i.e., to impede men's semen from getting in physical contact with the body of a partner during sexual intercourse. Every condom is a single-piece device, but they differ in task. For the male condom, the task is to put it on a rigid penis, and for the female condom, the task it is to place one ring at the end of the condom into the woman's vagina and then guide the penis into the condom through the other ring. The following figures show the procedural task analysis of using male and female condoms. They indicate the sequence of steps (square shapes) users must go through to successfully complete the tasks, and the decisions (diamond shape) they have to make during the procedure. These are presented as questions with Yes/No answers. Depending on the outcomes of the decision, users may need to return to previous steps, carry on the next step, or even take other

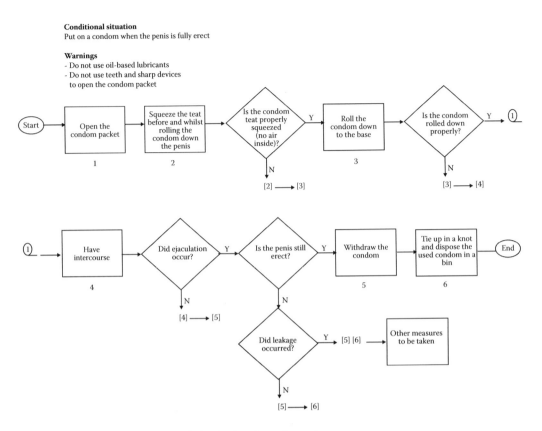

FIGURE 23.1 Procedural task analysis of using the male condom.

paths related to the procedure. For male condoms, warnings and conditional situations to initiate the task were added to the diagram as they are frequently found in the instructions available to users. The tasks can be considered similar in complexity as the number of steps and decision points for both male and female condoms are similar. A closer look at the diagrams leads to the assumption that these are "decision-making oriented tasks." For almost all steps, the outcomes demand decisions from users: the procedure for using male condoms presents four decision points out of six steps, whereas female condoms require five decision points out of seven steps (Figures 23.1 and 23.2).

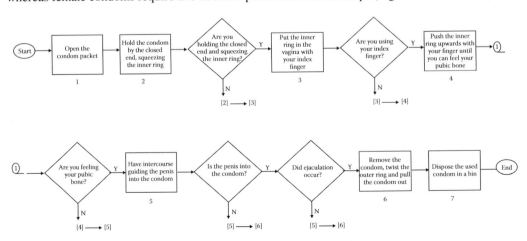

FIGURE 23.2 Procedural task analysis of using the female condom.

Considering the aspects mentioned above, one might wonder how successful these products really are and what difficulties users may face when performing those tasks. Furthermore, as condoms should present instructions to users—like any product for consumers—one may also wonder how effective such instructional materials are. Considering these questions, the following section presents some concerns on the design of PPSs on how to use male and female condoms regarding their graphic representation of information.

23.3 DESIGN OF PROCEDURAL PICTORIAL SEQUENCES ON HOW TO USE CONDOMS

Product instructions for users are mandatory and regulated by national and/or international institutions to guarantee consumers' rights. Thus, the condom industry has to provide users with information on how to put condoms on, as well as on possible hazards regarding the task, which may affect users and/or the product. In Brazil, the national consumer code (www.portaldoconsumidor.gov.br/) and the Ministry of Health regulate instructional materials addressed to users of condoms. In such material, PPSs are commonly employed to represent the tasks of using male and female condoms. They can be displayed in leaflets inserted in the packages or printed on them.

The pictorial representation of the tasks of using condoms involves conveying procedural and non-procedural information to users. The former regards the steps and the latter other information, such as warnings. As the tasks require the representation of a man's or a woman's body and of sexual intercourse, culture and users' level of visual literacy may pose constraints.

In order to reduce the impact of cultural values on the acceptance of pictorial representations of using condoms, some PPSs employ visual metaphors to represent the tasks and even cartoon-like representations. Examples are found in PPSs produced in Brazil. In one of them, the procedure of putting on a male condom is demonstrated using a banana (Figure 23.3). Other PPSs addressed to low-income and low-literacy audiences use funny characters to represent the penis and the male condom in a textless instruction (see Figure 23.3). In this example, the condom initially appears as a character and then becomes a "hat" to be worn by the penis character (!). In addition to the poor pictorial depiction of this PPS, the use of humor as a graphic strategy may be questionable.

FIGURE 23.3   Examples of PPSs on using the male condom available in Brazil. (Authors' private collection.)

Previous studies on educational and instructional materials show that humor may produce disapproving responses from audiences (e.g., Pettersson 1999; Spinillo 2006). They may consider humor in such materials a patronizing or even infantile approach to communication, particularly when information concerns themes taken seriously by users, such as health issues.

Although the use of visual rhetoric may promote the acceptance of PPSs on using condoms within certain audiences, it may hinder users' cognitive processes. Users have to interpret and infer various meanings from the pictures, i.e., "wear the hat" should mean "put the condom on." They also have to fill in information gaps, which may occur due to the impossibility of representing some contents of the task for cultural reasons (e.g., sexual taboo). For instance, sexual intercourse was not depicted in either of the given examples. Thus, users should infer that it occurs *after* the pictorial representation of putting the condom on and *before* removing the condom. Otherwise, the function of the condom may be critically compromised and the safe approach to sex jeopardized.

Regarding the depiction of procedural content, the absence of pictorial context may make visualizing some steps difficult due to incomplete representation of information. For instance, in the cartoon-like PPS of using a male condom, only partial depiction of fingers is employed to convey actions, such as wearing (putting on) the condom and squeezing its closed end. They demand a cognitive effort to infer context and therefore meaning from these images.

In a similar way, the use of drawing codes may make the interpretation of images in procedural material difficult. They are employed to allow users to visualize certain information that cannot be seen with the naked eye. For instance, sectional views of a woman's body can be found in pictorial sequences describing the procedure of using female condoms. They show how the condom should be placed inside the body and how it should be withdrawn (Figure 23.4). Although they facilitate task performance, they require a sophisticated visual repertoire from users. If they are not acquainted with such visual codes, they may not be able to interpret the pictures, or even, be afraid of performing the step/task as they may (mis)judge it as complicated/difficult.

In a study by Spinillo, Azevedo, and Benevides (2004), several drawbacks in the graphic representation of using male and female condoms were found in a sample of 54 PPSs available in Brazil. Among the results, it is worth highlighting those regarding sequences' reading direction, representation of warnings and of actions. For the former, the authors found an absence of reading guides (numbers, letters) in the majority of PPSs. According to the authors, this makes the identification of sequences' reading direction difficult. Users could follow the sequences either from left to right or from top to bottom, depending on how they interpret the pictures. This is likely to occur in audiences with a low level of literacy, who will rely on the meaning of images to determine the sequence flow. In relation to warnings, they were not found in most PPSs, and when they occurred they were generally represented by text rather than by pictures. This was seen by the authors as a weakness of the design, since some prohibitive warnings are of prime importance to task performance and users' safety. For instance, warning users about not using their teeth or a sharp object (e.g., scissors,

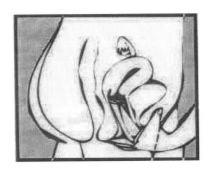

FIGURE 23.4 Image of sectional view of woman's body from a PPS on using the female condom. (Authors' private collection.)

cutters) to take the condom out of the packet to avoid damaging/tearing it. Other drawbacks found in the sample regard the lack of arrows to represent the sequence of the actions in most pictorial instructions. This could lead to misinterpretation of the steps depicted.

Unfortunately, weaknesses in the pictorial representation of the procedures for using male and female condoms are not restricted to Brazil. Instructional materials on this topic produced in Europe, Africa, and the United States presented problems similar to those discussed above. For example, in a PPS on how to use female condoms displayed in instructional materials in the United States and England, sectional views are employed to show the interior of the female body in three out of six pictures of the sequence. In Namibia, Africa, a study on visual messages intended to prevent HIV-AIDS carried out by Etezel (2006) show pictorial instructions lacking graphic devices conveying actions, and omitting intercourse (when the condom should be used). Because of the magnitude of this topic, studies on the effectiveness of instructional materials for task performance and on acceptability of information are of prime importance worldwide.

23.4 PREVIOUS STUDIES ON INSTRUCTIONAL PICTORIAL MATERIAL FOR USERS

Studies on cognition and information ergonomics agree that the use of pictures to convey instructions is beneficial to comprehension, facilitates information processes, and reduces cognitive load, particularly when used together with text (e.g., Wright 1999; Ganier 2001, 2004; Wogalter 2006). Moreover, Ganier (2004) stresses that their combined use optimizes the development of mental maps for action plans when a task has to be performed. The satisfactory performance of an instructional task relies on the completeness of the information provided and the quality of the graphic representation (e.g., Wright 1999; Spinillo, Padovani, and Lanzoni 2009).

Nevertheless, several aspects may influence the understanding of pictorial instructions, such as users' familiarity with the graphic presentation of information. The level of visual literacy plays an important part in communication. People not acquainted with pictorial sequences have difficulties in perceiving (or even do not perceive) the time lapse implied in a series of pictures, leading to communication failure (e.g., Maia 2008; Maia and Spinillo 2008). Comprehension of a pictorial message also depends on users' characteristics, such as age, culture, and level of literacy (Wogalter 2006; Pettersson 2007). Furthermore, a study conducted in England found that literate adults when reading wordless PPSs seem to expect the initial picture to represent a step (Spinillo and Dyson 2001), rather than other kinds of information. In the health field, such aspects are of prime importance, as misinterpretation of pictorial sequences in instructional material may compromise the effectiveness of the product in preventing diseases and infections, thereby possibly endangering people's lives and well-being.

In relation to how content should be communicated in instructional material, affirmative sentences are considered to be better comprehended than negative/prohibitive sentences (Wright 1999). Nevertheless, prohibitions are recommended when users should be aware of possible hazards related to the task or product (Wogater 2006). When using images to convey prohibitive warnings, semantic marks are employed, such as a slash or a cross over a picture. However, they may obstruct the visualization of what is depicted or may even be misinterpreted, if users are not acquainted with these marks (Wogalter 2006).

Another important aspect is the representation of actions in pictorial instructions that refer to the steps to be undertaken by users. Wanderley (2009) claims that an action when depicted provides readers with conceptual and graphic information. Conceptual information refers to the represented participants, movements, frequency, trajectory, and velocity of the depicted action (e.g., run and walk). Graphic information, on the other hand, refers to the ways that actions can be represented visually, such as postural depiction, pictorial context, and schematic/multiple images. The author highlights that a successful depiction of an action relies on the clarity of its representation allied to readers' visual repertoire to grasp its meaning.

When discussing people's attitudes toward product instructions, Wright (1981, 1999) suggests that factors other than the graphic quality and the appropriateness of an instruction may prevent users from reading it. This refers to users' pre-disposition or willingness to approach instructional material prior to carrying out a procedure, which is influenced by the degree of complexity associated with a product or task. Products considered of low complexity may lead users to believe that they can easily perform the task. Consequently, reading instructions is seen as unnecessary. Likewise, instructions may not be read if a product or a task is considered familiar to users. Such attitudes toward instructional material will certainly lead to misconception of the product and/ or of the task. In relation to male and female condoms, because they are seen as products of low complexity, users may think that the procedures for using them are obvious, and therefore, they may intuitively carry the tasks out. In this case, certain relevant information to users' safety can be missed, e.g., "squeeze the closed end of the male condom" or "do not use the teeth or a sharp object to take the condom out of the packet."

Considering the relevance of the correct use of condoms to health safety worldwide, their instructional materials should be carefully read by users. Thus, condom manufacturers as well as health governmental organizations ought to seek ways of encouraging users to read such instructional materials. Providing effective as well as attractive pictorial instructions on how to use male and female condoms plays a part in this scenario. Despite the relevance of pictorial instructional material on how to use condoms in the prevention of STIs/AIDS and unplanned pregnancies, little seems to have been investigated on this subject from an information ergonomics perspective. This is a serious matter, particularly when considering developing countries and low-literacy-level audiences, to whom pictures are the only source of information.

With this in mind, an investigation was conducted in Brazil, the purpose of which was to answer the question: *How effective are PPSs on using male and female condoms when addressed to users with a low level of literacy?* The next sections present the outcomes of this study and some recommendations for the design of pictorial instructions.

23.5   EXPERIMENTAL STUDIES ON USING MALE AND FEMALE CONDOMS

In order to investigate the effect of PPSs on task performance of using male and female condoms, two experimental studies were conducted in Brazil. Seventy-eight people took part in these studies. They were adults (40 men and 38 women) with a low level of literacy and with little or no experience of using condoms. The participants were divided into two groups for the experiments with each type of condom: (a) a control group and (b) an experimental group. For female condoms, the two groups consisted of women only, whereas for male condoms, both men and women took part. The task was performed in a simulated manner on male and female body models without the aid of the PPSs for the control group and with the aid of the PPSs for the experimental group. Figure 23.5 shows how participants were divided across groups for female and male condoms.

	Female condom	Male condom	
Control group (without PPS)	19 women	Same 19 women 20 men	39
Experimental group (with PPS)	19 women	Same 19 women 20 men	39
	38 women	38 women 40 men	78 participants

FIGURE 23.5   Participants division across groups for female and male condoms.

23.5.1 MATERIAL

For both groups, the experiment material consisted of male or female condoms and body models. For the experimental group, participants were also provided with a wordless PPS on how to use male condoms or female condoms. Figure 23.6 shows male and female condoms and their PPSs used in the experiments.

23.5.1.1 Procedural Pictorial Sequence on Male Condom

This was a sample-type from a previous study on a graphic presentation of pictorial instructions for using male condoms produced in Brazil (Spinillo, Azevedo, and Benevides 2004). It was in accordance with the graphic pattern found in a sample of 54 pictorial sequences analyzed in this study: vertical arrangement of pictures, drawing style, absence of reading guides and visual codes for representing actions, space and borders separating the pictures, and use of partial depiction.

The male condom PPS begins with a warning about how not to open the condom packet, followed by four steps: (1) squeeze the teat and place the condom on the penis fully erect; (2) squeeze the teat and unroll the condom down to the base of the penis; (3) withdraw the condom after sexual intercourse (with the penis still erect); hold it around the edge so as not to spill the sperm; (4) tie the condom in a knot and dispose of it.

23.5.1.2 Procedural Pictorial Sequence on Female Condom

This pictorial sequence was the one distributed by the Brazilian government to the population, and it is the official instructional material on this topic available to women in the country. The original PPS is displayed on the back of the condom package and presents numbers and arrows as reading guides. These features were removed from the testing PPS in order to make it a picture-only instruction, as for the PPS on male condom.

The PPS presents an introductory picture showing the female condom, followed by five steps: (1) squeeze the inner ring with your thumb, index finger, and middle finger; (2) put the inner ring into your vagina; (3) push the inner ring upward with your finger until you can feel your pubic bone; (4) during sexual intercourse, hold the outer ring with one hand and, with the other, guide the penis into

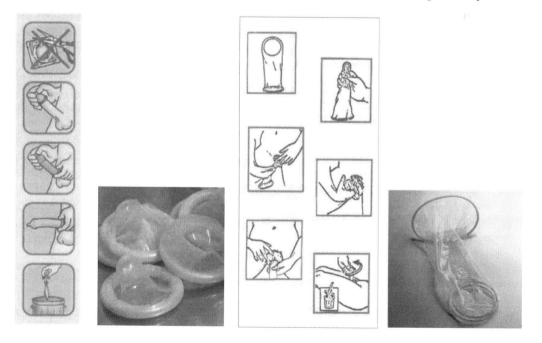

FIGURE 23.6 Male and female condoms and their PPSs used in the experiments.

the condom; (5) after sexual intercourse, twist the outer ring so as not to spill the sperm. Remove the condom by pulling it from the vagina and dispose of it in a bin.

23.5.2 PROCEDURES

In all experiments, the participants performed the tasks individually. A male or female condom, still in its package, was given to each participant. The participants were then asked to open the package, put the condom on and withdraw the condom from the body model. Those in the experimental group were given a PPS and allowed to read it whenever they wanted to. A semi-structured interview was conducted with each participant after task completion in order to find out their impressions of putting on and withdrawing the condoms. The questions asked were about the comprehension of the PPSs. Data were collected through videotape (tasks), audio recording (interviews), and note taking by the interviewers.

In both groups, the independent variable (X) measured was the presence of PPS, whether for the male or female condoms. The dependent variable (Y) was participants' task performance. In the experimental group, the other dependent variable (Y2) considered was the comprehension of the PPS.

23.5.3 RESULTS AND DISCUSSION

The results were analyzed in a qualitative manner and the numbers presented here are intended to show patterns and/or suggest trends in the findings. The qualitative approach was adopted as it allows the discussion of particular aspects that are of relevance to understand the scope of graphic representation of using condoms and its effects on task performance. Initially, the findings are presented for each type of condom. This is followed by a general discussion.

23.5.3.1 Male Condom

The general results on task performance on male condom in the control group, considering both men and women, show approximately two correct steps per participant, out of the four steps necessary to carry out the task. The number of errors was slightly higher than the number of steps (2.5 errors per participant). An interesting figure relates to the actions performed by participants to get acquainted with the condom and to infer on the steps ($N=21$). They were mostly attempts to identify the correct side to unroll the condom. These figures were higher in the experimental group, where participants could consult the PPS, either for getting the steps right (three out of the four) or for their efforts in identifying the condom ($N=26$). Although the performances were slightly better in the experimental group, these results seem to indicate a drawback in condom design regarding unrolling, which confused users, and which was not accounted for in the pictorial instruction.

When interviewed, most participants of both the control and experimental groups believed they performed the task successfully ($N=31$), and considered it an easy task ($N=30$ in the control group and $N=33$ in the experimental group). These impressions were generally based on their opinion that the task was simple and quick to carry out. Participants in the experimental group also found that the PPS helped in the task performance ($N=38$). These results suggest that participants are not aware of the difficulties in putting on and withdrawing male condoms, therefore, their assumptions about the task and product are misleading.

The results on the comprehension of the pictorial instruction on using male condoms were then a matter for concern. The majority of participants ($N=23$ out of 39) could not understand that the cross over the first picture in the warning meant prohibition. Also, the depiction of the step "squeeze the condom teat" was not properly perceived by nearly all participants ($N=35$). Similarly, the following steps (unroll and withdraw) were not fully understood ($N=32$). This was due to the lack of pictorial cues, such as arrows, to represent movement/action and the time/event lapse between putting on and withdrawing the male condom during sexual intercourse (steps 2 and 3). On the other hand, the depiction of disposing of the condom in a bin was noticed by participants, although to tie

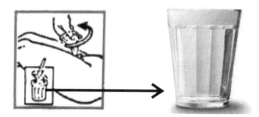

FIGURE 23.7   Small-scale depiction of a bin in the PPS and a glass commonly used in Brazil, which were mistakenly associated.

the condom in a knot was not ($N=28$). This may be due to the absence of graphic emphasis showing details (i.e., the knot).

23.5.3.2   Female Condom

Thirty-eight women took part in the experiments on how to use the female condom, and the results on task performance show similar figures in both the control and experimental groups. The figures increased slightly in the experimental group (with PPS). When we consider the figures per participant across groups, we can see that about three of the four steps were performed satisfactorily, and approximately 2.5 errors occurred when putting on and withdrawing the female condom from the body model. Similar to the results for the male condom, attempts to get acquainted with the condom also occurred during the task, and they increased in the experimental group ($N=48$ and $N=24$ in the control group). This suggests that, as with most products, participants need to be familiar with the condom to use it. Moreover, these figures further indicate that the initial picture showing the female condom was confusing. Nevertheless, this introductory image produced the highest level of comprehension ($N=18$ out of 19). This suggests that although the image enabled the recognition of the female condom, it did not provide the necessary visual features to properly perform the task. As a result, the participants did not have enough information about the product and consequently were not confident enough to put on the female condom.

Although the female condom's PPS produced satisfactory responses on comprehension across participants ($N=19$), it presented weaknesses that jeopardized task performance. About two-thirds of the participants did not fully understand the picture on how to hold the condom prior to inserting it in the female body (step 1). Six out of nineteen participants were not able to work out if the condom was being inserted or removed (step 3). The close depiction of the male body showing how to guide the partner during intercourse (step 4) was not acknowledged by some women. The last step, which showed that the female condom should be twisted while being withdrawn and that it should be disposed of in a bin, was only understood by two participants. This was because many different actions were depicted in a single picture. These outcomes suggest that some depictions lack pictorial context. They do not offer enough, if any, information about the product and/or task, and thus they represent the actions ambiguously.

In relation to the interviews, although they considered the task easy, most women admitted having difficulties in putting on and/or withdrawing the female condom. In the control group, 14 out of 19 participants believed they did not perform the task successfully due to their unfamiliarity with the female condom. In the PPS, the depiction in small scale/size of a waste bin for disposing of the condom led the participants in the experimental group to associate it with a glass shape common in Brazil (Figure 23.7). The poor quality of the PPS was also mentioned by some participants ($N=6$) as the main reason for not performing the task satisfactorily.

23.6   GENERAL DISCUSSION

The outcomes of the experiments, in general, indicate that both PPSs were not well understood by the participants, particularly when more than one action/step was depicted in a picture. The PPS did not strongly affect the task performance of the male condom, and it had a slightly positive effect on the task performance of the female condom. However, none of the participants carried out the tasks satisfactorily. It was also noticed that participants expected the PPSs to begin with a step, instead of a warning (male condom) or an introductory picture (female condom). Also, with the female condom, the women who looked at the PPS, performed the task better compared to those who did not.

Those results are in accordance with previous research on instructional material, particularly with regard to audiences with a low level of literacy in developing countries. They ratified the idea that users' level of familiarity with the product (condoms) affects task performance. Task performance is also influenced by how users view the complexity of the product (Wright 1981, 1999), which can make users misjudge their performance. The outcomes also confirmed that prohibitive statements conveyed through pictures may not be fully understood in instructional material if users are not familiar with semantic marks for negation, such as a cross over an image (Wright 1999; Wogalter 2006). These results further endorse previous findings, which suggested that participants may not notice that a moment/event has been omitted in a task, and thus not infer the time/event lapse in-between pictures in an instructional sequence (Maia 2008). Since this study did not look at participants' cognitive load regarding the reading of the instruction and task performance, its outcomes cannot be said to support (or not) the idea that images in instructional material reduce cognitive load and promote understanding (Ganier 2004). However, the results suggest that pictures may only facilitate comprehension in instructional material if properly designed. Thus, developers of instructional material should consider not only the use of pictures, but also the quality of their depiction.

23.7 CONCLUSIONS AND RECOMMENDATIONS

The present study has considered only two pictorial instructions on how to use male and female condoms, thus its findings cannot be over generalized. However, as previously mentioned, the PPS on the male condom presented the graphic characteristics common to the pictorial instructions on this topic available in Brazil, and the PPS on the female condom was the one largely distributed by the Ministry of Health. Thus, by considering the representative role that the PPSs investigated here may play in the prevention of STIs/AIDS in Brazil, it seems pertinent to draw some conclusions, attempting to answer the question earlier posed: *How effective are PPSs on using male and female condoms when addressed to users with a low level of literacy?*

According to the outcomes of this study, PPSs on how to use male and female condoms do not seem to successfully communicate their messages to illiterate audiences. Their poor design appears to make little difference in task performance. Hence, they may not contribute to decreasing the figures on contamination of STIs/AIDS in Brazil.

With the purpose of contributing to the improvement of the design of pictorial instructions on using male and female condoms in Brazil, some recommendations are proposed based on the study findings and on the literature on the topic (e.g., Pettersson 2007; Spinillo, Padovani, and Lanzoni 2009). For male condoms: (a) a picture conveying warnings at the beginning of the sequence should strongly differ from those depicting steps (e.g., distinct shape, frame, size, color) in order to prevent the message from being misunderstood; (b) a picture should be added to show the correct side to unroll the condom; (c) sexual intercourse should be represented to avoid an unnecessary time lapse between the frames of the PPS, which may lead to misinterpretations. For female condoms, designers should also consider: (a) an extreme close-up in the depiction should be avoided, as it may lead to loss of important context information for the task, especially when showing the male body during intercourse; (b) the elements depicted should be part of the audience's visual repertoire in order to ease recognition, i.e., when representing a waste bin; (c) pictures should not depict more than one action/step to avoid an excess of information in a picture, as this may affect users' discernment of the elements. Finally, for both male and female condoms: (a) their pictorial instructions should cater for emphatic devices, such as color and/or shapes to show details of the product or task that are relevant to users, such as "to tie the condom in a knot" before disposing of it; and (b) visual cues, such as arrows, should be used to convey movement/action in the pictures to avoid ambiguity.

Finally, we hope that the aspects discussed here and the outcomes from the experimental studies contribute not only to enhance the design of PPSs for using condoms, but also stress the relevance of a user-centered design approach to instructional material. For this to occur, users should be involved in the design process so that the material can cater for their information needs, characteristics, and cultural constraints. We also hope that condoms will be manufactured as an environmentally friendly product, and designed to be intuitively used by literate as well as illiterate audiences. Only then, pictorial instructions may become unnecessary.

ACKNOWLEDGMENT

We would like to thank all the participants who volunteered for the experiments, thereby making this chapter possible; Veronica Freire and Daniel Benevides for collaborating in this study; and CNPq – the National Council for Scientific Development in Brazil for funding this research.

REFERENCES

Acosta, F., Barker, G. 2003. *Homem, violência de gênero e saúde sexual e reprodutiva: Um estudo sobre homens do Rio de Janeiro/Brasil.* Rio de Janeiro: Instituto Promundo/NOOS.

Brazilian Ministry of Health. 2010. http://www.aids.gov.br/ (accessed February 18, 2010).

Brito, A.M., Castilho, E.A., and Szwarcwald, C.L. 2001. AIDS e infecção pelo HIV no Brasil: Uma epidemia multifacetada. *Revista da Sociedade Brasileira de Medicina Tropical* 34/2:207–17.

CBS News Health. 2010. http://www.cbsnews.com/stories/2010/02/11/health/ (accessed February 17, 2010).

Etezel, G. 2006. Visual communications and media methods for education, awareness and behavioural change. Unpublished research. Namibia.

Fonseca, M.G., Bastos, F.I., Derrico, M., Andrade, A.T., Travassos, C., and Szwarcwald, C.L. 2000. AIDS e grau de escolaridade no Brasil: evolução temporal de 1986 a 1996. *Caderno Saúde Pública* 16:77–87.

Ganier, F. 2001. Processing text and pictures in procedural instructions. *Information Design Journal* 10/2:143–53.

———. 2004. Les apports de la psychologie cognitive a la conception d'instructions procedurales. *InfoDesign-Revista Brasileira de Design da Informação* 1/1:16–28. http://www.infodesign.org.br/conteudo/artigos/10/port/art02_Les_apports.pdf

Geluda, K., Bosi, M.L.M., Cunha, A.J.L.A., and Trajman, A. 2006. Quando um não quer, dois não brigam: Um estudo sobre o não uso constante de preservativo masculino por adolescentes do Município do Rio de Janeiro, Brasil. *Cadernos de Saúde Pública* 8/22. http://www.scielosp.org/scielo.php?pid=S0102-311X2006000800015&script=sci_arttext.pdf

Gomez, C.A., and Marín, B.V. 1996. Gender, culture, and power: Barriers to HIV-prevention strategies for women. *The Journal of Sex Research* 33/4:355–62.

IBGE – Instituto Brasileiro de Geografia e Estatística. 2010. http://www.ibge.gov.br/ibgeteen/pesquisas/educacao.html (accessed February 18, 2010).

Maia, T.C. 2008. Estudo analítico da representação de dimensões temporais em instruções de produtos de consumo. In *Anais do P&D Design 8 - 8° Congresso Brasileiro de Pesquisa e Desenvolvimento em Design.* eds. R. Tori, E. P. Puftzenreuter, e P. L. Farias, 120–132. São Paulo: AEND-Associação Ensino e Pesquisa em Design.

Maia, T.C., and Spinillo, C.G. 2008. How are time related concepts pictorially represented in instructional material? In *Selected Readings of the International Visual Literacy Association Annual Conference 2007* (1st ed), eds. M.D. Avgerinou, R. Griffin, and C.G. Spinillo, 137–46. Norristown, PA: OmniPress.

Pettersson, R. 1999. Attention: An information design perspective. In *Proceedings of the Vision Plus 6, Drawing the Process: Visual Planning and Explaining.* ed. P. Simlinger, 145–51. Viena: IIID-International Institute for Information Design.

———. 2007. *It depends.* Tullinge. http://www.iiid.net/

Spinillo, C., Azevedo, E.R., and Benivides, D. 2004. Visual instructions in health printed material: An analytical study in PPS on how to use male and female condoms. In *Selected Readings of the Information Design International Conference*, eds. C.G. Spinillo, and S. Coutinho, 90–104. Recife: SBDI.

Spinillo, C., and Dyson, M. 2001. An exploratory study of reading procedural pictorial sequences. *Information Design Journal* 10/2:154–68.

Spinillo, C.G. 2006. Information design and cultural understanding. In *Knowledge Media Design. Theorie, Methodik, Praxis*, ed. M.P.F. Eibl, H. Reiterer, and F. Thissen, 319–29. Munique/Viena: Oldenbourg Wissenschaftsverlag.

Spinillo, C.G., Padovani, S. and Lanzoni, C. 2009. Patient Safety: Contributions from a Task Analysis Study on Medicine Usage by Brazilians. In *Human Interface and the Management of Information*. ed. G. Salvendy and M. Smith, vol. 5618, 604–8. Heidelberg: Springer.

Vystar. 2009. *FDA Clearance for Envy Natural Rubber Latex Condom Made with Vytex NRL* (http://www.vytex.com/BizDocs/R-EnvyFDA_Clearance.pdf). Press release. Retrieved 2010-02-05.

Wanderley, R.G. 2009. As imagens dinâmicas dos manuais de produtos eletroeletrônicos conseguem instruir seu leitor a realizar a ação necessária? In *Pesquisa Científica em Design da Informação: Anais do 4 Congresso Internacional de Design da Informação* eds. P. L. Farias, C. G. Spinillo, R. Tori, e L. A. L. Coelho, 57–63. Rio de Janeiro: SBDI-Sociedade Brasileira de Design da Informação.

Wogalter, M. (ed). 2006. *Handbook of Warnings* (1st ed). New York: Lawrence Erlbaum Associates.

Wright, P. 1981. The instructions clearly state…' Can't people read? *Applied Ergonomics* 12/3:131–41.

———. 1999. Printed instructions: Can research make a difference? In *Visual Information for Everyday Use: Design and Research Perspectives*, ed. H.J.G. Zwaga, T. Boersema, and H.C.M. Hoonhout, 45–66. London: Taylor & Francis.

Youssef, H. 1993. The history of the condom. *Journal of the Royal Society of Medicine* 86/4:226–28.

24 Special Needs in Pleasure-Based Products Design: A Case Study

Erminia Attaianese

CONTENTS

24.1 INTRODUCTION

As life expectancies rise, the number of elderly and disabled people living in ordinary homes increases, requiring domestic products that they can use easily and pleasurably in order to live independently. Several studies have been conducted on disabled people's products and systems design, centered on their functional dimensions and performances, without considering users' satisfaction characteristics, such as aspect and likeability. These kinds of products are usually considered as tools able to help users with special needs to overcome their physical limitations, thereby increasing their own capabilities. It is not necessary for them to be pleasant or user friendly. They only have to be effective and efficient, to work as useful devices, without considering the quality perceived in products. This chapter presents predicting needs criteria in the design of products for people with special needs, overcoming the efficacy and efficiency product requirements, and paying particular attention to the identification of product characteristics for users' satisfaction. A case study is presented concerning the study of special needs of people with movement disorders, in order to identify design principles and products requirements. Subsequently, a user-centered design process of a domestic telephone for people affected by Parkinson's disease is shown, in which the component acceptability has been previously emphasized. Design criteria for telephone likeability and ease of use have been defined.

24.2 DESIGN CRITERIA IN PLEASURE-BASED PRODUCTS FOR PEOPLE WITH SPECIAL NEEDS

24.2.1 MERGING PLEASURE AND SPECIAL NEEDS IN PRODUCT DESIGN

The ergonomic approach to product design for people with special needs is aimed at realizing objects and systems that can allow users to overcome their physical or cognitive limitations, identifying technical solutions that finally improve product usability for all. This point reflects a substantial change in the thought of users with special needs, replacing the idea of disability with the conviction that each one has different abilities. This may allow individuals, with temporary or permanent functional limitations, to benefit from products that can be used effectively by all, without assistance and without the stigmatization that is frequently associated with the use of assistive devices. To design a really inclusive product, seven design principles have been developed, concerning physical and cognitive aspects. Universal design principles actually suggest that products have to be flexible in use, to accommodate a wide range of individual preferences and abilities; simple and intuitive, giving perceptible information to address any ambient conditions or user's sensory abilities; and tolerant for error, minimizing the adverse consequences of failure actions. Products have to be used with a low physical effort and be sized and spaced appropriately for their approach and use. After all, products have to be designed to allow their equitable use (Beecher and Paquet 2005).

The set of above-mentioned principles forgets to point out the attention on the desirability of an inclusive product, and on its pleasure in use. On the contrary, perceived satisfaction by people with special needs using these products is particularly important, more than for other categories of users. Ease of use and likeability are two conditions that strongly influence product usability in terms of acceptance or refusal by disabled people, who often have to use effective products that are difficult, uncomfortable, and usually ugly—in one word "unpleasant." The products appearance underlines the fundamental role that this requirement has for people with special needs. In fact, the aesthetic characteristics of these kinds of products are usually neglected, on the basis that something useful, such as aids for disabled people, doesn't have to be likeable, just effective. This view leads to the design of products and systems whose aspects are not controlled enough, letting them appear as reserved only for disabled people. These products seem to highlight more their limitations, resulting in frequent refusal of the product. Perceived ease of use is conditioned by product appearance and acceptability by users. In fact, we know that, usually, attractive things work better (Norman 2004), because if the final product is aesthetically pleasing, this has a psychological effect on user' actions and behaviors. Moreover, user satisfaction, expressed in terms of comfort and acceptability, is one of the fundamental components of product usability together with effectiveness and efficiency (Green and Jordan 2002). The actual aim of the proposed methodology is trying to explain their implied components in special needs products design.

24.2.2 DESIGN METHODOLOGY

The presented methodology is founded on a detailed identification of users' needs, gathered from the analysis of the context of the product use. The objective of designing usable products and systems for people with physical limitations is to enable users to achieve goals usually accomplished through difficult actions considered simple for others. Psychological components highly influence the way in which the intended achieved by disabled people, more than for other categories of users.

Particular attention should be paid to satisfaction, for the role that product acceptability plays in the quality perceived and for the effects on the successful use of the product. Considering specific use conditions of products for people with special needs, ease of use and likeability appear to be the

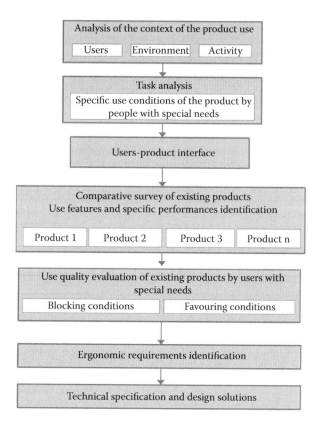

FIGURE 24.1 Methodology flow.

most significant goals to achieve in the design process. The aim of the process has been oriented to identify all the technical choices that make the product usable with pleasure, in terms of simplicity, friendliness, effortlessness, and fun in use.

The methodology we have tested was founded on the iterative process of user-centered design in which the components related to user satisfaction were deeply investigated, in order to identify product requirements that influence more user acceptance (Figure 24.1 depicts user-centered applied methodology). Starting from a detailed investigation of existing products use conditions by consumers with special needs, a series of objective and subjective evaluations have been conducted to identify the level of efficacy preformed by standard products and their ability to be perceived by users as effective, simple, and pleasant to use.

24.3 A PARTICULAR CATEGORY OF SPECIAL NEEDS: PEOPLE WITH MOVEMENT DISORDERS

24.3.1 DEFINITIONS AND GENERAL CONSIDERATIONS

Movement disorders are commonly considered as neurological conditions in which there is production of anomalous movements or their qualitative alterations rather than loss of muscular force. Main disorders concern chorea and jerky body movements, tics and tremors. Choreiform and jerky body movements are involuntary, irregular, and asymmetric movements, characterized by a variable width, which can affect one or both sides of the body. They are completely uncontrolled, purposeless, and rapid motions that interrupt normal movement or posture, but in some cases they are incorporated into deliberate movement patterns. Typical movements of chorea include facial grimacing, shoulder raising and lowering, and bending and extending the fingers and toes. They appear at rest, increasing during voluntary motions. Choreiform movements may occur in association with certain neurodegenerative diseases, including Wilson's disease and Huntington's disease, or systemic disorders. Symptoms generally start when patients are 25–50 years old, and may appear in association with slow, long, and not rhythmic motions affecting people both during movement and at rest, caused by the alteration of muscular tone. Choreiform movements disappear during sleep. Additionally, tics are rapid and involuntary movements. They can be classified as simple if they show fast and stereotyped motions of the face, shoulders, and limbs, and as complex when they are characterized by a sequence of movements. Tics reproduce movements and gestures of everyday life and are partially controlled by patients. They may be considered as a pathological sign or represent the typical symptom of the Gilles de la Tourette disease. Tremors are regular, rhythmic oscillations affecting a single part, a side, or the whole body at rest, during voluntary movements or maintaining a fixed posture. Tremors are commonly observed in Parkinsonism and Parkinson's disease patients, with several differences. Parkinson's disease tremor usually affects people at rest, showing a frequency of 4–5 Hz, but it may also occur during an action or when patients maintain their posture. In this case, it shows a variable frequency of 6–12 Hz. Tremor can also be defined as "essential," when it is not related to other symptoms and is not caused by an underlying disease. It is caused by abnormalities in areas of the brain that control movement while it does not occur as the result of a specific pathology. Tremor appears when maintaining a posture and worsens with voluntary movements, with a variable frequency. Other movement disorders are myoclonus- and dystonic-related movements. Myoclonus are rapid and uncontrolled contractions of muscles. Dystonia is a general term used to describe failure of regulation of muscle tone, usually a sustained increase in muscle tone, and shows slow and long torque movements. When opposing muscles of an extremity have increased tone, there is a limitation of voluntary activity. When the increased tone of opposing muscles is not symmetrical, the extremity or trunk may be drawn into a distorted posture that cannot be overcome voluntarily. The most common form of dystonia is dystonia musculorum deformans.

Movement disorders occur as a result of damage or disease in a region located at the base of the brain (basal ganglia). The basal ganglia is comprised of clusters of nerve cells (neurons) that send and receive electrical signals which are responsible for involuntary movement. Common causes of movement disorders may be attributed to many different conditions such as age-related changes, environmental toxins, genetic disorders, medications (e.g., antipsychotic drugs), and metabolic disorders (e.g., hyperthyroidism). They are also related to neurological pathology (e.g., Parkinson's disease or as an effect of multiple sclerosis) and as a consequence of cardiovascular disorders (e.g., stroke). Tremors, in particular, may be post-traumatic, as an outcome of a trauma to the brain, but they can also sometimes develop from conditions of fatigue, stress, and anxiety (Hallett 2003).

24.3.2 MAIN ASPECTS OF UPPER LIMB TREMORS IN MOVEMENT DISORDERS

Tremors are the most common movement disorders worldwide; incidence of essential tremor ranges from almost 4 to 36 cases per 1000 persons, to as high as 50 per 1000 persons older than 60 years. These data are actually very underestimated if we consider the prevalence of the other movement disorders affecting the upper limbs. Tremors are involuntary shaking movements, usually appearing as rhythmic oscillations caused by the alternate contractions (sometimes also rhythmic) of antagonist muscles. In medicine, tremors are classified on the basis of their frequency, occurring conditions, etiology (e.g., physiologic or pathologic), body part involved, and therapy responses. Tremor effects on the upper limbs can be observed in relation to voluntary movement execution, at rest or during muscular activation, and the physical characteristics of motions, in terms of frequency and amplitude.

In order to acquire the relevant data for the product design process, we can distinguish three kinds of tremors:

- *Resting or static tremors*: Occur when the limb is at rest and usually lessens with action.

- *Intention tremors*: Occur when patients are moving their limbs and disappear at rest. They are associated with a voluntary movement.
- *Postural tremors*: Occur when patients are holding their hands or legs in a particular position for a period of time. In some cases, it can be defined as *action tremors*, when it intensifies with action.

Essential tremor is a postural tremor of the hands and forearms: it is absent with rest and worsens with activity. As already stated, it is a disease characterized by uncontrolled trembling in part of the body. In relation to the upper limbs, essential tremor is associated with purposeful movement (e.g., holding a glass to drink, shaving, writing, buttoning a shirt) and occurs most often in the hands, but may also affect the arms. In fact, essential tremor mainly involves the distal part of the extremity, but it may also involve the wrist and beginnings of the fingers, with the appearance of flapping of the hands. It is usually present on both sides of the body, but when it occurs on one side, the dominant arm is usually involved. In the both sides form, the tremor affects mainly the dominant limbs. In essential tremor, the upper limbs are affected by pronation/supination and extending/flexion movements with a frequency of 2–12 Hz (5–12 Hz in the high frequency cases and 8–12 Hz in the low frequency cases). Generally, tremors worsen when the patient wants to execute fine movements. During adulthood or middle years, essential tremor may become progressively intense to the point of disability. Parkinsonism is a syndrome characterized by the presence of resting tremor, rigidity, bradykinesia (slow movements), and loss of postural reflexes. Although classically seen in Parkinson's disease, Parkinsonism may have other causes such as Alzheimer's disease. Parkinsonism tremor of the upper limbs has a frequency of 4–5 Hz and is usually asymmetric, prevalent on one side of the body. Tremor dynamic of the upper limbs is typical: the patient is affected by trembling motions primarily when his/her arm is at rest, but the tremor may decrease or end when the upper limb is voluntarily moved. This typology of tremor is activated by emotions; in fact, it occurs when the patient is solicited to answer a question or if he/she is only observed by someone. Among the intention tremors of the upper limbs, we can consider movements associated with dystonia. Dystonic typology of tremor affects patients with a sequence of rhythmic, repetitive, involuntary, twisting or writhing movements of the upper extremities; they can be rapid or slow, so they could have a duration of a few minutes to some hours (Findley 1995).

24.3.3 Movements Alterations in the Elderly

Tremors can happen at any age, but tend to be more common in older people. Together with tremors, the elderly usually present a set of movement disorders, due to the complex of physiological changes universally associated with aging rather than a specific pathology. Moreover, the number of neurological and cardiovascular diseases affecting body movements increases with age (Bhagwath 2001). Aging can also be defined as a progressive functional decline, or a gradual deterioration of physiological function with age, limiting peoples' normal performances. Physical changes largely affecting elderly motility, which gradually become limited and stereotyped, are a gradual reduction in height and weight loss due to loss of muscle and bone mass, a lower metabolic rate, lower reaction times, a decline in the nervous system and in certain cognitive functions, and a functional decline in audition, olfaction, and vision, together with other sensory alterations. Posture and gait change, weakness and slowed movement increase. Some elderly people have reduced reflexes and loss of balance. Walking may become unsteady, fatigue occurs more readily, and overall energy may be reduced (Graf et al. 2005). In relation to the upper limb movements, aging alterations bring a general reduction in arm and hand coordination. Movement slows and may become limited because of the loss of muscle mass reducing the strength and speed of motor stimulus along the nerves. Fine touch and fine skills may lessen, but involuntary movement, caused by muscle tremors, may increase. Sensory changes may influence upper limb movements in the elderly. In fact, senses become less acute, and older people may have trouble distinguishing details, showing difficulties in perceiving object borders.

24.3.4 SPECIAL NEEDS OF PEOPLE AFFECTED BY UPPER LIMB MOVEMENT DISORDERS

Tremors and other upper limb movement disorders may interfere with people's ability to perform daily activities. In particular, tremors, deeply involving the hands and forearms, may become very disabling because an affected person is trying to do something and, sometimes, as in the case of essential tremors, such tremors are not present when the limbs are not being used. Movement alterations are usually painless, although frequently simple actions, such as writing or eating, can quickly become painful for a person with hand tremors, gripping the pen or the fork tightly to maintain control in order to avoid their fall. Moreover, fatigue, stress, and emotions, in general, worsen the problem, especially when the affected person is with other persons, and feel that they are being observed. So, they have trouble enjoying activities, communication, and social interactions, suffering a general sense of isolation.

The physical and sensorial effects of movement disorders on upper limb motility may cause difficulties in people's elementary tasks, such as feeding, dressing, or executing personal hygiene actions, so that some of these activities can't be often done in complete autonomy. Trembling of one or both arms and forearms together with muscle rigidity, influence the fluidity sequence of movements and generate unsteady, to-and-fro motion of the upper limbs, with problems in fixing a stable position. For this reason, people may have trouble handling and using objects, grasping and gripping hand tools or just holding a glass or a cup to enjoy a beverage. A decreasing sense of balance and loss of muscle coordination may cause troubles in fine motors skills, with the consequent inability to successfully perform actions requiring fine movements of one or both hands and fingers. People can't effectively use small objects or switches, knobs and little buttons, and have trouble when setting up something with more pieces, and even when they just have to cork a bottle. Lessening of both voluntary and automatic motions brings an increase in reaction time from planning to executing movements, that appear slow, uncertain, and tentative, requiring more attention and concentration to be accomplished. To execute automatic motions, a person has to exert a voluntary control of the limb. Decreasing motion speed and amplitude causes troubles in raising and stretching the arms, so that the reaching area is reduced in people with movement disorders. Slowness of voluntary movements and a decrease in muscle strength, in association with sensorial disturbances of the arms, causes heaviness and fatigue, which is problematic when lifting objects, even they are not heavy, gradually increasing a general loss of motor performance.

Typical motor effects of upper limb movement can be listed as follows:

- Trembling of one or both arms and forearms
- Decreasing sense of balance
- Loss of muscle coordination
- Muscle rigidity
- Difficulty maintaining a fixed position
- Decreasing of motion speed (slowed movement increase)
- Decreasing of muscle strength
- Increasing sense of fatigue
- Increasing sense of limbs heaviness
- Loss of fine motor performance
- Decreasing movement amplitude
- Decreasing of movement self-control

Sensorial effects of upper limb movement disorders include:

- Anxiety and stress performing an action with decreasing action effective performance
- Loss of attention and concentration performances

- Decreasing or loss of autonomy
- Decreasing or loss of self-esteem

To design products fitting the needs of this special category of users, we can refer to the following typology of requirements:

- Usability features have to be assured to use the product easily
- Safety features have to be assured to protect the user from involuntary movements injuries
- Pleasure features have to be assured for product acceptance and equitable use

Table 24.1 shows the correlation of special needs of people affected by tremors, design principles, and requirements of products.

TABLE 24.1
Special Needs, Design Principles and Product Requirements

Special Needs	Design Principles	Product Requirements
Low weight	Reduce weight of tool or object Minimize weight held in the hand	Weight control
Low hand force	Minimize pinch force Minimize the effort to lift, push, and pull Prefer pushing rather than pulling whenever possible. *Pushing is generally preferred over pulling because a person is able to use their body weight to apply force to the load to get it to move.*	Force required control
Low actions speed	Reduce the speed of the sequence of actions necessary to use the product Reduce the speed of the motions required	Action speed control
Good grip	Providing appropriately sized and positioned grips or handles Providing appropriate shape or texture of the object so that it is easy to grip. *Roughness treatments to increase hand friction on grip are useful solutions.*	Slipperiness resistance Handling stability Morphological and dimensional adequacy
Possibility to use it with right or left hand	Design the object so that it can be used successfully with one or both hands Design the object so that it can be used with the right or the left hand	Right or left handling
Good access to the functions	Minimize the number of actions that require fine motions Minimize the number of actions that require fixing motions for a long time Providing appropriate sizes and shapes of knobs and buttons so that they are easy to manipulate. *Many times it means that commands have to be reciprocally well spaced.*	Movement tolerance Capacity to stress absorb Resistance to uncontrolled and accidental bumps Size and space for approach and use
Good visibility of commands	Evidencing with shape, colors, and position all the elements necessary to be manipulated to use the product	Perceptible information
Few actions required	Minimize the number of actions necessary to use the product	Simple and intuitive use
Few functions allowed	Minimize the number of functions that the product can perform with different procedures	One-purpose *Reduced number of functions*
Accurate styling	Providing care in aesthetic features of products, forecasting and designing sensorial and emotional reaction to the product	Not discriminant Beauty and appearance control

24.4   AN EXAMPLE OF PLEASURE-BASED DESIGN: A HOME TELEPHONE FOR PEOPLE AFFECTED BY PARKINSON'S DISEASE

24.4.1   DESIGN AIMS AND RATIONALE

Parkinson's disease is the classic cause of a resting tremor and is often accompanied by slowness of movement, muscle rigidity, and an abnormal gait. It is very frequent, its incidence is almost 2 cases per 1000 persons. The effects of Parkinson's disease on the movement abilities of the patients' upper limbs are very disabling: tremor provokes an increasing resistance to the articulations of the passive movements of flexion and extension, difficulty and slowdown at the beginning and during the fulfilment of movements (bradycinesia and akinesia), a great lengthening of reaction times, and a complete reduction in movement. These effects increase stressful moments, when a condition of concentration is required. Moreover, the monotony of the tone of voice decreases (hypophonia) and inattention takes place.

The objective of the project was to create a domestic telephone designed specifically to meet the needs presented by the conditions for use of this particular user target—people affected by Parkinson's disease, motor limitations to the upper limbs and a reduction in cognitive abilities. Nevertheless, this telephone is also usable for people who present with other disabling pathologies and difficulties in the fulfilment of precise movements of the arms and hands. This has occurred through the identification of specific morphological, dimensional, and materials characteristics ascribed to the components of the basic apparatus—base, micro telephone, keyboard—allowing people affected by Parkinson's diseases to perform comfortably, easily, and pleasantly the normal operations required for receiving and making a telephone call. The disabling upper limb symptoms of people affected by Parkinson's disease make the fulfilment of many daily activities, such as phoning, very difficult and not precise. In fact, the motor impediments make the use of a telephone complex for these users, by creating stressful conditions, often leading patients to give up the idea of making a call. These conditions are very serious, if we consider that the use of a telephone is an important need for these people, are usually at risk for depression, and put a high value on conversations and phone relationships in general (Attaianese, Caterina, and Manzi 2001).

24.4.2   PREDICTING NEEDS ANALYSIS

The design project has been based on the analysis of the users' needs and oriented toward defining the specific conditions for use of a standard telephone by people affected by Parkinson's disease. The analysis has been conducted on a sample of ten users, who have been asked to call and reply, using four existing telephones. The observations have been directed to pointing out the perceptive and motor effects of the pathology in relation to the use of telephones, in order to identify the critical conditions for use that these telephones present in relation to the particular user category. The report has been conducted on two fronts, dealing with the efficacy of use and the perceived quality of the sample telephones, used by experimental subjects. The usefulness performed, that is to say, the products ability to be used successfully, has been evaluated through the survey and the consideration of the mistakes in use and any other difficulties encountered by people in using the proposed products (Figure 24.2 shows some telephones tested during the study).

The quality of use really perceived by users, has been intended as product's ability to be used with ease and pleasure and surveyed by a questionnaire presented to a panel of users. The users have also been asked to establish the importance ascribed to the various components for satisfaction. This has indicated that simplicity, immediacy, and user friendliness in the dynamics of acceptance of this kind of product are very important. Pleasure has been interpreted by users mostly as an acceptance of the aesthetic components of the telephone, in terms of the appearance of the product, the form, materials, and colors. The survey has indicated that the preferred features for the users

FIGURE 24.2 Existing telephones tested by people affected by Parkinson's disease.

considered is that of a traditional telephone, characterized by a few, but clear, functions which are well known by users, so that the aesthetic and functional connotations desired are not different from those of a normal home telephone, used by all members of a family, with or without physical limitations. In whole, the study conducted for the identification of design requirements has pointed out that telephones produced nowadays don't have the characteristics useful for efficient and satisfactory use by people affected by tremors and limitations to the upper limbs, because they present some features not fitting these special users.

24.4.3 SPECIFIC USE CONDITIONS IDENTIFICATION: USE REQUIREMENTS AND PLEASURE-BASED CHARACTERISTICS

The analysis of the specific use conditions of the existing telephones has emphasized that upper limb limitations have critical effects on user efficacy and satisfaction during their utilization. They essentially concern the reduced ability of holding and handling the receiver in a fixed position; also typing the number on the keyboard became a critical task, evidencing increasing difficulties to hear the interlocutor voice and perceive the calling message. Moreover, the incidence of a lot of available telephone functions, without effective feedback confirming their efficacy, can be seen as an obstacle to the successful use of the telephone.

A telephone for people affected by Parkinson's disease should be designed to fit some of the special needs of users:

- Holding the receiver with a steady grip, despite unsteady hands and forearms
- Holding the receiver close to the face, despite its possible slipping due to the unsteady grip
- Typing exactly the number on the keyboard, despite the hand motions impediment and the difficulty in fulfilling fine movements
- Speaking and hearing the interlocutor on line, despite the difficulties in holding the receiver in a fixed position, close to the mouth and the ear
- Setting the receiver correctly on the telephone base, despite the risk of putting it down incorrectly because of arms tremors, keeping the line busy and remaining isolated
- Rolling the telephone cord, despite the difficulties in fulfilling precise movements and inattention, which could provoke unintentional shifting of the receiver on the base
- Perceiving visually the calling message, despite hypoacusia, inattention, and low concentration of the users
- Having few functions and being immediately recognizable through a helpful layout
- Obtaining the immediate perception of the efficacy of the used function
- Using the available functions through a low number of operations
- Having a telephone that doesn't seem different from those used daily in the home

Critical use conditions identification provides the possibility to define the telephone use requirements in relation to people with Parkinson's disease, and generally to all users affected by upper

FIGURE 24.3   Telephone prototype.

limb movement disorders. The main requirements are connected to the morphology of each telephone component and to the whole set of the apparatus.

Identified requirements fitting specific modalities in tasks accomplishment are

- Handling, with particular reference to comfort and safety
- Comfort of typing
- Control of the correct typing
- Control of the numbers scanning speed
- Stability on the horizontal plane
- Stability in the transmission of voice
- Immediate connection to the telephone call
- Auto-winding of the cord

Figures 24.3 through 24.5 provide views of the telephone prototype during its use.

Furthermore, some other general requirements connected with the telephone as an interactive device are

- Self-explanatory
- Recognizability

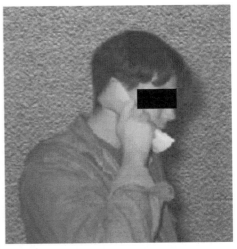

FIGURE 24.4   Receiver handling.

FIGURE 24.5   Push buttons and typing.

- Control
- Navigation and orientation among all different functions
- Ease of use
- Consistency between features of systems' components and their functions

24.4.4   Features and Details of the Final Design

The final project presents a telephone with a traditional appearance, few very simple functions, in accordance with the features required for domestic purpose.

Morphological and dimensional characteristics have been designed to assure simple and satisfactory use in order to overcome the critical points met by specific users. Handiness and comfort, which are fundamental requirements because of the user's grip difficulties and uncertainty, have been emphasized when designing a particular receiver morphology. Shaped and proportioned to allow a steady, comfortable, and secure hold, the receiver features are consistent with the anthropometric measures of the hand and the dynamics of the involuntary movements, in order to reduce the high risk of slipping during the approaching and removal movements to and from the face. Friction grip is also improved by a lightly wrinkled superficial treatment. The stability in the voice transmission has been obtained by shaping an asymmetrical profile of the telephone receiver. It is stretched toward the mouth, with the microphone area larger than usual, in order to overcome involuntary shifting due to the tremors and the consequent difficulty of keeping the receiver in a fixed position. The base of the telephone presents a space dimensioned for an immediate correct receiver reposition after the call and reply use.

The comfort of typing has been obtained by adequate shaping, sizing, and spacing of push buttons, and giving a light inclination to the keyboard. Buttons are concave, with a rough surface to increase finger friction, in order to avoid typing mistakes due to unsteady motions of the hands. Buttons are also spaced with a distance that prevents any typing mistakes due to the fingers imprecise motion.

Moreover, a visual display on the keyboard and a confirmation key allows verification of the accuracy of the numbers typed before entering the call, prevent on time all the occurring faults.

The stability of the apparatus, related to the risk of slipping on the horizontal plane due to the limbs uncontrolled tremors, is assured by mean of two small linear feet, placed under the telephone. The base presents a high friction surface and is covered with silicone material.

The immediate perceptibility of a telephone call, necessary to overcome the loss of attention and concentration characterized by people affected by Parkinson's disease, is assured by a linear blinking luminous LED, put along the sides of the telephone base, adding to the current acoustic signal of the telephone ring, a visual one for arriving calls. The visual led of the phone call also indicates the uncorrected receiver repositioning. The telephone is also provided with a system for cord auto-winding.

24.5 CONCLUSIONS

Today, the market doesn't offer products specifically designed for people with movement limitations of the upper limbs. Also, in the case of the telephones tested, standard solutions concern non-specific models. The new telephone presents a traditional appearance, even though particular attention has been given to the styling features, with a sober line and pleasant chromatic and tactile features. The most significant differences between the final project and the previously tested telephones concern, above all, the receiver's morphological and dimensional characteristics, proportioned and shaped in order to minimize the effects of involuntary movements. Its asymmetric profile, its lower part drawing near the mouth, and the wide surface of the microphone area allow users to hold the receiver comfortably and safely. Moreover, the tactile features of the grip surface increases hand friction, reducing the slipperiness conditions of trembling arms.

The morphological and dimensional characteristics of the push buttons are another new characteristic, also the distance between the button keys is greater than on the standard telephone, thereby reducing the risk of incorrect typing due to motor impediment and tremor.

The button keys are characterized by a very concave quadrangular shape, and have a particularly wrinkled surface. This choice characterizes the final project because it suggests a radical change from the current morphology of telephones, usually realized with convex buttons, not fitting the fingers' anatomy.

The presence of control systems for the typed number, thanks to a display on the top of the keyboard with a key for confirmation, aren't properly original elements, although in existing telephones we haven't the presence of such effective control devices. Also, the luminous LED for the arriving call, which can also indicate the uncorrected repositioning of the receiver, is quite a recurring system in standard models, but its allocation along the lower profile of the apparatus makes the perception of the calling message more efficient.

Even if the presented telephone has been designed to meet some users' special needs, it has been featured pursuing the goal to enhance its pleasantness and ease of use, compared with the current domestic telephone, for effective, efficient, and satisfactory use by everyone.

Features and functionalities of the telephone for people affected by Parkinson's disease were patented in Italy in 2001.

REFERENCES

Attaianese, E., Caterina, G., and Manzi, P. 2001. Predicting needs for disabled people pleasure-based products. A case study. In *Proceedings of the International Conference on Affective Human Factor Design*, eds. M.G. Helander, H.M. Khalid, and M.P. Tham, 304–8. London: Asian Academic Press.

Beecher, V., and Paquet, V. 2005. Survey instrument for universal design of consumer products. *Applied Ergonomics* 36:363–72.

Bhagwath, G. 2001. Tremors in elderly persons: Clinical features and management. *Hospital Physician* 49:31–37.

Findley, L. ed. 1995. *Handbook of Tremor Disorders*. New York: Marcel Dekker.

Graf, A., Judge, J.O., Ounpuu, S., and Thelen, D. 2005. The effect of walking speed on lower-extremity joint powers among elderly and adults who exhibit low physical performance. *Arch Phys Med Rehabil* 86:2177–83.

Green, W.S., and P.W. Jordan. 2002. *Pleasure with Products: Beyond Usability*. London: Taylor & Francis.

Hallett, M. 2003. *Movement Disorders*. London: Elsevier Health Sciences.

Norman, D. 2004. *Emotional Design*. New York: Basic Books.

25 Applied Anthropometry in Ergonomic Design for School Furniture

Luis Carlos Paschoarelli and José Carlos Plácido da Silva

CONTENTS

25.1 INTRODUCTION

The socio-historical evolution of developed countries is based on several factors, and we can highlight the basic education of the population among them. Politics and institutional activity have involved all socioeconomic levels, and physical infrastructure issues are tangible examples of the educational achievements observed in such communities.

Emerging nations wishing for the same socioeconomic status should also focus on this sort of strategy, giving special attention to the development of technology for physical infrastructure, which is important for educational praxis. A good example can be observed in Brazil, an emerging economy that seems to have contradicted the demand for schools, equipment, and didactic instruments more adapted to the needs of the students.

Of special importance is school furniture, which can be defined as basic equipment needed for classroom activities whose design should be informed by usability issues, especially the appropriate anthropometry of the product.

This chapter deals with the principals of anthropometry as they relate specifically to children, as well as some parameters for the use of the anthropometric data in the design of pre-school furniture. Finally, an example of the use of anthropometry in the ergonomic design of a school desk is presented.

25.2 ANTHROPOMETRIC PRINCIPLES

According to Pheasant (1996, 6), anthropometry "is a ramification of the human science that deals with measures of the human body: particularly measures such as size, shape, strength and occupational capacity."

419

Anthropometry is directly related to ergonomics, because without anthropometric references the sizing of objects or work surfaces is impossible. It is the science responsible for measuring the body's characteristics and functions, including linear dimensions, weight, volume, and types of movement.

Panero and Zelnik (1989) claim that the size and dimension of the body are the most important factors for design because of their relation to the ergonomic adjustment (ergofitting) of the user to the environment, an aspect of the human–machine interface of particular importance for ergonomists.

The Biomechanics Institute of Valencia (1992) defines anthropometrics as a science that presents "the measures and dimensions of the human body" and can be divided into two distinct classes:

1. Structural dimensions of the body, or static anthropometry, in which measurements are taken from an individual in either fixed or static positions. Such measurements are useful in general terms.
2. Functional dimensions of the body, or dynamic anthropometry, in which measurements are based on the position that the body occupies as a result of a certain movement. These measurements are more useful for design, seeing that the body is in a constant state of motion.

Historically, measurements of the human body have served as dimensional references (e.g., the foot as a unit of length), especially before the adoption of a decimalized system of measurement by the French Academy of Sciences in 1791 (Neufert 2002). The oldest documentation of a system of measurement based on human proportions was found in one of the pyramids in Memphis, Egypt, and dates from approximately 3000 BC.

In his first-century Roman architectural treatise, Vitruvius describes a system of proportionality of the human body and its implications on the metrology of his time. In the fifteenth century, the Italian painter Cennino Cennini wrote that human height and scale could be understood as the same distance, and in 1492, Leonardo da Vinci studied the movements and proportions of the body segments, initiating the concept of biomechanics.

In modern times, the use of anthropometric parameters has intensified due to industrialization and the mass production of goods, mainly since 1940. Anthropometry has become very important in aerospace projects and in complex work systems where human performance is critical (Iida 2005).

Because of the increase in international trade and the launch of increasingly segmented product lines, the use of anthropometric parameters has become essential for the development of safe and well-adapted projects for specific populations. A consideration, therefore, of some of the criteria for these parameters is necessary.

According to the descriptions of both Dul and Weerdmeester (2004) and Pheasant (1996), anthropometric parameters are always related to a specific population of users; when the parameters for one group are applied to products for a different population, adverse results can occur.

Thus, the anthropometric parameters used in product design must be based on the biological and sociocultural differences of the researched population (Roebuck, Kroemer, and Thomson 1975) and must focus on inter- and intragroup differences (Boueri Filho 2008), which according to Iida (2005), can include biotype, gender, age, and ethnicity.

Regarding age, it has been demonstrated that the dimensional adjustment of products for children is highly important for their comfort and security, since, according to Panero and Zelnik (1989), there is a causal link between unsuitable furniture and accidents.

Anthropometric parameters for children are available in the form of tables and graphics, most notably in the database "Childata" (Norris and Wilson 1995), which pools data from studies on American, British, German, and Dutch populations, and also for Brazilian children and adolescents in the research of Paschoarelli and Silva (1995), Silva (1997), Paschoarelli (1997), and Soares (2001), among other authors.

For specific literature on this subject, Croney (1978) and Iida (2005) stand out. Their studies show the lack of established anthropometric patterns for the populations under consideration. Hira (1980), Bosoni (1994), and Schianchi (1995) discuss the need for anthropometric data in order to ergonomically adjust the pre-school desk.

The application of anthropometric principles is fundamental to the ergonomic design of a product. Anthropometric differences should be considered attentively, especially in the standardized production of industrialized goods. The results of this process can be considered positive when they harmonize the "human-machine" system. For this reason, when such a system is in an educational context, "we must consider the dimensional features of the users" (Biomechanics Institute of Valencia 1992) in the design of its furnishings.

Important anthropometric studies have been undertaken both by normalizing agencies and by research centers around the world. In Brazil, both the army and the National Institute of Technology have spearheaded studies of this type, but, so far, there has been little development in the area of child anthropometry.

Nevertheless, important references relating to child anthropometry can be found in developed countries. Australia is one example, where research on child anthropometry has been conducted since 1908 (Oxford 1969). The development of such references has also been intensified in other countries in recent decades, with notable research being carried out by Hira (1980) in India, Mandal (1982) in Denmark, Panero and Zelnik (1989) in the United States, and the Biomechanics Institute of Valencia (1992), Spain, as well as the application of certain norms in many other countries.

25.3 APPLICATION OF ANTHROPOMETRIC DATA TO THE SCHOOL DESK

The gender, age, and national origin of the target population will define either the choice of the pre-existing data to be taken as a reference or the parameters of the sample to be used for the measurements.

The dimensional variation within populations, however, is so wide that, often, we cannot rely on the measures to function for the totality of individuals. Generally, a rating on the variation index equivalent to 90% coverage of the population is considered acceptable, including all but the smallest 5% and the largest 5%.

In order to verify the aptness of a work surface's profile, two-dimensional mannequins corresponding to the size groups of the selected population (mid-sized, small, and large individuals) are used. Based on this technique, it is possible to meet the requirements for the design intervention, which are defined in the following sections.

25.3.1 BASIC REMARKS ON THE DESK KIT—A

A.1: When school desks aren't adjustable for dimensional adequacy, it is necessary to provide them in different sizes for children of different ages, or even for those of the same age who have different anthropometric features.

A.2: The school desk kit must be built as if the seat and the work plane were separate systems, so that each can be adjusted separately to the needs and anthropometric demands of individual users.

A.3: Although the work plane and seat should be considered as separate systems, they nevertheless must be designed as a set due to the dimensional correlation essential for the adjustment of the work position itself.

25.3.2 REMARKS ON THE DESK SEATS—B

B.1: It is imperative that the user is allowed to change posture while seated; for this reason, the use of sculpted seats was excluded as a design possibility and we opted for flat seats.

B.2: The use of a pillow would provide added comfort to a flat seat; however, this would elevate costs as well as create hygiene issues and more frequent repair and replacement. The choice of hard seats isn't necessarily an ergonomic inconvenience.

B.3: It is preferable that the school desk seat be quadrilateral with rounded edges and with the front edge having a larger rounding radius in order to avoid unnecessary pressure on the bottom of the thighs and poples.

B.4: The seat back is an essential element of the school desk, permitting greater stability, contributing to a lower mechanical load in the trunk, and helping to maintain the physiological curvature of the spine, thereby providing support for the lumbar region, whose configuration must be slightly convex in the vertical axis and slightly concave in the horizontal axis.

The seat back position must mirror the elbow's distance from the seat since, when seated, the child's lumbar curve presents the same height as the elbow. Based on the population average for this variable (148.9 mm), the height was specified as 150 mm.

B.5: There must be enough room for free movement of the user's legs under the seat.

B.6: The seat height is the main factor that provides efficiency and comfort for the work surface; its determination has as a basic aim to avoid incorrect pressure on the thigh region. Moreover, when the seat is too high and the feet can't touch the ground, insensitivity of the feet can occur and, as a result, discomfort and fatigue.

In order to alleviate pressure on the thigh, however, the knee joint must be maintained at a right angle with the entire foot placed on the ground.

On the other hand, seats that are too low affect an improper sitting posture, bending the hip and creating a sharp angle between the femur and the trunk. As a consequence, the individual can't keep a concave position in the lumbar region.

Therefore, it is recommended that the popliteal height variable (08-B) is used to determine this specification, taking due precautions about the use of anthropometric data in the sizing of the work surface.

The average measurements for each category of the target population—3, 4, 5, and 6-year-olds (270, 290, 315, and 335 mm, respectively)—were used to determine the seat height.

B.7: The seat depth in school chair is as important as the last item. A shallow seat is normally uncomfortable, while an overly deep seat prevents proper use of the seat back. An important related parameter is the distance from the sacral plane to the popliteal angle. Again, the careful use of anthropometric data will prevent improper sizing of work surfaces.

For this particular project, we decided to determine three seat depth possibilities according to the results for the 5th, 50th, and 95th percentiles of the population: 250, 300, and 350 mm, respectively.

B.8: Seat width must be determined by the variable hip width, again with the goal of including the highest possible proportion of the population. In this case, based on the 95th percentile of the population (245 mm), the ideal width was determined as 250 mm.

B.9: The inclination of the seat plane is closely related to the angle of the seat plane with the seat back. Consequently, these measures determine user hip flexion. An inclination of 4° for the back in the horizontal plane is recommended, ranging from 3° to 5°.

B.10: The angle between the seat and the seat back must allow good posture, i.e., lumbar support by the seat back and weight distribution over the seat plane; the recommended angle is between 90° and 100°.

B.11: The seat back must always provide satisfactory lumbar support. Its upper edge must be low enough to allow for normal movement of the arms and its lower edge can begin from 100 to 150 mm above the seat plane.

25.3.3 Remarks on the Work Plane of School Desks—C

C.1: There must be enough room for the free movement of the legs under the desk. This space must not be used for storage, since any items placed there would restrict free movement.

The most important anthropometric parameter for determining seat height is the variable thigh height. In our target population, the 50th percentile was 95 mm, thus the distance chosen was 100 mm. Foot support is also a viable option, when very well planned.

C.2: The work plane must be designed so that it allows connection with other work planes for group activities. Thus, it has to be flat and parallel to the floor.

C.3: As previously mentioned, desks and chairs need to match, principally with respect to their height. The height of the desk must allow forearm support without requiring either forward trunk inclination or shoulder lift. Applying this parameter to the target population, for which seat height measurements and elbow-seat distances have already been presented, the resulting height for the work plane was determined to be 420, 440, 465, and 485 mm for the 3, 4, 5, and 6-year-olds, respectively.

C.4: The work plane must permit the user to recline (to a maximum inclination of 15°) in order to allow better performance for reading and writing activities, etc.

C.5: The height and depth of the work plane must be wide enough to accommodate every possible activity relating to the classroom learning process. We estimated that an area 500 mm long and 600 mm wide would be enough.

25.3.4 Other Remarks

D.1: The use of resistant materials was considered mandatory due to the expected stringent conditions of use and the possibility of damage and deterioration caused by accidents and mistreatment.

D.2: Light materials were also considered desirable because they would enable user transport of the furniture when necessary for classroom activities.

D.3: We decided on hygienic materials that would be easy to clean.

D.4: The many possibilities of use had to be taken into account during the design phase in order to avoid accidents and product failures.

D.5: Security was another essential aspect of the design process; it was necessary to avoid sharp edges and corners as well as other configurations that could lead to accidents or uncomfortable posture.

D.6: The use of color had to be suitable for the psychological and pedagogical circumstances, and was also used for dimensional tagging of each anthropometric level. It was determined that the seat and work plane should be neutral in color, either light gray or white, and that the base should be a different color for each size, i.e., blue, yellow, red, and green.

D.7: Space for textbooks and supplies was built into the structure of the school desk, though it is important to observe that the previously explained ergonomic conditions had to be obeyed. A complementary space doesn't necessarily represent an ergonomic problem.

After broaching the worrying reality in Brazil about educational infrastructure, especially classroom ergonomics, we have reviewed anthropometric theory, touching on significant research and the development of anthropometric standards, which are a fundamental tool for resolving these problems through design intervention.

Only through the design process will it be possible to accomplish more than just a mere change of scenery in Brazilian institutional infrastructure and arrive at a technological balance by taking advantage of opportunities where scientific theory and practice can work together in a concrete way.

25.4   APPLIED ANTHROPOMETRY IN THE PRE-SCHOOL DESK

In order to put into practice the concepts mentioned thus far, the methodology selected was an ergonomic design process, initially characterized by the framing and investigation of the problem and followed by project definition and briefings, with product design as the practical response.

The creation of design alternatives was characterized by the concomitant use of creative and logical techniques, resulting in four distinct concepts of pre-school desks (desk and chair) (Figure 25.1), which, after being assessed in terms of the design requisites, resulted in a single model. This model progressed through alterations and adjustments via technical representation and the development of prototypes.

Each set is composed of 65 parts, including ferules, interchangeable legs defined for each anthropometric pattern, a seat with correct backward tilt, a backrest providing proper lumbar support, an inclinable work surface for artistic activities, writing, and reading, and a supporting table that permits freedom of lay-out, as well as storage space.

Color-coded legs corresponding to each pre-school stage, and thus to different anthropometric patterns, were featured in this set. Red indicates the model for 3-year-olds; yellow for 4-year-olds; blue for 5-year-olds; and green for 6-year-olds (Figure 25.2). Versatility of assembly and disassembly allows children and teachers to rapidly create free space in the classroom.

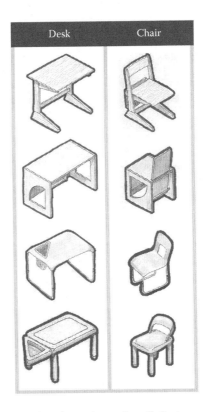

FIGURE 25.1   Creation process of design alternatives—four distinct concepts of pre-school desks.

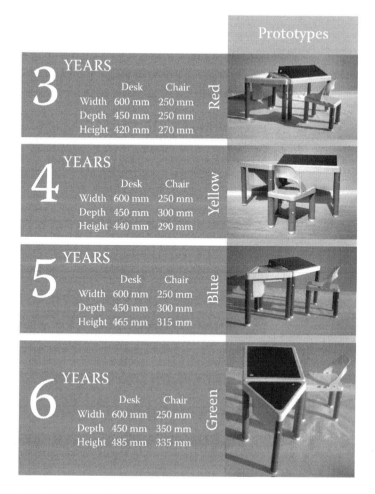

FIGURE 25.2 Results of the pre-school desk.

Other technical aspects regarding ecological questions were studied for this project, including the use of recyclable material. The prototypes were tested and achieved satisfactory results (Figure 25.3).

In this project, problems of pre-school desk design were observed, noted, and registered in a thorough investigation, resulting in parameters for the design process. The result of the design process was a furniture design proposal that met user-verified needs.

The finished product was awarded the "Seal of Quality" by the Brazilian Home Museum, an honorable mention at the 1997 Herman Miller International Competition for Industrial Design in Uruguay, and the "Good Form Certificate" from the Third National Awards for Furniture Design of the Abimovel – Brazilian Furniture Design show.

The importance of this research lies in the objective achieved, where a model for the conversion of science into technology is presented, demonstrating that the ergonomic design allows the development of products that represent a real improvement in quality of life. The adoption of scientific procedures in the process for developing pre-school furniture led to very satisfactory results, where aesthetic, technological, and utility aspects converged in a practical and viable way for an emerging economy.

FIGURE 25.3   Test of the pre-school desk prototypes.

REFERENCES

Biomechanics Institute of Valencia. 1992. *Guide of Recommendations for the Design of Ergonomic Furniture.* Biomechanics Institute of Valencia: Valencia (in Spanish).

Bosoni, G. 1994. Il moderno in piccolo. Appunti per una storia del disegno del mobile per l'infanzia (1850–1960). *Ottagono* 112:9–18 (in Italian).

Boueri Filho, J.J. 2008. Applied anthropometry to architecture, urbanism and industrial design. 1st ed. E-book. Sao Paulo: Estação das Letras e Cores (in Portuguese).

Croney, J. 1978. *Anthropometrics for Designers.* Barcelona: Gustavo Gilli (in Spanish).

Dul, J. and Weerdmeester, B. 2004. *Practical Ergonomics.* 2nd ed. Sao Paulo: Edgard Blücher (in Portuguese).

Hira, D.S. 1980. An ergonomic appraisal of educational desks. *Ergonomics* 23 (3): 213–21.

Iida, I. 2005. *Ergonomics: Project and Production.* Sao Paulo: Edgard Blücher (in Portuguese).

Mandal, A.C. 1982. The correct height of school furniture. *Human Factors* 24 (3): 257–69.

Neufert, E. 2002. *Architects Data.* 3rd ed. Oxford: Blackwell.

Norris, B., and Wilson, J.J. 1995. *Childata – The Handbook of Child Measurements and Capabilities.* Nottingham: Institute of Occupational Ergonomics.

Oxford, H.W. 1969. Anthropometric data for educational chairs. *Ergonomics* 12 (2): 140–61.

Panero, J. and Zelnik, M. 1989. *Las dimensiones humans en los espacios interiors: Estándares Antropométricos.* Mexico City: Gustavo Gili (in Spanish).

Paschoarelli, L.C. 1997. The school desk (workplace) as object of development of the children education: A contribution of ergonomic design. MSc Diss. Paulista State University (in Portuguese).

Paschoarelli, L.C., and Silva, J.C.P. 1995. Anthropometric survey with preschool children of the Bauru city. *Estudos em Design* 3 (2): 94–114 (in Portuguese).

Pheasant, S. 1996. *Bodyspace: Anthropometry, Ergonomics and Design of Work.* 2nd ed. London: Taylor & Francis.

Roebuck, J.A., Kroemer, K.H.E., and Thomson, W.G. 1975. *Engineering Anthropometry Methods.* New York: John Wiley.

Schianchi, F. 1995. Il banco di scuola. *Ottagono* 116:50–53 (in Italian).

Silva, J.C.P. 1997. Anthropometric data-collecting of the preschool to college students in Bauru – Sao Paulo city. Habilitation thesis. Univ. Estadual Paulista (in Portuguese).

Soares, M.M. 2001. Contributions of the ergonomics to product design of school furniture: University school desk, a case study. In *Ergodesign: Products and Process*, eds. A. de Moraes and B.C. Frisoni, 138–68. Rio de Janeiro: 2AB (in Portuguese).

26 Anthropometric Fitting of Office Furniture for Mexican Users

Lilia R. Prado-León

CONTENTS

26.1 INTRODUCTION

Ergonomics, with its research techniques and methodology, its growing body of descriptive data on human users and satisfactory design solutions, offers a science-based approach to user-centered design.

User-based design starts in the planning stage. It is necessary to begin by recognizing user characteristics and limitations, and determining their needs. If an object, system, or environment is intended for human use, then its design should be based on the physical and mental characteristics of its human users (Pheasant 1996).

The objective is to create a place where the user may perform his or her work in a comfortable, efficient, and safe manner. User-centered design is generally cyclic. It is frequently based on an initial task analysis; that is, research that deals with obtaining an operational description of user actions for performing an activity with the indicated object, and within a specific setting. This analysis may be conducted by means of user trials, in which a representative sample assesses the usability of a new or existing product. Results of this research suggest design modifications that are subsequently tested, and the cycle repeats itself. Traditional crafting implements, such as hand tools, may evolve over very long periods of time (centuries) following a similar cyclic process of trial and error.

In today's industrial societies, the majority of consumer products, such as clothing, tools, machines, furniture, and even residential buildings, are mass produced. Production demands also compel the designer to consider the "user" to be large groups of users, in such a way that an object with a standardized shape, size, and function may be accommodated for use by hundreds or thousands of different people (Prado-León and Ávila-Chaurand 2006).

These ergonomic fittings consist of a compatible relationship between elements of the component object and its corresponding human factor. The object is any man-made element or component with useful aims; that is, with the purpose of facilitating the performance of a practical activity. The human factor, for its part, is simply the human being interrelating with objects and surroundings to affect any sort of task (Prado-León and Ávila-Chaurand 2006).

Anthropometric fit is the optimal relationship between an object's physical dimensions (height, width, depth, etc.) and the structural or functional dimensions of those parts of the user's body that make direct contact with it while the system operates. It is thus necessary to have data available for the target user population in order to carry out anthropometric fitting. Given that these dimensions vary by age, sex, race, geographic region, and even from person to person with the same such characteristics, specific anthropometric databases are needed for the user population.

Anthropometric fitting is very important, since it is the basis for achieving physiological and biomechanical fittings (Prado-León, Ávila-Chaurand, and Herrera-Lugo 2005). A product's ergonomic properties are generally based on its comfort, safety, and usefulness, and these qualities are related in large measure to the user population and its characteristics. In the particular case of a computer workstation, the ergonomic properties are even more important because the worker may continuously perform tasks at this piece of furniture for many hours at a time.

Evidence for the relevance of considering anthropometric aspects in designing these stations may be found in the Liberty Mutual report (Kroemer and Kroemer 2001), which addresses the rise in disabilities due to hand, wrist, and shoulder disorders occasioned by ever-increasing computer use. Also, a large number of published works (Tullar et al. 2007; Montreuil et al. 2006; Tepper et al. 2003; Kroemer and Grandjean 2001; Anshel 1999; Bernard 1997) indicate that these disorders are related to workstation layout and arrangement, mainly including failures of anthropometric fitting and deficient lighting conditions. It has been suggested that other factors, such as organizational factors and psychosocial stress, may also influence musculoskeletal upper extremity disorders (American Industrial Hygiene Association 1994; Moon 1996).

Unfortunately, there is little data in Mexico regarding the effects of computer work. Támez-González et al. (2003) carried out a study of 218 computer workers in Mexico City and found a greater prevalence of musculoskeletal disorders in women. Computer use was associated with greater risk of visual fatigue, upper extremity disorders, and dermatological problems.

On the other hand, manufacturers of most consumer products, including computer workstations, do not consider data on the Mexican population when establishing furniture dimensions or their range of adjustment, but go by foreign, particularly North American, standards.

In August 2009, the Ergonomics Research Center (CIE, for its abbreviation in Spanish) conducted an evaluation of 70 workstations at the University of Guadalajara, by means of a checklist that encompassed chair, keyboard tray, monitor, work area, personal and work habits, and organization of tasks. The following are descriptions of these 70 evaluations' most relevant results, as they concern possible required anthropometric fitting.

To begin, a generalized problem was observed at many of the workstations (approximately 70%), in that the furniture used for computer work was not specifically designed for such tasks, and pencil-and-paper writing desks were being used instead. This fact meant that initial evaluation results were negative from an ergonomic standpoint.

Since the thigh-calf angle was not at approximately 90°, more than 87% of the seat heights were found to be outside the appropriate range. Although 91.4% had adjustable-height seats, 87.1% did not lower to a height of 36 cm.

The majority of the chairs had armrests (95.7%), but these could not be adjusted. Only 14% included these movable components. Additionally, 67.1% of the computer stations had no keyboard tray, and of the 32.1% that did, only 10% offered the possibility of height adjustment. In addition, at 68% of the evaluated stations, the keyboard tray and/or work surface height did not allow maintenance of an approximate 90° angle for the forearms. When analyzing the location of workstation monitors, it was found that 57% of the personnel had to bend in some way to focus on the monitor, generally because it was placed too high (71.4%).

The present chapter offers a simple methodology for conducting anthropometric fitting, taking as its example a computer workstation for administrative personnel at an educational institution and using anthropometric data from the Mexican population generated by the CIE at the University of Guadalajara, Mexico. It also includes a comparison with dimensions from another specially selected workstation, as calculated for the Mexican population and as recommended in foreign sources. Finally, it presents recommendations for improving the workstation using available furnishings, and explores some final considerations.

The finance department's "technical support" workplace was specially selected for study, because serious deficiencies in terms of anthropometric fitting were found at computer workstations there.

The following section will present a description of steps toward an anthropometric fitting methodology, summarizing examples from the abovementioned workstation.

26.2 METHODOLOGY FOR ANTHROPOMETRIC FITTING

26.2.1 ANALYSIS OF SYSTEM COMPONENTS

This step includes descriptions of users and an analysis of tasks to be performed with the product or products, in surroundings set up for achieving certain objectives.

Four people perform their work activities at this station—three women and one man. The women's ages are between 31 and 44, and the man is 28 years old. For anthropometric fitting of the furniture, I considered a broader age range than that represented by the current users: from 18 to 65 years of age, for both male and female users. This range is appropriate given Mexico's official working age of 18, with 65 being the retirement age at most institutions: it is thus feasible that personnel within that age range would be working at these stations.

Tasks in this workplace consist of entering financial data, making photocopies, and reviewing and correcting documents. The time dedicated to each of these tasks during the 8-hour workday varies: with between 1 and 7 hours of data entry and approximately 1 hour spent on document photocopying. The present analysis will only take into consideration data entry performed at the computer workstation itself; with the operator assigned to that station, a 31-year-old female, exemplifying the human–workstation relationship. As can be seen in Figure 26.1, the workstation is not built for a computer; it is a desk for pencil-and-paper work and thus has no keyboard tray, which makes for incorrect height. For this operator, the observed arm-forearm angle is increased to approximately 125°. Also, although the chair has armrests, these are not adjustable nor are they used for support, since their height is too low in relation to the work surface. The operator reports discomfort and annoyance as well, from supporting her wrists on the edge of the desk.

Taking into account the evaluation results and considering that the user has three points of physical contact with her workstation: the keyboard, seat, and screen (Pheasant 1996), the anthropometric

FIGURE 26.1 Technical support staff person at her workstation.

fitting will address: (a) seat height, (b) armrest height, (c) keyboard surface height, and (d) monitor height.

26.2.2 ESTABLISHING ERGONOMIC GUIDELINES

Once the tasks that users perform within the analyzed system are described, it is necessary to identify the ergonomic principles that apply to them. These guidelines are the result of decades of ergonomic studies concentrated on identifying the factors that influence discomfort and the development of musculoskeletal symptoms and illness at workstations, and consequently the establishment of guidelines for workplace furniture design. For example, in 1982, Drury and Coury began applying anthropometric, biomechanical, and orthopedic principles to arrive at basic principles for seat design. Today, although these principles continue to be respected in terms of ideal posture, it has also been widely recognized that the best furniture is that which permits freedom of movement and a variety of postures for alleviating the static effort of sedentary work (Kroemer and Kroemer 2001; Kroemer and Grandjean 2001).

It is important to note that at a seated workstation, where work surface height, thigh clearance, and chair and monitor height are necessarily considered as related and having dimensions that are critical for comfort, the interrelation between the different system components should be taken into account. The guidelines applicable to computer workstation dimensions, established in the previous step, are the following (Ontario Ministry of Labour Area Offices 2004):

a. To avoid ischemia from thigh compression, the subject's legs should not be dangling from the seat, and feet should be supported on the floor; thus, an approximate 90° angle between thighs and calves is recommended.

b. Shoulders should be relaxed (not raised), with arms hanging naturally at the sides of the body, resting on the armrest at an arm-forearm angle of approximately 90°.

c. For keyboard work, wrists should maintain an optimal skeletal configuration (straight, without flexion, extension, and ulnar/radial deviation exceeding 10°) and there should be sufficient space beneath the work surface for thigh clearance.

d. The head-neck should maintain an optimal skeletal configuration (head aligned with spine or bent slightly forward) when working at the monitor.

26.2.3 Specific Design Recommendations

The above principles should be translated into specific design recommendations or requirements, in which the referent anthropometric dimension and particular criteria are already established (Pheasant 1996). To continue with the previous example:

a. Seat height should be less than (50 mm) or equal to user's sitting popliteal height.

b. The armrest should be located slightly below the sitting elbow height (25 mm).

c. Height of the keyboard work surface is 30–50 mm below the elbow, but also considers thigh clearance, by means of a maximum thigh height plus 50 mm.

d. The highest point of the screen is at, or slightly below, sitting eye height.

26.2.4 Determining the Type of Anthropometric Fitting

The following are the principal approaches for implementing anthropometric fitting through design.

26.2.4.1 Design for the Average User

One very common but erroneous idea is that of basing anthropometric fittings on the average user. This derives from a mistaken concept of normal distribution statistics. From a technical point of view, the average is a theoretical datum obtained by dividing the sum of all data in a sample by the total data; which is to say that it is not an actual datum, and those actual data that happen to match it are extremely few.

Additionally, this average datum only indicates that all the rest are grouped around it, and never that the majority of the data match it. And most importantly, fitting to average data leaves at least 45% of the population, those above and below the average, in serious difficulties. Let us take the first of the fittings established for this exercise: seat height. If based on the average (according to the present data) for women aged 18–24 years, the value would correspond to 399 mm, which would mean that for approximately 45% of the population the edge of the seat would press into the thigh, since the seat would be up to 43 mm higher than the popliteal height for people in the 5th percentile (see Figure 26.2).

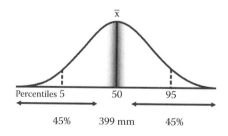

FIGURE 26.2 Design for the average, where approximately 90% of the population remains outside such anthropometric fitting. The average shown here is popliteal height for women 18–24 years old in the Guadalajara, Jalisco, México metropolitan area.

Lastly, due to evolutionary tendencies, there will in future be greater anthropometric variability, and fewer people tending toward the average. Modern times have encouraged a reduction of natural selection via improvements to health care and living conditions, increasing the range of individuals in a given population. In other words, a population will show a rise in its variability according to a relaxation of natural selection pressures.

This implies that the concept of "Mr. and Mrs. Average," or an ideal body type, will be less and less valid, and that designers should expect users to be different from themselves. This incongruence suggests that it will continue to be important to measure increased human variability in order to provide designers with necessary information so that a diverse user population might have fit with products available on the market (Bridger 1995).

26.2.4.2 Individualized Design

One common approach is to design the same product in different sizes. The Aeron Chair from Herman Miller is an example of this type of design, as it is sold in three measurements or sizes (small, medium, and large), although it should be clarified that this does not mean its design lacks adjustability, only that it is linked to the concept that comfort will be optimized if there is a "size" from which different parts of the chair may be adjusted.

A very popular example is the bicycle, which is found commercially in different sizes of the same style; its diversity corresponds to specific user gender, age, or size; while within the same structure and system, adjustments (like seat and handlebar height) permit further optimization. It is also possible to find objects and accessories of various sizes for children, according to the youngster's developmental stage, e.g., spoons, cups, plates, seats and tables, toys, crayons, etc. (Prado-León, Ávila-Chaurand, and Herrera-Lugo 2005).

26.2.4.3   Made-to-Measure Design

An extreme case of individualized design is personalization, or made-to-measure design. This approach would obviously meet the needs of 100% of the population, but that is not to be. The cost entailed in made-to-measure design puts it beyond the economic reach of most of the population, because it must be processed outside traditional production lines.

There are surely special needs where the cost–benefit relationship determines selection of this type of anthropometric fitting. The cases of high-performance sports accessories, or astronauts' suits, are examples of this (Prado-León, Ávila-Chaurand and Herrera-Lugo 2005).

26.2.4.4   Design for a Broad Range (5th to 95th Percentile)

In this approach, it is necessary to consider extreme values in the statistical distribution, first making a comparison between the effects of fitting a product feature, e.g., seat height, to one end of the curve and then observing the effects on a user at the other end; concluding by determining which seat height dimension will best benefit both ends, and thus 90% of the population. See Table 26.1, where it may be noted that the lowest value for popliteal height is 312 mm at the 5th percentile of females 55–65 years old, and the highest value is 488 mm for the 95th percentile of males between the ages of 18 and 24. If seat height is calculated based on the highest measurement, only 5% of the user population of this sex and age range (from the 95th to 100th percentile) will comply with the ergonomic principle of thigh-calf posture in a way that avoids pressure against the seat, and supports the feet on the floor. If the other end is taken, and seat height is considered based on the value of 312 mm, 95% of the user population, of both sexes and every age range, will comply with the principle and design recommendation that the seat height be equal to or less than their popliteal height, thereby avoiding pressure from ischemia (see Figure 26.3). Although seat height for individuals at higher percentiles will be much less than the specific recommendation of 50 mm from their popliteal height, this difference will not cause ischemia. Taller individuals will just have to lift their thighs, reducing the thigh-calf angle, which is less damaging than the other way around, where the legs dangle. Still, recalling that this is a system where diverse components are interrelated, the effect a lower seat may have on work surface height should be taken into account.

TABLE 26.1
Anthropometric Dimensions of Men and Women 18–65 years Old, in the Metropolitan Area of Guadalajara, Jalisco, México

Dimension	Women										Men									
	18–24 (n = 209)		25–34 (n = 195)		35–44 (n = 202)		45–54 (n = 260)		55–65 (n = 81)		18–24 (n = 196)		25–34 (n = 202)		35–44 (n = 176)		45–54 (n = 116)		55–65 (n = 79)	
	Percentiles																			
	5	95	5	95	5	95	5	95	5	95	5	95	5	95	5	95	5	95	5	95
Popliteal height, sitting	356	439	345	422	345	422	337	401	312[a]	408	400	488[b]	383	456	373	450	368	455	375	444
Elbow height, sitting	216	256	236	265	228	266	224	266	197[a]	255	212	268	224	279	236	274	209	280[b]	201	265
Thigh clearance height	122	164	124	168	130	168	127	174[b]	122	172	121	164	125	165[b]	126	162	126	153	122	154
Eye height, sitting	695	778	689	778	682	771	669	755	643[a]	755	736	841[b]	733	835	723	823	709	802	693	811

Note: All dimensions in millimeters.
[a] Minimum value.
[b] Maximum value.

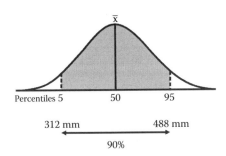

FIGURE 26.3 Design for a broad range, where anthropometric fitting accommodates approximately 90% of the population. Value of the 5th percentile is for popliteal height of women 55–65 years old, and value of the 95th percentile for popliteal height of men 18–24 years old in the Guadalajara, Jalisco, México metropolitan area.

26.2.4.5 Adjustable Design

A better alternative for design with anthropometric fit is to manufacture products whose critical use dimensions may be adjusted by the users themselves. In this way, a perfect fit for the majority of the population is assured. Adjustability makes it possible for the product to adapt to body measurements. It may be considered one of the best options given human variability, since it "matches" the specific design to each individual and generally encompasses 90% of the population, as the range of adjustment is almost always from the 5th to the 95th percentile.

Still, complex mechanical or electrical systems are often needed to attain this adjustability, which raises the cost compared with a product having fixed dimensions; but depending on the cost–benefit relationship, the approach has been useful in various cases, such as that of computer workstation chairs.

Along with cost, another problem with this approach is the extra work involved in having the user make adjustments, and also the fact that he or she generally has little idea of what constitutes the best fit.

26.2.4.6 Universal Design

In recent years, prospects for universal design have become more widespread. Universal design is an intentional process that includes all user groups in a product or environmental design. Particularly, in terms of anthropometric features, it refers to providing the appropriate size and space for reach, manipulation, and use, regardless of the user's body size (Beecher and Paquet 2005). This may be technically possible, as anthropometric variability is not infinite. For example, focusing on my data, the range of adjustment for seat height should be from the minimum value for women aged 55–65, to the maximum value for men between 18 and 24, or from 288 to 517 mm (see Figure 26.4).

In Mexico, this range of adjustment does not exist on the market, but it would be useful and very beneficial if manufacturers considered universal design for those products where its implementation may indeed be feasible.

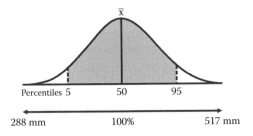

FIGURE 26.4 Universal design accommodating 100% of the user population in the anthropometric fitting. Minimum (women aged 55–65) and maximum (men aged 18–24) values are for popliteal height in the Guadalajara, Jalisco, México metropolitan area.

There are cases of universal design being applied to include disabled users within the population, for what is known as "accessibility." An example of this is widening elevators to accommodate wheelchairs. If a wheelchair's dimensions can fit into an elevator, a user at the 100th percentile of maximum body breadth will also be able to enter it on foot, and this then becomes universal design, or design for all.

The decision concerning which type of approach to use includes, first, an analysis of the need for an adjustable design or not; according to the task effected with the furniture or product, task duration, and effects when using one or other design type. It also includes considerations that have to do with possibilities for mass production and cost benefits.

In this case, it was determined that the work surface, chair seat, and armrest heights should all be adjustable. This took into account the following considerations: personnel at this station can work on data entry for up to 7 continuous hours, so furniture that is adjustable to user population anthropometric ranges is essential; otherwise, discomfort, pain, and even musculoskeletal injury may occur. Also, Verbeek (1991; cited by Bridger 1995) mentions that in practice there is no strong correlation between popliteal and elbow height, so it is necessary that seat and desk heights be adjustable. Lastly, anthropometric differences found in data for this user population are considerable, and furniture that does not adjust could not be adapted to those differences. Ranges of adjustment were thus established for the chair in terms of seat and armrest height, as well as for the height of the keyboard tray and monitor. Minimum and maximum ranges were determined for the latter, though locating height adjustment devices is not very feasible in Mexico, and the monitor is generally elevated by stacking books or other objects beneath it.

26.2.5 Percentile Selection

As shown in the previous step, percentile selection largely depends on the type of approach that has been put into effect.

We must not lose sight of how an effective anthropometric fit ensures that the product can be used according to the ergonomic principles required by at least 90% of its users, and that it has been common practice for many years to base any kind of fitting on the 5th and/or 95th percentiles.

In this case, the 5th and 95th percentiles were selected for the sex and age range in which values were, respectively, minimum and maximum, to encompass 90% of the possible user population, as may be seen in Table 26.1.

26.2.6 Extracting the Anthropometric Data for the Target User Population

As has been previously mentioned, anthropometric data for the population that will use the furniture must be made available. In the present work's study, unpublished data were used that had been generated by means of a 2004 anthropometric field study done by the CIE in the Guadalajara, Jalisco, Mexico metropolitan area, with a total sample of 2103 subjects of both sexes, arranged in the following age groups: 15–17, 18–24, 25–34, 35–44, 45–54, and 55–65.

In Table 26.1, we may observe the anthropometric data used in fitting the proposed dimensions. On occasion, a dimension does not exist in the consulted database, but it is possible to make a very approximate calculation based on other dimensions. For example, if the elbow height-seat dimension is missing, this can be obtained by subtracting the standing elbow height from the subject's stature, and then subtracting this difference from the sitting height (see Figure 26.5).

26.2.7 Considering Clearances

Clearances are the tolerances that permit accommodation of different parts of the body, and often even the whole body, as well as movements and considerations of user clothing or special equipment (Prado-León, Ávila-Chaurand, and Herrera-Lugo 2005).

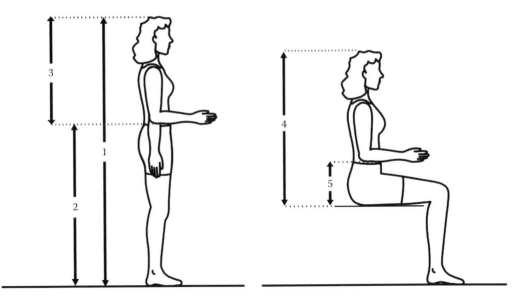

FIGURE 26.5 Graphic showing how to derive the sitting elbow height (5), from other dimensions: from stature (1) is subtracted standing elbow height (2) and the resulting difference is elbow-head height (3); from sitting stature (4) is subtracted the former height (3) and the difference is equivalent to measurement for the armrest-seat height dimension (5).

For seat and work surface height, the clearance to consider is the heel. For this clearance, it has been suggested that 25 mm be added to the height for men and 75 mm for women (Tayyari and Smith 1997). However, given that the case being analyzed presents its minimum height value at the 5th percentile of women in the 55–65 year age range, and that in this social context users generally wear a low heel, it was decided to consider a 25 mm clearance for women as well. Thigh clearance must also be considered for work surface height.

26.2.8   APPLICABLE ARITHMETIC OPERATIONS

a. Seat height* = popliteal height* – ergonomic guideline + heel clearance
   Seat height* = 312 mm – 50 mm + 25 mm = 287 mm
   Seat height† = popliteal height† – ergonomic guideline + heel clearance
   Seat height† = 488 mm – 50 mm + 25 mm = 463 mm

b. Armrest height* = elbow height* – ergonomic guideline
   Armrest height* = 197 mm – 25 mm = 172 mm
   Armrest height† = elbow height† – ergonomic guideline
   Armrest height† = 280 mm – 25 mm = 255 mm

c. Keyboard surface height* = seat height* + armrest height*
   Keyboard surface height* = 287 mm + 172 mm = 459 mm
   Keyboard surface height† = seat height† + armrest height†
   Keyboard surface height† = 463 mm + 255 mm = 718 mm

d. Monitor height* = seat height* + eye height
   Monitor height* = 287* mm + 643 mm = 930 mm
   Monitor height† = seat height† + eye height
   Monitor height† = 463 mm + 841 mm = 1304 mm

---

* Minimum value, 5th percentile.
† Maximum value, 95th percentile.

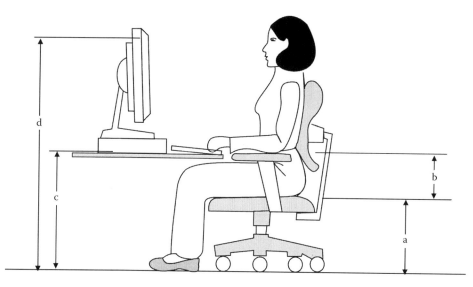

FIGURE 26.6   Two-dimensional simulation of anthropometric fitting with minimum range of adjustment for (a) seat height = 287 mm, (b) armrest-seat height = 172 mm, (c) keyboard tray surface height = 459 mm, and (d) upper edge of screen height = 930 mm (1:10 scale).

26.2.9   Workspace Simulation

This combines all design values obtained from a simulation based on a computer-generated scale drawing, mock-up, or three-dimensional (3D) computer model to assure that they will be compatible.

Figure 26.6 shows a two-dimensional 1:10 scale simulation of measurements determined for the 5th percentile of women aged 55–65 years, and Figure 26.7 for the 95th percentile of men aged 18–24 years.

Simulation is a very important step since this is where we may take note of errors, as well as how the various elements are interrelated. In this simulated example, it was found that using the keyboard surface guideline, which gives the ideal as 30–50 mm below the elbow (40 mm considered as average in the 30–50 mm range, with a 25 mm heel) would yield:

- Keyboard surface height[*] = sitting elbow height – 40 mm + 25 mm
  Where sitting elbow height may be obtained from the sum of popliteal height + elbow-to-seat height: 312 mm – 197 mm = 509 mm
  Keyboard surface height[*] = 509 mm – 40 mm + 25 mm = 494 mm
- Keyboard surface height[†] = sitting elbow height – 40 mm + 25 mm
  Where sitting elbow height may be obtained from the sum of popliteal height + elbow-to-seat height: 488 mm + 280 mm = 768 mm
  Keyboard surface height[†] = 768 mm – 40 mm + 25 mm = 753 mm

Placing these dimensions into the two-dimensional simulation, the surface height was still 35 mm higher than the armrest height for both men and women, which could imply wrist deviation.

It was thus decided to calculate keyboard surface height based on the previously calculated armrest height, as may be seen in step 8 of the anthropometric fitting.

---

[*] Minimum value, 5th percentile.
[†] Maximum value, 95th percentile.

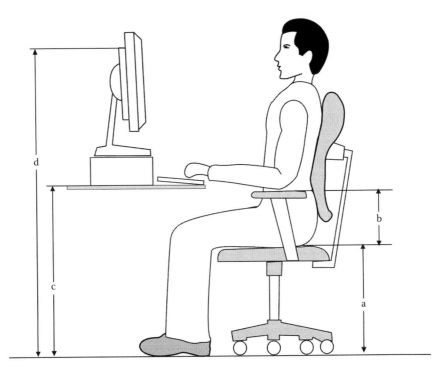

FIGURE 26.7   Two-dimensional simulation of anthropometric fitting with maximum range of adjustment for (a) seat height = 463 mm, (b) armrest-seat height = 255 mm, (c) keyboard tray surface height = 718 mm, and (d) upper edge of screen height = 1304 mm (1:10 scale).

The minimum range of adjustment for keyboard height should thus be from 459 mm to a maximum of 753 mm, for the user population of both men and women aged 18–65 years.

Figure 26.8 shows a woman in the 5th percentile of the 55–65 age range, at a workstation with dimensions set for the 95th percentile of men aged 18–24 years. What stands out is a significant mismatch between the dimensions of the person and the furniture, confirming the need for adjustable furniture.

Another important element arising during the simulation was that for a very short but very obese person (a possible extreme situation), established dimensions for the minimum workstation value provided insufficient thigh clearance. In this example, the space between the seat and work surface was 172 mm, allowing no thigh room and little freedom of movement for a person in the 5th percentile for popliteal and elbow height with 95th percentile thigh height (172 mm), in the female 55–65 age range. In this case, a special adjustment would be made, using the guidelines' flexibility ranges and taking care not to induce an awkward posture. For example, raising the armrest height to the highest possible range without lifting the arms, but opening the arm-forearm angle slightly (without exceeding the 120° limit); raising the keyboard surface height, although thigh clearance would then be less than the recommended 50 mm; and if necessary raising the seat height and using a footrest to keep the feet from dangling.

Once furniture dimensions were established using anthropometric measurements for the Mexican population, these data were compared with foreign standards.

26.3   COMPARISON WITH FOREIGN DATA

26.3.1   Seat Height

As may be observed in Table 26.2, both minimum and maximum values in the range established for seat height are much lower when using data from the Mexican population to calculate the

FIGURE 26.8 Woman in the 5th percentile of the 55–65 age range, at a workstation for a man in the 95th percentile of the 18–24 age range (1:10 scale).

measurement, compared with all foreign data. The maximum value presents no problem when Mexican users have chairs with heights based on foreign data, as in the case of the "technical support" station where the operator has her seat height at 470 mm with a possibility of raising it up to 520 mm; since the maximum height is simply not used. The problem is with minimum value, where no chair on the market can be lowered to the measurement (287 mm) calculated in this work.

TABLE 26.2
Measurement Comparisons of Computer Workstation Furniture

Dimension	(1)	(2)	(3)[a]	(4)[a]	(5)[a]	(6)
Seat height (mm)	430–525	287–463	500–570	–	430–570	420–510
Armrest height (mm)	220	172–255	–	–	–	190–240
Keyboard tray height (mm)	745	459–718	700–800	700–850	710–870	705–755[b]
Monitor height (mm)	1190[c]	930–1304[c]	1070–1150[d]	800–1100[d]	920–1160	–

Note: Column headings are as follows: (1) "Technical support" station; (2) Calculated dimension; (3) Heiden and Krueger (1984); (4) Lueder and Noro (1995); (5) Granjean et al. (1983); and (6) Ministry of Labour (2004).

[a] In Kroemer and Grandjean (2001).
[b] Recommended range for a fixed height.
[c] From floor to upper edge of monitor.
[d] From floor to center of monitor.

26.3.2 ARMREST HEIGHT

When comparing the fixed armrest measurement (220 mm) of the chair in the example workstation with the Canadian standard (190–240 mm) and the range established in this analysis (172–255 mm), it was observed to be within both ranges; however, the operator in Figure 26.1 could not use it because of its very low level in relation to the work surface. Effects from lack of forearm support when working at the computer have been reported by Cook, Burgess-Limerick, and Papalia (2004), who found that supporting the forearm significantly reduces ulnar deviation of the wrist as well as reported discomfort in computer users. Supporting the wrist was associated with less trapezius and anterior deltoid muscular activity.

26.3.3 WORK SURFACE HEIGHT

Also, the keyboard tray at the selected station had an intermediate height (745 mm) regarding ranges in foreign data and approached the maximum value of the range for established dimensions based on Mexican anthropometry (753 mm), which represents a serious problem for users not in the high percentiles. This may be confirmed by figuring out the difference between the calculated minimum value (459 mm) and the height value of the evaluated station, as well as the foreign data minimum (745 and 700 mm, respectively): difference = 286 and 241 mm, which is a very large difference. That is to say, if a short user's keyboard tray is 286 mm higher than recommended, this mismatch has very significant negative effects on his or her comfort, health, and efficiency.

In this operator's case, although the keyboard height is above elbow level, it is not such an extreme difference (see Figure 26.1), since her stature (1689 mm) places her very close to the 95th percentile of the data's 25–34 year age range.

However, even a 50 mm difference may have considerable effects, as Zennaro et al. (2004) noted when they carried out a study, which found that adding 50 mm to the appropriate keyboard height significantly increased muscular activity.

26.3.4 MONITOR HEIGHT

The majority of values in the ranges for monitor height are lower than in foreign data, somewhat exceptional according to previous comparisons, but explicable because they are taken from the floor to center screen and in this work from the floor to the upper edge of the screen. In the station selected for anthropometric analysis, the height was found to be within range (1190 mm), and the operator did not need to incline her head in order to see the monitor.

26.4 ADDITIONAL WORKSTATION RECOMMENDATIONS

Since, both for financial reasons and because these products simply do not exist on the market, it is not possible in the short term to replace existing furnishings with those meeting dimensional requirements as calculated in this work, a decision was made to include a series of suggestions that would use the same furniture while creating better ergonomic conditions.

One of the suggestions for the "technical support" workstation (Figure 26.1) was to elevate the seat height slightly (the operator had it at 470 mm), to bring the level of the elbow closer to the work surface, and to use a small cushion at the edge of the work surface as a wrist support. When raising chair height, use of a footrest is vital. Also, since chairs with the minimum value that this work calculated (287 mm) do not exist on the market, footrests are needed for shorter women.

Considering other important ergonomic features, use of a document holder is recommended. This should be placed beside the screen and at the same distance from the user: when an operator sets documents on the work surface beside the keyboard, it requires the head to incline downward.

The following guideline supports this recommendation.

The viewing area requiring frequent visual contact should be located within the central visual field. If the screen or material are placed outside this field, frequent visual contact will require excessive head and eye movements, causing neck, shoulder, and eye discomfort; thus, it is recommended that these items be placed to the side of the body's central line and angled to be perpendicular to the eyes (Anshell 1999).

It has also been shown that it is not enough to establish appropriate ranges for adjustable computer workstation furniture: training is necessary so that the furniture will truly be used in the correct fashion. For this, it is important that the user understands the objective of the adjustment, the correct way to do it, and the ramifications of not doing it. One example of the effect that training has on anthropometric computer station adjustment appears in a study by Verbeek (1991; cited by Bridger 1995). Before the program, a sampling of chair-desk fit in an office revealed deviations from the ideal heights of 71 mm for the seat and 70 mm for the desk. After the training program, participants reduced these deviations by 11 and 18 mm, respectively.

It must be noted that computer training should not only include anthropometric aspects, but all the ergonomic concerns that users may oversee and modify at their computer workstations, such as lighting conditions, task duration, breaks and exercise, etc.

Thus, the last recommendation is that training be provided to computer operators in all the above areas, with the objective of letting them know the reasons for, and the importance of, ergonomic guidelines to their work at the computer.

26.5 FINAL COMMENTS

The development of anthropometric fitting has been extended to encompass two-dimensional simulation, but it is often necessary to also include 3D analysis, which may use a mock-up or a 3D computer-generated model, as was mentioned in 26.2.9.

Also, preparation and testing of prototypes are very important parts of the design development process. These not only include product manufacture, but such special enhancements as photographs, biomechanical models, electromyographic evaluations, strength testing, psychophysical evaluations, user comfort measurements, etc.

As may be observed, the anthropometric fitting process is complex and demands a good grasp of ergonomic design principles. These principles derive in large measure from biomechanics and knowledge of muscular work, which allow us to judge physical and mechanical stress or load on the human body, primarily due to incorrect postures and static muscular effort. New findings in these areas permit development of new furniture designs and subsequent evaluations, dealing with improved postures, reduced effort, and increased efficiency for computer tasks; such as the vertical keyboard (Van Galen, Liesker, and de Haan 2007) and a device that slants the work surface and moves the keyboard closer to the screen (Tepper et al. 2003).

As has been shown throughout this chapter, a product's anthropometric fit is just one very important part of its ergonomic fitting process, which also entails consideration of anatomical, physiological, biomechanical, psychological, and sociocultural factors.

Throughout this work, evidence has also arisen for the need to generate anthropometric data on user populations for ergonomic design. The CIE has conducted various field studies with this objective (Ávila-Chaurand, Prado-León, and González-Muñoz 2001; Prado-León, Ávila-Chaurand, and González-Muñoz 2007). The most recent anthropometric study took place in 2004 and used a 3D scanner to obtain measurements. However, these studies were conducted in very localized geographic zones (Guadalajara's metropolitan area) and a national study is required to enable anthropometric fitting for all Mexican users.

The dearth of data, and Mexico's limited recognition and practice of ergonomics, obviously carry with them a lack of nationally legislated ergonomic standards for workstations. And yet, globalization has greatly impacted computer use in Mexico, spreading to many office and even home workstations, but with little consideration of ergonomics. While the effects of such proliferation have not been widely evaluated, this does not keep them from manifesting themselves.

REFERENCES

American Industrial Hygiene Association. 1994. *An Ergonomics Guide to VDT Workstations*. Arlington, VI: AIHA.

Anshel, J. 1999. *Visual Ergonomics in the Workplace*. London: Taylor & Francis.

Ávila-Chaurand, R., Prado-León, L.R., and González-Muñoz, E. 2007. *Dimensiones antropométricas. Población Latinoamericana*. México: Universidad de Guadalajara.

Beecher, V., and Paquet, V. 2005. Survey instrument for the universal design of consumer products. *Applied Ergonomics* 36:363–72.

Bernard, P.B. ed. 1997. *Musculoskeletal Disorders and Workplace Factors. A Critical Review of Epidemiologic Evidence for Work-related Musculoskeletal Disorders of the Neck, Upper Extremity and Low Back*. Cincinnati, OH: U.S. Department of Health and Human services. National Institute for Occupational Safety and Health.

Bridger, R.S. 1995. *Introduction to Ergonomics*. New York: McGraw-Hill.

Cook, C., Burgess-Limerick., R., and Papalia, S. 2004. The effect of upper extremity support on upper extremity posture and muscle activity during keyboard use. *Applied Ergonomics* 35:285–92.

Drury, C.G., and Coury, B.G. 1982. Methodology for chair assessment. *Applied Ergonomics* 13:195–202.

Kroemer, K.H.E., and Grandjean, E. 2001. *Fitting the Task to the Human. A Textbook of Occupational Ergonomics*. London: Taylor & Francis.

Kroemer, K.H.E., and Kroemer, A.D. 2001. *Office Ergonomics*. London: Taylor & Francis.

Montreuil, S., Laflamme, L., Brisson, C., and Teiger, C. 2006. Conditions that influence the elimination of postural constraints after office employees working with VDU have received ergonomics training. *Work* 26:157–66.

Moon, S.D. 1996. A psychosocial view of cumulative trauma disorders. Implications for occupational health and prevention. In *Beyond Biomechanics. Psychosocial Aspects of Musculoskeletal Disorders in Office Work*, eds. S.D. Moon and S.L. Sauter, 109–143. Bristol, PA: Taylor & Francis.

Ontario Ministry of Labour Area Offices. 2004. *Computer Ergonomics: Workstation Layout and Lighting. Health and Safety Guidelines*. Canada: Ministry of Labour.

Pheasant, S. 1996. *Body Space. Anthropometry, Ergonomics and the Design of Work*. London: Taylor & Francis.

Prado-León, L.R., and Ávila-Chaurand, R. 2006. *Ergonomía y diseño de espacios habitables. Factores antropométricos y socioculturales*. México: Universidad de Guadalajara.

Prado-León, L.R., Ávila-Chaurand, R., and González-Muñoz, E. 2001. Anthropometric study of Mexican primary school children. *Applied Ergonomics* 32:339–45.

Prado-León, L.R., Ávila-Chaurand, R., and Herrera-Lugo, E. 2005. *Factores ergonómicos en el diseño. Antropometría*. México: Universidad de Guadalajara.

Támez-González, S., Ortiz-Hernández, L., Martínez-Alcántara, S., and Méndez-Ramírez, I. 2003. Riesgos y daños a la salud derivados del uso de Videoterminal. *Salud Pública de México* 43 (3): 171–80.

Tayyari, F., and Smith, J.L. 1997. *Occupational Ergonomics. Principles and Applications*. Norwell, MA: Kluwer Academic.

Tepper, M., Vollenbroek-Hutten, M.M.R., Hermens, H.J., and Baten, C.T.M. 2003. The effect of an ergonomic computer device on muscle activity of the upper trapezius muscle during typing. *Applied Ergonomics* 34:125–30.

Tullar, J., Amick, B.C., Robertson, M.M., Fossel, C., Coley, A.H., Hupert, N., Jenkins, M., and Katz, J.N. 2007. Direct observation of computer workplace risk factors of college students. *Work* 28:77–83.

Van Galen, G.P., Liesker, H., and de Haan, A. 2007. Effects of a vertical keyboard design on typing performance user comfort and muscle tension. *Applied Ergonomics* 38:99–107.

Zennaro, D., Lâubli, T., Krebs, D., Krueger, H., and Klipstein, A. 2004. Trapezius muscle motor unit activity in symptomatic participants during finger tapping using properly and improperly adjusted desks. *Human Factors* 46 (2): 252–66.

27 User-Centered Approach for Sailing Yacht Design

Giuseppe Di Bucchianico and Andrea Vallicelli

CONTENTS

27.1 INTRODUCTION

Yacht design is the applicative sector of industrial design that deals with the planning of pleasure boats. It is a complex field of industrial design, in which highly specialized branches of learning converge with as many specific planning contributions that the yacht designer finds himself/herself having to coordinate to establish the formal values of the artifact-boat, full of symbolic and functional aspects.

In particular, sailing yacht design deals with difficulties of a superior nature, because of its need to integrate the technological and technical innovation and figurative and cultural codification aspects with the demands of livability of the areas in continual movement and of the overall efficiency of the system-boat expressed by the end users.

Therefore, yacht design finds a valid operative support in the methods, techniques, and instruments developed by applied ergonomics to direct planning action with awareness toward a morphological and spatial research that is both elegant in its language and fluid in its articulation, and functionally efficient and pleasant in its use. In fact, the user-centered design (UCD) approach, with its theoretical and applicative scope, allows the individual and his/her requirements to be placed at the center of yacht design, which reflects on the main levels of the requirements of the nautical product: from those regarding the safety of equipment and areas, to their functionality and ease of use, to the requirement for pleasantness of the environment and finishing.

To describe the possible intersections and interactions between ergonomics and sailing yacht design, this chapter is articulated in three parts.

Section 27.2 outlines the peculiarity and multidimensional complexity of this design field, particularly sailing yacht design, underlining the aspects on which the UCD ergonomic approach can provide planning indications that are useful to the anthropocentric management of the project.

Section 27.3 starts off from a known classification of the ergonomic requirements of products (safety-functionality-usability-pleasantness) to outline a synthesis from it attributed to the nautical product, with the aim of identifying the various levels of a possible contribution of applied ergonomics to yacht design.

Section 27.4 reports two experiences of ergonomic research applied to sailing yacht design that show how ergonomics can contribute to identifying highly innovative solutions oriented toward the well-being of the user even in such a sector as pleasure boat sailing, highly connoted by a complex system of semantic and technological ties.

27.2   YACHT DESIGN AND ERGONOMICS

The multidimensional complexity of yacht design, in particular of sailing pleasure crafts, passes through a system of ties that involve several branches and areas of competence, at times highly specialized: from structural engineering to material engineering, from fluid dynamics to marketing. Add to this is the foreshadowing complexity of a product that must have living characteristics often halfway between a passenger compartment and a house and that, furthermore, is for use in trim conditions in continual movement.

The UCD approach, together with the complex system of methods, instruments, and control and assessment techniques of the project, which ergonomics has developed in the last decades, offers the designer a useful contribution to the anthropocentric management of yacht design.

27.2.1   Characteristics of Yacht Design

Yacht design is the applicative sector of industrial design that deals with the planning of pleasure boats, which have an internal living area and are for open-sea, sports, or cruiser navigation. It is a planning sector characterized by a particular complexity, for it must coordinate highly diversified and interacting multidisciplinary areas of competence, which cover the scientific fields of engineering, architecture, ergonomics, marketing, and ecology, with their respective specialized disciplinary articulations. For example, considering sailing yachts only from an engineering point of view, it is a matter of highly complex aeolian-propelled machines that move between two fluids (air and water), stressed by static and dynamic loads. Therefore, their study involves several branches of learning among which there is naval architecture (study of resistance to the movement of the hull), aerodynamics (efficiency of the appendages and sails), structural engineering, and material technology.

As far as the genesis of the project is concerned, however, it can be said that the greater complexity derives from the fact that the pleasure craft embodies both the internal symbolic and functional values of the house, of the "refuge" (stability, strength, safety, privacy), and the external ones of the "vehicle" (lightness, dynamicity, maneuverability). In this, even the dimensions of the object have an important role in the definition of the relationships between the concepts of house and vehicle: in general, it can be assumed that so much bigger the boat is, especially on the inside, so much the relationship with anthropometric constrains and with proxemic and figurative "dimensions" of housing becomes tight (also as a consequence of the greater effective "stability" of the bigger sailing yachts), and, therefore, its relationship with the sea becomes weaker. Thus, if the comparison with civil architectural areas and its more typical furnishing systems is generally immediate and spontaneous in big ships, in small crafts, in particular sailing ones, one often witnesses compromises between the spatiality of the living area and that of the binnacle. Furthermore, in contrast with other means of "habitable" locomotion, crafts are forced to resolve the antinomy between movement and stop with a greater planning effort: "A boat is represented as an object which moves even when it is still. When a camper van is still, it is static like a house; a craft is always in movement even when it is still" (Spadolini 1987). A boat, therefore, is represented as an unstable object that moves inside an element that is also unstable. Therefore, the external "shell" takes upon itself the arduous task of carrying out the role of a boundary, limit, edge, real and symbolic, between an internal, finite, static, and domestic world, and the sense of infinity conveyed by the external marine context, which is variable, unstable, wild, and indomitable. In a sailing yacht, these relations become complicated because the "rolling" point of sailing and the continual movement are almost a rule.

Furthermore, in planning a sailing craft it is necessary to consider the need to integrate the technological and technical innovation aspects, in continual evolution, with a specific figurative and cultural codification, at times with a millennia-long tradition, particularly present in this field of the project.

The first image that probably comes to everybody's mind when thinking of a sailing craft is that of a "deck surface," of a rather complex morphology, cluttered with specialized equipment placed in an apparently casual way and without a visible aesthetic sense, on which members of the crew carry out specific tasks, at times very rapidly, without apparent order. In reality, by studying the tasks and equipment of a sailing craft in a more analytical way, it transpires how it is, above all, a place of highly organized and disciplined activity, based on rigorous formal, personal, and functional relations between artifacts and individuals, with expressly hierarchical roles, which operate in limited areas for the same common aim, which is the overall performance of the craft.

An analogous consideration could be made for the "underdeck" areas: whether it is a sailing craft or a motor maxi yacht, the nautical internal areas are, however, variously recognizable because of certain distinctive elements that are present in them, which are an expression of functional results or simple formal heritage tied in with nautical tradition.

Therefore, for the designer it is a matter of dealing with an articulated system of historically represented morphological and spatial relations, in which the multiplicity of human activities, the spatial areas, and the equipment present on the craft, which, if, on the one hand, continues even today to relate to a strong tradition of nautical practice, on the other hand, is called on to deal with the evolution of roles and tasks on board, which tend to follow morphological, functional, and technological modifications in positions and equipment. In the evolutionary panorama to which we refer, which is constantly oriented toward research and experimentation, the nautical designer carries out a primary and coordinating role between the various areas of competence involved, for a typological and functional redefinition, both of the spatial areas and of positions and single equipment, which take into consideration, on the one hand, the evocative-cultural result of tradition and, on the other hand, the technological progress obtained in the field of materials, building, and production techniques, and electronic miniaturization.

27.2.2 User-Centered Design Applied to Sailing Yacht Design

Yacht design is a particularly complex applicative sector of industrial design, both because it operates on highly variable representative scales, from the architectonic scale to that of single equipment, and because every project must deal contemporarily with the diversified system of limitations imposed by the user (from physical-dimensional ones to psychosocial ones) and with the overall efficiency of the "system-boat" (from engineering ones to organizational or simply functional ones of the various maneuvering positions and equipment).

Thus, the methods and techniques developed by ergonomics and applied to design represent a valid operative and instrumental support for yacht design, in order to direct the planning action with awareness toward morphological and spatial research that is both elegant in its language and fluid in its articulation, and functionally efficient and pleasant to use. In particular, the UCD ergonomic approach, which is based on the idea of planning artifacts that can be utilized by users with maximum efficiency and minimum physical and mental discomfort, provides the most meaningful theoretical and methodological contribution to the development of some aspects of yacht design. It is an approach that allows the acquisition and assessment of users' requirements by means of structured and verifiable methods, to turn them into planning instruments. It proposes a cyclical course, articulated in a series of activities (Tosi 2003):

- Identifying the basic functions and aims of the product (principal, secondary, and accessory)
- Specifying the use context and its components (characteristic of users, task, equipment, and physical and social environment in which the interaction with the product is carried out)
- Identifying the demands of the users and organization, thereby defining the ergonomic requirements of the product
- Producing planning solutions and prototypes, to assess them, with the users' contribution, in relation to the requirements identified

Within this course, the process of identification and acquisition of the users' demands takes on a central role: it can come about both on an objective basis (by referring to measurable parameters and establishing for each of them, by appealing to regulations, the most suitable acceptability thresholds), and on a personal basis, when one is interested in identifying even latent emotions, preferences, expectations, and desires. The users' demands and the consequent requirements of the products may, in turn, be expressed according to different priority levels, from the lowest (safety and accessibility) to the highest (pleasantness and gratification).

Because of the complexity of yacht design, it tends to be faced naturally in a "particle" way: a clear division of tasks and roles among different, extremely specialized areas of competence tends to approach the project in parts, often with a distinct division between the aspects of a "productive" nature (economic, building, and systemic) and the so-called "human" factors (regarding the individual or social use of the product). The role of the yacht designer is precisely to plan the shape of the nautical product by coordinating, integrating, and articulating such factors. Therefore, he/she finds a valid aid in the UCD methods and instruments to assess and plan, most especially, the man–boat interaction with greater awareness: he/she manages to do this at all levels referable to the users' demands and regarding the different scales of the project, from the single pieces of equipment present on board to the more complex spatial areas.

27.3 VARIOUS APPLICATIVE FIELDS OF USER-CENTERED DESIGN TO SAILING YACHT DESIGN

The UCD places at the center of its identification process, by means of the users' demands, some ergonomic requirements for the product. The ergonomic literature, on the other hand, has recently placed four different requirement levels referable to the users' demands in a close hierarchical relationship (Jordan 2000). The requirements of the nautical product refer to various aspects. It is possible to identify an analogous articulation for the nautical product too. Thus, the requirements concerning the safety of equipment and the environment are identified—those regarding their functionality, ease, and practicalness of use, including the requirements that concern their pleasantness, the physical and mental "pleasure" that is experienced in interacting with the product and the environment. For each one of them, ergonomics is already able to offer a precious contribution in relation to the nautical project of environment, positions, equipment, and finishing, as well as, in a figurative sense, to establishing all those tasks and activities, even apparently secondary, which are carried out on the craft.

In particular, on the safety on board plane, ergonomics provides the most suitable instruments to analyze the characteristics of individuals and the limits of their psychophysical abilities, even in extreme environmental and postural conditions; on the functionality plane, ergonomic research can relate directly to the typological evolution of the equipment and the various parts of the craft to allow a critical, objective, well-pondered reading of it; on the plane of usability of the equipment and postural comfort, ergonomic practice allows the identification of more innovative solutions also by means of observation of organization and structure of tasks on board; on the pleasantness plane, ergonomics has defined the most useful instruments and methods to assess the psychosensory interactions of individuals with components, equipment, and the environment.

Furthermore, in all cases, the multidisciplinary approach of ergonomics to the design and the availability of methods, intervention procedures, and operative instruments that it offers, allows the study of the requirements of the user's well-being to be faced whether in relation to the single product/equipment or to the task/position or to the environment/context in which he/she finds himself/herself.

27.3.1 SAFETY ON BOARD

The "safety" requirement of an artifact may be considered both generic and defined and regulated by a complex and articulated system of national and international standards, even in a specific planning sector such as yacht design.

In general terms, safety may be defined as the set of conditions regarding the safety of users as well as the defense and prevention of damage depending on accidental factors. In the nautical field, some sources of danger are obviously correlated directly to the technical-productive aspects of the craft: therefore, there are numerous safety standards oriented toward the control and testing of the aspects of structure, building, buoyancy, fire resistance, etc.

One may add that, in the nautical field, almost all safety standards are attributed "only" to regulating such aspects. For example, even in the international normative picture, just a few technical safety standards are applied to the crew. Among them is the ISO 15085: 2003 standard, entitled "Small craft: man-overboard prevention and recovery," whose salient points refer particularly to the differences in the level of the deck, the requirements of foot stops, gunwales, manropes and stays, non-slip elements, and boarding means usable without assistance. Furthermore, this standard highlights another problematic aspect concerning the safety requirements of pleasure boats. However, in this case, too, the various aspects are not considered with the completeness and organicity that the subject would require.

Thus, it happens that the variability of the physical and perceptive characteristics of the users as a fundamental aspect to define the accidental sources of danger are not considered. Nor, least of all, are the numerous "disturbance" effects that interactions with the context can produce on the individual considered, which, in actual fact, reduce his/her physical abilities. On the contrary, ergonomic planning teaches that observation and analysis of the "system of interactions" between individuals, contexts, activities, and equipment represent the fundamental principle of every "safe" and "comfortable" project. This is a concept that can also be considered to be at the root of the difference between the "safety" and "ergonomics" of a product: in fact, by means of ergonomic planning it is possible to obtain "also" better safety conditions, while the simple application of adequate safety measures may be attributed strictly to single pieces of equipment or activities, do not in themselves guarantee overall conditions of well-being for the individuals who have to carry out particular tasks with particular equipment in specific environmental conditions.

Unfortunately, on many craft, as happens in many other "workplaces," the realization of maneuvering positions is often the result of a simple "assembly" process of the single elements that make it up: perhaps they correspond individually to specific safety standards, but they are not the subject of a real, combined, and coordinated planning activity, with the inevitable risk of accidents.

On the contrary, to reduce the risks, to optimize the efficiency of the "boat" system and, at the same time, to pursue the well-being of the individuals involved in the activities on board, it is advisable to make reference above all to the so-called ergonomic principle of "totality of interventions," which calls for turning one's planning attention to general, wide-ranging subjects, leaving the single aspects of the problem to a second stage of in-depth study.

Starting from such a consideration, the multidisciplinary wealth of ergonomics is already able to offer the yacht designer many operative instruments and suggestions useful for planning equipment, positions, and "safe" areas on pleasure boats in an ergonomically correct way. Among these, it is advisable to remember, for example, that:

- The correct use of anthropometric data (percentile) and the dimensioning of the accessible (zcr) and visibility (ovz) areas, are useful for an initial definition of the dimensional requirements of areas and positions, and for the placing of components and equipment.
- The assessment of risk factors to the muscular-skeletal structure (by appealing to the many methods defined clearly by ergonomic research, such as OWAS, RULA, NIOSH, etc.) is fundamental to define the functional requirements attributed to posture, movements, and muscular efforts, often taken on in particularly uncomfortable and tiring postural conditions.
- Knowledge of the sensorial abilities and the most elementary mental mechanisms of individuals seems advisable to facilitate the comprehensibility of the use of equipment and commands, reducing, for example, the stress caused by mental fatigue or accidents due to the monotony and repetitiveness of tasks, which are always lying in wait during long navigations.
- Furthermore, analysis of the organization of activities on board allows the reduction of risks of accidents caused by human error in coordinated and collective activities, which exist, for example, on sailing boats during maneuvers to change the points of sailing.

The knowledge defined clearly by ergonomics, therefore, already seems particularly useful at this primary definition level of product requirements regarding safety to widen the range of the variables tied in with the planning subject, allowing contemporarily the objective checking also of many aspects concerning the safety of individuals.

27.3.2 Functionality of Equipment and Environment

We said that the craft itself sums up the typical values of a dwelling and a vehicle. We can say that, above all for the smaller craft, the design is in search of continual compromises precisely between the spatiality of the living area and those of the "binnacle." This means that in a planning sense the proxemic and perceptive relations must be taken into consideration with which the users relate to in the "reduced" areas of the craft, which are obviously different in comparison with the more traditional architectonic ones.

Furthermore, the sailing boat's space is, by definition, a space that moves and tilts continually and noticeably in all directions. It is constantly in motion, even when the boat is "still" at its moorings.

The extent of the movements that the craft is subject to depends prevalently on its dimensions, speed, angle of contact with the waves and, naturally, meteomarine conditions. Such motion affects the conditions of life and work on board in a meaningful way. In particular, the rolling (transversal rotatory motion) and pitching (longitudinal rotatory motion) can take on a meaningful role in the definition of the conditions of well-being of the passengers, causing the classic malaise commonly called "sea sickness," to real loss of balance. Therefore, the need for users to learn to live with the incessant movement of the craft without being excessively disorientated by it is another factor that greatly affects many aspects of planning.

These initial considerations involve some initial, substantial, typological, distributive, and dimensional variations of the nautical environment compared to the structure of dwellings on dry land. Therefore, the difference reaches well beyond the simple morphological and stylistic characterization of the design or choice of materials for the finishing, which could refer directly to nautical tradition. The need for living and work areas that allow the most common activities on board to be carried out, living both with the narrowness of the spaces, and with the even violent movements of the craft, and with a transversal listing of the whole hull as regards the horizon, which can last for relatively long periods, has caused every component of the craft, both outside and inside, from the single equipment to the positions and micro environment, to envisage planning solutions full of strictly functional values that often derive from a nautical tradition consolidated by centuries of history. Therefore, case records of planning in the nautical field are a rich source of ideas and considerations attributed to typological, morphological, and functional solutions, often extremely original, general, and detailed. Some are of a general nature, attributed to all parts of the craft, such as those that tend toward a generalized reduction of corners, in search of a morphological continuity between surfaces, never interrupted and connected as much as possible; or those that tend toward eliminating mobile furnishings; or, finally, those that deal with the continual lateral listing of the craft, by adopting planning solutions now consolidated ("nautical" ladders with sunken steps, horizontally pivoted elements, shelves and surfaces with raised edges, etc.). Other specific solutions, subsequently added, are attributed to single environmental units (wardroom, chartwork, cabins, bathrooms, cockpits, etc.). Systems are involved, widely discussed in the literature of the sector, in which a compromise is found between the requirements of the user, attributed to the activities they carry out, and the adaptable and movable characteristics of the minimum spaces assigned to them.

In these cases too, ergonomic research offers a planning approach and all the necessary instruments to identify and verify particularly innovative solutions even starting from a critical-dismantling or historical-typological analysis of the requirements for the use of parts and components of the "boat" product, from the most consolidated planning solutions to the experimental ones.

27.3.3   USABILITY OF EQUIPMENT AND POSTURAL COMFORT

In recent years, ergonomics has developed many methods and operative instruments that are useful to the design and assessment of the usability aspects of the products. Procedures are involved, about which detailed literature is now available, which are able to quantify and, above all, provide objective data on the qualitative aspects attributed to the apparent subjectivity of judgment of the ease of use of a product. In particular, the so-called "user trials," carried out with the direct involvement of the users, provide reliable indications about the requirements, expectations, and possible problems that they reveal regarding the use of the product that is the object of the study (Rubin 1994). Methods are involved that are obviously also applicable to the equipment present on pleasure boats, although few cases of this type are reported on in the literature.

The planning complexity of certain aspects of yacht design concerning the man–boat–sea interaction, on the contrary, requires ergonomic research regarding the assessment of usability to deal with two specific planning conditions of the sector, which in a certain sense would entail a clear definition of new methods oriented, so to speak, toward a greater "evaluative multidimensionality."

The first condition refers to the minimum dimension of certain spatial areas on board (bathroom, kitchen, chartwork, control cockpit, etc.). Real living units are involved, often nearer to the dimensions of the "binnacle" than of the "living area" in which the synchronicity with which certain tasks have to be carried out prevails: these aspects would require the need to assess the overall usability of the system rather than to carry out tests regarding the single pieces of equipment.

The second condition, concerns some highly specialized equipment present on board, such as winches, grinders, rudders, bridges, etc., for which the posture takes on a fundamental role. This posture, in turn, can be extremely variable, as it is tied in with the nature of the tasks, the environmental conditions (listing boat, climatic conditions, etc.), the areas available and, naturally, the various equipment that must be used, even contemporarily. It is well known that an incorrect posture can cause a feeling of weight, irritation, numbness, and stiffness of parts of the body, pain, and even serious musculoskeletal ailments or damage. On the contrary, correct posture can notably improve the efficiency and effectiveness of the action and the user's satisfaction. Many postures on a boat are classified as "special", are hard to classify and continually modified to maintain balance during the execution of the task. This means that the effectiveness of the action and the overall efficiency of the system cannot leave out of consideration an assessment associated with the equipment, their operating position, and the variability of posture that the user may assume while using the equipment to carry out a certain task.

This last consideration offers an idea for further careful thoughts regarding the assessment of usability of nautical equipment. Nor should the study of their correct use, in fact, leave out of consideration an analysis of the overall organization of activities on board and, above all, the hierarchical structure of the tasks that each individual is called on to carry out, e.g., during a single maneuver of the craft, when every single activity must be coordinated with those of all the other members of the crew. In this sense, planning every single piece of equipment in view of its use means taking into account both the demands imposed by the overall organization of the activities on board and the possibility of the equipment adapting itself to a subjective attitude of the single user, which is also a variable regarding the task and environmental conditions of use.

Therefore, it is a matter of verifying the possibility of using more than one assessment method of the usability together to plan the equipment-cockpit system of maneuver, maybe in association with other analytical instruments, as for example those concerning the hierarchical task analysis (Shepherd 2001), which allows breaking down, in an objective and analytical way, the activities and tasks of each member of the crew, including even single, elementary actions, to identify planning limitations in an analytical manner, even of a postural type.

27.3.4   PLEASANTNESS OF SURROUNDINGS AND FINISHING

The relationship established between users, products, and the environment often reaches far beyond the simple use relationship. This is valid also for equipment whose functional result seems at first sight to be the only one to condition its aspect, as in the case of the equipment or areas of pleasure navigation. Pleasantness concerns the positive sensations and emotions produced by the use of an object or an environment. It is well known how its subjectivity is strongly tied in with aspects of variability of individuals of a cultural, sensorial, and temporal type. This represents one of the most recent research and experimentation areas of ergonomics applied to design: the study, design, and assessment of the sensorial qualities of artifacts is possible today also by means of the analysis of desires, aspirations, emotions, dreams, attitudes, personality, and behavior of the users. It deals with apparently subjective elements of judgment, which ergonomics manages to bring back to an objectivity scale with particular methods and operative instruments.

Considering the declaredly "recreational" nature of yachting, it is easy to understand how important it is for navigation producers also, at a strategic and competitive level, to get to know, interpret, and control in an objective planning way the sensations (tactile, prehensile, functional, thermal, chromatic, acoustic, and, at times, even olfactory) experienced by the users who interact with objects, systems, and the environment.

In recent years, there have been various theoretical and methodological elaborations on the theme of emotionality and pleasantness of products and its assessment. The concept of emotional design

(Norman 2004), namely the possibility of interacting emotionally with a product, even though on a deep-rooted, behavioral, and reflective triple dimension, is indirectly related to the preceding conception of the "four pleasures" (Jordan 2000), with which the generic concept of pleasantness in the four pleasures—physio, socio, psycho, and ideo—is articulated.

Such theoretical considerations follow the empirical approach to the theme of assessment of pleasantness temporally. Among the many methods that started off the experimentation, Kansei Engineering (Nagamachi 1989) proposes a "consumer-oriented" procedure with which an assessment can be made of the sensations produced on individuals by the various solutions of shapes, materials, surfaces, and colors, by using a database built on the key words and adjectives used by extended samples of users; the Sequam method (Bandini Buti, Bonapace, and Tarzia 1997), useful in assessing the emotional aspects of the products by means of the sensations experienced by the users when they come into direct contact with the objects; and the more recent Citarasa Engineering (Khalid 2005), a novel user-centered approach to design that integrates the cognitive and affective requirements of customers.

Such theoretical considerations and assessment methods are evidently applicable also to the products for yachting, to assess the sensorial pleasantness of the environment (coloring, luminosity, formal choices, sonority, etc.) and, above all, of the materials used for the finishing of the surfaces (from the non-slip surfaces on deck to the visual or tactile effects of certain finishing materials used for the exterior or interior, etc.). It is, of course, the opening of research that is still at the start in this applicative field, which promises, though, to give interesting results to increase the awareness of the designer and the objectivity of his/her choices attributed to aspects that, until now, were the exclusive prerogative of his/her individual sensitiveness and perceptive abilities.

27.4 APPLICATIVE EXPERIENCES OF USER-CENTERED DESIGN TO SAILING YACHT DESIGN

This section reports the results of two different researches carried out on the theme of ergonomics applied to sailing yacht design. It deals with assessment tests whose experimental results have allowed the definition of certain guidelines useful for orienting project choices on several aspects regarding the shape of pleasure sailing boats, facilitating in actual fact their identification of innovative solutions at a functional and aesthetic level.

In particular, the first experience (Section 27.4.1) reports the results of an observation test attributed to the tasks and posture of the tailer, a figure that has the decisive role of maneuvering and regulating the sails on crafts. In this case, the analysis has highlighted its principal elements of postural discomfort, defining some guidelines regarding both the various operative conditions of the tailer (postural, organizational, and localizing) and the various equipment used, to facilitate the identification of an innovative solution for the control cockpit.

The second experience (Section 27.4.2) refers to the development and execution of an assessment test of the tactile pleasantness of the finishes of the deck surfaces. The experience has provided designers with some planning guidelines that have been shown to be useful in avoiding a simply "decorative" approach to design and the placing of the "macro-grips."

27.4.1 ANALYSIS OF THE TASKS AND POSTURE OF THE TAILER OF A 17-METER SAILING YACHT

Here, the results are reported of an evaluation test referred, in particular, both to the tasks and to the postures of one single role on board a sailing yacht: the "tailer," who is the subject taking care of furling and setting the sails. The target of the evaluation test was to point out the most important aspects of a tailer's postural discomfort, in order to define some guidelines referring to different operative conditions (postural, organizational, and positional) as well as to different sets of rigging.

27.4.1.1 Introduction

In every organized work system, the joint approach of ergonomics and work organization is a formidable means of increasing system productivity and, at the same time, improving the quality of life. Moreover, together they contribute to reducing injuries, errors, stress, and strains and to enhancing the satisfaction and care of those engaged in the group activity (Corlett and Di Martino 1998).

Methods allowing a hierarchical evaluation of the tasks, or an analysis of the postures of subjects accomplishing the same tasks, are numerous. However, there are few methods evaluating both these aspects.

Nevertheless, the integrated use of different methods of task and work analyses may reveal the best way to analyze the different postures of subjects acting in organized work systems, especially in the more complex ones, in order to guide design choices on new, aesthetic, comfortable, and functional station solutions.

A sailing yacht, for example, can be considered a particularly complex "organized work system," which is defined by an interrelated set of elements (activities, riggings, and people), working in special environmental conditions, pursuing the main object of "sailing" with efficiency, effectiveness, and satisfaction. On board, each hand has several defined roles, duties, and tasks: these are carried out by interacting with specific riggings in specific "areas" of the sailing yacht. However, these "stations" do not always derive from a UCD: they often derive from compromises, which are usually aesthetical or simply hierarchical, in regard to other emplacements on board. As a matter of fact, the most efficient design, from both an organizational and a postural point of view, does not always match to a "good-looking cockpit," that is "aesthetically" interesting.

27.4.1.2 Objectives

Among the most important roles on board a sailing yacht, the "tailer" is the determinant subject taking care of furling and setting the sails.

Here, the results of an evaluation test regarding the tasks and the postures of a tailer are outlined. The target of the evaluation test was to point out the most important aspects of a tailer's postural discomfort, in order to define some guidelines referring to the design of different operative conditions (postural, organizational, and positional), as well as to different sets of rigging.

27.4.1.3 Method

The research was experimentally organized into three different phases: during the first phase, data and information were collected concerning both in general concerning the navigation with a sailing yacht similar to the studied one, and the role, the riggings, and the tasks of a tailer; in the second phase, theories and methods were further examined in order to analyze the tasks and postures of subjects working together in organized systems and to apply these methods in an original way to the theoretical evaluation of a tailer's tasks and postures; in the third phase, a direct observation was planned in order to verify the rightness of the theoretical data of the preview steps and to point out any critical points. The study was completed with a comparison of data collected during the different phases: this allowed some guidelines referring to the design of a tailer's emplacement inside a sailing yacht cockpit to be defined.

27.4.1.3.1 Tailer's Role and Riggings

The study started collecting data and information mostly from the nautical literature, concerning both the operating principles of the propelling system of a sailing yacht (such as the reaches, the riggings, and the most common practices), and the different roles of the crew on a sailing yacht similar to the studied one.

Since the study was concerned with group activity in reduced spaces, this fact was important in beginning to understand the kind of direct and hierarchical relations between the tailer and the rest of the crew and between the tailer and the different parts and riggings of the sailing boat interacting with him.

Moreover, the first phase of the research was useful to collect initial information on the tailer's most usual postures compared to the reaches, the intensity of strains, and the kind of cockpit (Figure 27.1).

27.4.1.3.2 Tasks and Postures Analysis

Using methods and protocols such as the hierarchical task analysis (HTA) and the Ovako work analysis system (OWAS), an integrated and original use of both was set, in order to analyze the different postures that the tailer takes up in rapid succession, in comparison with the tasks connected to the most important riggings of a sailing yacht. In particular, from the nautical literature, information was collected on the tailer's usual tasks and schemes during the principal riggings (bearing away, turning, luffing, and gybing); later, thanks to the experience of old-timer yachtsmen, it was possible to piece together the riggings temporal sequences, according to the HTA approach. Four different "fixed plan sequences" were derived, in which each next task is due to the completion of the previous. Moreover, for each step of the tailer's task sequences a length of time was supposed, and an OWAS postural evaluation was carried out, in regard to the different kinds of cockpits and the tailer's typical postures (Figure 27.2).

27.4.1.3.3 Experimental Evaluation

Afterward, a direct observation test was planned and developed. The purpose was to verify the correctness of the theoretical data and hypotheses collected in the previous phases (task sequences, postures, times of performances), and to point out any practical skills and unusual postures that were not mentioned in the specific literature.

In particular, six users were observed (three skilled and three not skilled) while acting as a tailer during the four riggings previously analyzed (bearing away, turning, luffing, and gybing): times of performances were measured and postures were observed, using original observation assessment surveys. On the whole, 24 postures were observed and 24 times of performances were measured.

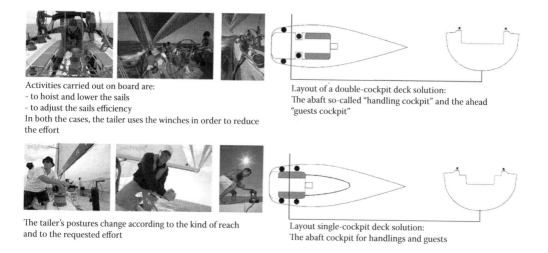

Activities carried out on board are:
- to hoist and lower the sails
- to adjust the sails efficiency
In both the cases, the tailer uses the winches in order to reduce the effort

Layout of a double-cockpit deck solution:
The abaft so-called "handling cockpit" and the ahead "guests cockpit"

The tailer's postures change according to the kind of reach and to the requested effort

Layout single-cockpit deck solution:
The abaft cockpit for handlings and guests

FIGURE 27.1 Different tailer's tasks (using winches), postures, and kinds of cockpits.

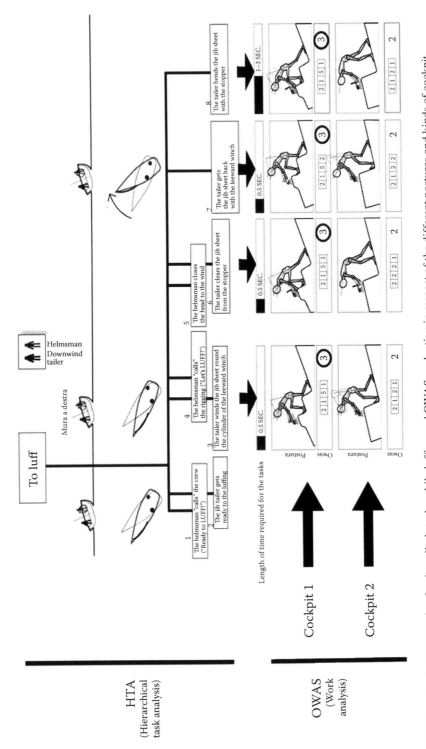

FIGURE 27.2 Fixed sequence plan for the tailer's tasks while luffing and OWAS evaluation in respect of the different postures and kinds of cockpit.

27.4.1.4 Results

The experimental results collected in the third phase basically allowed the validation of the previous theoretical setting regarding the task sequences.

Regarding some critical levels of the tailer's postures derived from the theoretical application of the OWAS method, a fundamental reduction was noted because of the very short length of time of many tasks (about 1–3 seconds), which basically reduced their level of discomfort.

For some tasks and postures, referring in particular to the use of winches, direct observation confirmed the same critical levels as the OWAS analysis. Therefore, regarding the use of winches, further careful considerations were developed to better clarify some problems observed in the field.

Subsequently, it was possible to define some guidelines to favor and guide design choices on postures, stations, and riggings, referring to the tailer's role when using winches (Figure 27.3).

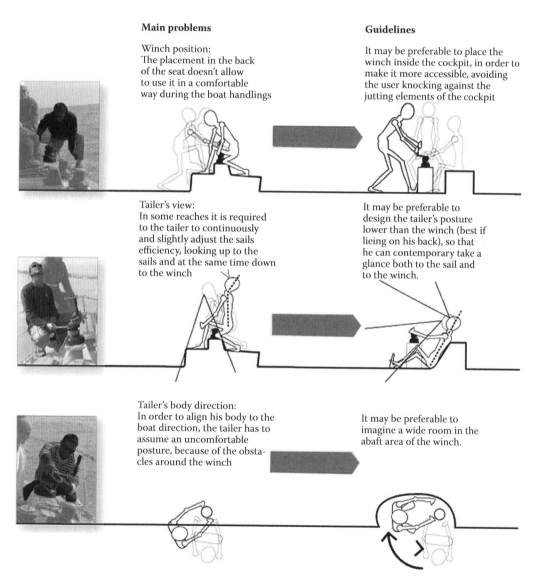

FIGURE 27.3 The main problems deriving from using the winches and the resulting guidelines.

The research ended with possible application of design guidelines, enforced in an interesting and innovative aesthetic solution for a cockpit (Figure 27.4).

27.4.2 Assessment of the Tactile Pleasantness of the Antiskid Deck Surfaces of a Sailing Yacht

In this section, the results of a study on the development and evaluation test of the tactile pleasantness of antiskid deck surfaces are reported. The test has given yacht designers some guidelines that are useful in avoiding a simple "decorating" approach to the "macro-grip" design of deck surfaces on sailing boats.

27.4.2.1 Introduction

The pleasantness of a product can be related to several factors, ranging from physiological and sensorial feelings to factors referred to as social or "ideological" fields (Jordan 2000). But within these fields, the consciousness and control capability of designers has not developed in the same way. Regarding the simple sensorial pleasure, for example, the historical supremacy of sight over the other senses has not allowed a harmonic and conscious development of aesthetic languages. As a matter of fact, sensations such as odors, noises, tactile qualities, and aesthetic feeling cannot be controlled and measured by design tools. Western culture has developed a mature critical conscience related to the "visual" design of artifacts, but, at the same time, the result is not the same regarding those aspects connected to the experience of product perception. Consequently, designers are usually not able to rule objectively the aspects of the project regarding other sensations as much as they can rule the visual sensations. This situation can also be seen in the field of some antiskid deck surfaces design. As the users on board sailing yachts are constantly in contact with these surfaces with their naked bodies, the sensations and feelings aroused by the material are particularly important.

The evaluation test here reported, aims to provide some guidelines on antiskid deck surfaces design, in order to avoid a simple "decorating" approach.

27.4.2.2 Objectives

The general complexity of nautical design and the large number of variables related to it, makes it difficult to verify objectively all the factors implied in the multi-sensorial perception of a sailing yacht. This is true also for the aspects determining the tactile pleasure of some antiskid deck surfaces with which the users on board are constantly in contact.

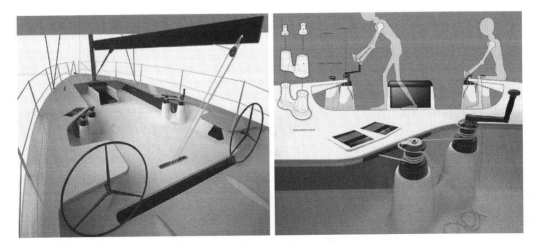

FIGURE 27.4 Application of guidelines to design an innovative cockpit solution.

In order to objectively define these aspects, test designers started their study by individualizing two main aspects regarding the antiskid performances of surfaces:

- The so-called "micro-grips," depending on the chemical-physical features forming the material (toughness, friction factor, etc.)
- The so-called "macro-grips," depending on the morphological features of surface finishing (height, dimension, distribution, and design of relief elements, etc.)

The evaluation test focused on the evaluation of macro-grips because of their influence on designer's choices.

The experimental results allowed the provision of some design suggestions to guide design choices, in order to avoid a simple "decorating" approach in "macro-grip" design.

27.4.2.3   Method

The evaluation test was set up and conducted referring to the sensorial quality assessment method (Sequam) (Bandini Buti, Bonapace, and Tarzia 1997; Bonapace 2002), though with some methodological adaptations necessary for the feature of the object to be surveyed and for the limits imposed by the application context. The Sequam starts by separating the morphological features of an antiskid surface: this compared to its main physical properties ("product elements"), which are quantifiable and objectively measurable. Then, the Sequam associates these properties together with the individual sensations perceived by the users, which can be also measured by applying some well-established techniques (like structured questionnaires with bipolar scales of assessment, etc.). The idea is that by associating the individual sensations with the physical features of the product, suggestions could be given to designers regarding the conscious development of the different design solutions. That's the reason why the whole test design was articulated in three phases, although the experiment was later restricted to the first two phases.

The first phase (Section 27.4.2.3.1), in which pleasure parameters were explored, allowed to define sensorial (tactile) parameters which were required to be analyzed. This was possible by examining some products whose tactile characteristics were similar to those meant to be assessed. It also allowed the product elements of antiskid surfaces to be classified, which were decisive in the assessment of the tactile parameter investigated.

The second phase (Section 27.4.2.3.2), in which the evaluation test on tactile pleasure was carried out, was performed using maquettes made for that purpose, thus isolating and assessing the single tactile product features. In this phase, 24 users were requested to answer questionnaires with bipolar scales, while sitting or standing on the 18 maquettes.

According to the Sequam, in a third phase the pleasantness of products has to be verified by means of working prototypes in real use conditions. That's why the third stage was not performed in this evaluation test, due to the evident difficulties regarding both the product (prototyping of a deck surface of a 10 m sailing yachting) and the real use conditions (in navigation).

27.4.2.3.1   Parameters Exploration

A careful analysis of the tactile parameters of antiskid surfaces has shown test designers how the antiskid performance takes place mainly by means of "macro-grip" or relief elements (which are characterized by their height, dimension, toughness, distribution, density, design, etc.) and by "micro-grip" connected to the physical features of the material itself (friction, heat, etc.). In particular, test designers noticed that the design of relief elements can be so determining in defining other overall performance aspects of surfaces (e.g., the draining features) that it plays a fundamental role in determining both the capability and efficacy of the antiskid effect, and those aspects connected to the pleasure of using deck surfaces. That's why test designers classified the characteristics of relief elements (Figure 27.5): it allowed the stages of the following test to be organized, and further to give useful suggestions in order to realize 18 maquettes (Figure 27.6).

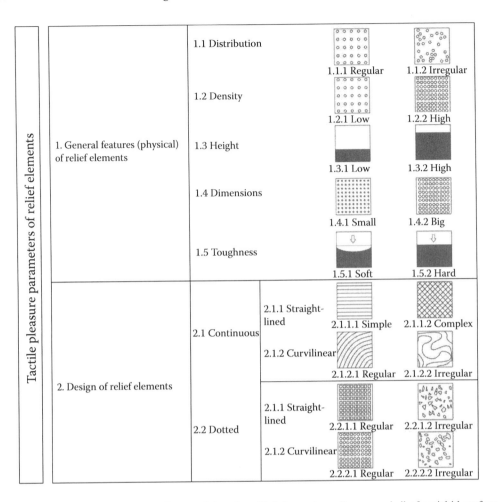

FIGURE 27.5 Tactile parameters characterizing the relief elements, or "macro-grip," of antiskid surfaces, producing objective and measurable qualities.

27.4.2.3.2 Test Performance

Maquettes were submitted for individual evaluation by means of 24 subjects (eight owners of a sailing boat, eight people who had occasional navigation experience on a sailing boat, and eight people who had no navigation experience on a sailing boat). It is useful to underline that, in order to limit the assessment on tactile aspects, during the test the maquettes were covered with a black cloth so that the users could not be influenced, as far as possible, by other important aspects of visual type. The questionnaire included questions on the bipolar scale. The test, performed in the laboratory, was carried out with users wearing bathing costumes only, letting them stand and sit on the maquettes while looking some pictures that made them conscious of the context that the tested maquettes were coming from (Figure 27.7).

27.4.2.4 Results

Correlations between "product elements" and the "reactions" of single users were collected during the test by means of a structured questionnaire with several questions related to bipolar scales of assessment. This led to the first synthesis of data processing reporting on similar assessment scales, the morphological features of the surfaces and the levels of tactile pleasantness felt by the subjects (Figure 27.8).

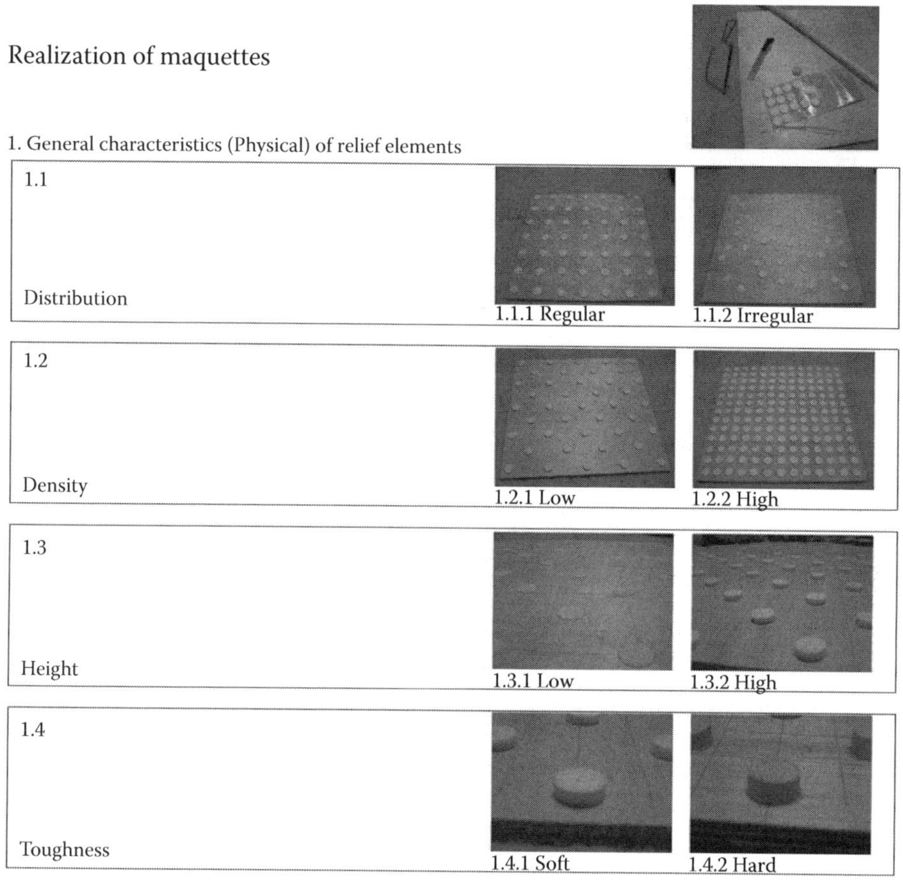

Realization of maquettes

1. General characteristics (Physical) of relief elements

1.1 Distribution	1.1.1 Regular	1.1.2 Irregular
1.2 Density	1.2.1 Low	1.2.2 High
1.3 Height	1.3.1 Low	1.3.2 High
1.4 Toughness	1.4.1 Soft	1.4.2 Hard

FIGURE 27.6 Maquettes useful to investigate some tactile parameters.

A later critical reading of these data Allowed to focus on some guidelines useful to the design of antiskid deck surfaces for sailing boats. Two groups of guidelines were distinguished, particularly referring to the morphological features of the grip in two different types of use:

- As a trample plane, which is connected to the dynamic activity of walking
- As a supporting plane for the body, which is connected to sitting, lying, or simply standing positions

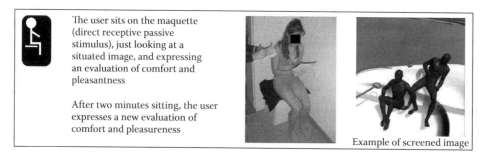

The user sits on the maquette (direct receptive passive stimulus), just looking at a situated image, and expressing an evaluation of comfort and pleasantness

After two minutes sitting, the user expresses a new evaluation of comfort and pleasureness

Example of screened image

FIGURE 27.7 During each assessment, the user watches an evoking picture from the context where the tested surfaces are potentially used.

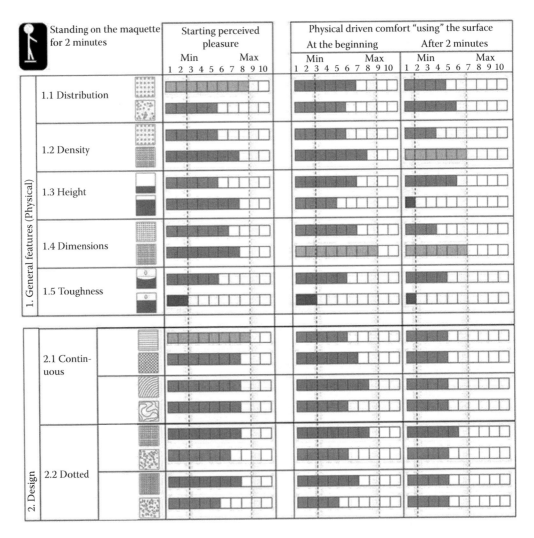

FIGURE 27.8   An analytical synthesis of data collection.

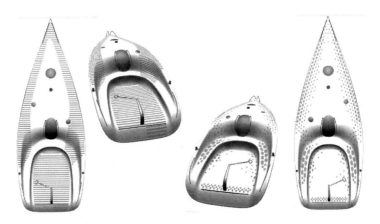

FIGURE 27.9   Two different macro-grip design solutions of deck surfaces according to guidelines.

Later, the guidelines obtained at the end of the test were used to make hypotheses on some possible macro-grip design solutions of the deck surfaces on a sailing boat (Figure 27.9).

CREDITS

Giuseppe Di Bucchianico is the author of Sections 27.2 and 27.3. Andrea Vallicelli is the author of Section 27.1. Section 27.4 was written by both the authors. In particular, Andrea Vallicellli wrote Sections 27.4.1.1 and 27.4.2.1. Giuseppe Di Bucchianico wrote the remaining parts. Furthermore, Section 27.4 reports the results of two researches carried out in the field of two different degree theses in industrial design carried out at the Faculty of Architecture in Pescara. In particular, in Section 27.4.1, reference is made to the thesis entitled "Naked 55. 17-metre open-sea sailing craft" (supervisor Prof. A. Vallicelli, co-examiner Prof. G. Di Bucchianico, graduand Marco Scuderi). In Section 27.4.2, reference is made to the thesis entitled "10-metre day cruiser sailing craft" (supervisor Prof. A. Vallicelli, co-examiner Prof. G. Di Bucchianico, graduands Francesco Merla and Grazia Patruno).

REFERENCES

Bandini Buti, L., Bonapace, L., and Tarzia, A. 1997. Sensorial Quality Assessment: A method to incorporate perceived user sensations in product design. Application in the field of automobiles. Paper presented at the IEA 1997 Meeting, Helsinki.

Bonapace, L. 2002. Linking product properties to pleasure: The Sensorial Quality Assessment Method – SEQUAM. In *Pleasure with Products: Beyond Usability*, eds. W.S. Green and P.W. Jordan, 187–217. London: Taylor & Francis.

Corlett, N., and Di Martino, V. 1998. *Work Organization and Ergonomics*. Geneva: International Labour Office.

Jordan, P.W. 2000. *Designing Pleasurable Products. An Introduction to the New Human Factors*. London: Taylor & Francis.

Kahlid, H.M. 2005. Computerized automotive technology reconfiguration (CATER) system for reverse engineering and mass customization. Paper presented at EuroMalaysia IST Call 5 FP6 Proposal, Damai Sciences, Kuala Lumpur.

Nagamachi, M. 1989. *Kansei Engineering*. Tokyo: Kaibundo.

Norman, D.A. 2004. *Emotional Design*. New York: Basic Books.

Rubin, J. 1994. *Handbook of Usability Testing*. New York: John Wiley.

Shepherd, A. 2001. *Hierarchical Task Analysis*. London: Taylor & Francis.

Spadolini, P.L. 1987. Le analisi delle funzioni d'uso e loro relazioni, gli spazi minimi ed i problemi dimensionali nelle imbarcazioni da diporto a vela e a motore. In *Architettura imbarcazioni da diporto*, ed. A. Vallicelli. Vol. 1, p. 24, Firenze: Cesati.

Tosi, F. 2003. Il progetto centrato sull'utente. In *Ergonomia e ambiente. Progettare per i cinque sensi*, ed. M. Beccali, M. Gussoni, and F. Tosi, 95–116. Milano: Il Sole 24 Ore.

28 Design for Driver's Comfort: Discomfort Assessment and Discomfort Manifestation among Professional and Amateur Car Drivers

David Ravnik

CONTENTS

28.1 INTRODUCTION

It is a fact that the modern population has become more *homo sedens* and therefore has more problems with the locomotor system than in the past. As our reliance on cars increases due to long-distance job commutes, driving has become a significant part of our daily routine. By spending more and more time in cars, driving can be considered a major source of physical and psychological stress in day-to-day living. With prime ergonomic elements being posture, force, and repetition, it is posture that is the most important for a driver (Anon 2010). The concepts of comfort and discomfort experienced while driving a car are under debate. There is no widely accepted definition, although it is beyond dispute that comfort and discomfort are feelings that are subjective in nature (De Looze, Kuijt-Evers, and Van Dieen 2003). Discomfort has to be validated from the aspects of intensity, quality, and body location—where it is felt and its appearance over time (Straker 1998).

Several subjective assessment methods have been developed to measure human responses, ranking from mild discomfort to pain (Corlett and Bishop 1976; Ebe and Griffin 2000). Among car drivers, the most commonly used method of discomfort evaluation has been by self-reported questionnaire. An example of findings is provided by Myers and Schierhout (1996), who acknowledged the validity of self-reported questionnaires when applied to large test groups. One of the most frequently encountered self-reported questionnaires is the Standardized Nordic Questionnaire (Kurioka, Jonsson, and Kilborn 1987). Another widespread questionnaire is the Questionnaire

Body part Discomfort Scale by Corlett and Bishop (1976). Various different modifications have been made (Porter, Gyi, and Tait 2003; El Falou et al. 2003; Ravnik, Otahal, and Fikfak 2008).

Several researchers have suggested possible factors that affect human discomfort while driving. Personal factors identified by scientific research include: body dimensions (McFadden et al. 2000), age (Jensen, Tuchsen, and Orhede 1996), gender (Parkin, Mackay, and Cooper 1995), driving experience (El Falou et al. 2003), and biochemistry and metabolism (Pope et al. 1998). Factors related to the driving environment include: possibilities for seat adjustment (Harrison et al. 1999), driving posture (Aaras et al. 1997), distribution of pressure and body anthropology (Porter, Gyi, and Tait 2003), progression of muscle fatigue and the duration of driving (El Falou et al. 2003), forces exchanged with the vehicle (Mourant and Sadhu 2002), postural shifts (Liao and Drury 2000), and the possible presence of vibration (Ebe and Griffin 2000). El Falou et al. (2003) admitted that it is difficult to evaluate a driver's discomfort from the accessible methodology listed above.

Several objective methods (e.g., posture analysis, pressure measurements, and electromyography—EMG) are used to assess comfort or discomfort while sitting (Westgaard 1988). However, these methods are rarely used to assess car driving. Pressure distribution appears to be the objective measure with the clearest association with subjective ratings (De Looze, Kuijt-Evers, and Van Dieen 2003). For other variables, i.e., spinal profile or muscle activity, the reported associations are less clear and usually not statistically significant (De Looze, Kuijt-Evers, and Van Dieen 2003). Gyi and Porter (1999) hold opposing opinions. They stated that levels of pressure in the prediction of discomfort are unsatisfactory. Despite measurements by Gyi and Porter (1999) and Porter, Gyi, and Tait (2003), who could not find a relationship between discomfort and values measured by pressure, they still hold the opinion that compression on the contact interface of a human to the seat could be the prime agent in the prediction of discomfort. Attempts to correlate postural angles and distances derived from photographs with subjective judgments of physical discomfort reported in a questionnaire were unsuccessful (Starr, Shute, and Thompson 1985). It is assumed that discomfort increases with mechanical strain in joint areas (Straker 1998). The appearance of discomfort does not require the presence of muscular fatigue (El Falou et al. 2003). According to Liao and Drury (2000), one distinguishable signal of discomfort is a shift in posture. The feeling of subjective fatigue does not always correlate to objective measurements of fatigue (Mabbott and Newman 2001). Fountain (2003) compared the validity of results gained by rapid upper limb assessment (RULA) with records gained from EMG and a subjective evaluation of discomfort. Positions assessed by RULA as high risk for a rise in musculoskeletal problems were, at the same time, those that also found by individuals to be the most uncomfortable. According to Yamazaki (1992), the results of the relation between the characteristics of surface deformation, anthropometry, sitting posture, and comfort perception showed that the comfort of each morphological fitting did not correspond to one special and single parameter of these physical factors, but was represented by a function with many parameters related to deformation, posture, and body anthropology. It can be concluded that a questionnare and CORLETT are excellent at predicting the location of discomfort, the Borg CR10 scale is a good indicator of the level of discomfort, while OWAS and RULA are useful to appraise body posture to predict the appearance of discomfort (Ravnik, Otahal, and Fikfak 2008).

The human is both an active and passive participant in traffic. The body of a driver and co-driver or passenger is exposed to vibration and other forces. A research into problems in the musculoskeletal system reported by drivers and co-drivers/passengers was made primarily on a group of 118 amateurs (Ravnik 2005) and secondarily on a group of 200 rally competitors (Ravnik 2009). Agents that contribute to the discomfort and appearance of pain in the musculoskeletal system are mainly incorrect posture, exposure to vibration, and other mechanical forces (Harrison et al. 2000). Posture during driving and forces that influence the human body can increase exposure to postural problems and can contribute significantly to the appearance of discomfort (Harrison et al. 2000), mainly by a mechanism of accelerated degenerative changes on intervertebral joints (Cheung, Zhang, and Chow 2003). Emotional stress, such as rally driving, could also be a great influence on degenerative changes of the spine (Hirano et al. 1988). The lumbar spine has its own resonant frequency of between 4 and 5 Hz and this is the frequency that appears during driving (Hedge 2003). Posture during driving enables a transfer of different forces to the spine and causes problems in the locomotive system, back pain, neck troubles, stress, etc.

28.2 POSTURE AND DRIVING

Posture is usually meant the distribution of body mass with respect to gravity (Irvin 1998). It could be said that any one sitting position is like another sitting position, but at the same time, small differences play a big role. The first of these is how the person is sitting. Lower limbs are active while driving a car and cannot be used at the same time for support. Drivers were found to adapt to changes in vehicle geometry primarily by changes in limb posture, whereas torso posture remains relatively constant. Stature accounts for most of the anthropometrically related variability in driving posture, and gender differences appear to be explained by body size variation. Large intersubject differences in torso posture, which are fairly stable across different seat and package conditions, are not closely related to standard anthropometric measures. These findings can be used to predict the effects of changes in vehicle and seat design on driving postures for populations with a wide range of anthropometric characteristics (Reed et al. 2000). According to driver body dimensions and the sitting distance from the steering wheel, women sit closer to the steering wheel (Parkin, Mackay, and Cooper 1995) than men do and this difference is accounted for by variations in body dimensions, especially height (McFadden et al. 2000).

Based on research, the National Institute of Occupational Safety and Health (USA) stated in their report a manifesting relationship between a compulsory and unsuitable driving posture and musculoskeletal disorders (Lueder 2004). Similar conclusions are contained in studies by other authors (Aaras et al. 1997), who also declare that compulsory maintenance of a specific posture increases discomfort and becomes a health hazard. Leboeuf-Yde (2004) found that physical work is connected with abnormalities of the spine, degeneration of intervertebral discs, osteoarthritis of facet joints of the spine, etc., but there is no evidence for a causal link between a sedentary lifestyle and lower back pain. A forced position can cause degenerative changes in the cervical, thoracic, and lumbar spine (Graf, Guggenbuhl, and Krueger 1993). A kyphotic curve of the spine increases stress on the intervertebral discs, stretches the posterior ligaments, and decreases nerve supply. This leads to the sensation of discomfort and pain in the back. A static sitting position over a long time causes tension in the muscles, nerves, vessels, ligaments, and joint capsules, and compression of these structures causes local chemical changes connected to fatigue in muscles, decreased blood flow, and partial ischemia, at worst leading to an impaired nervous response resulting from pressure, and secondary to inflammation (Straker 1998). The contact area of the chair causes deformation of soft tissue, leading to a decreased flow of blood and nutritious materials, and hence to the sensation of discomfort (Christie, Kumar, and Warren 1995). The construction of the body structure influences pressure in the area of the tuber ischiadicus (TI) and the thighs during sitting (Porter, Gyi, and Tait 2003). Slimmer individuals have higher pressure in the TI area and heavier individuals have more pressure under the thighs. The position of the body during driving was examined by Porter and Gyi (2002) and Ravnik, Otahal, and Fikfak (2008).

Both myoelectric activity and disc pressure decrease when the back is supported. The backrest inclination is the most important of the support parameters; myoelectric activity and disc pressure both decrease with an increase in inclination. Disc pressure is also considerably reduced when lumbar support is increased and when armrests are used (Andersson et al. 1975). Electromyographic studies have shown that the erector spinae muscles are resonant with sinusoidal vibration, and that maximum muscle activity is recorded at a vibration frequency of 4–6 Hz, resulting in noticeable muscle fatigue (Wilder et al. 1982).

Lumbar lordosis is affected by both the trunk–thigh angle and knee angle. Subjects in seats with seat back inclinations of 110° to 130°, with concomitant lumbar support, have the lowest disc pressures and lowest EMG recordings of spinal muscles. Seat back inclination is always measured from the front edge

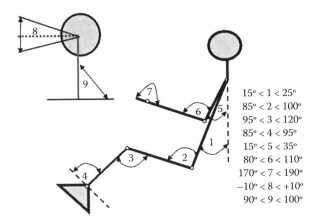

FIGURE 28.1   Ergonomic angles of a car driver in the saggital plane.

(horizontal) to the seat back. A seat bottom posterior inclination of 5° and armrests can further reduce lumbar disc pressure and EMG readings while seated. To reduce forward tilted head postures, the seat back inclination of 110° is preferable over higher inclinations (Harrison et al. 1999). A seat back inclined to 110° or more and with lumbar support will reduce disc pressure (Magnusson and Pope 1998).

The ergonomic angles of the car driver in the saggital plane (Wisner and Rebiffe 1963; Ravnik, Otahal, and Fikfak 2008) and a correct driving posture are shown in Figures 28.1 and 28.2, respectively. Numbers from 1 to 7 represent different joints. Their minimal ergonomic angle value can be seen on the left side of the number of the joint and maximal ergonomic angle on the right side.

1: Angle between the spine and vertical line;
2: Hip flexion;
3: Knee flexion;
4: Ankle angle;
5: Angle between upper arm and vertical line;
6: Elbow angle;
7: Wrist angle;
8: Angle of the middle line between the meatus acusticus externus and lip line and the meatus acusticus externus and eye line; and
9: Angle between the horizontal line and line from the acromion to meatus acusticus externus.

FIGURE 28.2   Correct driving posture.

A significant number of occupants were seated in an unsuitable position while traveling (Dinas and Fildes 2002). Driving positions that create the most health problems are sitting with a bent back, hyperlordotic neck, upper arms too high, elbows more or less straight, and flexed and ulnar-deviated wrist positions (Tostrup 1994). Posture is vital to reducing back pain for both the driver and passenger.

A driver's behavior while driving could be influenced by many stimuli. According to Novak and Votruba (2004), it is possible to divide stimuli into two main groups: inner and external, and natural and artificial stimuli. Not only physical, organizational, or psychosocial work environment characteristics, but also cognitive task characteristics are important for the assessment of postural effects (Karwowski et al. 1994), in our case while driving a car. The frequency of reported discomfort also increased with higher annual mileage (Porter and Gyi 2002) and a stressful lifestyle, leading to muscle tension in different body parts before even starting the journey, which leads to incorrect posture while driving.

Ergonomics can play an important role in the reduction of risk of injury to the driver (Lyons 2002). Ergonomic factors were found to have evident effects on musculoskeletal discomfort (Sauter, Schleifer, and Knutson 1991). According to ergonomic criteria, basic aspects of anthropometry. Vehicle designers must guarantee that the car seat is suitable for all kinds of people (Andreoni et al. 2002). Seats were found to be very important in impulse attenuation leading to a higher transmissibility (Pope et al. 1998). Both static seat characteristics (seat stiffness) and dynamic seat characteristics (vibration magnitude) can influence judgment of seat comfort (Ebe and Griffin 2000).

The challenge for every manufacturer is to strike a balance between safety and comfort. A series of requirements for optimal car seat design have been developed. Ideally, the optimal car seat should have (Hedge 2003):

- Adjustable seat back incline (100° from horizontal is optimal)
- Changeable seat bottom depth (from seat back to front edge)
- Adjustable seat height—there should be room for your fist between the crown of your head and the roof of the car
- Adjustable seat bottom incline
- Seat bottom cushion with firm (dense) foam
- Adjustable lumbar support (horizontally and vertically adjustable)—a seat with an inflatable lumbar support is ideal
- Depth pulsating lumbar support, to reduce static load
- Adjustable bilateral armrests
- Adjustable head restraint with lordosis pad—it is easy to adjust the head restraint so the top is even with your eyes
- Seat shock absorbers to dampen frequencies between 1 and 20 Hz
- Linear front-back seat adjustment to allow differently sized drivers to reach the pedals

In addition, drivers of cars with more adjustable driving packages had fewer reported musculoskeletal troubles (Porter and Gyi 2002).

28.3   EFFECTS OF VIBRATION ON THE HUMAN BODY

Prolonged exposure to whole-body vibration is often connected with driving in traffic (Jandak 1994). Lower back symptoms among bus drivers occurred at whole-body vibration exposure levels that were lower than the health-based exposure limits proposed by the International Standard ISO 2631/1 (Bovenzi and Zadini 1992). It is almost impossible to say how much vibration exposure has to accumulate before people are affected, because sensitivity to vibration varies from person to person. The body being much the same as a machine, can tolerate certain levels of vibrational energy, but eventually starts to deteriorate and fail as long-term damage is experienced, and the natural processes and systems of the body are disrupted. Whole-body vibration has many more widespread and varied effects, but these effects are not particularly clear as the body does not have one receptor for this energy.

The resonant frequency of the body ranges from 2 Hz for the lower limbs, 4–8 Hz for the trunk and shoulders, up to 50–200 Hz for the hands (Chaffin and Andersson 1991). The most important in this context is the resonant frequency of the lumbar spine, which is approximately 4 Hz (Panjabi et al. 1986). The effect of vibration on the body varies from segment to segment; a woman driver is affected by vibration more than a woman passenger. Horizontal vibration affects body segments (lower arm, upper arm, head, thorax, torso, cervical spine, and lumbar spine) more than vertical vibration, while the thorax is affected by vertical vibration more than horizontal vibration (Qassem and Othman 1996).

Studies have shown that subjects exposed to vibration, such as drivers, are more likely to have locomotor problems than other subjects not exposed to vibration (Bovenzi and Hulshof 1999; Dupuis and Zerlett 1987). The most pronounced and common effect of whole-body vibration exposure may be lower back pain, which is caused by various mechanisms of vibration on the musculoskeletal system of the body, namely, the degeneration of the intervertebral discs, which leads to an impairment of the mechanics of the vertebral column, allowing tissues and nerves to be strained and pinched, leading to various back problems (Harrison et al. 2000; Hedge 2003). But, according to Videman et al. (2000), significant effects of lifetime driving on disc degeneration were not evident. Results do not support driving and its associated whole-body vibration as a significant cause of disc degeneration and question the theory that the higher incidence of back pain among drivers is due to accelerated disc degeneration (Pope et al. 1998). Other driving-related factors, such as postural stress, may deserve more attention (Videman et al. 2000). The nutrition of the discs is also affected by long periods of sitting, aggravated by exposure to vibration. The vertebrae are also damaged by vibration energy, which leads to an accumulation of micro fractures at the end plates of the vertebrae and associated pain. Muscle fatigue also occurs as the muscles try to react to the vibrational energy to maintain balance and protect and support the spinal column, but these are often too slow as the muscular and nervous system cannot react fast enough to the vibrational shocks and forces being applied to the body (Anon 2010).

Vibration in combination with driving posture causes a measurable grade of discomfort, especially during long-term driving (Ravnik 2005). During stage rallying, musculoskeletal injuries may be provoked by the high magnitude of vibration and shock to which the driver and co-driver are exposed. Drivers and co-drivers experience similar exposure to whole-body mechanical shocks and vibration, but different exposure to hand/wrist stressors (Mansfield and Marshall 2001). The occurrence of low back symptoms increases with increasing whole-body vibration exposure expressed in terms of total (lifetime) vibration dose (years per square meter per second to the power of 4), equivalent vibration magnitude (meter per second, squared), and duration of exposure (years of service) (Bovenzi 1996).

Long-term driving vibration is among the highest risk factors for neck and back injuries and pain. Professional driving is a risk factor for prolapsed cervical intervertebral discs (Jensen, Tuchsen, and Orhede 1998). There is a higher prevalence of cervical spine discomfort for co-drivers (62%) than for drivers (46%) (Mansfield and Marshall 2001). It is suggested that professional drivers are at an increased risk of lower back pain and injury due to a variety of factors, such as vibration of the whole body, prolonged sitting, awkward posture, lifting and carrying, and psychosocial issues (Lyons 2002). Suspected health effects of whole-body vibration include: blurred vision, a decrease in manual coordination, drowsiness (even with proper rest), low back pain/injury, insomnia, headaches, and upset stomach. Bus and lorry drivers are more likely to experience a range of respiratory, gastrointestinal, and musculoskeletal disorders than a population matched for age and social class that does not drive professionally. Drivers can also expect a higher incidence of cancers, particularly lung and bowel, than a matched population (Whitelegg 1995).

Vibration could effect (Martin 2010):

- Physiology: cardiac rhythm, respiratory rhythm, blood circulation, vasoconstriction, endocrine secretions, central nervous system, paresthesia
- Pathology: musculoskeletal injuries, back disorders, hand-arm vibration syndrome
- Comfort: subjective effects, pain, nausea
- Performance: vision, posture
- Movement and coordination: strain, perception, illusion
- Neurophysiology: sensory information, alteration of nervous messages, perception, motor response, tonic vibration reflex, spinal reflexes

28.3.1 BODY CHEMISTRY DUE TO VIBRATION

The effect of vibration on body chemistry was investigated by Kamenski and Nosova (1989). According to them, exposure to whole-body vibration caused noticeable shifts in the metabolic and neurohormonal status of subjects. There was a tendency for a reduction in the acetylcholine content in the blood and lactic acid concentration was increased by 25.2% immediately after the end of exposure and by 30.8% after 30 minutes. There was a significant drop in the AMF level of 50.3%. Adrenaline excretion after exposure was unchanged while that of noradrenalin was somewhat reduced. These changes were interpreted by the authors as being related to the adaptive regulatory function of the body under exposure to whole-body vibration. The increase in lactic acid is due to muscle activity in the spine directed at maintaining posture under exposure to vibration. The reduction of AMF was considered to be the most significant result of exposure because this nucleotide is involved in the energy production processes of the body and modifies many physiological reactions in response to vibration exposure. The authors concluded that changes in body metabolism may explain the fatigue effects. The von Willebrand factor (vWf) is a complex protein whose release is a marker for endothelial damage; serum levels of its antigen (vWFAg) can be used as a marker for such changes. Back discomfort appeared and vWf levels were significantly increased following sitting upright, compared with lying flat, and increased further following vibration. These results demonstrate that vibration has a significant effect in increasing back discomfort and the serum levels of vWFAg, and it is possible that vibration may induce vascular damage within the spine (Pope et al. 1998).

28.4 DISCOMFORT STANDARDS AND DETECTION OF DISCOMFORT

To describe discomfort fully, we need to validate it from four subjective standpoints: intensity, quality, body parts, and their behavior over time (time diagram) (Straker 1998).

Besides several subjective methodologies, several objective methods (e.g., posture analysis, pressure measurements, and EMG) are used to assess comfort or discomfort while sitting (Westgaard 1988; Bluthner, Seidel, and Hinz 2001). According to de Looze, Kuijt-Evers, and Van Dieen (2003), distribution of contact pressure could be objectively measured, indicating a relation to subjective evaluation. Gyi and Porter (1999) have the opposite opinion, writing that levels of pressure in the prediction of discomfort are unsatisfactory. It was assumed that pressure on the interface between the body and seat during vibration exposure is the most important contributing factor to the feeling of discomfort during driving (Wu, Rakheja, and Boileau 1998).

Attempts to correlate postural angles and distances derived from photographs with subjective judgments of physical discomfort reported in a questionnaire were unsuccessful (Starr et al. 1985). It is assumed that discomfort increases with mechanical strain in joint areas (Straker 1998). The feeling of subjective fatigue does not always correlate to objective measures of fatigue (Mabbott and Newman 2001). El Falou et al. (2003) found that discomfort does not require the presence of muscular fatigue. According to Liao and Drury (2000), shifts in posture are a distinguishable signal of discomfort. Among the physiological values that could be used are heart rate, blood pressure, breath frequency, conductivity of skin, perspiration of skin, and temperature. According to Straker (1998), we cannot measure discomfort directly, but we can understand measured values and their relation to discomfort.

Subjects were compared using a valid, reliable, computer-assisted slide digitizing system called the postural analysis digitizing system (PADS) (Braun 1991). Fountain (2003) compared the validity of results gained by the RULA questionnaire, with the record gained from EMG and subjective valuation of discomfort. Statistically, a relationship was found between the perceived amount of discomfort and RULA. Positions supported by RULA, such as high risk for an increase in musculoskeletal problems, were at the same time those perceived by individuals as causing the most discomfort. Vergara and Page (2002) analyzed the causes of lumbar discomfort during sitting on a chair, by analyzing the relationship of lumbar curvature, pelvis inclination, and their mobility with discomfort. Greater changes of posture are a good indicator of discomfort, and lordotic postures with a forward leaning pelvis and low mobility are the principal causes of the increase in discomfort (Vergara and Page 2002). Commonly, where pain of the musculoskeletal system is present, the pain is without objective evidence of disease, trauma, or disorder (Irvin 1998).

The results of the interrelation between the characteristics of surface deformation, parameters of body build, sitting posture, and feeling of comfort show that the comfort of each morphological fitting does not correspond to one special and single parameter of those physical factors, but is represented by a function with many parameters related to the deformation, posture, and body build (Yamazaki 1992). A technique for postural strain on the upper body assessment (LUBA) is based on experimental data for the composite index of perceived discomfort (ratio values) for a set of joint motions, including the hand, arm, neck, and back, and the corresponding maximum holding time in static postures. This was used to assess subjects' discomfort for varying joint motions. The proposed scheme can be used for evaluating and redesigning static working postures in industry (Kee and Karwowski 2001).

28.4.1 DOCUMENTED DISCOMFORT FROM EXPOSURE TO DRIVING

It seems that mental fatigue and monotony influence the sensation of discomfort. Discomfort increased significantly during the trial, regardless of the experimental condition (El Falou et al. 2003). According to Porter, Gyi, and Tait (2003), the weight would have to be distributed in the gluteal area with minimum pressure below the thighs, since hypertension in the soft tissue of the thighs produces discomfort. Ebe and Griffin (2000) found that individuals with lower general pressure in the TI area feel less discomfort than those with larger general pressure. El Falou et al. (2003) admitted that with the accessible methodology it is difficult to evaluate a driver's discomfort. The authors use all the above methods of subjective questionnaires, based on imparted sensations of an individual during testing. They often use other measurements, independent statements from individuals that are related to discomfort and try to find some correlation. For example, concerning the distribution of pressure on the contact area between the body and seat (Porter, Gyi, and Tait 2003; Wu, Rakheja, and Boileau 1998), progression of muscle fatigue and driving performance (El Falou et al. 2003), body posture while driving (Porter, Gyi, and Tait 2003; Ravnik, Otahal, and Fikfak 2008; Starr, Shute, and Thompson 1985; Andreoni et al. 2002), vibration (Ebe and Griffin 2000; El Falou et al. 2003), postural shifts (Liao and Drury 2000), etc.

28.5 BIOMECHANICAL MODEL OF A DRIVER AND RESONANCE

The human body is a non-linear system and fluid-driven mechanical system that exhibits a number of natural rhythms and resonances. The rhythms vary from regularities of sleeping patterns, alpha brain waves, heart rhythms, hormonal regularities, walking and running, blinking, speaking, singing, etc. Many humans are sensitive to external low-frequency vibration. The constituents of the standing human body can be excited into resonance at frequencies between 3 and 100 Hz. The abdominal viscera can be set into (uncomfortable) oscillation at the lowest of these frequencies. The torso and pelvis have excitable modes of oscillation in the 5–7 Hz range. The upper torso and spine begin to oscillate between 10 and 14 Hz, while the head and shoulders can be excited between 20 and 30 Hz. Finally, there is eyeball resonance in the 60–90 Hz range. Because of the relative displacement of body part resonances below 10 Hz, they are particularly uncomfortable and can lead to serious damage if sustained (Ravnik 2002). According to Kitazaki and Griffin (1998), a vertical principal resonance has been found between 4 and 6 Hz or 4 and 8 Hz (Bilban 1999) in the driving point impedance or apparent mass, in seat-to-head transmissibilities, in seat-to-spine transmissibilities, and in abdominal pressure. Eight modes of vibration response were obtained below 10 Hz (1.1, 2.2, 3.4, 4.9, 5.6, 8.1, 8.7, and 9.3 Hz). A principal resonance of the human body at about 5 Hz consisted of an entire body mode, in which the skeleton moved vertically due to axial and shear deformation of buttock tissue, in a phase with a vertical visceral mode, and a bending mode of the upper thoracic and cervical spine. A bending mode of the lumbar and lower thoracic spine was found with a pitching mode of the head in the next higher mode located close to the principal mode. The second principal resonance at about 8 Hz corresponded to pitching modes of the pelvis and a second visceral mode. When subjects changed posture from erect to slouched, the natural frequency of the entire body mode decreased, resulting in a decrease in the principal resonance frequency. Shear deformation of buttock tissue increased in the entire body mode due to the same change of posture. The complex body motions suggest that any force causing injury from whole-body vibration will not be well predicted by biodynamic models incapable of representing the appropriate body motions and the effects of body posture. It seems likely that the greatest risk for back problems will arise from bending deformation of the spine (Kitazaki and Griffin 1998).

A crucial state, in which exceedingly important processes proceed under the pressure of vibration, is resonance (characteristic quantity—resonant frequency). It has a direct relation to their mechanical features (mechanical impedance). The ratio between stiffness and mass designates "tune" into the system of resonant frequency (Otahál 1997). An organism that is exposed to forces near the resonance has a tendency to quiver "*ad infinitum.*" This causes risk suspense construction sequences inside the tissues (Otahál 1997). It was ascertained that the natural frequency of the human body during vibration exposure turns completely independently, maintaining that the person wants to remain as far as possible from the frequency of vibratory stimulus. In the area of proprioception, interoception, and exteroception, it could lead to distorted perception, devious sensation (phantoms, cold hands, contact acatamathesia, diplopia). The human body is constituted of a system of masses, among which are elastic and viscous connections (springs and absorbers) (Ravnik 2002). The masses of cavus organs depend on their content and blood supply. Elastic characteristics also change according to different muscle tone, arm position, etc. (Bilban 1999). The resonance of the pelvis–seat system lies in the range between 4 and 15 Hz with two peaks. The system of head–pelvis embodies resonance in the area of 9 Hz; the system of head–shoulder vibrates at higher frequencies (in this case 28–32 Hz). From the abovementioned example, it is only possible to generalize about definite toleration, because the impedance of the structures (especially elasticity) is highly dependent on muscle tone, intrathoracal pressure, etc., on factors determined by the level of the "tune" system–adaptability (Otahál 1997). Under a frequency of 2 Hz, the human body behaves as a firm body. The basic resonant frequency of vertical oscillations of the human body is among 4–6 Hz in a sitting position and in the area of 5 and 12–15 Hz during standing. The effects of vibration become more localized with an increase of vibration frequency. Vibration has a cumulative effect, which means that structural and also functional effects are directly dependent on the exposure history, the time of exposure, or the general exposition to the vibration (Ravnik 2002). Resonant frequencies of different body parts are presented in Table 28.1.

TABLE 28.1
Resonant Frequencies of Different Body Parts

Body Segment	Resonant Frequency
Eyes	12–21 Hz (Griffin 1996), 15–60 Hz (Dupuis and Zerlett 1987), 12–40 Hz (Bilban 1999)
Head, neck	20 Hz (Dupuis and Zerlett 1987)
Shoulder	4–8 Hz (Bilban 1999)
Upper extremities	20–40 Hz (Dupuis and Zerlett 1987), 8–30 Hz (Bilban 1999), 16–30 Hz (Chaffin and Andersson 1991)
Hand-wrist	50–200 Hz (Chaffin and Andersson 1991)
System thorax—abdomen	3 Hz (Dupuis and Zerlett 1987; Bilban 1999), 4–8 Hz (Chaffin and Andersson 1991)
Spine	5 Hz (Dupuis and Zerlett 1987; Bilban 1999), 4–5 Hz (Hedge 2003; Panjabi et al. 1986), 10–12 Hz (Chaffin and Andersson 1991)
Pelvis	4–15 Hz (Chaffin and Andersson 1991)
Lower extremities	2 Hz (Chaffin and Andersson 1991)—knees are flexed, >20 Hz (Chaffin and Andersson 1991)—knees are extended

28.6 DISCUSSION

The appearance of discomfort and pain among competitors (drivers and co-drivers) is high. Discomfort is most frequently mentioned in the cervical spine, thoracic and lumbar spine area, and in the buttocks (Ravnik, Otahal, and Fikfak 2008). Among professional drivers, discomfort in the lumbar spine area appeared in 52% of competitors (Ravnik 2009), or 70% as stated by Mansfield and Marshall (2001), and 89% as stated by Videman et al. (2000). In the group of 118 amateur drivers (Ravnik 2005), discomfort in the lumbar spine appeared in 36%. The appearance of discomfort and pain in the lumbar spine area among rally drivers and co-drivers is higher (Ravnik 2009) than in the amateur driver group (Ravnik 2005) and also higher than in the basic population (29% appearance) as stated by Bovenzi and Hulshof (1999), but similar to industrial employees who were exposed to vibration, and vibration was considered problematic. It is interesting to note that discomfort in the neck area appeared among 38% of competitors in comparison to 37% of amateurs (Ravnik 2009). The data are opposite to that expected, as it was expected that oscillations of the head during rally driving would have a higher influence on the appearance of discomfort.

A research made on a group of professional bus drivers (Netterstrom and Knud 1989) has shown that 57% of participants had problems in the area of the spine. The appearance of discomfort in the spine area in the group of professional drivers was 59% (Ravnik 2009) while the research on amateurs (Ravnik 2005) stated 71% of appearance. Discomfort that appeared during and after a competition was mainly discomfort in the spine area (60%), leg area (54%), and arm area (40%) (Ravnik 2009). Statistically significant differences in the appearance of discomfort among drivers and co-drivers were mainly in the thigh area, stomach area, and the lower arms. The reason for the difference in the thigh area could be that drivers legs are active while co-drivers' legs are in a stabilized posture and are in isometric muscle work. Drivers' lower arms are more strained because of the activity and also because of the vibration transmission from the wheel. This also correlates with the results of Mansfield and Marshall (2001).

Women drivers are more exposed to the appearance of discomfort, especially in the neck and lumbar spine area (Ravnik 2005). Driving for longer than 1 hour is found to be critical in the group of amateur drivers, because discomfort appeared in 10% after 1 hour (Ravnik 2005). On the other hand, 86% of people in the group of rally drivers noticed the appearance of discomfort after 1 hour. This confirms the statement that rally driving and extreme strain, vibration, and sitting in the car causes discomfort to appear faster in the musculoskeletal system in contrast to amateur drivers (Ravnik 2009).

There is no single posture that can be comfortably maintained for long periods of time. Any prolonged posture will lead to static pressure on the muscles and joint tissues and, consequently, can cause discomfort. The majority of the modern population spends a lot of time in cars. A lot of people drive long distances daily to and from work and many of them don't or even can't adjust their car seat. Correct adjustment of the seat can decrease strain on the locomotor system. Concerning seat adjustment, the distance from the steering wheel is important. The recommended distance is when the wrists of the extended arms are placed on top of the steering wheel. Female drivers sit closer to the steering wheel than male drivers, because of different body dimensions. According to the questionnaire, more women than men reported experiencing discomfort in the locomotor system during driving, and female drivers were found to report greater levels of discomfort in the locomotor system during driving. Vibration disturbs women more than men. A statistically important difference between men and women in localization of discomfort was found in the neck and lumbar area (Ravnik, Otahal, and Fikfak 2008). Among bus drivers, 57% of subjects (Nettestrom and Knud 1989) were found to have a problem in the spine area. The percentage of discomfort appearing in the spine area in the group of subjects, where all were non-professional drivers, was 72.1% (Ravnik 2005). We should be concerned about the results of Evans (1994), where only 12% of professional drivers worked until they retired.

Drivers were found to adapt to changes in the vehicle geometry primarily by changes in limb posture, whereas torso posture remained relatively constant (Reed et al. 2000). The height of the seat is important, chiefly because of the driver's vision and the roof shouldn't restrict the head. Legs and arms shouldn't be extended. In the event of an accident, they should act as a spring so that the share of force will be absorbed rather than in extended extremities, where the force can be transferred to the pelvis and shoulders, which could cause injuries. Seats that have lumbar support maintain the correct curve of the lumbar spine. According to Hedge (2003), the seat back should have an angle of 100° (upright). According to Ravnik (2005), 60% of those who felt discomfort in the neck, have the seat back position at an angle (backward slant) of between 110° and 120°. Sitting causes the pelvis to rotate backward, which causes a reduction in lumbar lordosis, trunk-thigh angle, and knee angle, and an increase in muscle strain and disc pressure (Harrison et al. 1999). Subjects in seats with seat back inclinations of 110° to 130°, with concomitant lumbar support, have the lowest disc pressure and lowest EMG spinal muscle recordings (Harrison et al. 1999). It is important that the position would not provoke a forward head position. Although almost all subjects in the amateur group (Ravnik 2005) thought that their seat was correctly adjusted, the distance between the occiput and head restraint was on average 7.7 cm (SD±38 cm). The top of the head restraint should be under the line of the top of the driver's ears and the back of the head should be as close as possible to the head restraint. Research on whiplash indicates that the greater the gap between the head and the head restraint, the greater the injury (Lawrence and Siegmund 2000). Reed et al. (2000) supposed that the hip position and the height of the eyes are important. If someone is sitting on a correctly adjusted seat, it could still be problematic, because the spine is fixed in one position for a longer period of time and the spine is constructed to move. The effects on the human body of vibration and sitting are complicated (Ravnik, Otahal, and Fikfak 2008). They act on organisms directly (mechanically), which act on the structure of the tissues, and indirectly (e.g., over receptors), which act on the function of the locomotor system. Vibration is perceived very subjectively, because it can provoke responses in receptors, for which they are not adequate. This is the reason that the effects of vibration on the human body are underestimated. Problems always appear first as functional disturbances and therefore if we don't recognize them and react in time, they can become morphological or structural disturbances. Vibration acts on both the mechanical and biological characteristic of the organism (Otahál 1997). Exposure to vibration should be congruent with normatives, chiefly with International Standard ISO 2631-1 from 1997 and Australian Standard AS 2670.1 from 1990. The truth in practice is that problems even appear where exposure to vibration is not, according to the classification, in the danger area (Bovenzi and Zadini 1992). For young drivers, it is important to realize that postural faults established in adolescence can cause pain syndrome in adulthood.

The results of research made by Ravnik, Otahal, and Fikfak (2008) confirm the hypothesis that different regions of the human body experience different levels of discomfort due to the activity of driving among non-professional drivers. The regions associated with the highest levels of self-reported discomfort, according to the Borg CR10 scale, were the neck, the thoracic spine, and the lumbar spine while, according to the questionnaire and CORLETT, discomfort was mostly noted in the neck and lumbar spine area. It can be concluded that the questionnare and CORLETT are excellent at predicting the location of the discomfort, the Borg CR10 scale is a good indicator of the level of discomfort, and OWAS and RULA are valuable for appraising body posture to predict the appearance of discomfort.

People who sit incorrectly in car seats are asking for trouble. There is almost twice as much pressure on the back during sitting incorrectly than during standing up. Those most at risk are people who not only spend long periods of time in the car, but also those who make infrequent short journeys in the car, because it can be compared to an unaccustomed form of exercise (Hutchful 2003). It is concluded that recommendations for car interiors that are listed in terms of static angles and distances are currently unsubstantial and, thus, not yet ready to be codified as formal standards. Currently, there are no exposure limits for fatigue and vibration that are accepted by experts in the field.

Discomfort, a subjectively perceived sensation, appears to be a significant detector in the rise of musculoskeletal problems. It is influenced not only by mechanical tension of the soft tissues and local chemical changes, but also by mental and social factors. Many authors do not consider the experiment complexly in the light of driving, because such experiments last just a few minutes. It is proven that discomfort increases over time. When considering vibration, it seems unlikely that single-axis mechanical impedance data can be directly transferred to a multi-axis environment (Holmlund and Lundstrom 2001). For a sufficient valuation of discomfort, four standpoints are necessary—intensity: detected with scales; quality: verbal characteristic of sensation; body part: where discomfort is felt, usually by mapping to a body map; and their behavior over time. The authors used all the above methods of subjective questionnaires. Another measurement of quantity frequently used is independent individual statements that relate to discomfort and then it is possible to find some correlation. For example, pressure distribution, progression of muscle fatigue, driving performance, body posture, and vibration. No clear relation between objective measurements and sensation of discomfort was found.

Driving postures, vibration, impact from the driving surface, and handling are the main reasons for health problems, as well as cars that are not designed according to ergonomic principles (Tostrup 1994). The results clearly showed that exposure to car driving is associated with reported sickness absenteeism due to lower back trouble, and that those who drive as part of their job appear to be more at risk from lower back trouble than those whose jobs primarily involve sitting (not driving) and standing activities (Porter and Gyi 2002). The sitting position and exposure to vibration were identified as the most likely causal factors associated with this high proportion of back trouble. Exposure to whole-body vibration during vehicle use is frequently underestimated as a contributor to musculoskeletal injuries, particularly back pain. Involuntary muscle activation can occur not only in previously vibrated muscles (or its antagonists), but can also appear in muscles more distant from the vibration (Gurfinkel et al. 1998). It is thought that accidents where the driver suddenly and unexpectedly veers to the left (Europe, USA) or right (United Kingdom, Japan) could be the result of vibration exposure.

A forward head position and elevated shoulders have been implicated in the development of or discomfort felt among drivers. According to Lawrence and Siegmund (2000), the major contributors of neck pain while driving are insufficient headroom and inadequate seat positioning. A forward head posture, which is very common among drivers, can affect important postural joints, such as the atlanto-occipital joint, cervical spine, scapulothoracic joint, and glenohumeral joint (Christman 2000). Direct and associated pain, discomfort, and dysfunction in the aforementioned joints can be directly attributed to the effects of a forward head posture. When muscles are placed under additional stress, the vertebral joints and discs are placed under additional physiological strain. According to Christman (2000), for every 2.5 cm that the head moves forward from neutral, an additional 6.8–13.6 kg of tension is placed on the supporting neck muscles.

28.7 CONCLUSIONS

Practicing a healthy posture is like holding a defence shield against future problems in the locomotor system. It is a vital requirement when training drivers in the importance of developing measures to reduce or avoid problems, e.g., the selection of an individual's car with respect to comfort and postural criteria. Discomfort and pain can be prevented. It is concluded that recommendations for drivers are phrased in terms of static angles; distances are currently unsubstantiated and, thus, are not yet ready to be codified as formal standards. The human being's natural behavior is to change posture frequently. The seated posture is determined by both the design of the seat and the task to be performed.

Vibration has a cumulative effect, which means that the functional and structural consequences are directly dependent on complete exposure to vibration. Vibration in combination with sitting causes discomfort to appear earlier than just sitting. We are less aware of vibration that is dangerous. According to our research, people change their position during driving rather than stopping more frequently and being physically active. Discomfort during driving mostly appears in the spine region and in the leg and shoulder areas, which can also be caused by discomfort in the spine region. After exposure to whole-body vibration, the muscles are fatigued and the discs compressed, and are therefore less capable of absorbing and distributing weight. It would seem reasonable to recommend the avoidance of heavy lifting immediately after exposure to vibration. The solution to decrease concentrating forces on the lower lumbar spine area is to add foam wedges to the seat-bottom to elevate the pelvis. Driver height may provide a good surrogate for sitting distance from the steering wheel when investigating the role of driver position in actual crash situations (McFadden et al. 2000).

Car driving has an undisputed influence on human perception. Vibration and driving a car cause discomfort in the locomotor system and appear faster than in other forms of sitting. Most often, discomfort appears in the spine region and women were found to be more sensitive to discomfort. Correct car seat adjustment, awareness of posture and vibration, getting in and out of the vehicle correctly and, most importantly, more frequent rests and movement can contribute to maintaining our health. The health of drivers is an important issue in public health, occupational health, transport policy, and employment conditions. No concerted consideration has been given to these factors, which are the cause of poor health and this is an area of neglect that needs urgent attention. Measures to protect and improve the health of drivers should be pursued in a way that maximizes advantages for all sectors of society.

The height and inclination of the seat, the position and shape of the seat back, and the presence of armrests influence the seated posture. Since all postures become uncomfortable and may be a risk factor for discomfort and pain if maintained too long, the seat should permit alterations in posture. Drivers are exposed to a number of health problems as a direct result of the posture adopted while driving. Sitting in the driving position exerts considerable strain on the spine and can cause a number of problems with the musculoskeletal system, in particular backache, neck problems, pulled muscles, and general stiffness. The driving posture also causes problems for the digestive system. It is very important in understanding the demanding working environment of drivers that there is a complex interaction between the physical aspects of that environment (posture, ventilation, noise) and the psychological stress that will contribute to or amplify physical ailments (Whitelegg 1995). The health effects of whole-body vibration on female reproductive organs and the vertebral column should be carefully investigated (Bovenzi 1996). Women, and most likely men, would benefit greatly from international harmonization in European head restraint standards. Until then, both women and men should be encouraged to adjust their adjustable head restraint, when possible, to behind their head's center of gravity and sit with the back of their head as close as possible to their head restraint (Chapline et al. 2000). Drivers are very susceptible to health-related problems and this situation will deteriorate over the next few years. An extensive experimental study on a possible relationship between vibration and fatigue could be considered, although it is likely that this would be expensive to conduct. Such a study is necessary to establish whether the effects of vibration would be noticeable among all other contributors to fatigue and each factor known to contribute to fatigue would need to be controlled for (the amount of time awake, time taken on the task, rest and sleep, circadian factors).

From the literature reviewed, it is clear, that discomfort while driving a car depends on many applied factors, and that are not covered by one valid test. The main problem could be the fact that discomfort is a subjective experience and is therefore very individually recognized and always depends on the subjective statements of the person. There is a poor correlation between the subjective sensation of discomfort and the objective records. According to the results of the CORLETT method, breaks and time spent outside the car can decrease the symptoms of discomfort (Ravnik, Otahal, and Fikfak 2008). It could be assumed that a history of exposure is very important and also the physical condition of the body. There must be an exposure line, where symptoms can be reversible or irreversible. For now, we have only standards, but the literature reviewed found deficiencies. The results confirm that among non-professional drivers, different areas of the human body experience different levels of discomfort due to the activity of driving. The regions associated with the highest levels of self-reported discomfort were, according to the Borg CR10 scale, the neck, the thoracic spine, and the lumbar spine, while according to the questionnaire and CORLETT, discomfort was mostly noted in the neck and lumbar spine area. According to these results, the conclusion could be that the whole spine (different levels of neck, thoracic spine, and lumbar spine) is more at risk of the appearance of discomfort (Ravnik, Otahal, and Fikfak 2008). Currently, research among non-professional drivers is still rare. Almost all research specializes in professional drivers' problems.

The purpose of this chapter is also to elaborate on the procedure of driver's risk assessments. This procedure was also later tested among rally drivers (Ravnik, Otahal, and Fikfak 2008; Ravnik 2009). There are three suggested stages (I, II, III). The following stages, I–III, could be developed during a person's involvement in "discomfort and vibration during driving" to assess the risk of physical symptoms associated with driving. Stage I is the initial risk assessment (questionnaire) for all drivers. Stage II is a detailed risk assessment (interview) for drivers with a high exposure to driving (more than 4 hours per day) and/or are already experiencing driving-related discomfort. Stage III could be urgent action for drivers with severe discomfort or recurring pain, with a medical history of back or neck injury, drivers with an inappropriate car, high driving exposure, or other risk factors. Information from the initial and detailed risk assessments (I and II) should be considered as part of an integrated approach involving, where necessary, additional training, medical input, reduced exposure to driving, a change of car, a change of daily tasks, a change of lifestyle, or specialist advice (e.g., doctor, ergonomist, physiotherapist, psychologist, biomedical specialist, etc.). Physiotherapy and ergonomy are two of the most important agents in the reduction of discomfort. All team leaders can use this therapeutic knowledge positively because they can directly influence the competitor's being and indirectly on results. Discomfort and pain can be prevented.

REFERENCES

Aaras, A., Fostervold, K.I., Ro, O., Thoresen, M., and Larsen, S. 1997. Postural load during VDU work: A comparison between various work postures. *Ergonomics* 40 (11): 1255–68.

Andersson, B.J., Ortengren, R., Nachemson, A.L., Elfstrom, G., and Broman, H. 1975. The sitting posture: An electromyographic and discometric study. *Orthop Clin North Am* 6 (1): 105–20.

Andreoni, G., Santambrogio, G.C., Rabuffetti, M., and Pedotti, A. 2002. Method for the analysis of posture and interface pressure of car drivers. *Appl Ergon* 33 (6): 511–22.

Anon 2010. http://www.teutonicsales.com/pdf/Exposure/Workplace%20Vibration%20Research.pdf (accessed January 8, 2010).

Bilban, M. 1999. Medicina dela: Obremenitve in škodljivosti pri delu. Vibracije in ravnotežje. Závod za varstvo pri delu, Ljubljana: 66–67, 401–409.

Bluthner, R., Seidel, H., and Hinz, B. 2001. Examination of the myoelectric activity of back muscles during random vibration – methodical approach and first results. *Clin Biomech* 16 Suppl 1:S25–30.

Bovenzi, M. 1996. Low back pain disorders and exposure to whole-body vibration in the workplace. *Semin Perinatol* 20 (1): 38–53.

Bovenzi, M., and Hulshof, C.T.J. 1999. An updated review of epidemiologic studies on the relationship between exposure to whole-body vibration and low back pain (1986–1997). *Int Arch Occup Environ Health* 72 (6): 351–65.

Bovenzi, M., and Zadini, A. 1992. Self-reported low back symptoms in urban bus drivers exposed to whole-body vibration. *Spine* 17 (9): 1048–59.

Braun, B.L. 1991. Postural differences between asymptomatic men and women and craniofacial pain patients. *Arch Phys Med Rehabil* 72 (9): 653–56.

Chaffin, D.B., and Andersson, G.B.J. 1991. *Occupational Biomechanics* (2nd ed). New York: John Wiley.

Chapline, J.F., Ferguson, S.A., Lillis, R.P., Lund, A.K., and Williams, A.F. 2000. Neck pain and head restraint position relative to the driver's head in rear-end collisions. *Accid Anal Prev* 32 (2): 287–97.

Cheung, J.T., Zhang, M., and Chow, D.H. 2003. Biomechanical responses of the intervertebral joints to static and vibrational loading: a finite element study. *Clin Biomech* 18 (9): 790–99.

Christie, H.J., Kumar, S., and Warren, S.A. 1995. Postural aberrations in low back pain. *Arch Phys Med Rehabil* 76 (3): 218–24.

Christman, J. 2000. Posture analysis 101: PowerPosture System. Ventura Design.

Corlett, E.N., and Bishop, R. 1976. A technique for assessing postural discomfort. *Ergonomics* 19 (2): 175–82.

De Looze, M.P., Kuijt-Evers, L.F., and Van Dieen, J. 2003. Sitting comfort and discomfort and the relationships with objective measures. *Ergonomics* 46 (10): 985–97.

Dinas, A., and Fildes, B.N. 2002. Observations of seating position of front seat occupants relative to the side of the vehicle. *Annu Proc Assoc Adv Automot Med* 46:27–43.

Dupuis, H., and Zerlett, G. 1987. Whole-body vibration and disorders of the spine. *Int Arch Occup Environ Health* 59 (4): 323–36.

Ebe, K., and Griffin, M.J. 2000. Qualitative models of seat discomfort including static and dynamic factors. *Ergonomics* 43 (6): 771–90.

El Falou, W., Duchene, J., Grabisch, M., Hewson, D., Langeron, Y., and Lino, F. 2003. Evaluation of driver discomfort during long-duration car driving. *Appl Ergon* 34 (3): 249–55.

Evans, G.W. 1994. Working on the hot seat: Urban bus operators. *Acc Anal Prev* 26 (2): 181–93.

Fountain, L.J. 2003. Examining RULA's postural scoring system with selected physiological and psychophysiological measures. *Int J Occup Saf Ergon* 9 (4): 383–92.

Graf, M., Guggenbuhl, U., and Krueger, H. 1993. Investigations on the effects of seat shape and slope on posture, comfort and back muscle activity. *Int J Ind Ergon* 12 (1–2): 91–103.

Griffin, M.J. 1996. *Handbook of Human Vibrations*. London: Academic Press.

Gurfinkel, V.S., Levik, Y.S., Kazennikov, O.V., and Selionov, V.A. 1998. Locomotor-like movements evoked by leg muscle vibration in humans. *Eur J Neurosci* 10: 1608–12.

Gyi, D.E., and Porter, J.M. 1999. Interface pressure and the prediction of car seat discomfort. *Appl Ergon* 30 (2): 99–107.

Harrison, D.D., Harrison, S.O., Croft, A.C., Harrison, D.E., and Troyanovich, S.J. 1999. Sitting biomechanics part I: Review of the literature. *J Manipulative Physiol Ther* 22 (9): 594–609.

———. 2000. Sitting biomechanics, part II: Optimal car driver's seat and optimal driver's spinal model. *J Manipulative Physiol Ther* 23 (1): 37–47.

Hedge, A. 2003. Back care behind the wheel. *ErgoSolution Magazine* 34–36.

Hirano, N., Tsuji, H., Ohshima, H., Kitano, S., Itoh, T., and Sano, A. 1988. Analysis of rabbit intervertebral disc physiology based on water metabolism. II. Changes in normal intervertebral discs under axial vibratory load. *Spine* 13 (11): 1297–1302.

Holmlund, P., and Lundstrom, R. 2001. Mechanical impedance of the sitting human body in single-axis compared to multi-axis whole-body vibration exposure. *Clin Biomech* 16 (1): 101–10.

Hutchful, T. 2003. Bad back linked to driving posture. www.medicalnewstoday.com/articles/4467.php (accessed May 15, 2003).

Irvin, R.E. 1998. The origin and relief of common pain. *J Back Musculoskelet Rehabil* 11 (2): 89–130.

Jandak, Z. 1994. *Kritéria pro diagnostiku poškození lidského organismu v důsledku expozice hluku nebo vibracím*. Praha: Iga MZČR.

Jensen, M.V., Tuchsen, F., and Orhede, E. 1996. Prolapsed cervical intervertebral disc in male professional drivers in Denmark, 1981–1990. A longitudinal study of hospitalizations. *Spine* 21 (20): 2352–55.

———. 1998. Prolapsed cervical intervertebral disc in male professional drivers in Denmark, 1981–1990. *Ugerskr Laeger* 160 (26): 3913–16.

Kamenski, Y., and Nosova, I.M. 1989. Effect of whole body vibration on certain indicators of neuro-endocrine processes. *Noise Vibr Bull* 205–6.

Karwowski, W., Eberts, R., Salvendy, G., and Noland, S. 1994. The effects of computer interface design on human postural dynamics. *Ergonomics* 37 (4): 703–24.

Kee, D., and Karwowski, W. 2001. LUBA: An assessment technique for postural loading on the upper body based on joint motion discomfort and maximum holding time. *Appl Ergon* 32 (4): 357–66.

Kitazaki, S., and Griffin, M.J. 1998. Resonance behaviour of the seated human body and effects on posture. *J Biomech* 31: 143–49.

Kurioka, I., Jonsson, B., and Kilborn, A. 1987. Standardized Nordic Questionnaire for the analysis of musculoskeletal symptoms. *Appl Ergon* 18: 233–37.

Lawrence, J.M., and Siegmund, G.P. 2000. Seat back and head restraint response during low-speed rear-end automobile collisions. *Accid Anal Prev* 32 (2): 219–32.

Leboeuf-Yde, C. 2004. Back pain – individual and genetic factors. *J Electromyogr Kinesiol* 14 (1): 129–33.

Liao, M.H., and Drury, C.G. 2000. Posture, discomfort and performance in a VDT task. *Ergonomics* 43 (3): 345–59.

Lueder, R. 2004. Ergonomics of seated movement. A review of scientific literature. Humanics ErgoSystems, Inc. Calabasas, California.

Lyons, J. 2002. Factors contributing to low back pain among professional drivers: A review of current literature and possible ergonomic controls. *Work* 19 (1): 95–102.

Mabbott, N.A., and Newman, S.L. 2001. Safety Improvements in Prescriptive Driving Hours. AUSTROADS Report AP-R182, Sydney, New South Wales.

Magnusson, M.L., and Pope, M.H. 1998. A review of the biomechanics and epidemiology of working postures. *J Sound Vib* 215 (4): 965–76.

Mansfield, N.J., and Marshall, J.M. 2001. Symptoms of musculoskeletal disorders in stage rally drivers and co-drivers. *Br J Sports Med* 35 (5): 314–20.

Martin, J.B. 2010. An overview of human vibration. www.coursehero.com/file/5820093/Lecture-14-Vibration-10-29 (accessed April 15, 2010).

McFadden, M., Powers, J., Brown, W., and Walker, M. 2000. Vehicle and driver attributes affecting distance from the steering wheel in motor vehicles. *Hum Factors* 42 (4): 676–82.

Mourant, R., and Sadhu, P. 2002. Evaluation of force feedback steering in a fixed based driving simulator. Proceeding of the Human Factor and Ergonomics Society 46th Annual Meeting, 2202–2205.

Myers, J.E., and Schierhout, G.H. 1996. Is self-reported pain an appropriate outcome in ergonomic-epidemiological studies of work related musculoskeletal disorders? *Am J Ind Med* 30: 93–98.

Netterstrom, B., and Knud, J. 1989. Low back trouble among urban bus drivers in Denmark. *Scand J Med* 17 (2): 203–6.

Novak, M., and Votruba, Z. 2004. Challenge of human factor influence for car safety. *Neural Netw World* 1: 37–47.

Otahál, S. 1997. Biomechanika tkání pohybového systému: Účinek vibrací. In: Kolektiv autorů. *Pohybový systém a zátěž*, 80–83. Prague: Grada Publishing.

Panjabi, M.H., Andersson, G.B.J., Jorneus, L., Hult, E., and Mattsson, L. 1986. In vivo measurements of spinal column vibration. *J Bone Joint Surg* 68-A: 695–702.

Parkin, S., Mackay, G.M., and Cooper, A. 1995. How drivers sit in cars. *Accid Anal Prev* 27 (6): 777–83.

Pope, M.H., Magnusson, M., Broman, N.H., and Hasson, T. 1998. The dynamic response of human subjects while seated in car seats. *Iowa Orthop J* 18: 124–31.

Porter, J.M., and Gyi, D.E. 2002. The prevalence of musculoskeletal troubles among car drivers. *Occup Med* (Lond) 52 (1): 4–12.

Porter, J.M., Gyi, D.E., and Tait, H.A. 2003. Interface pressure data and the prediction of driver discomfort in road trials. *Appl Ergon* 34 (3): 207–14.

Qassem, W., and Othman, M.O. 1996. Vibration effects on setting pregnant women – subjects of various masses. *J Biomech* 29 (4): 493–501.

Ravnik, D. 2002. The influence of vibrations on the local hemodynamics of blood. In *Komplexita biomaterialu a tkanovych struktur* eds. K. Jelen, S. Kušova, M. Chalupova, and J. Otahal, 309–72. Univerzita Karlova v Praze, FTVS, ČSB; Prague.

———. 2005. The phenomenon of discomfort in the locomotor system during car driving. *Fizioterapija* 13 (1): 31–36.

————. 2009. Physical forces and stress as a risk factor with developing problems in the locomotor system during car driving. Proceedings of XV. Symposium of Slovenian physiotherapists, ed. F. Kresal, 30–37. Chamber of physiotherapists of Slovenia, Ljubljana.

Ravnik, D., Otahal, S., and Fikfak, M.D. 2008. Using different methods to assess the discomfort during car driving. *Coll Antropol* 32 (1): 267–76.

Reed, M.P., Manary, M.A., Flannagan, C.A., and Schneider, L.W. 2000. Effects of vehicle interior geometry and anthropometric variables on automobile driving posture. *Hum Factors* 42 (4): 541–52.

Sauter, S.L., Schleifer, L.M., and Knutson, S.J. 1991. Work posture, workstation design, and musculoskeletal discomfort in a VDT data entry task. *Hum Factors* 33 (2): 151–67.

Starr, S.J., Shute, S.J., and Thompson, C.R. 1985. Relating posture to discomfort in VDT use. *J Occup Med* 27 (4): 269–71.

Straker, L.M. 1998. Body discomfort assessment tools. In *The Occupational Ergonomics Handbook,* 1st ed., eds. W. Karwowski, and W. S. Marras, 1239–52. Salem, USA: CRC Press LLC.

Tostrup, B. 1994. Ergonomic aspects on snowmobile driving. *Arctic Med Res* 53 Suppl 3: 45–54.

————. http://www.scribd.com/doc/21260095/Ergonomics-Project-Report (accessed June 30, 2010).

Vergara, M., and Page, A. 2002. Relationship between comfort and back posture and mobility in sitting-posture. *Appl Ergon* 33 (1): 1–8.

Videman, T., Simonen, R., Usenius, J., Osterman, K., and Battie, M. 2000. The long-term effects of rally driving on spinal pathology. *Clin Biomech* 15 (2): 83–86.

Westgaard, R.H. 1988. Measurement and evaluation of postural load in occupational work situations. *Eur J Appl Physiol Occup Physiol* 57 (3): 291–304.

Whitelegg, J. 1995. Health of professional drivers. A Report for Transport and General Workers Union. Eco-Logica Ltd., White Cross, Lancaster.

Wilder, D.G., Woodworth, B.B., Frymoyar, J.W., and Pope, M.H. 1982. Vibration and human spine. *Spine* 7: 243–54.

Wisner, A., and Rebiffe, R. 1963. L'utilisation des donnees anthropometriques dans la conception du poste de travail. *Trav Hum* 26: 193–217.

Wu, X., Rakheja, S., and Boileau, P.É. 1998. Study of human-seat interface pressure distribution under vertical vibration. *Int J Ind Ergon* 21 (6): 433–49.

Yamazaki, N. 1992. Analysis of sitting comfortability of driver's seat by contact shape. *Ergonomics* 35 (5–6): 677–92.

29 Ergonomics as a Strategy Tool: An Approach through Design

Jairo José Drummond Câmara and Róber Dias Botelho

CONTENTS

29.1 INTRODUCTION

The emergence of new production processes, the successful use of new materials, and the opening of new markets globally, have influenced how ergonomics can benefit our society. In the past, ergonomics was incorporated primarily into strategies that could offer reductions in the cost of materials and the cost of production, and strategies that could increase corporate efficiency and profitability. Similarly, business partnerships and distribution systems have been created for the purpose of reducing the overall costs to consumers. Such strategies by themselves are no longer sufficient. New approaches are needed that can add competitive value to products as well as to processes.

Souza (1999) proposes that competitiveness of organizations can be achieved through the adoption of new strategies, which include features that go beyond simple costs reductions. These can differentiate products and services from the competition. One such variable includes aesthetics.

In the past, the concept of "added value" was limited to a simple stereotype within the industrial/commercial world. Vargas (2002) defined the aggregate value of a product or process as the ratio between the actual costs incurred and the work performed within a given period of time. The focus was directed toward the performance achieved in comparison to what was spent achieving it.

With the evolution of new manufacturing processes and the emergence of new technologies after the 1980s, many new products took on shapes that conformed to some "standards" in function and form. Therefore, the development of new products needed to take new directions. Subsequently, design teams within organizations began to take on a more critical role and were no longer restricted to working only with products and the materials used in making them. Design teams were beginning to become involved in designing the language of a "corporate image."

Câmara (1993) divided the word "design" into its Anglo-Saxon origins of "de + sign," which means "giving meaning." In French-speaking countries, design is understood to imply a project design or a visual expression of an idea. In English-speaking countries, the word "design" can refer to a wide range of design activities. It becomes clear that the activities carried out by a designer can produce a product differentiation that may offer the basis for a corporate message to a target audience. In this sense, the designer can give the product a "value" that can be understood by the user.

In Value Analysis, the actual value of a product, process, or a system, depends on the person who evaluates it, just as on the degree of acceptability by the costumer. A product position's over the competition is determinated by its largest real value in comparison to another product, with the same function, based on local economic conditions. Therefore, a combination of specific types of values determinates the real value of a product, service or system, which increases with higher use and esteem and decreases with an increase in the cost of the product. As for Gurgel (2001, 431), "the economic value of a product or service increases as the product costs are reduced while not reducing the economic value of the product or service itself. The value of a product will be equal to the value of its basic and secondary functions that are exercised by the various components supporting the parts that meet the product's principal function."

The concept of "value" has evolved over time. In the early twentieth century, a consumer had the option to choose a car in any color ... provided it was black (Henry Ford and the Model T). However, with the advances in new manufacturing technologies and communication, and rapid expansion in international information flow (also known as globalization), cultural factors became major considerations in defining "value."

To Pereira Filho (1994), a value analysis (VA) attempts to reconcile the value envisioned by the entrepreneur with the value perceived by the user or consumer. A VA assigns a product a fair price while, at the same time, assuring some profit for the seller. Csillag (1995, 59) notes that, "Value Engineering is an organized effort, directed to analyze the functions of goods and services to achieve the necessary functions and essential features in the most profitable way."

Reaching new markets through differential marketing has become a major challenge in the design of new industrial products. Depending on each company, product, and marketing strategy, a particular product feature will have to receive specific attention. To illustrate the business/financial benefits of a product and its associated social and cultural implications, the impact of design of new products will be illustrated through three case studies: (1) design outcome of a "hyper-economic" vehicle, (2) conversion of a Brazilian pick-up truck, and (3) arena and convention hall. These examples illustrate the role of design in macro-ergonomics considerations and the principles of "value creation" through product design.

29.2 STRATEGIES

29.2.1 METHOD

This chapter offers selected research results, including a summary of references and products as well as an analysis of the data. Three case studies are included.

29.2.2 MACRO-ERGONOMICS

The International Ergonomics Association (IEA; *apud* Carayon 2006, 3) defines ergonomics as "A scientific discipline concerned with the understanding of interactions among humans and other elements of a system." Ergonomics is also a profession that applies theory, principles, data, and methods to design in order to optimize human well-being and overall system performance. In this definition, the system represents the physical, cognitive, and organizational artifacts that people interact with. The system can be a technology or a device; a person, a team, or an organization; a procedure, policy, or guideline; or a physical environment. Interactions between people and the systems are tasks. Ergonomic professionals are concerned with the design of systems to make the systems fit the needs, abilities, and limitations of human beings.

In Hendrick and Kleiner (2002), macro-ergonomics is defined as the analysis, design, and evaluation of work systems. The term "work" refers to any form of human effort or activity, including recreation and leisure pursuits. A system refers to socio-technical systems. These systems may be as simple as a single individual using a hand tool or as complex as a multi-national organization. A work system consists of two or more persons interacting with some form of: (1) job design, (2) hardware and/or software, (3) internal environment, (4) external environment, and (5) an organizational design that includes all of the work system's structure and processes.

In Brown (*apud* Bezerra 1998), macro-ergonomics is seen as socio-technical systems and the concepts and procedures of socio-technical systems are applied to the field of ergonomics. Bezerra (1998) points out that macro-ergonomics views organizations as open systems in constant interaction with the environment and, through processes of adaptation, may present organizational "dysfunctions," which are reflected in their performances and particularly in their social subsystems. The method used in ergonomics is the "ergonomic analysis," which provides an analysis of the work and promotes improvement of the interface between the machine, the human, and the organization.

Macro-ergonomics represents an efficient strategy for consolidating human issues with issues of survival in the world today. The key to a successful macro-ergonomics intervention is to consider socio-technical parameters in conjunction with input from a person who will be impacted by the ergonomic interventions. Conceptually, macro-ergonomics is a socio-technical approach that addresses four aspects: the technological issues, personnel needs and requirements, the work that needs to be performed, and the external environment that impacts the entire process, including the structure of the organization and the processes by which the organization operates (Guimarães and Hendrick, *apud* Bitencourt, Guimarães, and Santos 2006). To achieve improvements in efficiency, it will be important to select a "market differential" that provides the foundation for a strategy that will add intrinsic value to the product during the development of that product. The ergonomic concepts, when properly dealt with by design, not only allow the generation of a more organized space, but will also create interfaces that are more efficient and secure. These concepts also allow for the presence of an aesthetic language in the design of the product.

29.2.3 VALUE ANALYSIS

The Industrial Revolution served as a foundation for new solutions that now in modern times can satisfy many of our daily needs. Products designed during the Industrial Revolution exceeded the minimal design requirements. They served as a design transition from the nineteenth century "possibility" to our "actual" products, services, and systems in place today. Prior to the twentieth century, product development focused primarily on the use of materials as specified by the product design requirements.

Military conflicts, regional as well as global, have greatly impacted the pace of development of new products, services, and technologies. Of these, World War II played an important role. World War II fostered the emergence of human factors in product and systems design. Piske (2003) notes that the special needs of World War II forced the U.S. government to allocate important raw materials, including rare metals, only to military applications. This policy forced a search for alternative materials. Many of the alternative materials that were subsequently developed exhibited better characteristics than the "original" materials. Examples include new polymer materials used for building construction, the manufacturing of automobiles, aircraft, clothing, and novel fabric materials. This process gave birth to the field of VA.

Duchamp (1988) and Lawrence D. Miles (Duchamp and Miles (1998: 16)) explain that in 1947, as a result of suggestions made by executives at the General Electric Company (GE), the first studies of VA were undertaken. Prior to this, design methodologies focused only on the replacement of materials, optimization of the use of materials, and overall cost reduction. As a result of the successes at GE, the U.S. Navy adopted this strategy in 1954. However, debate about the advantages and disadvantages of VA and value engineering (VE) continued. However, according to Souza (1999), VA/VE received international recognition only after the Society of American Value Engineers (SAVE) was established in 1959. Currently, the term "value management" is also used to describe VA/VE. Thus, both terms have become synonymous with the methodology. Pereira Filho (1994) argues that the expansion of this methodology occurred globally only during the latter part of the 1960s. In Brazil, AV was also adopted during the same period when Singer of Brazil SA, GE, and other large companies such as Philips, Mercedes-Benz, Volkswagen, General Motors (GM), etc., embraced this methodology.

According to Pereira Filho (1994), the word "value" can take on several different meanings and it is often used to denote "cost" or "price." This confusion is not new. Around the year 350 BC, Aristotle identified seven types of "value" that are applicable today: (1) economic, (2) political, (3) moral, (4) aesthetic, (5) social, (6) juridical, and (7) religious. Of all seven definitions, the economic value is the only one that can be quantified; the others can only be assessed subjectively. Thus, an economic value can be divided into four categories as described by Csillag (1995): (1) "value of use," including a monetary measure of the properties or qualities exhibited by the performance and use of a product or service; (2) "estimated value," including the properties and characteristics of an item that make it a desirable possession; (3) "cost value," which is characterized by the money or other resources required to produce or obtain a product; and (4) "exchange value," which is related to the monetary measure of the properties or qualities of an item that makes it possible to exchange for another item. Csillag also argues that a "value" can be assigned through a cash equivalent based on the personal opinion of the evaluator.

In the context of management and development of new products, Toledo and Machado (2008) note that "value" can be defined as the ability to provide customers with a product they want at the right time, at a fair price as defined by the customer in each case.

The four values highlighted define the goal of project design teams. Using these goals, it is possible to develop strategies for the design of a new product or to proceed with the redesign of an existing one. The authors cited above illustrate the importance of conducting a review not only with regard to the design and performance standards of products and processes, but also to review the context in which the products will be marketed. To date, the standards for the "value" of esteem are understudied. For companies and for the consumer market alike, "values" remain subjective because they are based on a cultural context and individual life experiences.

29.3 CASE STUDIES

29.3.1 HYPER-EFFICIENT VEHICLE

The first case entitled "Project Sabiá" involved a series of concept cars that were developed during a 16-year period by researchers at the Design and Ergonomics Research and Development Center in the School of Design at the State University of Minas Gerais in Brazil.

Allowing students to conceptualize a design, and develop, manufacture (build), and test the product under actual field conditions provided a unique learning opportunity for the students enrolled in a course of Industrial Design. This project called for an approach that required for the design and development of an experimental vehicle using an interdisciplinary perspective. With this philosophy in mind, the project offered an opportunity for a select group of students and professors to pursue intellectual goals and challenges that, in general, were not part of the normal degree program.

Project Sabiá was conceived as a strategy to foster thinking among students about novel applications of new materials and processes in the development of new cars. Project Sabiá was part of an international competition program sponsored by the petroleum industry to promote the design and development of new and novel fuel-efficient vehicles. Project Sabiá entered six competitions, submitting a unique design for each of the six events taking place in the time period between 1993 and 2009.

The focus of Project Sabiá's strategy was to enhance the esteem "value" of the product (vehicle) using the elements that, according to the concepts of VA, would help promote its exchange value.

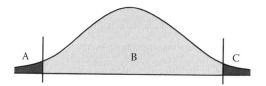

FIGURE 29.1   Gaussian curve. (Courtesy: Câmara.)

Since the vehicle represented a "one-time" product, the costs of construction became irrelevant. To maximize the value of the vehicle, the Sabiá team emphasized the aesthetics of the vehicle. Design elements embedded in Brazilian culture, especially those in the State of Minas Gerais, were highlighted. In addition to incorporating a cultural language into the design, ergonomic principles (associated with engineering materials) were used extensively to promote a high level of driver comfort, high fuel efficiency, and an outstanding shape. These characteristics were achieved through an iterative design process where improvements were incorporated into the design with each new version of the vehicle.

Ergonomics played a central role in the development of all Sabiá vehicles. Use of new materials, weight considerations, driver comfort, and anthropometric limitations with regard to vehicle controls, emergency egress, component use, and assembly considerations, etc., were all central to the design process. To better evaluate driver comfort and safety, a special ergonomic chair was built to obtain accurate measurements for use in the design of the vehicle cockpit. Similarly, pedals, steering wheel, speed controls, and emergency equipment were arranged to optimize the driver–vehicle interaction. Thus, it was possible through the application of design and engineering of materials, to consider form/material/process/function in the overall design and development process of the vehicle. It must also be noted that not only were ergonomic principles used in the design and construction of the vehicle, but human factors engineering was also incorporated into the logistics of transporting the vehicle around the globe and "marketing" the Sabiá vehicle to the community once it arrived at its destination and on its return to Brazil.

According to Rebiffe and Guillien (1991), the first step in evaluating vehicle ergonomics must be to accurately define the morphological characteristics of the "client." In an experimental vehicle such as the Sabiá vehicle, the "client" was the driver who needed to perform important functions inside the vehicle cockpit during the competition. The designers of the Sabiá vehicles were able to excel in applying human factors principles in the final version of the vehicle that was ultimately entered in the competition.

When considering all aspects of product design and development, one must take a strategic approach (vision) toward achieving design excellence. This is especially important when competing for recognition internationally. Good ergonomics implies that one must receive good reviews from a design jury.[*] Members of a design jury evaluate the mechanical, functional as well as the aesthetic aspects of a product (the vehicle). Since the Sabiá entries were developed at a school of design and the development budget was limited, it was decided to focus on the strength offered by the student designers. Therefore, the project strategy was to focus on design. The objective was to have the Sabiá vehicle placed in the "middle" in terms of fuel efficiency rating (mechanics) while placed top in the rating area of design and aesthetics.

Such a strategy is illustrated by a Gaussian curve involving two different scenarios (Figure 29.1). In the first scenario, the vertical axis represents the participating vehicles (approximately 250 per competition) and the horizontal axis represents the percentage of: A, low performance, B, medium performance, and C, high performance vehicles. Performance means fuel economy. In the second scenario, the horizontal axis represents design excellence, where A represents low excellence, B is medium excellence, and C is high excellence. The objective for the Sabiá vehicle entries was to be placed in the extreme right of the design excellence curve while only having to reach the B category for performance (fuel efficiency). This strategy resulted in a total of three design awards for the Sabiá vehicles that were entered in six separate Eco-Marathons.

---

[*] The Eco-Marathons have awards for best fuel efficiency, best design, best ecological concept, and so on.

FIGURE 29.2   Sabiá 1, winner of the Design Award. (Courtesy: Câmara.)

The Sabiá team based much of its design strategy on information and data collected in laboratory studies. A special adjustable ergonomic chair was used to determine optimal driver positioning, and electromyographic measurements of selected muscle groups were used to obtain data on driver comfort while positioned inside various cockpit designs. The outcome was a state-of-the-art vehicle that clearly differentiated itself from all of the other vehicles competing in the Eco-marathons. The unique design solution provided "added value" to the vehicle as a product. The focus on good ergonomics coupled with non-conventional shapes (avoidance of the traditional tear-drop shape used by most vehicles to achieve better aerodynamics), reduced overall costs and also helped make the vehicle stand out so that the public (news media) and the design jury paid more attention to this vehicle.

During a 15-year period when the Sabiá project actively generated innovative prototype design solutions, projects received international recognition from the design profession. In 1994, Sabiá 1 (Figure 29.2), which was inspired by the shape of a sailplane without wings, received the French "Prix d'Honneur du Design" (Design Award). This recognition was received despite the fact that mechanical problems prevented the prototype from completing the required number of laps around the track of the Paul Ricard Speedway located in southern France.

In the following year (1995), Sabiá 2 won the French jury award, again based on the vehicle's unconventional design inspired by the Belugas marine mammal. The team was considered the most elegant because of their uniforms, which were designed to complement the vehicle shape and colors.

In 2000, the Sabiá project based its new design on features seen in aircraft and automobile racing cars of the 1930s. Sabiá 3 (Figure 29.3) returned to France at the Paul Armagnac Speedway located in the southwest of France. This time, the team was awarded the "Prix Spécial du Design pour une

FIGURE 29.3   Sabiá 3, winner of the Special Design Award for a Foreign Team. (Courtesy: Câmara.)

FIGURE 29.4   Sabiá 6, winner of the 2009 Innovation Design Award. (Courtesy: Câmara.)

Équipe Étrangère"—Special Design Award for a Foreign Team. The prototype vehicle achieved a fuel efficiency of 83.78 km/L.

Two years later, in 2002, Sabiá 5 achieved a fuel efficiency of 480 km/L. As a result of the art deco-inspired shape, the vehicle was selected from more than 250 entries to represent the marketing image for the 2003 eco-marathon. The vehicle design was used in all subsequent advertising material (posters, releases, videos).

In 2004, Sabiá 5 was redesigned, exhibiting a simplified morphology. The design highlighted the mechanical components used, the driver seating arrangement inside the cockpit, and the materials used according to their ecological properties. Again, based on its innovative design features, Sabiá 5 was used as the marketing image for the 2005 competition.

In 2009, Sabiá 6 (Figure 29.4) offered a design strategy that featured the application of light-weight materials, mechanical simplicity,[*] and the use of materials that exhibited a low environmental impact. The morphology included elements used in previous Sabiá vehicles and associated with the automotive styles and trends of the 1920s and 1930s. Additionally, an emphasis on value was included: esteem, exchange, and use. The ergonomics used in the design and development of Sabiá 6 was described in scientific papers presented in 2009 at the ErgoDesign Forum, in Lyon, France, and at the IEA Congress in Beijing, China. Sabiá 6 was entered in the 2009 eco-marathon competition, which was held at the Autoclub Speedway in Fontana, California. The vehicle achieved a fuel efficiency of 98.3 km/L and received the Innovation Design Award.

It should be remembered that the six Sabiá prototype vehicles were developed by different groups of students. The objective in developing the different types of vehicles was to help train future professionals in the design process with a view toward the automotive field. It is evident that the various Sabiá projects encouraged design students to adopt a scientific method of thinking while incorporating ergonomic and human factors principles into their design projects.

The "value" to the university and the community during the past 15 years can be measured in terms of the total number of students trained, scholarly papers published, and awards received. For all three categories, the benefits clearly outweigh the "costs." Project Sabiá was clearly a value-added proposition.

29.3.2   Double-Cabin Pick-up Truck

In the late 1980s, a team of designers in Brazil was presented with the opportunity to modify the design of a pick-up truck that was manufactured by GM of Brazil. The project was called Nellore. The availability of imported cars in Brazil was limited at that time because of stringent government import controls. GM of Brazil truck models D-20 and A-20 were available but designed primarily

---

[*] As the majority of the prototypes, it used a 35cc, four-stroke engine, air refrigerated.

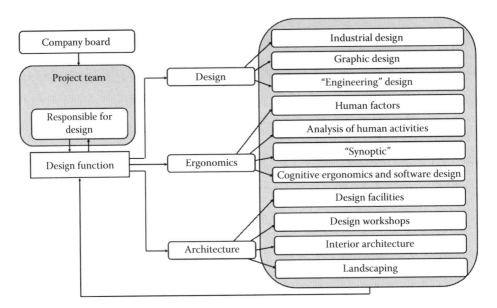

FIGURE 29.5   Integration of the design function in an industrial project. (Courtesy: Câmara, 1993: 196.)

for rural use. The goal of the design project was to develop a new version with an extended cab feature that could accommodate three more passengers in comfort. The design modification was to add value to the original design and add prestige to the product.

To reduce costs, the new design maintained its original two-door configuration while extending the chassis and frame. However, this approach required the use of steel to maintain corrosion resistance of the frame. Fiberglass was introduced selectively as a cost-saving feature. To maintain "value," however, materials similar to those used in the original model were used. In contrast to the design philosophy promoted by the Sabiá project, the pick-up design emphasized complexity in order to maintain value and prestige. Sales performance was good. This redesigned vehicle reached a sales number of 2000 units during a 6-year period, generating a profit equivalent to US$8 million.

29.3.3   CONCERT HALL AND SPORTS COMPLEX

Using an ergonomics-based design strategy similar to the strategy used in the Sabiá project, a team of designers, architects, and engineers developed a concert hall and sports arena complex. The first step required a designer to identify the focus of the "product" within the given location and "marketplace" and determine potential problems that could be associated with specific architectural design solutions.[*]

Design goes beyond the creation of graphic elements and products. Design must be seen as a component of a larger system that serves as an integrator within a management process. It has the power to unite the tangible and intangible, such as the brand and the physical product. It can also consolidate user needs and product features, value and cost, technology and human factors, ergonomics and production, ergonomics and the effective use of products, materials, and perceptions (Martins 2004). In addition, Câmara (1993) suggests that a design supervisor be appointed as project team leader who can identify the intended goals (Figure 29.5).

---

[*] Similarly, a group of investors in the aluminum production sector developed a project to deploy an innovative power plant, producing aluminum, supported on three fronts: economic profitability; organization and way of operation; and a well-conceived plant to optimize the environment because of its expected useful life of 30 years. The actions used by the group adopted the architecture and the ergonomics sought for a new brand image. To this end, they explored three main areas: the conception, the performance, and future operations. This project was a reference that supported the realization of a PhD thesis (Câmara 1993).

FIGURE 29.6 Concert hall and sports complex. (Courtesy: Câmara.)

After identifying the most important needs of the Brazilian city of Belo Horizonte, the goals and objectives of the project were identified and presented to a group of investors. A design concept for a multi-purpose cultural and sports complex emerged after extensive discussions between the investors, community leaders, architects, engineers, and designers.

Once again, ergonomics played a central role in the development strategy of the project. It was agreed that the new "product" would have to meet the needs of a wide range of end users, such as city administrators, theater performers, musicians, actors, athletes, spectators, security personnel, and maintenance staff. A common aesthetics language was selected, including aspects related to use and a sense of appreciation. Similarly, it was agreed to acknowledge in the design the pedagogical practices of the school using the facility by adopting a unique brand-image within the socio-cultural context.

Recently, a poll conducted by the newspaper *Estado de Minas*, asked the residents of Belo Horizonte which architectural works in the city would be remembered in the future as an icon. Even though the multi-purpose complex was already 5 years old, the facility (of the Marist School Dom Silvério) was selected over all other buildings (Figure 29.6).

The architect Bruce Goff (*apud* Norman 2008) says that in architecture, there is a reason why you do something and then there is the real reason. In the case of souvenir buildings, despite their ostensible functions (though irrelevant), their real reason remains to induce the human memory.

In addition to the community benefits provided by the facility in terms of promoting social and cultural activities, the project sparked local urban renewal through private and government investments. This included construction of a new shopping center and a new roadway, and upgrading of pedestrian walkways. The multi-purpose facility represented an investment of approximately US$24 million. In addition to the benefits realized by the investors, the project had a significant positive impact on the socio-cultural context of the entire city.

29.4 CONCLUSION

Much has been written about "value" and "value adds." However, those who are asked to generate new values are challenged by individuals and groups who have a vested interest in the value of the product, service, or system. These groups expect values that are comparable to the values of the products, services, or systems from which they originate. Ultimately, the value will be determined by how well the new product, service, or system will serve the requirements of the interest groups.

The case studies presented in this chapter highlight three different projects: the design for a one-of-a-kind competition vehicle, the design of a new truck adapted from an existing one, and the design of a multi-use public space. In each of the three projects, the primary strategy was to use ergonomics as a differentiator in order to increase the value through aesthetics, efficiency, prestige, or flexibility. The common feature among the three projects is the fact that all were conceived from the inside out. The three projects considered all functions of the product, directly or indirectly, through the eyes of the end user.

ACKNOWLEDGMENTS

The authors wish to thank the CNPq and FAPEMIG, Brazilian Research Foundations. Appreciation is also extended to Livia Galvão Fiuza for her contributions to this chapter. Finally, a special thanks to Prof. Dr. Uwe Reischl, Boise University, for the final revision of the text.

REFERENCES

Bezerra, L.A.H. 1998. *O Estudo da Biografia de uma Empresa como Apóio à Intervenção Ergonômica: um Estudo de Caso – Proposta para Implementação da Ergonomia em uma Empresa de Saneamento*. MSc Diss. Federal University of Santa Catarina, Brazil. http://www.eps.ufsc.br/disserta98/bezerra/ (accessed May 14, 2009).

Bitencourt, R.S., Guimarães, L.B.M., and Santos, P.H. 2006. *Uma Aplicação Inclusiva da Macroergonomia no Setor Industrial Calçadista*. ABERGO 2006: 14° Congresso Brasileiro de Ergonomia/4° Fórum Brasileiro de Ergonomia. Curitiba, PR. Out. Nov. 2006. Departamento de Engenharia de Produção, Universidade Federal do Rio Grande do Sul – UFRGS. http://www.producao.ufrgs.br/arquivos/publicacoes/69_art11.pdf (accessed May 14, 2009).

Câmara, J.J.D. 1993. *Le Design et l'Ergonomie dans les Investissements Industriels: un Approche Intégrative pour l'Usine du Futur*. These du Centre Projets et Produits Nouveaux – École des Mines de Paris. Spécialité: Management de Projets et Génie Industriel. Paris, France.

Carayon, P. 2006. *Handbook of Human Factors and Ergonomics in Health Care and Patient Safety*. 1008. Mahwah, EUA: Lawrence Erlbaum Associates. [in English].

Csillag, J.M. 1995. *Análise de Valor: Metodologia do Valor; Engenharia do Valor; Gerenciamento do Valor; Redução de Custos; Racionalização Administrativa*, 4th ed. São Paulo: Atlas.

Duchamp, R. 1988. *La Conception de Produits Nouveaux: Tecnologies de Pointe*. Paris: Hermes.

Gurgel, F.A. 2001. *Administração do Produto*. 2a ed., 537. Sao Paulo: Atlas.

Hendrick, H.W., and Kleiner, B.M. 2002. *Macroergonomics: Theory; Methods and Applications*, 1st ed. London: CRC Press.

Machado, M.C., and Toledo, N.N. 2008. *Gestão do Processo de Desenvolvimento de Produtos: uma abordagem baseada na criação de valor*. São Paulo: Atlas.

Martins, R.F.F. 2004. A Gestão de Design como uma Estratégia Organizacional – um Modelo de Integração do Design em Organizações. PhD Diss. Federal University of Santa Catarina, Brazil. http://teses.eps.ufsc.br/defesa/pdf/10827.pdf (accessed September 8, 2005).

Norman, D.A. 2008. *Design Emocional*. Rio de Janeiro: Rocco.

Pereira Filho, R.R. 1994. *Análise de Valor*. São Paulo: Nobel.

Piske, I. 2003. *Análise de Valor*. Competitiva Consultoria e Treinamento Ltda. http://www.competitiva.com.br/analise.html (accessed September 12, 2003).

Rebiffe, R., and Guillien, J. 1991. *Ergonomie du poste de conduite des voitures particulières et variabilité anthropometrique de la clientele* (article). Design for Everyone, Published Proceedings of the 11th Congress of the International Ergonomics Association. 1507, 1509, Edited by Y. Quéinnec and F. Daniellou.

Vargas, R.V. 2002. *Análise de Valor Agregado em Projetos*. Rio de Janeiro: Brasport.

Index